Feature Papers in Biological Factors

Feature Papers in Biological Factors

Collection Editor

Tracey Martin

Basel • Beijing • Wuhan • Barcelona • Belgrade • Novi Sad • Cluj • Manchester

Collection Editor
Tracey Martin
School of Medicine
Cardiff University
Cardiff
UK

Editorial Office
MDPI AG
Grosspeteranlage 5
4052 Basel, Switzerland

This is a reprint of the Topical Collection, published open access by the journal *Biomolecules* (ISSN 2218-273X), freely accessible at: https://www.mdpi.com/journal/biomolecules/topical_collections/Biological_Factors.

For citation purposes, cite each article independently as indicated on the article page online and as indicated below:

Lastname, A.A.; Lastname, B.B. Article Title. *Journal Name* **Year**, *Volume Number*, Page Range.

ISBN 978-3-7258-6882-7 (Hbk)
ISBN 978-3-7258-6883-4 (PDF)
https://doi.org/10.3390/books978-3-7258-6883-4

Contents

About the Editor

Tracey A. Martin

Tracey A. Martin is a Senior Lecturer in Cell and Tumour Biology and Director of Postgraduate Research in the School of Medicine at Cardiff University, where she leads a research programme focused on cancer biology and metastasis. Her work has centred on the role of tight junctions — the protein complexes that connect adjacent cells — in cancer progression, particularly how changes in these structures facilitate tumour cell dissociation, invasion and metastasis. She has a longstanding interest in angiogenesis and the mechanisms by which tumour cells breach vascular and blood–brain barriers, with implications for brain metastasis and potential therapeutic targeting. Her group aims to identify molecular regulators of metastasis that could serve as diagnostic markers or treatment targets. Dr. Martin completed her PhD in Molecular Biology at Cardiff and after post-doctoral research on the hepatocyte growth factor (HGF) and its antagonist in cancer, she has published extensively on tight junction proteins, cancer cell signalling and metastatic mechanisms. She has presented over 70 papers at national and international conferences and contributed invited reviews and book chapters. Dr. Martin also plays key leadership roles in postgraduate training and equality, diversity and inclusion within the School of Medicine. She supervises many PhD and MD students and contributes to undergraduate and postgraduate teaching.

biomolecules

MDPI

Review

Decoding the Spectrum of Anorexia Nervosa: Clinical Impact, Molecular Insights, and Therapeutic Perspectives

Dimitris C. Kounatidis [1] and Natalia G. Vallianou [2,*]

1 Diabetes Center, First Department of Propaedeutic Internal Medicine, Medical School, National and Kapodistrian University of Athens, Laiko General Hospital, 11527 Athens, Greece; dimitriskounatidis82@outlook.com
2 First Department of Internal Medicine, Sismanogleio General Hospital, 15126 Athens, Greece
* Correspondence: natalia.vallianou@gmail.com

Abstract: Anorexia nervosa (AN) is a severe psychiatric disorder with the highest mortality rate among mental illnesses, characterized by an intense fear of weight gain, persistent restriction of energy intake, and a distorted perception of body image. Despite decades of investigation, the pathogenesis of AN is only partially understood and is recognized as multifactorial, involving genetic, sociocultural, and neurobiological determinants. Beyond its core psychopathological features, AN leads to a wide spectrum of systemic complications, including cardiovascular, renal, skeletal, and endocrine dysfunctions. Increasing evidence implicates autophagy and oxidative stress as key molecular mechanisms underpinning its pathophysiology, while growing attention has been directed toward immune dysregulation and alterations in the gut–brain axis as potential mediators of disease onset and progression. Therapeutic advances, however, remain limited. Current management relies primarily on nutritional rehabilitation and psychotherapeutic interventions, while treatment outcomes are constrained by high relapse rates and the lack of pharmacological agents with proven efficacy. In this context, a more comprehensive understanding of the clinical spectrum and molecular substrates of AN is essential to improving prognosis and guiding the development of novel therapeutic strategies. This narrative review synthesizes current evidence on the non-psychopathological dimensions of AN, encompassing its clinical manifestations, systemic complications, and implicated molecular pathways. It also appraises existing treatment modalities and examines emerging interventions with translational potential. Overall, this review aims to provide clinicians and researchers with an updated and integrative overview of AN, shedding light on novel directions in ongoing research.

Keywords: anorexia nervosa; autophagy; gut–brain axis; hormones; immunity; neuromodulation; novel therapies; oxidative stress; systemic complications

1. Introduction

Anorexia nervosa (AN) is a complex and potentially life-threatening eating disorder (ED) defined by persistent restriction of energy intake, intense fear of weight gain, and a distorted body image, resulting in clinically significant underweight and widespread systemic consequences. AN carries the highest mortality rate among psychiatric illnesses, primarily due to the medical sequelae of malnutrition. Mortality risk is further increased in the presence of psychiatric comorbidities, particularly within the first decade after diagnosis, with suicide representing a leading cause of death [1,2]. The disorder disproportionately affects females, with a lifetime prevalence of up to 4%, compared with 0.3% in males. While overall incidence has remained relatively stable, diagnoses among individuals younger than 15 years are rising, likely due to earlier onset and improved detection [3].

According to the Diagnostic and Statistical Manual of Mental Disorders, Fifth Edition (DSM-5), AN is defined by three cardinal features: (i) restriction of energy intake resulting in significantly low body weight relative to age, sex, developmental stage, and health status; (ii) intense fear of weight gain or persistent behaviors that prevent weight restoration; and (iii) disturbance in body image, excessive influence of body weight or shape on self-evaluation, or failure to recognize the severity of underweight [4]. Two primary subtypes of AN are delineated, the restricting type and the binge-eating/purging type, distinguished by the presence of recurrent binge eating or purging behaviors [4,5]. The DSM-5 further recognizes atypical AN, in which the core psychopathological features are present despite the individual maintaining a weight within or above the normal range, potentially exhibiting distinct patterns in the frequency and severity of medical complications [4,6].

AN pathogenesis remains only partially understood, yet converging evidence supports a multifactorial etiology encompassing genetic, epigenetic, sociocultural, psychiatric, and biological influences. Recent molecular studies have implicated the *WDR6* gene in disease susceptibility, and differential DNA methylation patterns have also been observed in women with AN compared with both healthy controls and remitted cases. These motifs are particularly notable in genes such as *SYNJ2*, *PRKAG2*, and *STAT3*, which are involved in psychiatric, metabolic, and immune pathways [7,8]. Emerging research data also highlight autophagy, oxidative stress, immune dysregulation, and alterations in the gut microbiota as pivotal mechanisms contributing to disease onset and progression [9,10].

Treatment of AN remains particularly challenging due to patients' frequent resistance to weight restoration and the uncertainty often reported by non-specialist physicians managing complex inpatient cases. These barriers underscore the necessity of multidisciplinary care, specialist involvement, and targeted physician training to enhance clinical confidence and decision-making [11,12]. Integrating advances in the molecular and systemic understanding of AN with its clinical manifestations is essential for optimizing current management and informing the development of novel therapeutic strategies.

This narrative review provides a comprehensive synthesis of current knowledge on the clinical spectrum of AN, highlighting its systemic repercussions and associated laboratory correlates. It also incorporates emerging insights into the disorder's molecular mechanisms, including the roles of the immune system and gut microbiome, while examining contemporary therapeutic strategies. Finally, the review identifies key areas for future research aimed at improving treatment outcomes.

2. Literature Search

For the purposes of this narrative review, a comprehensive literature search was conducted in the PubMed (NIH) database using the keywords "anorexia nervosa", "pathogenesis", and "treatment". The search primarily focused on publications from the past five years, yielding a total of 441 records. Priority was given to research articles, systematic and narrative reviews, clinical trials, and meta-analyses relevant to the pathophysiology and therapeutic approaches to AN. Reference lists of included papers were further screened to identify additional pertinent studies. Given the breadth of the literature retrieved, it is acknowledged that not all studies could be discussed in exhaustive detail within the scope of the present review.

3. Neurological Manifestations in Anorexia Nervosa

In AN, central nervous system (CNS) alterations are characterized primarily by atrophy of both gray and white matter, often accompanied by deficits in higher-order cognitive functions. Neuroimaging studies have revealed ventricular enlargement and widening of cortical sulci, indicative of diffuse cerebral volume loss [13–17]. A recent study by Gupta et al. demonstrated that even in early-stage, acute restrictive-type AN, structural brain changes resembled those typically associated with accelerated aging. Using magnetic resonance imaging (MRI) coupled with machine learning–based brain-age analysis, this study identified marked deviations from normative brain-age profiles. Partial reversal of these alterations was observed following nutritional rehabilitation, although complete restoration was not achieved in all cases [18].

Neuropsychological investigations have identified cognitive rigidity as a common feature in AN, persisting in some individuals despite weight restoration [19]. Morphometric analyses have localized volume reductions to regions such as the thalamus, hippocampus, and cerebellum, areas integral to sensory integration, memory, and motor coordination. At the cellular level, chronic malnutrition deprives neural tissue of essential substrates, including polyunsaturated fatty acids (PUFAs), potentially impairing membrane integrity, synaptic transmission, and myelination, thereby contributing to the observed structural deficits [20–25]. Notably, individuals with AN, particularly those experiencing rapid weight loss, are at risk for Wernicke's encephalopathy due to severe thiamine malnourishment [26].

Evidence also indicates that women with AN exhibit diminished sympathetic nervous system activity, likely reflecting an adaptive response to chronic energy restriction [27]. Peripheral nervous system (PNS) involvement has further been reported, manifesting as distal weakness, paresthesia, or, in severe cases, foot drop. These deficits are thought to result from direct axonal injury secondary to nutrient depletion, compounded by connective tissue alterations that compromise neural support structures. Most PNS abnormalities improve with weight restoration, although recovery trajectories remain variable [28].

4. Cardiopulmonary Disease in Anorexia Nervosa

4.1. Cardiovascular Complications

Cardiovascular abnormalities represent some of the most frequent medical complications of AN, occurring in up to 87% of patients. Clinical manifestations encompass left ventricular atrophy, mitral valve prolapse, bradycardia, hypotension (often orthostatic), QTc prolongation, arrhythmias, pericarditis, ischemic heart disease, and heart failure (HF). Mortality due to cardiovascular causes is particularly elevated, with epidemiological data indicating a two- to threefold higher risk compared with the general population [29,30]. In a large Taiwanese cohort comprising 2081 patients with AN, Tseng et al. reported a 4.8% cumulative incidence of major adverse cardiovascular events (MACEs) over a five-year follow-up period. The risk of HF, structural cardiac abnormalities, and conduction disturbances was greatest shortly after diagnosis, whereas the incidence of ischemic heart disease rose significantly only by the fifth year. Older age and male sex were identified as independent risk factors for adverse cardiovascular outcomes [31].

Bradycardia, reported in 36–95% of patients, and hypotension, observed in up to 85%, correlate inversely with body mass index (BMI) and often necessitate hospitalization when BMI falls below 15 kg/m^2 [30,32–35]. Bradycardia likely reflects an adaptive decrease in basal metabolic rate, whereas hypotension arises from autonomic dysregulation characterized by increased vagal tone. QTc prolongation, observed in up to 40% of patients, becomes clinically significant when exceeding 600 ms due to the risk of ventricular tachycardia, torsades de pointes, and sudden cardiac death [34]. Electrolyte abnormalities, particularly hypokalemia and hypomagnesemia, are common precipitants, although QTc prolongation may also occur independently of electrolyte imbalance, implicating autonomic dysfunction. Certain antipsychotic agents may further exacerbate repolarization disturbances [36–38].

Structural myocardial alterations are also well documented. Echocardiography often reveals reduced left ventricular mass index proportional to the degree of undernutrition [34,35]. Histopathological findings include mitochondrial swelling, myxoid and lipofuscin accumulation, and interstitial fibrosis [39]. Thiamine deficiency contributes significantly to myocardial dysfunction, and beriberi cardiomyopathy may coexist [40]. Mitral valve prolapse, affecting up to 25% of patients, likely results from disproportion between ventricular chamber size and leaflet dimensions and typically resolves with weight restoration [30]. Pericardial effusion, observed in up to one-third of patients, is associated with thyroid hormone alterations and hypoalbuminemia and generally regresses following nutritional rehabilitation [30,34].

The refeeding phase is a particularly vulnerable period, as the metabolic transition from catabolism to anabolism can trigger refeeding syndrome, typically manifesting with hypokalemia, hypomagnesemia, and hypophosphatemia within the first 10 days of nutritional reintroduction. These shifts impair myocardial energy metabolism and oxygen delivery, predisposing to arrhythmias and HF even in the presence of preserved ejection fraction [32,34,41,42]. Thiamine deficiency may further exacerbate cardiac dysfunction. Proactive supplementation with electrolytes and thiamine during refeeding is therefore essential to prevent these potentially fatal complications [32].

4.2. Pulmonary Complications

Pulmonary complications in AN are less common but clinically relevant. Isolated reports describe spontaneous pneumothorax and pneumomediastinum [43]. Proposed mechanisms include malnutrition-induced bullae or bleb formation predisposing to alveolar rupture, as well as acute intrathoracic pressure changes from forceful vomiting [44]. Aspiration pneumonia may also occur in advanced stages, likely secondary to impaired pharyngeal muscle coordination from severe malnutrition. An increased frequency of emphysema has also been noted, with some authors proposing the entity of "nutritional emphysema," though its pathogenesis and validity remain debated [30,45].

5. Liver Disease in Anorexia Nervosa

Although AN is an uncommon cause of liver disease, abnormalities in liver function are frequently observed among hospitalized AN patients [46]. The predominant mechanism involves starvation-induced autophagy, triggered by endoplasmic reticulum stress. Initially adaptive, this process may become excessive with progressive malnutrition, resulting in hepatocellular injury and cell death [47,48]. Additional contributors include secondary hypoxia due to hypovolemia, particularly in the setting of hypotension or cardiac dysfunction, and, in some cases, iron overload [49,50]. In contrast, insulin resistance (IR) does not appear to play a significant role [51].

Elevated serum aminotransferases are the most common liver abnormality, occurring in more than half of patients. These elevations typically reflect increased hepatocyte membrane permeability from autophagy rather than frank necrosis [52,53]. Although usually mild, aminotransferase levels can rise substantially when BMI falls below 13 kg/m^2 [53,54]. The biochemical profile often resembles metabolic dysfunction-associated steatotic liver disease (MASLD), with alanine aminotransferase (ALT) exceeding aspartate aminotransferase (AST). Rarely, lactate dehydrogenase (LDH) may increase, usually in the context of hepatic hypoxia, and severe hypoglycemia may occur due to depleted glycogen stores [55]. Notably, serum albumin may remain within the normal range despite profound malnutrition, whereas hematocrit-adjusted albumin has been proposed as a more reliable marker of nutritional status and hepatic recovery [50,56,57].

The international normalized ratio (INR) is typically modestly elevated, though more pronounced increases may be observed in patients with marked transaminase abnormalities. These changes are thought to result primarily from impaired vitamin K absorption rather than reduced hepatic synthetic capacity [58]. Standard anticoagulant prophylaxis for venous thromboembolism (VTE) is generally safe, supported by thromboelastography data indicating preserved coagulation function [59]. Nonetheless, severe coagulopathy and persistent thrombocytopenia, even after refeeding, have been reported, highlighting the importance of regular monitoring of hematological and coagulation parameters. Acute liver failure is rare, while isolated cases of portal hypertension have been described [60–62].

Liver biopsy is seldom required but, when performed, typically demonstrates abundant autophagosomes with minimal inflammation and little or no necrosis. Hepatic steatosis is generally absent in starvation-related hepatitis unless the biopsy is obtained during refeeding, when glucose and fat accumulate in the liver [52]. Gradual nutritional rehabilitation remains the cornerstone of management. Serum aminotransferases usually begin to normalize within two weeks of refeeding, although full recovery may take longer. A mild, transient rise in aminotransferases is often observed during the early refeeding phase [51,52]. Figure 1 presents the key features and clinical management considerations of liver disease in patients with AN.

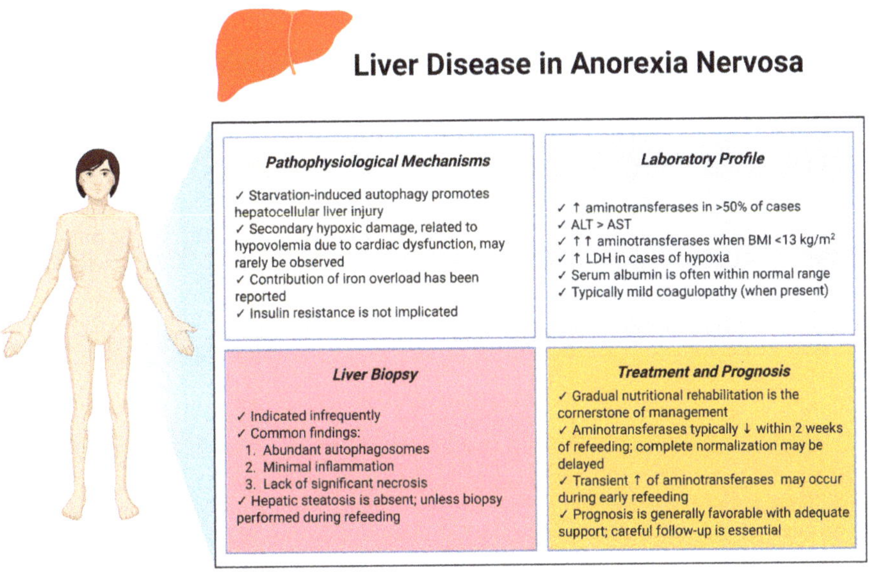

Figure 1. Main characteristics and key approaches to liver disease in anorexia nervosa. Abbreviations: ALT: alanine aminotransferase; AST: aspartate aminotransferase; BMI: body mass index; LDH: lactate dehydrogenase; ↑: increase; ↓: decrease. Created in BioRender. Kounatidis, D. (2025) https://BioRender.com/zs7ulab (assessed on 8 September 2025).

6. Gastrointestinal Disease in Anorexia Nervosa

Gastrointestinal (GI) symptoms are highly prevalent in individuals with AN and may reinforce restrictive eating behaviors, as patients often attribute reduced food intake to abdominal discomfort. Reported manifestations encompass the entire GI tract, including abdominal pain, constipation, and diarrhea, with documented cases of gastroesophageal reflux disease (GERD) and irritable bowel syndrome (IBS) [63,64]. More severe complications, such as gastroparesis and, less commonly, acute gastric dilatation (particularly following binge episodes) have also been observed. Pancreatitis may arise in severely malnourished patients, potentially as a result of hypoperfusion, duodenogastric reflux, or pancreatic atrophy [63,65]. Overly aggressive refeeding may exacerbate these GI complications. Notably, early-life GI disturbances, like persistent constipation or diarrhea, have been linked to subsequent development of AN, though the contribution of shared familial or genetic factors cannot be excluded [66]. While symptoms such as constipation and abdominal pain generally improve with nutritional rehabilitation, complaints of dyspepsia may persist, with psychological distress emerging as a significant predictor of ongoing GI morbidity [67].

7. Kidney Disease in Anorexia Nervosa

In a cohort of 395 adolescents with newly diagnosed AN, Gurevich et al. reported impaired kidney function in nearly 37% of patients [68]. Renal complications in AN are primarily driven by electrolyte disturbances and volume depletion, with hypokalemia representing the key pathogenic factor. Hypokalemia commonly arises from self-induced vomiting, laxative misuse, or diuretic abuse and is particularly prevalent in the binge-eating/purging subtype compared with the restricting subtype [68–70]. Severe, acute hypokalemia in the context of intravascular depletion may precipitate hypotension, reduced renal perfusion, and acute kidney injury (AKI). During refeeding syndrome, marked hypophosphatemia (serum phosphorus <0.3 mmol/L) can further exacerbate AKI through rhabdomyolysis [69–72]. Moreover, chronic hypokalemia may induce structural renal injury, termed hypokalemic nephropathy, histologically characterized by vacuolization of proximal tubular epithelial cells, with distal tubular involvement occurring less frequently. Additional reported abnormalities include non-specific glomerulosclerosis, interstitial nephritis, tubular atrophy, and medullary cyst formation. Notably, correction of hypokalemia may reduce cyst burden [69]. Figure 2 depicts the spectrum of histopathological changes associated with hypokalemic nephropathy in AN.

Other electrolyte disturbances further exacerbate renal dysfunction in AN. Hypomagnesemia, frequently coexisting with hypokalemia, often reflects diuretic or laxative misuse compounded by nutritional deficiencies [69–73]. Hyponatremia, resulting from excessive water intake or volume depletion, can trigger inappropriate vasopressin secretion, perpetuating water retention and electrolyte imbalance. Interestingly, this process may be aggravated by concomitant antipsychotic therapy [69]. Beyond electrolyte-mediated injury, nephrocalcinosis, defined by calcium oxalate and/or calcium phosphate deposition within the renal parenchyma, has been reported, with malnutrition and chronic volume depletion implicated in its pathogenesis. Nephrolithiasis may also occur, commonly as ammonium urate calculi promoted by dehydration. Collectively, these abnormalities place patients with AN at risk for both AKI and chronic kidney disease (CKD), with some cases progressing to end-stage renal disease (ESRD) [73,74]. Accurate assessment of renal function in AN remains challenging. Severe muscle wasting reduces creatinine generation, limiting the reliability of creatinine-based estimations of glomerular filtration rate (eGFR). The potential utility of cystatin C, either alone or incorporated into Chronic Kidney Disease Epidemiology Collaboration (CKD-EPI) equations, warrants further investigation as a more accurate biomarker in this population [69].

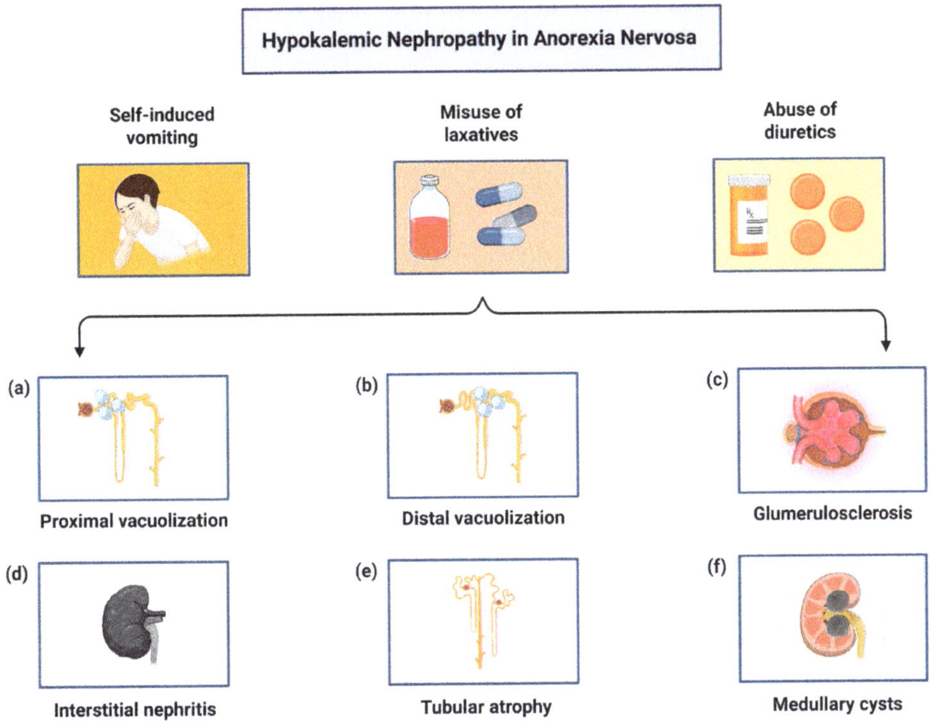

Figure 2. The histopathological spectrum of hypokalemic nephropathy in anorexia nervosa. Created in BioRender. Kounatidis, D. (2025) https://BioRender.com/4tnmh4k (assessed on 1 November 2025).

8. Hormonal Abnormalities in Anorexia Nervosa

8.1. Hypothalamic–Pituitary–Adrenal (HPA) Axis Dysregulation

Patients with AN, particularly those with the binge–purge subtype, frequently display sustained overactivity of the hypothalamic–pituitary–adrenal (HPA) axis, reflected by elevated levels of corticotropin-releasing hormone (CRH) and adrenocorticotropic hormone (ACTH) [75]. A rare case of ACTH deficiency with secondary adrenal insufficiency has also been documented [76]. Excessive CRH and ACTH secretion contributes to hypercortisolemia, with affected individuals often exhibiting increased salivary and 24 h urinary cortisol levels [77]. The cortisol awakening response is typically exaggerated, especially in the binge–purge subtype compared with the restrictive subtype [78]. In contrast, hair cortisol concentration, a biomarker of long-term cortisol exposure, does not appear elevated and shows weak correlation with urinary cortisol output, possibly indicating altered cortisol incorporation or enzyme dysregulation secondary to severe malnutrition [79]. Weight restoration generally leads to normalization of circulating cortisol levels; however, persistent HPA axis abnormalities have been reported. Schmalbach et al. observed that, despite normalized basal cortisol, individuals with AN demonstrated a blunted cortisol response to acute stress, a pattern more closely associated with residual eating-disorder psychopathology and psychological distress than with BMI [80]. Pharmacological suppression of cortisol is not recommended, given the absence of proven benefit and the potential risks of adrenal insufficiency and further weight loss [81].

8.2. Growth Hormone–Insulin-like Growth Factor 1 (GH–IGF-1) Axis Dysregulation

Prolonged starvation in AN is characterized by increased growth hormone (GH) secretion and pulsatility, accompanied by markedly reduced circulating insulin-like growth factor 1 (IGF-1) levels. This paradoxical pattern likely reflects a compensatory adaptation aimed at maintaining euglycemia through enhanced gluconeogenesis and promoting lipolysis. In AN, however, GH hypersecretion occurs in the context of GH resistance, mediated by reduced insulin concentrations and elevated fibroblast growth factor 21 (FGF21), both of which impair GH receptor signaling and hepatic IGF-1 synthesis [77,81]. Growth hormone secretagogue receptor (GHSR) signaling has also been implicated in circadian adaptation during periods of food restriction. In a murine model of AN, GHSR-deficient mice exhibited impaired circadian adjustment, with diminished daytime and pre-prandial activity [82]. Although premorbid stature in AN is generally normal, adult height often remains below midparental expectations. Partial catch-up growth may occur with nutritional rehabilitation; however, final stature depends on chronological age, bone age, growth velocity during hospitalization, and concurrent changes in luteinizing hormone (LH) secretion [83]. Growth hormone therapy has demonstrated potential benefits in patients with persistent AN-related growth failure, increasing linear growth velocity compared with placebo over a 12-month period [84].

8.3. Hypothalamic–Pituitary–Gonadal (HPG) Axis Dysregulation

Menstrual irregularities are highly prevalent in AN, with functional hypogonadotropic hypogonadism (FHH) most commonly manifesting as amenorrhea, reported in up to 85% of cases. Loss of adipose tissue leads to leptin deficiency, disrupting kisspeptin-mediated stimulation of hypothalamic gonadotropin-releasing hormone (GnRH) neurons. This impairs pulsatile GnRH secretion and subsequently reduces LH and follicle-stimulating hormone (FSH) output. As a result, ovarian estradiol and testicular testosterone production decline [77,81]. Weight restoration typically reverses FHH, though menstrual recovery may take up to one year, and approximately 20% of women fail to resume menses despite normalization of body weight [81,85]. Among weight-restored individuals, higher estradiol concentrations and greater LH/FSH responsiveness to GnRH predict menstruation resumption. Specific endocrine thresholds, such as the presence of at least two LH pulses within four hours and an LH/GnRH response $\geq$33 IU/L, have demonstrated strong predictive value independent of body weight [86]. Despite amenorrhea, conception remains possible; thus, early fertility counseling is recommended due to increased obstetric risks, including low birth weight, preterm delivery, and intrauterine growth restriction [87,88].

8.4. Hypothalamic-Pituitary-Thyroid (HPT) Axis Dysregulation

Typical clinical manifestations of AN, including dry skin, hypothermia, hypotension, and bradycardia, frequently reflect underlying thyroid dysfunction. Severe malnutrition often induces a non-thyroidal illness syndrome (NTIS), characterized by reduced triiodothyronine (T3), elevated reverse T3 (rT3), and thyroid-stimulating hormone (TSH) and free thyroxine (fT4) levels within the low-normal range [89,90]. Suppression of hypothalamic-pituitary-thyroid (HPT) axis activity, indicated by reduced fT3 and a decreased fT3/fT4 ratio, appears more pronounced in individuals with severe depressive symptoms [91]. An fT3/fT4 ratio below 2.0 has been proposed as a biomarker distinguishing NTIS from central hypothyroidism in adolescents [92]. Thyroid hormone alterations generally improve with nutritional rehabilitation, although complete normalization is not universal. Animal studies suggest that early nutritional support may prevent T3 decline during fasting by modulating deiodinase activity and leptin signaling [93]. Exogenous thyroid hormone replacement is not recommended, as it may disrupt adaptive hypometabolism and heighten cardiovascular risk [81].

8.5. Posterior Pituitary Hormone Dysregulation

Individuals with AN may exhibit impaired osmoregulation, as evidenced by lower plasma sodium and osmolality, elevated antidiuretic hormone (ADH) concentrations, and diminished urine-concentrating capacity. ADH responses appear dysregulated, with persistently dilute urine particularly in chronic illness [94]. In contrast, copeptin, used as a stable surrogate marker of ADH, does not differ between medically stable patients with AN and healthy controls, suggesting that ADH activity may not be elevated under normohydrated conditions [95]. Notably, central diabetes insipidus can emerge during the refeeding phase, with MRI findings of absent posterior pituitary high signal (PPHS) indicating ADH granule depletion [96,97].

Oxytocin dysregulation has also been implicated in AN pathophysiology, with reduced serum and cerebrospinal fluid levels reported [81,98]. Elevated postprandial oxytocin concentrations have been associated with heightened anxiety and depressive symptoms, independent of cortisol and largely unaffected by leptin status [99]. Conversely, other studies have described postprandial oxytocin declines, more pronounced in atypical AN, potentially contributing to appetite dysregulation [100]. Associations between oxytocin and psychopathology may vary by subtype, with restrictive AN showing broadly negative correlations across symptom domains, while binge–purge AN is more closely linked to depressive symptoms [101]. These findings have prompted investigation of intranasal oxytocin (IN-OT) as a therapeutic strategy. A pilot inpatient trial reported reductions in eating-related preoccupations, cognitive rigidity, and salivary cortisol responses. However, a subsequent multicenter randomized controlled trial (RCT) failed to replicate these results, suggesting that treatment effects may depend on baseline symptom severity and comorbid psychiatric profiles [102,103].

8.6. Adipokine Dysregulation

Fat mass depletion in AN may produce significant alterations in adipokine secretion. Leptin, a key regulator of satiety, neuroendocrine function, and metabolic homeostasis, is reduced, initiating starvation-adaptive processes that operate through both threshold-dependent "switches" and graded, tissue-specific modulation [104]. Low leptin contributes to hyperactivity, a relationship absent post-recovery, when activity aligns more closely with core psychopathology [105,106]. Ghrelin signaling is impaired via increased methylation of the growth hormone secretagogue receptor 1a (GHSR1a) promoter, consistent with "ghrelin resistance," explaining the lack of compensatory hyperphagia despite elevated circulating and exogenous ghrelin, while anxiety, food preoccupation, and altered reward processing persist [107]. Elevated liver-expressed antimicrobial peptide 2 (LEAP2), an endogenous GHSR antagonist, may reinforce this selective resistance, highlighting complex interactions between metabolic signals and neural reward circuits [108]. Circulating adiponectin, which is typically elevated during caloric restriction, is influenced by metabolic and inflammatory mediators such as insulin, resistin, triglycerides, and interleukin (IL)-6. Its concentrations correlate positively with GH pulsatility but remain unaltered by short-term GH elevations, suggesting regulation primarily by fat mass rather than by acute endocrine fluctuations [109,110].

9. Musculoskeletal Disease in Anorexia Nervosa

Malnutrition in subjects with AN is closely correlated with generalized fatigue and attenuated physical endurance, mainly resulting from skeletal muscle atrophy. Nutritional deficits, electrolyte imbalances, metabolic alterations, and hypercortisolemia contribute to the selective degeneration of type II muscle fibers [111,112]. This catabolic milieu can lead to functional impairments, including difficulties in maintaining posture, and a heightened risk of falls. Skeletal involvement in AN is predominantly characterized by a reduction in bone mineral density (BMD), which can progress from osteopenia to osteoporosis, thereby predisposing to fragility fractures. Dual-energy X-ray absorptiometry (DEXA) studies have demonstrated markedly reduced BMD in individuals with AN, often within the osteoporotic range [113,114]. The pathophysiology underlying these bone abnormalities is multifactorial. FHH favors estrogen deficiency, while changes in GH secretion and reduced IGF-1 impair osteoblastic bone formation. Additional contributing factors include vitamin D deficiency, depletion of calcium and phosphorus, persistent hypercortisolemia, and dysregulation of adipokines such as leptin and adiponectin. In addition, serum osteocalcin concentrations are decreased, reflecting reduced osteoblastic activity and a consequent imbalance between bone formation and resorption [115,116]. Histological and imaging studies further indicate a pathological shift in bone marrow composition from osteogenesis toward adipogenesis, thereby accelerating bone loss [117].

10. Cutaneous Manifestations in Anorexia Nervosa

Skin signs are common in advanced AN, particularly when BMI falls below 16 kg/m^2 [118]. A functional classification identifies four principal categories: (i) cutaneous signs of starvation, (ii) lesions related to self-induced vomiting, (iii) injuries from self-inflicted behaviors, and (iv) changes associated with substance misuse, including laxatives and diuretics. Often, multiple mechanisms contribute to individual clinical findings [119]. Xerosis, or dry skin, is among the most prevalent features, attributed to reduced sebaceous gland activity and altered lipid composition. Dehydration from purging behaviors, diuretic or laxative use, and excessive hygiene further exacerbates this condition [119,120]. Hair abnormalities are also frequent and include alopecia, brittle or thinning hair, and lanugo-like growth, with the latter potentially related to decreased 5-α-reductase activity or concurrent hypothyroidism. Additional cutaneous changes include eczema, acne, nail fragility, and pigmentary alterations such as hyperpigmentation, carotenoderma, and acrocyanosis [119,121,122]. Carotenoderma is typically linked to a high intake of carotenoid-rich vegetables, while acrocyanosis reflects underlying vascular dysregulation. Pruritus is frequently reported and may be generalized, influenced by hormonal imbalance, and impaired thermoregulation. Characteristic signs include Russell's sign, manifested as callused knuckles from repeated self-induced vomiting, and perimolysis, characterized by enamel erosion resulting from chronic exposure to gastric acid. Among the most common dermatological findings in AN are xerosis, alopecia, dental caries, brittle hair, and nail fragility, while features such as hypertrichosis, Russell's sign, perimolysis, and self-inflicted lesions are considered more specific to the disorder. Early recognition of these manifestations may facilitate identification of patients who do not openly disclose disordered eating behaviors [123,124]. Figure 3 summarizes the common clinical features observed in AN.

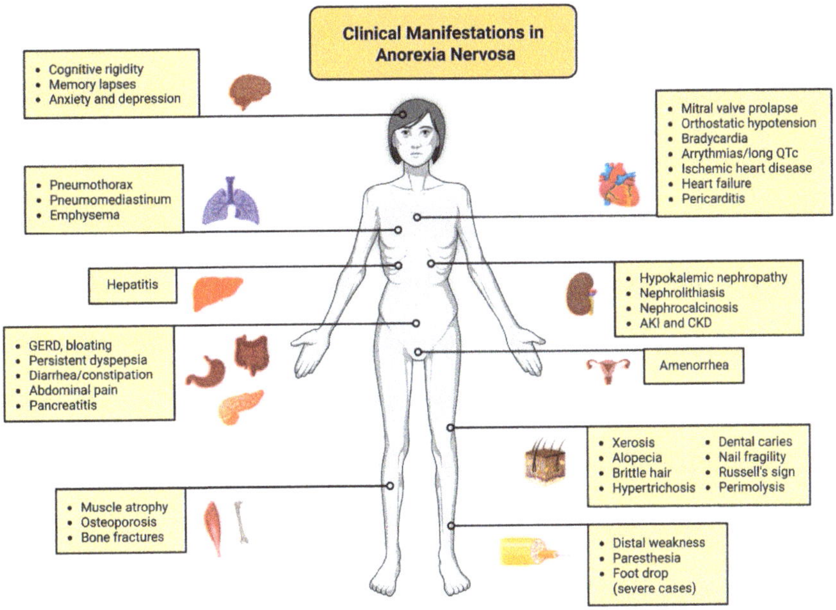

Figure 3. Clinical consequences of anorexia nervosa. Abbreviations: AKI: acute kidney injury; CKD: chronic kidney disease; GERD: gastroesophageal reflux disease; LV: left ventricular. Created in BioRender. Kounatidis, D. (2025) https://BioRender.com/3hfpuek (assessed on 6 September 2025).

11. Hematological Alterations in Anorexia Nervosa

Hematological assessment represents an essential component of the clinical evaluation of AN, as complete blood count (CBC) abnormalities commonly reflect the systemic effects of malnutrition. The most frequently observed alterations include anemia, leukopenia, and thrombocytopenia, all of which generally improve with nutritional rehabilitation [125]. The prevalence and severity of these changes correlate closely with the degree and duration of protein–energy malnutrition, as reflected by BMI and phase angle measurements [126]. Reported frequencies vary across studies, likely owing to differences in sample characteristics and nutritional status. Earlier investigations from Europe and the United States identified anemia as the most frequent abnormality, followed by leukopenia and, less often, thrombocytopenia [127,128]. More recent data, however, indicate that leukopenia predominates, while a global meta-analysis estimated the overall prevalence of anemia in AN at 44.8% [129,130]. Bone marrow biopsies typically reveal hypocellularity with adipose replacement and, in some cases, gelatinous marrow transformation (GMT) [131]. Despite these findings, bone marrow biopsy is not recommended in routine clinical practice [129].

During refeeding, hematological indices may transiently deteriorate, often reaching their nadir between days five and ten of hospitalization. This decline tends to be more pronounced in restrictive-type AN or in the presence of intercurrent infection and may necessitate transfusion support [132]. Anemia is usually normocytic and normochromic, though cases of macrocytosis have been reported [127,129]. The relationship between neutropenia and infection risk in AN remains uncertain. Individuals admitted with a BMI below 12 kg/m^2 appear particularly vulnerable, and earlier studies suggested that impaired bactericidal activity and reduced polymorphonuclear cell adherence may compound this risk [133,134]. Nevertheless, evidence for a higher incidence of infections in AN is inconclusive. Clinical vigilance is therefore critical, as infections may present atypically, with absent fever or blunted inflammatory responses [135].

In cases of severe neutropenia, the therapeutic role of granulocyte colony-stimulating factor (G-CSF) remains poorly defined but may be considered in protracted or complicated presentations [62,130,136]. Thrombocytopenia can also occur in a subset of patients and may be accompanied by increased platelet distribution width (PDW), a parameter potentially reflecting both illness duration and the rate of weight loss [137]. In instances of severe hepatic dysfunction, impaired thrombopoietin synthesis may further contribute to reduced platelet counts [138]. Figure 4 summarizes the principal hematological and laboratory abnormalities observed in AN.

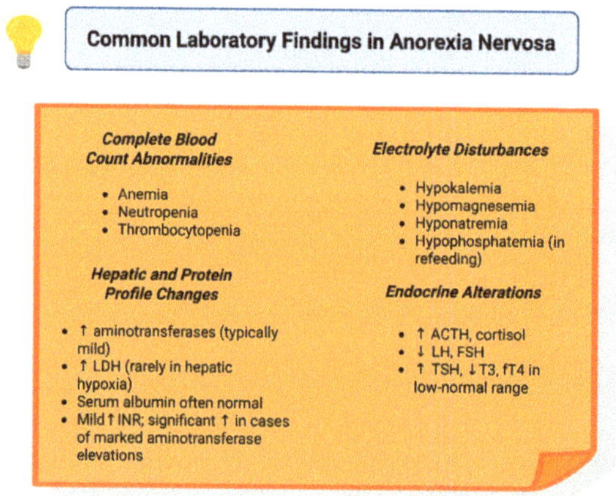

Figure 4. Common laboratory findings in anorexia nervosa. Abbreviations: ACTH: adrenocorticotropic hormone; fT4: free thyroxine; FSH: follicle-stimulating hormone; INR: international normalized ratio; LDH: lactate dehydrogenase; LH: luteinizing hormone; T3: triiodothyronine; TSH: thyroid-stimulating hormone; ↑: increase; ↓: decrease. Created in BioRender. Kounatidis, D. (2025) https://BioRender.com/ye572om (assessed on 7 September 2025).

12. Emerging Pathophysiological Pathways in Anorexia Nervosa: The Role of Immunity and the Gut–Brain Axis

Beyond the well-characterized systemic adaptations associated with AN, accumulating evidence implicates dysregulation of immune function and disturbances within the gut–brain axis as significant contributors to its complex pathophysiology. These emerging frameworks highlight the intricate interplay between peripheral metabolic and inflammatory processes and central neural regulation of appetite, mood, and cognition. A deeper understanding of these interactions may clarify how AN influences and is maintained by both immune and neurobiological mechanisms and could guide the development of more targeted prevention and treatment strategies. The following subsections synthesize current data on the role of immune alterations and gut–brain communication in the onset, development, and progression of AN.

12.1. Immunological Alterations in Anorexia Nervosa

Nutritional deprivation has long been implicated in immune dysregulation, and accumulating data suggest that AN is accompanied by distinct alterations in both innate and adaptive immunity [139,140]. However, the available evidence remains heterogeneous and is often limited by small cohort sizes. Components of innate immunity appear blunted, although the contribution of specific subsets such as macrophages and dendritic cells has yet to be fully characterized [141]. Circulating levels of macrophage inhibitory cytokine-1 (MIC-1) tend to rise during starvation and decline following refeeding, a fluctuation that may be potentiated by hyperinsulinemia [142,143]. Natural killer (NK) cell counts are generally reduced, while their cytotoxic function appears largely preserved [141,144].

Adaptive immune responses are likewise perturbed. Although the proliferative capacity of T cells is usually maintained, an elevated $CD4^+/CD8^+$ ratio is frequently observed, largely due to selective depletion of $CD8^+$ subsets, a shift that could impair antiviral defenses [145]. Overexpression of the C-X-C chemokine receptor type 4 (CXCR4) on $CD4^+$ T cells has been associated with both psychological distress and nutritional status in adolescents with AN [146], whereas regulatory T cell (Treg) function appears relatively preserved [147]. Findings regarding humoral immunity are ambiguous. B cell counts may be normal or elevated, with restrictive-type AN showing expansion of antigen-experienced subsets alongside depletion of immunoregulatory populations [148]. Antibody titers are typically reduced or within normal limits, yet vaccine responsiveness remains intact [141,149]. Reports also suggest aberrant T–B cell crosstalk, and a decreased neutrophil-to-lymphocyte ratio (NLR) has been proposed as a potential marker of disease severity [141,150].

Cytokine profiling has yielded similarly variable results, although most studies report elevated concentrations of pro-inflammatory cytokines, including IL-1β, IL-6, and IL-8 [151]. Prostaglandin E_2 (PGE_2) has been implicated in IL-1β-driven anorexic behavior, albeit with strain-specific variability [152]. Findings regarding tumor necrosis factor-alpha (TNF-α) remain inconsistent. For instance, one study demonstrated in vitro release comparable to controls, whereas lipopolysaccharide (LPS)-stimulated secretion appeared attenuated, possibly reflecting reduced monocyte availability. Other investigations reported increased IL-6 and transforming growth factor-beta (TGF-β) without significant TNF-α alterations, while some data support spontaneous TNF-α production [151,153]. Reduced interferon-gamma (IFN-γ) secretion by lymphocytes has also been described, though findings are not uniform across studies [151,154].

Importantly, immune abnormalities may emerge even in the early stages of AN, with improvements in cytokine profiles during treatment paralleling reductions in depressive symptoms [155]. Restrictive-type AN may not initially exhibit overt pro-inflammatory activation; nonetheless, disturbances in immune-related mediators such as C-X-C motif chemokine ligand 1 (CXCL1) and matrix metalloproteinase-8 (MMP-8) have been documented [156]. Complement activity remains underexplored, although reduced C3 levels have been correlated with disease severity and improve with weight restoration [157]. Oxidative stress also appears to modulate immune function, as indicated by reduced serum uric acid and total antioxidant capacity, both of which normalize following nutritional rehabilitation [158,159].

While some immune abnormalities in AN resemble those observed in primary malnutrition, the overlap is only partial, suggesting that caloric restriction alone does not fully account for the immunological phenotype of AN [141]. Endocrine factors, particularly adipokines such as leptin and adiponectin, likely exert additional modulatory effects on immune responses [160]. Collectively, the available evidence supports an immunological contribution to the pathophysiology of AN, although its precise clinical significance remains incompletely defined. The partially autoimmune potential of AN is further suggested by its co-occurrence with immune-mediated diseases, particularly those affecting the gastrointestinal tract, such as celiac disease (CeD) and inflammatory bowel disease (IBD). In the case of CeD, comparable patterns of immune dysregulation have been described, albeit mediated through distinct mechanisms of aberrant immune activation and inflammatory signaling [161]. On the other hand, the relationship between AN and IBD remains less consistent. Population-based data and genetic analyses yield divergent findings regarding both the strength and direction of association between AN and specific IBD subtypes, namely Crohn's disease (CD) and ulcerative colitis (UC). Overall, the mechanistic basis underlying the interplay between AN and immune-mediated GI disorders remains unclear and warrants further investigation [162–164]. Additional immune-related phenomena include hepatic cytolysis linked to anti-hypoxia-inducible factor 1-alpha (HIF-1α) autoantibodies, and reactivation of latent human herpesviruses 6 and 7 (HHV-6, HHV-7) leading to pityriasis rosea. Interestingly, reduced salivary immunoglobulin M (IgM) concentrations have been correlated with increased dental plaque accumulation in subjects with AN [50,165–167].

12.2. The Gut–Brain Axis in Anorexia Nervosa

The gut microbiota, comprising bacteria, viruses, fungi, and archaea, and its genomic counterpart, the gut microbiome, play a central role in maintaining host homeostasis. Disruption of this balance gives rise to gut dysbiosis, which in AN is marked by increased intestinal permeability, reduced microbial diversity, and distinct compositional alterations. These include elevated levels of *Enterobacterales*, *Proteobacteria*, and *Methanobrevibacter smithii*, alongside reductions in *Faecalibacterium prausnitzii* and *Firmicutes* [168–170]. Beyond bacterial shifts, patients with AN demonstrate diminished viral–bacterial interactions and enrichment of microbial pathways linked to neurotransmitter degradation, suggesting that dysbiosis may contribute to neurobiological processes underlying the disorder [171]. Within this context, the gut–brain axis (GBA), a bidirectional communication network integrating immune, hormonal, and metabolic signals, has become a major focus of contemporary research [172,173].

Loss of intestinal barrier integrity facilitates the translocation of LPS from Gram-negative bacteria into the circulation via disrupted tight junction proteins, a phenomenon commonly referred to as "leaky gut". This process elicits an immune response marked by the release of pro-inflammatory cytokines, including IL-6 and TNF-α, which can cross the blood–brain barrier and promote neuroinflammation [170]. The hypothalamus, enriched in cytokine receptors and central to appetite regulation, is particularly sensitive to these signals. Toll-like receptor 2 (TLR-2) and toll-like receptor 4 (TLR-4) activation of microglia, leading to NLR family pyrin domain-containing protein 3 (NLRP3) inflammasome activation, provides a key pathway through which peripheral immune disturbances may disrupt hypothalamic function [174–176]. The role of the hypothalamus in weakening immune defense and further destabilizing intestinal barrier integrity is underscored by the chronic hyperactivity of the HPA axis observed in AN, manifested in elevated cortisol secretion, particularly in the binge–purge subtype [177].

Metabolic signaling through short-chain fatty acids (SCFAs) constitutes another mechanism shaping neuroinflammation. SCFAs, such as butyrate, are generated by bacterial fermentation of dietary fibers, and patients with AN have been shown to display reduced fecal concentrations of butyrate and acetate compared with healthy controls [178]. Diminished butyrate production has also been associated with lower levels of brain-derived neurotrophic factor (BDNF), impairing neuronal growth, survival, and plasticity, and thereby contributing to disturbances of appetite regulation and mood [179]. Other microbiota-derived metabolites, including caseinolytic peptidase B (ClpB), can mimic the anorexigenic hormone α-melanocyte-stimulating hormone (α-MSH), further enhancing appetite suppression [172]. In addition, severely ill patients may demonstrate reduced circulating concentrations of diazepam-binding inhibitor (DBI). A decline in DBI weakens inhibitory control over anorexigenic signals such as growth differentiation factor 15 (GDF15) and lipocalin-2 (LCN2), thereby reinforcing restrictive eating behaviors [180].

Altered tryptophan metabolism represents an additional pathway through which the gut influences brain function in AN. Gut microbes divert tryptophan away from serotonin (5-HT) synthesis, reducing the substrate available for neurotransmitter production. Combined with dietary restriction, this leads to further 5-HT depletion, increasing vulnerability to mood disturbances and cognitive deficits [181]. Autonomic dysregulation may boost these effects, as the gut–vagus nerve pathway serves as a critical conduit for microbial signaling to the brain. Evidence from animal studies indicates that vagal input can modulate dopamine release, inflammatory responses, and emotional behavior, although human data remain mixed [182].

Overall, current evidence indicates that gut dysbiosis in AN disrupts neuroimmune, neuroendocrine, and neurotrophic processes through the GBA. This interplay positions the gut microbiome not only as a potential biomarker of disease progression but also as a promising therapeutic target. Figure 5 presents the role of the GBA in the development and progression of AN.

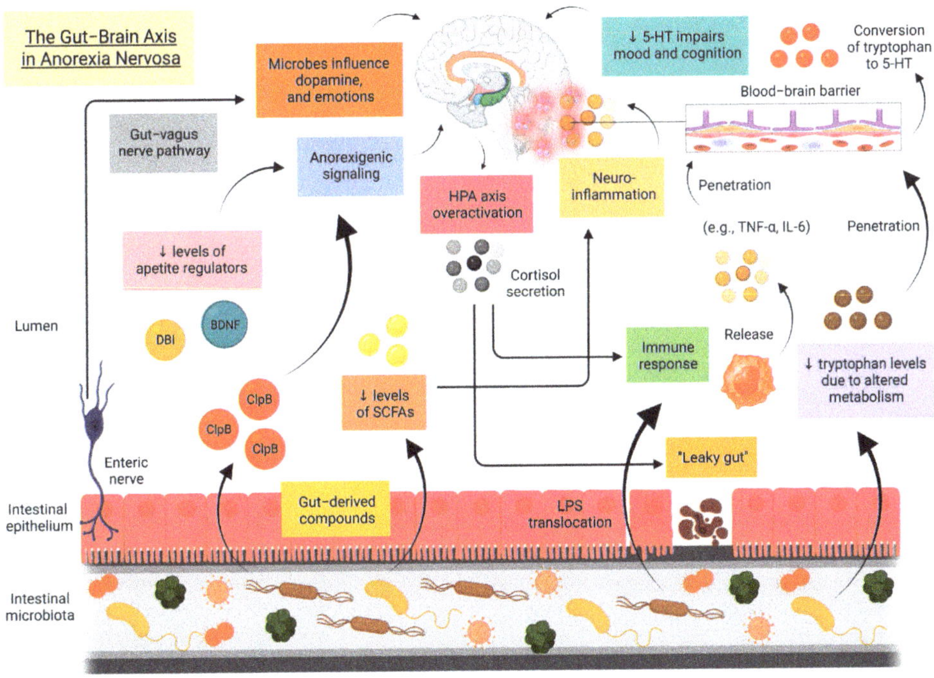

Figure 5. The "busy" gut–brain axis in anorexia nervosa. This figure depicts the role of the GBA in the development and progression of AN, through interconnected neuroimmune, neuroendocrine, and neurotrophic pathways. A central event in this cascade is the loss of intestinal barrier integrity, commonly referred to as "leaky gut", which permits the translocation of LPS into the systemic circulation, leading to peripheral immune activation and promoting neuroinflammation. These effects are exacerbated by reduced production of SCFAs and by sustained hypercortisolemia resulting from chronic hyperactivation of the HPA axis. The bacterial chaperone protein ClpB exerts anorexigenic effects via molecular mimicry with α-MSH, which are compounded by decreased circulating levels of appetite regulators, such as DBI and BDNF. Lastly, disturbances in tryptophan metabolism reduce serotonin bioavailability, while impaired autonomic regulation, particularly through altered vagus nerve signaling, further contributes to disease progression. Abbreviations: 5-HT: 5-hydroxytryptamine;

α-MSH: α-melanocyte-stimulating hormone; AN: anorexia nervosa; BDNF: brain-derived neurotrophic factor; ClpB: caseinolytic peptidase B; DBI: diazepam-binding inhibitor; GBA: gut–brain axis; HPA: hypothalamic–pituitary–adrenal; IL-6: interleukin-6; LPS: lipopolysaccharides; SCFA: short-chain fatty acid; TNF-α: tumor necrosis factor-alpha; ↓: decrease. Created in BioRender. Kounatidis, D. (2025) https://BioRender.com/o4pd1rv (assessed on 7 September 2025).

13. Anorexia Nervosa Management: Current Strategies and Therapeutic Perspectives

13.1. Current Strategies

The management of AN requires a coordinated, multidisciplinary approach integrating the expertise of psychiatrists, psychologists, and medical–nutrition specialists. Nutritional education should be initiated at the outset and reinforced throughout treatment, with the goal of establishing sustainable, health-promoting eating behaviors. The involvement of registered dietitians is essential to ensure safe and progressive weight restoration. Traditional refeeding protocols typically began with approximately 1200 kcal/day; however, more recent strategies employing 1400–1600 kcal/day have been shown to be safe when patients are closely monitored. Vigilance for refeeding syndrome remains paramount, given its potential lethality during early nutritional rehabilitation. Electrolyte levels must be assessed regularly, and supplementation with thiamine, riboflavin, vitamin D, calcium, and other essential micronutrients should be administered. A realistic weight-gain target is approximately 0.5 kg per week [183]. Table 1 depicts recent studies of nutritional management in patients with AN.

Table 1. Major studies from the past three years on medical nutrition management in patients with anorexia nervosa.

Author, Year	Type of Study	Findings	Remarks
Brennan et al., 2025 [184]	Semi-structured interviews were undertaken by 11 clinicians who had been working with adolescents aged 12–18 y.o. with anorexia nervosa (AN) who had been on a daily program.	The role of the dietician was pointed out as a crucial component of this intervention daily program.	The personalized nature of each nutritional interventionist turned out to be the most important determinant between the different phases of the nutritional program, especially between the weight restoration and maintenance phase.
Bohon et al., 2025 [185]	The authors have summarized key recommendations for AN management in the United States.	Early recognition of restrictive AN in the population together with family interventions and the help from a multidisciplinary team is of utmost importance.	The importance of screening tools for eating disorders in the general population has been stressed in this review manuscript.
Dozio et al., 2025 [186]	Observational study involving 79 patients with AN in Italy who had been admitted for management of AN.	The results after 6 months of medical management of these patients supported that early fat and protein restoration in these patients together with their close monitoring were very helpful.	The authors concluded that sustained medical nutritionists care and the implementation of a multidisciplinary team for as long as 6 months revealed encouraging results.
Brennan et al., 2024 [187]	A retrospective cohort study involving 92 patients with AN from Maudsley Center for Child and Adolescent Eating Disorders in London, United Kingdom.	The results of this study supported that there was no difference regarding weight gain among patients who received dietician's guidance compared to those who did not.	The authors concluded that more information is needed in order to timely intervene with the consultation of a registered dietician among patients with AN.

Table 1. *Cont.*

Author, Year	Type of Study	Findings	Remarks
Ayrolles et al., 2024 [188]	A scoping review from France regarding patients with early-onset pre-pubertal AN.	The authors have pointed out that there is lack of information regarding this particular group of early-onset AN.	The authors concluded that much more research is mandatory in this early-onset pre-pubertal group of patients with AN to better characterize and manage them.
Teo et al., 2024 [189]	A retrospective cohort study involving 47 patients with AN in Singapore.	Patients with AN can achieve weight restoration with a more dynamic management and for less duration of in-hospital stay with safety.	The authors concluded that in Asia and from their patient cohort, a shorter and more dynamic calorie administration could result in a lower duration of in-hospital stay. However, they have highlighted that the incidence of the refeeding syndrome should be further studied in this Asian group of patients.
Campos del Portillo et al., 2024 [190]	A consensus report from Spain.	The authors reported that outpatient management was most often preferred. However, there should be frequent re-assessment even at out of hospital care of these patients by a multidisciplinary team.	The authors point out the significance of alertness for the refeeding syndrome, especially among very malnourished patients with AN.
Hunter et al., 2024 [191]	A qualitative study interviewing 14 female patients with AN in the UK to evaluate the effects of a plant-based diet.	The authors proposed that a plant-based diet could be motivating for patients with AN.	The authors concluded that the participants found this plant-based diet motivating and that it was giving them good support for healthier eating habits.
Cinelli et al., 2023 [192]	An observational study including 8 outpatients with AN in Rome, Italy.	The patients were seen at baseline and after 6 months. The Mediterranean Diet was reported to be a good choice regarding weight management among these patients with AN.	The authors concluded that the Mediterranean Diet could offer a very promising dietary plan in patients with AN. However, the group consisted of only 8 patients.
Sthener et al., 2023 [193]	An editorial manuscript.	The authors reported that the initiation of nutritional management should be gradually increased during the first days due to the increased risk of refeeding syndrome.	The authors pointed out that the use of parenteral nutrition should be restricted among patients with AN due to increased risk for infections and metabolic adverse effects.

Psychotherapy remains the cornerstone of AN treatment. Evidence-based modalities include family-based interventions, cognitive behavioral therapy (CBT), and the Maudsley model for adults. Emerging approaches, such as the provision of calorie-dense foods in prepackaged formats, have been investigated to enhance dietary adherence. Acceptance and commitment therapy (ACT), which emphasizes restructuring maladaptive cognitions, fostering intrinsic motivation, and reinforcing value-driven behaviors, has also garnered increasing attention. On the contrary, pharmacological interventions have demonstrated limited efficacy. Antidepressants, in particular, yield minimal clinical benefit. To date, no effective pharmacological treatment for AN has been established, a therapeutic gap that has prompted intensified research into alternative strategies [194,195].

13.2. Challenges and Therapeutic Perspectives

13.2.1. Atypical Antipsychotics

The limited efficacy of antidepressants in AN has prompted growing interest in the therapeutic potential of atypical antipsychotics. Among second-generation agents, olanzapine and aripiprazole appear particularly promising in cases of severe or treatment-resistant illness. Although the evidence remains inconsistent, recent studies identify olanzapine as a potential candidate, with reported benefits including weight restoration and attenuation of obsessive food-related thoughts. Qualitative research has described olanzapine treatment as a liminal "rite of passage", offering adolescents with AN a sense of autonomy and empowerment, while simultaneously underscoring tensions around weight gain and communication with parents and clinicians [196].

13.2.2. Psychedelics, Dronabinol and Ketamine

Psychedelic compounds have recently gained attention as potential therapeutic strategies for AN. Psilocybin, a serotonergic agent derived from fungi, may modulate mood, perception, and cognition and has been proposed as a candidate treatment. In a phase 1 open-label feasibility study, women with AN received a single 25 mg dose of psilocybin accompanied by psychological support. The intervention was well tolerated, with only two brief, asymptomatic episodes of hypoglycemia, and participants generally considered the treatment acceptable [197]. Supporting these findings, a large observational study of 6612 individuals with EDs reported self-perceived symptom improvement specifically associated with cannabis or psychedelic use, although such anecdotal evidence requires confirmation in controlled trials [198]. Evidence is also available for the synthetic cannabinoid agonist dronabinol, derived from a randomized, double-blind, placebo-controlled crossover study involving twenty-five women aged over 18 years with AN of at least five years' duration. Over the four-week treatment period, dronabinol administration resulted in small but statistically significant weight gain ($p < 0.01$) and was generally well tolerated [199].

Research has likewise examined ketamine, an N-methyl-D-aspartate (NMDA) receptor antagonist with established antidepressant efficacy. Beyond its mood-regulating properties, ketamine may enhance neuroplasticity, neurogenesis, and hippocampal integrity while mitigating neuroinflammation and depressive symptoms that impede recovery in AN. When combined with psychotherapy, ketamine could further promote cognitive flexibility and therapeutic engagement. Early findings, derived primarily from case studies, are encouraging. Nevertheless, further research is required to establish optimal dosing regimens, clinical efficacy, and long-term safety, particularly given the ongoing concerns surrounding ketamine use and potential safety risks [200].

13.2.3. Anti-Inflammatory and Mitochondria-Targeted Strategies

Anti-inflammatory interventions, particularly those involving cyclooxygenase (COX) inhibition, have been investigated for their potential to mitigate anxiety and mood disturbances while promoting neural resilience and synaptic integrity [195]. Another attractive therapeutic target is growth differentiation factor 15 (GDF15), a stress-responsive cytokine implicated in mitochondrial oxidative stress and energy homeostasis. Emerging evidence indicates that subsets of patients with AN display elevated circulating GDF15 levels [201]. In conditions such as cancer cachexia and chemotherapy-induced anorexia, increased GDF15 concentrations have been associated with appetite suppression through the activation of emetic and nausea-related pathways [201,202]. These findings have led to the hypothesis that GDF15 neutralization might ameliorate weight loss in AN. However, experimental studies using genetic ablation or antibody-mediated neutralization of GDF15 in sepsis models have demonstrated negligible effects on appetite or body weight, complicating the translational relevance of this approach [203]. Consequently, it remains uncertain whether GDF15-targeted interventions could yield tangible benefits in promoting weight restoration. The paucity of large-scale, controlled studies highlights the need for further investigation into GDF15 modulation, as delineating its precise role in AN may offer new perspectives for metabolic and behavioral therapeutic strategies.

13.2.4. Recombinant Human Leptin Analogs

As previously discussed, hypoleptinemia in AN may exacerbate weight-related anxieties and obsessive food-related cognitions. Recombinant leptin analogs, particularly metreleptin, have therefore been proposed as potential therapeutic agents targeting both metabolic and affective dysregulation [204]. Originally approved for the management of metabolic complications in lipodystrophy, metreleptin has demonstrated diverse endocrine-regulatory properties, including the restoration of reproductive function in women with hypothalamic amenorrhea and the attenuation of weight-gain-related fear and anxiety [205,206]. A notable case report described a 16-year-old girl with atypical AN who experienced rapid improvement in mood, diminished weight-related phobia, and facilitated hospital discharge following an 11-day course of metreleptin administration [207]. Beyond AN, leptin signaling has also become a focus of investigation in oncology, where leptin antagonists have shown encouraging yet unconfirmed results in clinical settings [208]. Similarly, leptin-based interventions for AN remain largely experimental, as no RCTs have yet been conducted. Moreover, the modest weight reduction observed in some metreleptin-treated cases warrants cautious interpretation and vigilance regarding its catabolic potential. Notably, although metreleptin has been reported to transiently enhance hunger shortly after subcutaneous administration, this effect appears to wane over time and may even be followed by mild weight loss [209]. Taken together, while isolated case reports suggest that metreleptin may offer psychological and metabolic benefits in selected individuals with AN, its clinical application, even in short courses, remains contentious and requires rigorous validation through well-designed clinical trials.

13.2.5. Gut-Directed Therapeutic Interventions

The gut microbiome has emerged as an attractive therapeutic target in AN. Probiotics and prebiotics, acting synergistically, support microbial balance and promote immune, metabolic, and gastrointestinal health. They have demonstrated benefits in type 2 diabetes (T2D), obesity, and IBD [210]. Preliminary findings in AN suggest possible roles in modulating immune activity and alleviating comorbid depression and anxiety [211]. In a randomized, double-blind, placebo-controlled trial, *Lactobacillus reuteri* had no effect on constipation at three months but was associated with greater stool frequency and faster normalization of body weight at six months. Additional benefits included improved BMD and vitamin D levels, suggesting that probiotics may represent a safe adjunct to nutritional rehabilitation in pediatric AN [212]. However, results across studies remain heterogeneous, and larger trials are needed [195].

Fecal microbiota transplantation (FMT) has also been investigated. Transplantation of microbiota from AN patients into mice reduced food intake and locomotor activity while altering microbial profiles and appetite-regulating hormones. Conversely, FMT from healthy donors reversed these effects, while data from meta-analyses support its role in enhancing microbial diversity [213]. Findings to date remain mixed. For example, Kooij et al. reported that FMT from AN patients did not alter cognitive flexibility, anxiety, or dopaminergic signaling in animal models, highlighting the need for more refined experimental approaches [214]. Meanwhile, an ongoing RCT is investigating whether direct colonic administration of SCFAs over six weeks can influence stress reactivity, eating behaviors, BMI, and restrictive patterns in AN [215].

13.2.6. Neuromodulation with Intermittent Theta-Burst Stimulation (iTBS)

Intermittent theta-burst stimulation (iTBS) is a patterned form of non-invasive brain stimulation that delivers bursts of magnetic pulses to modulate cortical excitability and synaptic plasticity. While iTBS has demonstrated efficacy in major depressive disorder and other neuropsychiatric conditions, its use in AN remains exploratory and largely confined to pilot investigations [195]. A current feasibility study is enrolling 66 adolescents and young adults with AN who have shown suboptimal response to conventional care. Participants are randomized to receive either active or sham iTBS in conjunction with standard treatment. The primary outcome measure is four-month treatment retention, with secondary endpoints including changes in symptom severity, quality of life (QoL), and neurobiological indices. The results of this trial are anticipated to inform the methodological and clinical design of future large-scale studies evaluating iTBS as an adjunctive therapeutic strategy in AN [216].

Collectively, the emerging interventions discussed above represent innovative efforts to directly modulate the neurobiological underpinnings of AN. Although these strategies hold theoretical and preliminary clinical promise for improving outcomes and enhancing QoL, most remain at preclinical or early translational stages and are not yet suitable for routine clinical implementation. The existing body of research is limited by small sample sizes, lack of randomization, and reliance on uncontrolled or case-based data, thereby precluding firm causal inferences. Of note, the 2023 update of the World Federation of Societies of Biological Psychiatry (WFSBP) guidelines on the pharmacological management of EDs emphasized that olanzapine has not received marketing authorization from any medicinal regulatory agency for the treatment of AN, underscoring the continuing absence of an approved pharmacotherapy for this condition [217]. Figure 6 summarizes the range of emerging therapeutic strategies currently under investigation for AN management.

Figure 6. Emerging therapies in anorexia nervosa. Abbreviations: COX: cyclooxygenase; FMT: fecal microbiota transplantation; GDF15: growth and differentiation factor 15; iTBS: intermittent theta-burst stimulation; ROS: reactive oxygen species; SCFA: short-chain fatty acid. Created in BioRender. Kounatidis, D. (2025) https://BioRender.com/jqw1j8l (assessed on 3 November 2025).

14. Conclusions

AN remains one of the most challenging disorders at the interface of psychiatry, internal medicine, and neuroscience. Its complex pathophysiology, spanning genetic, immunological, and gut–brain interactions, underscores the need for an integrative approach that transcends traditional psychopathological frameworks. While substantial progress has been made in delineating its systemic complications and molecular mechanisms, translation into effective and durable treatments has lagged behind. Current therapeutic strategies, though essential, often fall short due to high relapse rates and limited pharmacological efficacy. Moving forward, research should prioritize the identification of novel biological targets, the refinement of multimodal interventions, and the personalization of treatment according to the individual patient's clinical and biological profile. Importantly, bridging basic science discoveries with clinical practice will be critical to improving outcomes and reducing the burden of this devastating disorder.

Author Contributions: D.C.K. and N.G.V. conceived the idea of the review, organized its plan, wrote, reviewed, edited and supervised the manuscript; N.G.V. performed the literature search and prepared all references and the table; D.C.K. made the figures. All authors have read and agreed to the published version of the manuscript.

Funding: This research received no external funding.

Institutional Review Board Statement: Not applicable.

Informed Consent Statement: Not applicable.

Data Availability Statement: Not applicable.

Conflicts of Interest: The authors declare no conflicts of interest.

Abbreviations

5-HT: serotonin; ACT: acceptance and commitment therapy; ACTH: adrenocorticotropic hormone; ADH: antidiuretic hormone; ALT: alanine aminotransferase; AN: anorexia nervosa; AST: aspartate aminotransferase; BDNF: brain-derived neurotrophic factor; BMD: bone mineral density; BMI: body mass index; CBC: complete blood count; CD: Crohn's disease; CBT: cognitive behavioral therapy; CeD: celiac disease; CKD: chronic kidney disease; CKD-EPI: Chronic Kidney Disease Epidemiology Collaboration; CNS: central nervous system; COX: cyclooxygenase; CRH: corticotropin-releasing hormone; DEXA: dual-energy X-ray absorptiometry; DBI: diazepam-binding inhibitor; ED: eating disorder; eGFR: estimated glomerular filtration rate; ESRD: end-stage renal disease; FGF21: fibroblast growth factor 21; FHH: functional hypogonadotropic hypogonadism; fT3: free triiodothyronine; fT4: free thyroxine; FSH: follicle-stimulating hormone; FMT: fecal microbiota transplantation; G-CSF: granulocyte colony-stimulating factor; GDF15: growth differentiation factor 15; GBA: gut–brain axis; GH: growth hormone; GHSR: growth hormone secretagogue receptor; GHSR1a: growth hormone secretagogue receptor 1a; GnRH: gonadotropin-releasing hormone; HIF-1α: hypoxia-inducible factor 1-alpha; HPA: hypothalamic–pituitary–adrenal; HPG: hypothalamic-pituitary-gonadal; HPT: hypothalamic–pituitary–thyroid; HF: heart failure; HLA: human leukocyte antigen; HHV-6: human herpesvirus 6; HHV-7: human herpesvirus 7; IBD: inflammatory bowel disease; IN-OT: intranasal oxytocin; INR: international normalized ratio; IR: insulin resistance; iTBS: intermittent theta-burst stimulation; IGF-1: insulin-like growth factor 1; LCN2: lipocalin-2; LEAP2: liver-expressed antimicrobial peptide 2; LH: luteinizing hormone; LPS: lipopolysaccharide; MACE: major adverse cardiovascular event; MASLD: metabolic dysfunction-associated steatotic liver disease; MIC-1: macrophage inhibitory cytokine-1; MRI: magnetic resonance imaging; MMP-8: matrix metalloproteinase-8; NMDA: N-methyl-D-aspartate; NK: natural killer; NLR: neutrophil-to-lymphocyte ratio; NLRP3: NLR family pyrin domain-containing protein 3; NTIS: non-thyroidal illness syndrome; PGE2: prostaglandin E2; PDW: platelet distribution width; PPHS: posterior pituitary high signal; PNS: peripheral nervous system; PUFAs: polyunsaturated fatty acids; QoL: quality of life; RCT: randomized controlled trial; rT3: reverse triiodothyronine; SCFAs: short-chain fatty acids; DSM-5: Diagnostic and Statistical Manual of Mental Disorders, Fifth Edition; T2D: type 2 diabetes; T3: triiodothyronine; TGF-β: transforming growth factor-beta; TLR-2: toll-like receptor 2; TLR-4: toll-like receptor 4; TNF-α: tumor necrosis factor-alpha; TRH: thyrotropin-releasing hormone; TSH: thyroid-stimulating hormone; UC: ulcerative colitis; VTE: venous thromboembolism; WFSBP: World Federation of Societies of Biological Psychiatry.

References

1. Auger, N.; Potter, B.J.; Ukah, U.V.; Low, N.; Israël, M.; Steiger, H.; Healy-Profitós, J.; Paradis, G. Anorexia nervosa and the long-term risk of mortality in women. *World Psychiatry* **2021**, *20*, 448–449. [CrossRef] [PubMed]
2. Søeby, M.; Gribsholt, S.B.; Clausen, L.; Richelsen, B. Overall and cause-specific mortality in anorexia nervosa; impact of psychiatric comorbidity and sex in a 40-year follow-up study. *Int. J. Eat. Disord.* **2024**, *57*, 1842–1853. [CrossRef]
3. Van Eeden, A.E.; Van Hoeken, D.; Hoek, H.W. Incidence, prevalence and mortality of anorexia nervosa and bulimia nervosa. *Curr. Opin. Psychiatry* **2021**, *34*, 515–524. [CrossRef]
4. Dalle Grave, R.; Sartirana, M.; El Ghoch, M.; Calugi, S. DSM-5 severity specifiers for anorexia nervosa and treatment outcomes in adult females. *Eat. Behav.* **2018**, *31*, 18–23. [CrossRef]
5. Hebebrand, J.; Gradl-Dietsch, G.; Peters, T.; Correll, C.U.; Haas, V. The Diagnosis and Treatment of Anorexia Nervosa in Childhood and Adolescence. *Dtsch. Arztebl. Int.* **2024**, *121*, 164–174. [CrossRef]
6. Walsh, B.T.; Hagan, K.E.; Lockwood, C. A systematic review comparing atypical anorexia nervosa and anorexia nervosa. *Int. J. Eat. Disord.* **2023**, *56*, 798–820. [CrossRef] [PubMed]
7. Adams, D.M.; Reay, W.R.; Cairns, M.J. Multiomic prioritisation of risk genes for anorexia nervosa. *Psychol. Med.* **2023**, *53*, 6754–6762. [CrossRef]
8. Steiger, H.; Booij, L.; Thaler, L.; St-Hilaire, A.; Israël, M.; Casey, K.F.; Oliverio, S.; Crescenzi, O.; Lee, V.; Turecki, G.; et al. DNA methylation in people with anorexia nervosa: Epigenome-wide patterns in actively ill, long-term remitted, and healthy-eater women. *World J. Biol. Psychiatry* **2023**, *24*, 254–259. [CrossRef]
9. Sirufo, M.M.; Magnanimi, L.M.; Ginaldi, L.; De Martinis, M. Anorexia nervosa and autoimmune comorbidities: A bidirectional route? *CNS Neurosci. Ther.* **2022**, *28*, 1921–1929. [CrossRef]
10. Butler, M.J.; Perrini, A.A.; Eckel, L.A. The Role of the Gut Microbiome, Immunity, and Neuroinflammation in the Pathophysiology of Eating Disorders. *Nutrients* **2021**, *13*, 500. [CrossRef] [PubMed]
11. Davidson, A.R.; Braham, S.; Dasey, L.; Reidlinger, D.P. Physicians' perspectives on the treatment of patients with eating disorders in the acute setting. *J. Eat. Disord.* **2019**, *7*, 1. [CrossRef] [PubMed]
12. Guinhut, M.; Godart, N.; Benadjaoud, M.A.; Melchior, J.C.; Hanachi, M. Five-year mortality of severely malnourished patients with chronic anorexia nervosa admitted to a medical unit. *Acta Psychiatr. Scand.* **2021**, *143*, 130–140. [CrossRef]
13. Stedal, K.; Ely, A.V.; Kurniadi, N.; Lopez, E.; Kaye, W.H.; Wierenga, C.E. A process approach to verbal memory assessment: Exploratory evidence of inefficient learning in women remitted from anorexia nervosa. *J. Clin. Exp. Neuropsychol.* **2018**, *41*, 653–663. [CrossRef]
14. Nickel, K.; Joos, A.; Tebartz van Elst, L.; Matthis, J.; Holovics, L.; Endres, D.; Zeeck, A.; Hartmann, A.; Tüscher, O.; Maier, S. Recovery of cortical volume and thickness after remission from acute anorexia nervosa. *Int. J. Eat. Disord.* **2018**, *51*, 1056–1069. [CrossRef]
15. De La Cruz, F.; Schumann, A.; Rieger, K.; Giuliano, M.D.; Bär, K.J. Fibre-specific white matter changes in anorexia nervosa. *Psychiatry Res. Neuroimaging* **2023**, *336*, 111736. [CrossRef] [PubMed]
16. Gaiaschi, L.; Priori, E.C.; Mensi, M.M.; Verri, M.; Buonocore, D.; Parisi, S.; Hernandez, L.N.Q.; Brambilla, I.; Ferrari, B.; De Luca, F.; et al. New perspectives on the role of biological factors in anorexia nervosa: Brain volume reduction or oxidative stress, which came first? *Neurobiol. Dis.* **2024**, *199*, 106580. [CrossRef] [PubMed]
17. Collantoni, E.; Madan, C.R.; Meneguzzo, P.; Chiappini, I.; Tenconi, E.; Manara, R.; Favaro, A. Cortical Complexity in Anorexia Nervosa: A Fractal Dimension Analysis. *J. Clin. Med.* **2020**, *9*, 833. [CrossRef]
18. Gupta, Y.; de la Cruz, F.; Rieger, K.; di Giuliano, M.; Gaser, C.; Cole, J.; Breithaupt, L.; Holsen, L.M.; Eddy, K.T.; Thomas, J.J.; et al. Does restrictive anorexia nervosa impact brain aging? A machine learning approach to estimate age based on brain structure. *Comput. Biol. Med.* **2025**, *194*, 110484. [CrossRef]
19. Mitchell, J.E.; Peterson, C.B. Anorexia Nervosa. *N. Engl. J. Med.* **2020**, *382*, 1343–1351. [CrossRef]
20. Brodrick, B.B.; Adler-Neal, A.L.; Palka, J.M.; Mishra, V.; Aslan, S.; McAdams, C.J. Structural brain differences in recovering and weight-recovered adult outpatient women with anorexia nervosa. *J. Eat. Disord.* **2021**, *9*, 108. [CrossRef]

21. De La Cruz, F.; Schumann, A.; Suttkus, S.; Helbing, N.; Zopf, R.; Bär, K.J. Cortical thinning and associated connectivity changes in patients with anorexia nervosa. *Transl. Psychiatry* **2021**, *11*, 95. [CrossRef]
22. Walton, E.; Bernardoni, F.; Batury, V.L.; Bahnsen, K.; Larivière, S.; Abbate-Daga, G.; Andres-Perpiña, S.; Bang, L.; Bischoff-Grethe, A.; Brooks, S.J.; et al. Brain Structure in Acutely Underweight and Partially Weight-Restored Individuals with Anorexia Nervosa: A Coordinated Analysis by the ENIGMA Eating Disorders Working Group. *Biol. Psychiatry* **2022**, *92*, 730–738. [CrossRef]
23. Asami, T.; Takaishi, M.; Nakamura, R.; Yoshimi, A.; Konishi, J.; Aoyama, K.; Fujita, J.; Miyazaki, H.; Aoki, Y.; Asanuma, K.; et al. Structural brain abnormalities in adolescent patients with anorexia nervosa at both the acute and weight-recovered phase. *Brain Imaging Behav.* **2022**, *16*, 1372–1380. [CrossRef]
24. Fuglset, T.S. Is set-shifting and central coherence in anorexia nervosa influenced by body mass index, anxiety or depression? A systematic review. *BMC Psychiatry* **2021**, *21*, 137. [CrossRef] [PubMed]
25. Fuglset, T.S. Set-shifting, central coherence and decision-making in individuals recovered from anorexia nervosa: A systematic review. *J. Eat. Disord.* **2019**, *7*, 22. [CrossRef]
26. Oudman, E.; Wijnia, J.W.; Oey, M.J.; van Dam, M.J.; Postma, A. Preventing Wernicke's encephalopathy in anorexia nervosa: A systematic review. *Psychiatry Clin. Neurosci.* **2018**, *72*, 774–779. [CrossRef] [PubMed]
27. Jenkins, Z.M.; Castle, D.J.; Eikelis, N.; Phillipou, A.; Lambert, G.W.; Lambert, E.A. Autonomic nervous system function in women with anorexia nervosa. *Clin. Auton. Res.* **2022**, *32*, 29–42. [CrossRef] [PubMed]
28. Puckett, L.; Grayeb, D.; Khatri, V.; Cass, K.; Mehler, P. A Comprehensive Review of Complications and New Findings Associated with Anorexia Nervosa. *J. Clin. Med.* **2021**, *10*, 2555. [CrossRef]
29. Tseng, M.M.; Chien, L.N.; Tu, C.Y.; Liu, H.Y. Mortality in anorexia nervosa and bulimia nervosa: A population-based cohort study in Taiwan, 2002–2017. *Int. J. Eat. Disord.* **2023**, *56*, 1135–1144. [CrossRef]
30. Cost, J.; Krantz, M.J.; Mehler, P.S. Medical complications of anorexia nervosa. *Cleve. Clin. J. Med.* **2020**, *87*, 361–366. [CrossRef]
31. Tseng, M.M.; Chiou, K.R.; Shao, J.Y.; Liu, H.Y. Incidence and Risk of Cardiovascular Outcomes in Patients with Anorexia Nervosa. *JAMA Netw. Open* **2024**, *7*, e2451094. [CrossRef] [PubMed]
32. Sachs, K.V.; Harnke, B.; Mehler, P.S.; Krantz, M.J. Cardiovascular complications of anorexia nervosa: A systematic review. *Int. J. Eat. Disord.* **2016**, *49*, 238–248. [CrossRef] [PubMed]
33. Burns, J.; Shank, C.; Ganigara, M.; Saldanha, N.; Dhar, A. Cardiac complications of malnutrition in adolescent patients: A narrative review of contemporary literature. *Ann. Pediatr. Cardiol.* **2021**, *14*, 501–506. [CrossRef] [PubMed]
34. Giovinazzo, S.; Sukkar, S.G.; Rosa, G.M.; Zappi, A.; Bezante, G.P.; Balbi, M.; Brunelli, C. Anorexia nervosa and heart disease: A systematic review. *Eat. Weight Disord.-Stud. Anorex. Bulim. Obes.* **2019**, *24*, 199–207. [CrossRef]
35. Goldfarb, M.; De Hert, M.; Detraux, J.; Di Palo, K.; Munir, H.; Music, S.; Piña, I.; Ringen, P.A. Severe mental illness and cardiovascular disease: JACC state-of-the-art review. *J. Am. Coll. Cardiol.* **2022**, *80*, 918–933. [CrossRef]
36. Cass, K.; McGuire, C.; Bjork, I.; Sobotka, N.; Walsh, K.; Mehler, P.S. Medical Complications of Anorexia Nervosa. *Psychosomatics* **2020**, *61*, 625–631. [CrossRef]
37. Benayon, M.; Latchupatula, L.; Kacer, E.; Shanjer, M.; Weiss, E.; Amar, S.; Zweig, N.; Ghadim, M.; Portman, R.; Balakrishnan, N.; et al. QTc Interval Prolongation and Its Association With Electrolyte Abnormalities and Psychotropic Drug Use Among Patients With Eating Disorders. *CJC Pediatr. Congenit. Heart Dis.* **2023**, *3*, 14–21. [CrossRef]
38. Mehler, P.S.; Anderson, K.; Bauschka, M.; Cost, J.; Farooq, A. Emergency room presentations of people with anorexia nervosa. *J. Eat. Disord.* **2023**, *11*, 16. [CrossRef]
39. Lamzabi, I.; Syed, S.; Reddy, V.B.; Jain, R.; Harbhajanka, A.; Arunkumar, P. Myocardial changes in a patient with anorexia nervosa: A case report and review of literature. *Am. J. Clin. Pathol.* **2015**, *143*, 734–737. [CrossRef]
40. Arnés-García, D.; Zamorano-García, E.; Gallego-Romero, I. Beri-beri: A rare and reversible cause of heart failure. *Med. Clin.* **2025**, *164*, 106893. [CrossRef]
41. Crook, M.A. Cardiac abnormalities in the refeeding syndrome. *Nutrition* **2017**, *35*, 146–147. [CrossRef]
42. O'Meara, D.; Sheward, L.; Hartman-Munick, S.; Addison, J.; Abu-El-Haija, A. Craving Answers. *N. Engl. J. Med.* **2022**, *386*, 880–886. [CrossRef]
43. Jensen, V.M.; Stoving, R.K.; Andersen, P.E. Anorexia nervosa with massive pulmonary air leak and extraordinary propagation. *Int. J. Eat. Disord.* **2017**, *50*, 451–453. [CrossRef]
44. Nitsch, A.; Kearns, M.; Mehler, P. Pulmonary complications of eating disorders: A literature review. *J. Eat. Disord.* **2023**, *11*, 12. [CrossRef] [PubMed]
45. Cook, V.J.; Coxson, H.O.; Mason, A.G.; Bai, T.R. Bullae, bronchiectasis and nutritional emphysema in severe anorexia nervosa. *Can. Respir. J.* **2001**, *8*, 361–365. [CrossRef]
46. Vootla, V.R.; Daniel, M. Abnormal Liver Function Tests in an Anorexia Nervosa Patient and an Atypical Manifestation of Refeeding Syndrome. *Case Rep. Gastroenterol.* **2015**, *9*, 261–265. [CrossRef]
47. Schuster, K.; Staffeld, A.; Zimmermann, A.; Böge, N.; Lang, S.; Kuhla, A.; Frintrop, L. Starvation in Mice Induces Liver Damage Associated with Autophagy. *Nutrients* **2024**, *16*, 1191. [CrossRef] [PubMed]

48. Kheloufi, M.; Boulanger, C.M.; Durand, F.; Rautou, P.E. Liver autophagy in anorexia nervosa and acute liver injury. *BioMed Res. Int.* **2014**, *2014*, 701064. [CrossRef]
49. Rautou, P.E.; Cazals-Hatem, D.; Moreau, R.; Francoz, C.; Feldmann, G.; Lebrec, D.; Ogier-Denis, E.; Bedossa, P.; Valla, D.; Durand, F. Acute liver cell damage in patients with anorexia nervosa: A possible role of starvation-induced hepatocyte autophagy. *Gastroenterology* **2008**, *135*, 840–848.e3. [CrossRef]
50. Yoshida, T.; Namiki, T.; Yamaga, M.; Onishi, S.; Takemoto, M. Iron overload may be critical for liver dysfunction in anorexia nervosa, and the role of haematocrit-adjusted albumin in assessing nutritional status: A case report. *BMC Pediatr.* **2023**, *23*, 547. [CrossRef]
51. Fanin, A.; Miele, L.; Bertolini, E.; Giorgini, A.; Pontiroli, A.E.; Benetti, A. Liver alterations in anorexia nervosa are not caused by insulin resistance. *Intern. Emerg. Med.* **2020**, *15*, 337–339. [CrossRef]
52. Rosen, E.; Sabel, A.L.; Brinton, J.T.; Catanach, B.; Gaudiani, J.L.; Mehler, P.S. Liver dysfunction in patients with severe anorexia nervosa. *Int. J. Eat. Disord.* **2016**, *49*, 151–158. [CrossRef]
53. Cuntz, U.; Voderholzer, U. Liver Damage Is Related to the Degree of Being Underweight in Anorexia Nervosa and Improves Rapidly with Weight Gain. *Nutrients* **2022**, *14*, 2378. [CrossRef] [PubMed]
54. Rosen, E.; Bakshi, N.; Watters, A.; Rosen, H.R.; Mehler, P.S. Hepatic Complications of Anorexia Nervosa. *Dig. Dis. Sci.* **2017**, *62*, 2977–2981. [CrossRef]
55. Fong, H.F.; Divasta, A.D.; Difabio, D.; Ringelheim, J.; Jonas, M.M.; Gordon, C.M. Prevalence and predictors of abnormal liver enzymes in young women with anorexia nervosa. *J. Pediatr.* **2008**, *153*, 247–253. [CrossRef]
56. Narayanan, V.; Gaudiani, J.L.; Mehler, P.S. Serum albumin levels may not correlate with weight status in severe anorexia nervosa. *Eat. Disord.* **2009**, *17*, 322–326. [CrossRef] [PubMed]
57. Giordano, F.; Arnone, S.; Santeusanio, F.; Pampanelli, S. Brief elevation of hepatic enzymes due to liver ischemia in anorexia nervosa. *Eat. Weight Disord.-Stud. Anorex. Bulim. Obes.* **2010**, *15*, e294–e297. [CrossRef] [PubMed]
58. Kurisu, K.; Sato, K.; Matsuoka, M.; Otani, M.; Yoshiuchi, K. Thrombocytopenia and PT-INR in patients with anorexia nervosa and severe liver dysfunction. *Biopsychosoc. Med.* **2023**, *17*, 9. [CrossRef]
59. Gaudiani, J.L.; Kashuk, J.L.; Chu, E.S.; Narayanan, V.; Mehler, P.S. The use of thrombelastography to determine coagulation status in severe anorexia nervosa: A case series. *Int. J. Eat. Disord.* **2010**, *43*, 382–385. [CrossRef]
60. Miyachi, Y.; Sakiko, K.; Yokoi, T.; Kaido, T. Rapidly progressing severe coagulopathy and thrombocytopenia in extreme anorexia nervosa patient with small bowel strangulation: A case report. *Int. J. Surg. Case Rep.* **2023**, *112*, 108985. [CrossRef]
61. Koga, A.; Toda, K.; Tatsushima, K.; Matsuubayashi, S.; Tamura, N.; Imamura, M.; Kawai, K. Portal hypertension in prolonged anorexia nervosa with laxative abuse: A case report of three patients. *Int. J. Eat. Disord.* **2019**, *52*, 211–215. [CrossRef]
62. Loisel, A.; Caunes, A.; Birsen, R.; Moro, M.R.; Boussaid, I.; Blanchet, C. When refeeding is not enough: Severe and prolonged pancytopenia in an adolescent with anorexia nervosa. *Eat. Weight. Disord.-Stud. Anorex. Bulim. Obes.* **2022**, *27*, 3797–3801. [CrossRef]
63. Norris, M.L.; Harrison, M.E.; Isserlin, L.; Robinson, A.; Feder, S.; Sampson, M. Gastrointestinal complications associated with anorexia nervosa: A systematic review. *Int. J. Eat. Disord.* **2016**, *49*, 216–237. [CrossRef]
64. Perkins, S.J.; Keville, S.; Schmidt, U.; Chalder, T. Eating disorders and irritable bowel syndrome: Is there a link? *J. Psychosom. Res.* **2005**, *59*, 57–64. [CrossRef] [PubMed]
65. Morris, L.G.; Stephenson, K.E.; Herring, S.; Marti, J.L. Recurrent acute pancreatitis in anorexia and bulimia. *JOP* **2004**, *5*, 231–234.
66. Wiklund, C.A.; Kuja-Halkola, R.; Thornton, L.M.; Hübel, C.; Leppä, V.; Bulik, C.M. Prolonged constipation and diarrhea in childhood and disordered eating in adolescence. *J. Psychosom. Res.* **2019**, *126*, 109797. [CrossRef] [PubMed]
67. Riedlinger, C.; Mazurak, N.; Schäffeler, N.; Stengel, A.; Giel, K.E.; Zipfel, S.; Enck, P.; Mack, I. Gastrointestinal complaints in patients with anorexia nervosa in the timecourse of inpatient treatment. *Front. Psychiatry* **2022**, *13*, 962837. [CrossRef] [PubMed]
68. Gurevich, E.; Steiling, S.; Landau, D. Incidence of Impaired Kidney Function Among Adolescent Patients Hospitalized with Anorexia Nervosa. *JAMA Netw. Open* **2021**, *4*, e2134908. [CrossRef]
69. Bouquegneau, A.; Dubois, B.E.; Krzesinski, J.M.; Delanaye, P. Anorexia nervosa and the kidney. *Am. J. Kidney Dis.* **2012**, *60*, 299–307. [CrossRef]
70. Puckett, L. Renal and electrolyte complications in eating disorders: A comprehensive review. *J. Eat. Disord.* **2023**, *11*, 26. [CrossRef]
71. Tseng, M.C.M.; Chien, L.N.; Tu, C.Y.; Zheng, C.M.; Liu, H.Y. Risk of dialysis and renal diseases in patients with anorexia nervosa in Taiwan. *Int. J. Eat. Disord.* **2023**, *56*, 991–1000. [CrossRef]
72. Ohne, M.; Watanabe, Y.; Oyake, C.; Onuki, Y.; Watanabe, T.; Ikeda, H. Pediatric early-phase anorexia nervosa without hypokalemia exhibiting acute tubular injury: A case report. *CEN Case Rep.* **2024**, *13*, 504–509. [CrossRef] [PubMed]
73. Khatri, V.; Bauschka, M.; Foley, M.; Lundberg, C.; Mehler, P. A Multi-Disciplinary Approach to Managing End-Stage Renal Disease in Anorexia Nervosa: A Case Report. *Clin. Med. Insights Case Rep.* **2023**, *16*, 11795476231169385. [CrossRef] [PubMed]

74. Zhu, Y.; Kawanishi, K.; Sawa, C.; Luo, P.; Chen, P.; Luo, M.; Kang, D.; Honda, K. Application of low-vacuum scanning electron microscopy and energy-dispersive X-ray spectrometry for detection of renal tubular crystals: A case of nephrocalcinosis in the setting of anorexia nervosa. *BMC Nephrol.* **2025**, *26*, 352. [CrossRef]
75. Schorr, M.; Miller, K.K. The endocrine manifestations of anorexia nervosa: Mechanisms and management. *Nat. Rev. Endocrinol.* **2017**, *13*, 174–186. [CrossRef] [PubMed]
76. Shunsuke, H.; Taishi, A.; Kohei, N. A Case of Hypoglycemia Associated with Anorexia Nervosa Revealing Isolated Adrenocorticotropic Hormone (ACTH) Deficiency. *Cureus* **2025**, *17*, e79383. [CrossRef]
77. Misra, M.; Klibanski, A. Endocrine consequences of anorexia nervosa. *Lancet Diabetes Endocrinol.* **2014**, *2*, 581–592. [CrossRef]
78. Monteleone, A.M.; Monteleone, P.; Marciello, F.; Pellegrino, F.; Castellini, G.; Maj, M. Differences in Cortisol Awakening Response between Binge-Purging and Restrictive Patients with Anorexia Nervosa. *Eur. Eat. Disord. Rev.* **2017**, *25*, 13–18. [CrossRef]
79. Ritschel, F.; Clas, S.; Geisler, D.; Haas, V.; Seidel, M.; Steding, J.; Roessner, V.; Kirschbaum, C.; Ehrlich, S. Is hypercortisolism in anorexia nervosa detectable using hair samples? *J. Psychiatr. Res.* **2018**, *98*, 87–94. [CrossRef]
80. Schmalbach, I.; Herhaus, B.; Pässler, S.; Runst, S.; Berth, H.; Wolff-Stephan, S.; Petrowski, K. Correction: Cortisol reactivity in patients with anorexia nervosa after stress induction. *Transl. Psychiatry* **2021**, *11*, 208. [CrossRef]
81. Haines, M.S. Endocrine complications of anorexia nervosa. *J. Eat. Disord.* **2023**, *11*, 24. [CrossRef]
82. Mattioni, J.; Duriez, P.; Aïdat, S.; Lebrun, N.; Bohlooly-Y, M.; Gorwood, P.; Viltart, O.; Tolle, V. Altered circadian pattern of activity in a chronic activity-based anorexia nervosa-like female mouse model deficient for GHSR. *Psychoneuroendocrinology* **2025**, *177*, 107453. [CrossRef]
83. Modan-Moses, D.; Yaroslavsky, A.; Pinhas-Hamiel, O.; Levy-Shraga, Y.; Kochavi, B.; Iron-Segev, S.; Enoch-Levy, A.; Toledano, A.; Stein, D. Prospective Longitudinal Assessment of Linear Growth and Adult Height in Female Adolescents with Anorexia Nervosa. *J. Clin. Endocrinol. Metab.* **2021**, *106*, e1–e10. [CrossRef]
84. Léger, J.; Fjellestad-Paulsen, A.; Bargiacchi, A.; Pages, J.; Chevenne, D.; Alison, M.; Alberti, C.; Guilmin-Crepon, S. One Year of GH Treatment for Growth Failure in Children with Anorexia Nervosa: A Randomized Placebo-Controlled Trial. *J. Clin. Endocrinol. Metab.* **2021**, *106*, e2535–e2546. [CrossRef]
85. Faust, J.P.; Goldschmidt, A.B.; Anderson, K.E.; Glunz, C.; Brown, M.; Loeb, K.L.; Katzman, D.K.; Le Grange, D. Resumption of menses in anorexia nervosa during a course of family-based treatment. *J. Eat. Disord.* **2013**, *1*, 12. [CrossRef]
86. Galusca, B.; Gay, A.; Belleton, G.; Eisinger, M.; Massoubre, C.; Lang, F.; Grouselle, D.; Estour, B.; Germain, N. Mechanisms and predictors of menses resumption once normal weight is reached in anorexia nervosa. *J. Eat. Disord.* **2023**, *11*, 172. [CrossRef] [PubMed]
87. Bulik, C.M.; Hoffman, E.R.; Von Holle, A.; Torgersen, L.; Stoltenberg, C.; Reichborn-Kjennerud, T. Unplanned pregnancy in women with anorexia nervosa. *Obstet. Gynecol.* **2010**, *116*, 1136–1140. [CrossRef]
88. Pan, J.R.; Li, T.Y.; Tucker, D.; Chen, K.Y. Pregnancy outcomes in women with active anorexia nervosa: A systematic review. *J. Eat. Disord.* **2022**, *10*, 25. [CrossRef] [PubMed]
89. Calcaterra, V.; Magenes, V.C.; Siccardo, F.; Hruby, C.; Basso, M.; Conte, V.; Maggioni, G.; Fabiano, V.; Russo, S.; Veggiotti, P.; et al. Thyroid dysfunction in children and adolescents affected by undernourished and overnourished eating disorders. *Front. Nutr.* **2023**, *10*, 1205331. [CrossRef] [PubMed]
90. Selvaraj, N.; Bobby, Z.; Sridhar, M.G. Is euthyroid sick syndrome a defensive mechanism against oxidative stress? *Med. Hypotheses* **2008**, *71*, 404–405. [CrossRef]
91. Wronski, M.L.; Tam, F.I.; Seidel, M.; Mirtschink, P.; Poitz, D.M.; Bahnsen, K.; Steinhäuser, J.L.; Bauer, M.; Roessner, V.; Ehrlich, S. Associations between pituitary-thyroid hormones and depressive symptoms in individuals with anorexia nervosa before and after weight-recovery. *Psychoneuroendocrinology* **2022**, *137*, 105630. [CrossRef]
92. Nomura, R.; Miyai, K.; Kuge, R.; Okura, T.; Goto, M.; Hasegawa, Y. Free T3 to free T4 ratio less than 2.0 suggests low T3 syndrome rather than central hypothyroidism from the age of two to eighteen years. *Endocr. J.* **2017**, *64*, 213–219. [CrossRef] [PubMed]
93. Mebis, L.; Eerdekens, A.; Güiza, F.; Princen, L.; Derde, S.; Vanwijngaerden, Y.M.; Vanhorebeek, I.; Darras, V.M.; Van den Berghe, G.; Langouche, L. Contribution of nutritional deficit to the pathogenesis of the nonthyroidal illness syndrome in critical illness: A rabbit model study. *Endocrinology* **2012**, *153*, 973–984. [CrossRef]
94. Evrard, F.; Pinto da Cunha, M.; Lambert, M.; Devuyst, O. Impaired osmoregulation in anorexia nervosa: A case-control study. *Nephrol. Dial. Transplant.* **2004**, *19*, 3034–3039. [CrossRef]
95. Goetze, J.P.; Støving, R.K. Copeptin in anorexia nervosa. *Brain Behav.* **2020**, *10*, e01551. [CrossRef] [PubMed]
96. Rosen, E.L.; Thambundit, A.; Mehler, P.S.; Mittelman, S.D. Central diabetes insipidus associated with refeeding in anorexia nervosa: A case report. *Int. J. Eat. Disord.* **2019**, *52*, 752–756. [CrossRef]
97. Sato, N.; Endo, K.; Ishizaka, H.; Matsumoto, M. Serial MR intensity changes of the posterior pituitary in a patient with anorexia nervosa, high serum ADH, and oliguria. *J. Comput. Assist. Tomogr.* **1993**, *17*, 648–650. [CrossRef]

98. Afinogenova, Y.; Schmelkin, C.; Plessow, F.; Thomas, J.J.; Pulumo, R.; Micali, N.; Miller, K.K.; Eddy, K.T.; Lawson, E.A. Low Fasting Oxytocin Levels Are Associated with Psychopathology in Anorexia Nervosa in Partial Recovery. *J. Clin. Psychiatry* **2016**, *77*, e1483–e1490. [CrossRef] [PubMed]
99. Lawson, E.A.; Holsen, L.; Santin, M.; DeSanti, R.; Meenaghan, E.; Eddy, K.; Herzog, D.; Goldstein, J.; Klibanski, A. Correction: Postprandial oxytocin secretion is associated with severity of anxiety and depressive symptoms in anorexia nervosa. *J. Clin. Psychiatry* **2015**, *76*, 666. [CrossRef]
100. Aulinas, A.; Plessow, F.; Pulumo, R.L.; Asanza, E.; Mancuso, C.J.; Slattery, M.; Tolley, C.; Thomas, J.J.; Eddy, K.T.; Miller, K.K.; et al. Disrupted Oxytocin-Appetite Signaling in Females with Anorexia Nervosa. *J. Clin. Endocrinol. Metab.* **2019**, *104*, 4931–4940. [CrossRef]
101. Plessow, F.; Galbiati, F.; Eddy, K.T.; Misra, M.; Miller, K.K.; Klibanski, A.; Aulinas, A.; Lawson, E.A. Low oxytocin levels are broadly associated with more pronounced psychopathology in anorexia nervosa with primarily restricting but not binge/purge eating behavior. *Front. Endocrinol.* **2023**, *13*, 1049541. [CrossRef] [PubMed]
102. Russell, J.; Maguire, S.; Hunt, G.E.; Kesby, A.; Suraev, A.; Stuart, J.; Booth, J.; McGregor, I.S. Intranasal oxytocin in the treatment of anorexia nervosa: Randomized controlled trial during re-feeding. *Psychoneuroendocrinology* **2018**, *87*, 83–92. [CrossRef] [PubMed]
103. Maguire, S.; Kesby, A.; Brownlow, R.; Hunt, G.E.; Kim, M.; McAulay, C.; Grisham, J.R.; McGregor, I.S.; Suraev, A.; Kevin, R.C.; et al. A phase II randomised controlled trial of intranasal oxytocin in anorexia nervosa. *Psychoneuroendocrinology* **2024**, *164*, 107032. [CrossRef]
104. Hebebrand, J.; Hildebrandt, T.; Schlögl, H.; Seitz, J.; Denecke, S.; Vieira, D.; Gradl-Dietsch, G.; Peters, T.; Antel, J.; Lau, D.; et al. The role of hypoleptinemia in the psychological and behavioral adaptation to starvation: Implications for anorexia nervosa. *Neurosci. Biobehav. Rev.* **2022**, *141*, 104807. [CrossRef]
105. Hebebrand, J.; Plieger, M.; Milos, G.; Peters, T.; Hinney, A.; Antel, J. Does hypoleptinemia trigger entrapment in anorexia nervosa? Etiological and clinical considerations. *Eur. Eat. Disord. Rev.* **2024**, *32*, 557–574. [CrossRef] [PubMed]
106. Ehrlich, S.; Burghardt, R.; Schneider, N.; Broecker-Preuss, M.; Weiss, D.; Merle, J.V.; Craciun, E.M.; Pfeiffer, E.; Mann, K.; Lehmkuhl, U.; et al. The role of leptin and cortisol in hyperactivity in patients with acute and weight-recovered anorexia nervosa. *Prog. Neuropsychopharmacol. Biol. Psychiatry* **2009**, *33*, 658–662. [CrossRef]
107. Batury, V.L.; Walton, E.; Tam, F.; Wronski, M.L.; Buchholz, V.; Frieling, H.; Ehrlich, S. DNA methylation of ghrelin and leptin receptors in underweight and recovered patients with anorexia nervosa. *J. Psychiatr. Res.* **2020**, *131*, 271–278. [CrossRef]
108. Tolle, V.; du Montcel, C.T.; Mattioni, J.; Schéle, E.; Viltart, O.; Dickson, S.L. To eat or not to eat: A role for ghrelin and LEAP2 in eating disorders? *Neurosci. Appl.* **2024**, *3*, 104045. [CrossRef]
109. Tural, U.; Iosifescu, D.V. Adiponectin in anorexia nervosa and its modifiers: A meta-regression study. *Int. J. Eat. Disord.* **2022**, *55*, 1279–1290. [CrossRef]
110. Makimura, H.; Stanley, T.L.; Chen, C.Y.; Branch, K.L.; Grinspoon, S.K. Relationship of adiponectin to endogenous GH pulse secretion parameters in response to stimulation with a growth hormone releasing factor. *Growth Horm. IGF Res.* **2011**, *21*, 155–159. [CrossRef] [PubMed]
111. Ramsey, K.A.; Meskers, C.G.M.; Trappenburg, M.C.; Verlaan, S.; Reijnierse, E.M.; Whittaker, A.C.; Maier, A.B. Malnutrition is associated with dynamic physical performance. *Aging Clin. Exp. Res.* **2020**, *32*, 1085–1092. [CrossRef]
112. Rosa-Caldwell, M.E.; Eddy, K.T.; Rutkove, S.B.; Breithaupt, L. Anorexia nervosa and muscle health: A systematic review of our current understanding and future recommendations for study. *Int. J. Eat. Disord.* **2023**, *56*, 483–500. [CrossRef]
113. de Araújo, I.M.; Rebolho, M.V.F.; Gomes, M.M.; Suen, V.M.; de Paula, F.J.A. Bone evaluation and relationship between body composition and bone mass in anorexia nervosa followed up by a multidisciplinary team. *Endocrine* **2025**, *89*, 259–266. [CrossRef] [PubMed]
114. DiVasta, A.D.; Stamoulis, C.; Rubin, C.T.; Gallagher, J.S.; Kiel, D.P.; Snyder, B.D.; Gordon, C.M. Low-Magnitude Mechanical Signals to Preserve Skeletal Health in Female Adolescents with Anorexia Nervosa: A Randomized Clinical Trial. *JAMA Netw. Open* **2024**, *7*, e2441779. [CrossRef]
115. Legroux, I.; Cortet, B. Factors influencing bone loss in anorexia nervosa: Assessment and therapeutic options. *RMD Open* **2019**, *5*, e001009. [CrossRef]
116. Gibson, D.; Filan, Z.; Westmoreland, P.; Mehler, P.S. Loss of Bone Density in Patients with Anorexia Nervosa Food That Alone Will Not Cure. *Nutrients* **2024**, *16*, 3593. [CrossRef]
117. Donat, A.; Knapstein, P.R.; Jiang, S.; Baranowsky, A.; Ballhause, T.M.; Frosch, K.H.; Keller, J. Glucose Metabolism in Osteoblasts in Healthy and Pathophysiological Conditions. *Int. J. Mol. Sci.* **2021**, *22*, 4120. [CrossRef]
118. Hediger, C.; Rost, B.; Itin, P. Cutaneous manifestations in anorexia nervosa. *Schweiz. Med. Wochenschr.* **2000**, *130*, 565–575.
119. Strumia, R. Skin signs in anorexia nervosa. *Dermatoendocrinology* **2009**, *1*, 268–270. [CrossRef]
120. Strumia, R. Eating disorders and the skin. *Clin. Dermatol.* **2013**, *31*, 80–85. [CrossRef]
121. Safer, J.D. Thyroid hormone action on skin. *Dermatoendocrinology* **2011**, *3*, 211–215. [CrossRef]

122. Lackner, S.; Meier-Allard, N.; Mörkl, S.; Müller, W.; Fürhapter-Rieger, A.; Mangge, H.; Zelzer, S.; Holasek, S. Hypercarotenemia in Anorexia Nervosa Patients May Influence Weight Balance: Results of a Clinical Cross-Sectional Cohort Study. *Front. Psychiatry* **2021**, *12*, 758300. [CrossRef]
123. Rallis, E.; Lotsaris, K.; Grech, V.S.; Tertipi, N.; Sfyri, E.; Kefala, V. The Nutrient-Skin Connection: Diagnosing Eating Disorders Through Dermatologic Signs. *Nutrients* **2024**, *16*, 4354. [CrossRef] [PubMed]
124. Stamu-O'Brien, C.; Shivakumar, S.; Messas, T.; Kroumpouzos, G. Through the looking glass: Skin signs that help diagnose eating disorders. *Clin. Dermatol.* **2023**, *41*, 67–74. [CrossRef] [PubMed]
125. Cleary, B.S.; Gaudiani, J.L.; Mehler, P.S. Interpreting the complete blood count in anorexia nervosa. *Eat. Disord.* **2010**, *18*, 132–139. [CrossRef]
126. De Filippo, E.; Marra, M.; Alfinito, F.; Di Guglielmo, M.L.; Majorano, P.; Cerciello, G.; De Caprio, C.; Contaldo, F.; Pasanisi, F. Hematological complications in anorexia nervosa. *Eur. J. Clin. Nutr.* **2016**, *70*, 1305–1308. [CrossRef]
127. Hütter, G.; Ganepola, S.; Hofmann, W.K. The hematology of anorexia nervosa. *Int. J. Eat. Disord.* **2009**, *42*, 293–300. [CrossRef] [PubMed]
128. Miller, K.K.; Grinspoon, S.K.; Ciampa, J.; Hier, J.; Herzog, D.; Klibanski, A. Medical findings in outpatients with anorexia nervosa. *Arch. Intern. Med.* **2005**, *165*, 561–566. [CrossRef]
129. Walsh, K.; Blalock, D.V.; Mehler, P.S. Hematologic findings in a large sample of patients with anorexia nervosa and bulimia nervosa. *Am. J. Hematol.* **2020**, *95*, E98–E101. [CrossRef]
130. Salari, N.; Razavizadeh, S.; Abdolmaleki, A.; Heidarian, P.; Rahimi, A.; Shohaimi, S.; Mohammadi, M. The global prevalence of anemia in patients with anorexia nervosa: A systematic review and meta-analysis. *BMC Psychol.* **2025**, *13*, 343. [CrossRef] [PubMed]
131. Takeshima, M.; Ishikawa, H.; Kitadate, A.; Sasaki, R.; Kobayashi, T.; Nanjyo, H.; Kanbayashi, T.; Shimizu, T. Anorexia nervosa-associated pancytopenia mimicking idiopathic aplastic anemia: A case report. *BMC Psychiatry* **2018**, *18*, 150. [CrossRef] [PubMed]
132. Funayama, M.; Koreki, A.; Mimura, Y.; Takata, T.; Ogino, S.; Kurose, S.; Shimizu, Y.; Kudo, S. Restrictive type and infectious complications might predict nadir hematological values among individuals with anorexia nervosa during the refeeding period: A retrospective study. *J. Eat. Disord.* **2022**, *10*, 64. [CrossRef] [PubMed]
133. Devuyst, O.; Lambert, M.; Rodhain, J.; Lefebvre, C.; Coche, E. Haematological changes and infectious complications in anorexia nervosa: A case-control study. *QJM Int. J. Med.* **1993**, *86*, 791–799.
134. Palmblad, J.; Fohlin, L.; Lundström, M. Anorexia nervosa and polymorphonuclear (PMN) granulocyte reactions. *Scand. J. Haematol.* **1977**, *19*, 334–342. [CrossRef]
135. Brown, R.F.; Bartrop, R.; Beumont, P.; Birmingham, C.L. Bacterial infections in anorexia nervosa: Delayed recognition increases complications. *Int. J. Eat. Disord.* **2005**, *37*, 261–265. [CrossRef]
136. Mashimoto, M.; Higuchi, F.; Okazaki, S.; Hoshii, Y.; Nakagawa, S. Decreased Volume of Bone Marrow Adipocytes With Sparse Gelatinous Marrow Transformation in a Patient With Pancytopenia With Anorexia Nervosa: A Case Report. *Cureus* **2024**, *16*, e58390. [CrossRef]
137. Oświęcimska, J.; Malczyk, Ż.; Szymlak, A.; Mikołajczak, A.; Ziora, K.; Zamlynski, J.; Machura, E.; Zajac, P.; Koczy, B.; Kasperska-Zajac, A. Changes in Platelet Count and Size Indices in Adolescent Patients with Anorexia Nervosa. *Clin. Appl. Thromb./Hemost.* **2017**, *23*, 562–566. [CrossRef]
138. Mitchell, O.; Feldman, D.M.; Diakow, M.; Sigal, S.H. The pathophysiology of thrombocytopenia in chronic liver disease. *Hepatic Med.: Evid. Res.* **2016**, *8*, 39–50. [CrossRef]
139. Bourke, C.D.; Berkley, J.A.; Prendergast, A.J. Immune Dysfunction as a Cause and Consequence of Malnutrition. *Trends Immunol.* **2016**, *37*, 386–398. [CrossRef]
140. Marcos, A.; Nova, E.; Montero, A. Changes in the immune system are conditioned by nutrition. *Eur. J. Clin. Nutr.* **2003**, *57*, 66–69. [CrossRef]
141. Gibson, D.; Mehler, P.S. Anorexia Nervosa and the Immune System—A Narrative Review. *J. Clin. Med.* **2019**, *8*, 1915. [CrossRef]
142. Dostálová, I.; Kaválková, P.; Papežová, H.; Domluvilová, D.; Zikán, V.; Haluzík, M. Association of macrophage inhibitory cytokine-1 with nutritional status, body composition and bone mineral density in patients with anorexia nervosa: The influence of partial realimentation. *Nutr. Metab.* **2010**, *7*, 34. [CrossRef]
143. Karczewska-Kupczewska, M.; Kowalska, I.; Nikolajuk, A.; Adamska, A.; Otziomek, E.; Gorska, M.; Straczkowski, M. Hyperinsulinemia acutely increases serum macrophage inhibitory cytokine-1 concentration in anorexia nervosa and obesity. *Clin. Endocrinol.* **2012**, *76*, 46–50. [CrossRef] [PubMed]
144. Omodei, D.; Pucino, V.; Labruna, G.; Procaccini, C.; Galgani, M.; Perna, F.; Pirozzi, D.; De Caprio, C.; Marone, G.; Fontana, L.; et al. Immune-metabolic profiling of anorexic patients reveals an anti-oxidant and anti-inflammatory phenotype. *Metabolism* **2015**, *64*, 396–405. [CrossRef]

145. Elegido, A.; Graell, M.; Andrés, P.; Gheorghe, A.; Marcos, A.; Nova, E. Increased naive CD4+ and B lymphocyte subsets are associated with body mass loss and drive relative lymphocytosis in anorexia nervosa patients. *Nutr. Res.* **2017**, *39*, 43–50. [CrossRef] [PubMed]
146. Freff, J.; Bröker, L.; Leite Dantas, R.; Schwarte, K.; Bühlmeier, J.; Kraft, I.; Hinney, A.; Buhlmann, U.; Arolt, V.; Dannlowski, U.; et al. Expression of CXCR4 on CD4+ T cells predicts body composition parameters in female adolescents with anorexia nervosa. *Front. Psychiatry* **2022**, *13*, 960905. [CrossRef]
147. Pászthy, B.; Svec, P.; Vásárhelyi, B.; Túry, F.; Mazzag, J.; Tulassay, T.; Treszl, A. Investigation of regulatory T cells in anorexia nervosa. *Eur. J. Clin. Nutr.* **2007**, *61*, 1245–1249. [CrossRef] [PubMed]
148. Freff, J.; Schwarte, K.; Bröker, L.; Bühlmeier, J.; Kraft, I.; Öztürk, D.; Hinney, A.; Arolt, V.; Dannlowski, U.; Romer, G.; et al. Alterations in B cell subsets correlate with body composition parameters in female adolescents with anorexia nervosa. *Sci. Rep.* **2021**, *11*, 1125. [CrossRef]
149. Xiao, K.; Gillissie, E.S.; Lui, L.M.; Ceban, F.; Teopiz, K.M.; Gill, H.; Cao, B.; Ho, R.; Rosenblat, J.D.; McIntyre, R.S. Immune response to vaccination in adults with mental disorders: A systematic review. *J. Affect. Disord.* **2022**, *304*, 66–77. [CrossRef]
150. Caldiroli, A.; La Tegola, D.; Affaticati, L.M.; Manzo, F.; Cella, F.; Scalia, A.; Capuzzi, E.; Nicastro, M.; Colmegna, F.; Buoli, M.; et al. Clinical and peripheral biomarkers in female patients affected by Anorexia: Does the Neutrophil/Lymphocyte Ratio (NLR) affect severity? *Nutrients* **2023**, *15*, 1133. [CrossRef]
151. Słotwińska, S.M.; Słotwiński, R. Immune disorders in anorexia. *Cent. Eur. J. Immunol.* **2017**, *42*, 294–300. [CrossRef]
152. Nilsson, A.; Elander, L.; Hallbeck, M.; Kugelberg, U.Ö.; Engblom, D.; Blomqvist, A. The involvement of prostaglandin E2 in interleukin-1β evoked anorexia is strain dependent. *Brain Behav. Immun.* **2017**, *60*, 27–31. [CrossRef] [PubMed]
153. Pomeroy, C.; Eckert, E.; Hu, S.; Eiken, B.; Mentink, M.; Crosby, R.D.; Chao, C.C. Role of interleukin-6 and transforming growth factor-beta in anorexia nervosa. *Biol. Psychiatry* **1994**, *36*, 836–839. [CrossRef]
154. Polack, E.; Nahmod, V.E.; Emeric-Sauval, E.; Bello, M.; Costas, M.; Finkielman, S.; Arzt, E. Low lymphocyte interferon-gamma production and variable proliferative response in anorexia nervosa patients. *J. Clin. Immunol.* **1993**, *13*, 445–451. [CrossRef]
155. Di Paolo, A.; Membrino, V.; Alia, S.; Nanetti, L.; Svarca, L.E.; Perrone, M.L.; Aquilanti, L.; Mazzanti, L.; Vignini, A.; Salvolini, E.; et al. Pro-inflammatory cytokine alterations in recent onset anorexia nervosa adolescent female patients before and after 6 months of integrated therapy: A case-control study. *J. Investig. Med.* **2024**, *72*, 522–531. [CrossRef]
156. Tyszkiewicz-Nwafor, M.; Jowik, K.; Paszynska, E.; Dutkiewicz, A.; Słopien, A.; Dmitrzak-Weglarz, M. Expression of immune-related proteins and their association with neuropeptides in adolescent patients with anorexia nervosa. *Neuropeptides* **2022**, *91*, 102214. [CrossRef] [PubMed]
157. Flierl, M.A.; Gaudiani, J.L.; Sabel, A.L.; Long, C.S.; Stahel, P.F.; Mehler, P.S. Complement C3 serum levels in anorexia nervosa: A potential biomarker for the severity of disease? *Ann. Gen. Psychiatry* **2011**, *10*, 16. [CrossRef]
158. Wagner-Skacel, J.; Haidacher, F.; Wiener, M.; Pahsini, K.; Marinschek, S.; Lahousen, T.; Wonisch, W.; Bengesser, S.; Butler, M.I.; Lackner, S.; et al. Oxidative Status in Adult Anorexia Nervosa Patients and Healthy Controls-Results from a Cross-Sectional Pilot Study. *Antioxidants* **2022**, *11*, 842. [CrossRef]
159. Amerio, A.; Martino, E.; Strangio, A.; Aguglia, A.; Escelsior, A.; Conio, B.; Sukkar, S.G.; Saverino, D. Autoantibodies, Oxidative Stress, and Nutritional State in Anorexia Nervosa. *Antibodies* **2024**, *14*, 1. [CrossRef] [PubMed]
160. Taylor, E.B. The complex role of adipokines in obesity, inflammation, and autoimmunity. *Clin. Sci.* **2021**, *135*, 731–752. [CrossRef]
161. Wei, Y.; Wang, Y.; Yuan, Y.; Chen, J. Celiac Disease, Gluten-Free Diet, and Eating Disorders: From Bench to Bedside. *Foods* **2024**, *14*, 74. [CrossRef]
162. Larsen, J.T.; Yilmaz, Z.; Vilhjálmsson, B.J.; Thornton, L.M.; Benros, M.E.; Musliner, K.L.; Eating Disorders Working Group of the Psychiatric Genomics Consortium; Werge, T.; Hougaard, D.M.; Mortensen, P.B.; et al. Anorexia nervosa and inflammatory bowel diseases—Diagnostic and genetic associations. *JCPP Adv.* **2021**, *1*, e12036. [CrossRef]
163. Su, Q.; Li, J.; Lu, Y.; Wu, M.; Liang, J.; An, Z. Anorexia and bulimia in relation to ulcerative colitis: A Mendelian randomization study. *Front. Nutr.* **2024**, *11*, 1400713. [CrossRef]
164. Ilzarbe, L.; Fàbrega, M.; Quintero, R.; Bastidas, A.; Pintor, L.; García-Campayo, J.; Gomollón, F.; Ilzarbe, D. Inflammatory Bowel Disease and Eating Disorders: A systematized review of comorbidity. *J. Psychosom. Res.* **2017**, *102*, 47–53. [CrossRef] [PubMed]
165. Joshkon, A.; Reytier, C.; Bertin, D.; Castets, S.; Reynaud, R.; Bat, F.; Mege, J.L.; Blot-Chabaud, M.; Darmon, P.; Bardin, N. Identification of autoantibodies against HIF1a in patients with anorexia nervosa and their potential role in hepatic cytolysis. *Sci. Rep.* **2025**, *15*, 21347. [CrossRef] [PubMed]
166. Ciccarese, G.; Guadagno, A.; Herzum, A.; Amicarelli, V.; Drago, F. Purpuric pityriasis rosea in patients with anorexia nervosa. *Dermatol. Rep.* **2025**, *17*, 10053. [CrossRef]
167. Paszynska, E.; Hernik, A.; Slopien, A.; Boucher, Y.; Tyszkiewicz-Nwafor, M.; Roszak, M.; Bilska, K.; Dmitrzak-Weglarz, M. Expression of salivary immunoglobulins and their association with analgesic neuropeptide opiorphin in anorexia nervosa during adolescence. *J. Eat. Disord.* **2022**, *10*, 118. [CrossRef]

168. Vallianou, N.; Christodoulatos, G.S.; Karampela, I.; Tsilingiris, D.; Magkos, F.; Stratigou, T.; Kounatidis, D.; Dalamaga, M. Understanding the Role of the Gut Microbiome and Microbial Metabolites in Non-Alcoholic Fatty Liver Disease: Current Evidence and Perspectives. *Biomolecules* **2021**, *12*, 56. [CrossRef] [PubMed]
169. Terry, S.M.; Barnett, J.A.; Gibson, D.L. A critical analysis of eating disorders and the gut microbiome. *J. Eat. Disord.* **2022**, *10*, 154. [CrossRef]
170. Monteleone, A.M.; Troisi, J.; Serena, G.; Fasano, A.; Dalle Grave, R.; Cascino, G.; Marciello, F.; Calugi, S.; Scala, G.; Corrivetti, G.; et al. The Gut Microbiome and Metabolomics Profiles of Restricting and Binge-Purging Type Anorexia Nervosa. *Nutrients* **2021**, *13*, 507. [CrossRef] [PubMed]
171. Fan, Y.; Støving, R.K.; Berreira Ibraim, S.; Hyötyläinen, T.; Thirion, F.; Arora, T.; Lyu, L.; Stankevic, E.; Hansen, T.H.; Déchelotte, P.; et al. The gut microbiota contributes to the pathogenesis of anorexia nervosa in humans and mice. *Nat. Microbiol.* **2023**, *8*, 787–802. [CrossRef]
172. Vallianou, N.G.; Kounatidis, D.; Panagopoulos, F.; Evangelopoulos, A.; Stamatopoulos, V.; Papagiorgos, A.; Geladari, E.; Dalamaga, M. Gut Microbiota and Its Role in the Brain-Gut-Kidney Axis in Hypertension. *Curr. Hypertens. Rep.* **2023**, *25*, 367–376. [CrossRef]
173. Shahid, F.; Afridi, Z.; Ali, M.; Saddique, M.N.; Iqbal, J.; Jha, A. Beneath the surface: Unveiling the gut-brain axis in anorexia and depression. *Eat. Weight Disord.-Stud. Anorex. Bulim. Obes.* **2025**, *30*, 14. [CrossRef]
174. Rousseau, L.; Salaün, C.; Langlois, L.; Achamrah, N.; Coëffier, M. Burden of negative energy balance in anorexia nervosa: Role of circadian rhythm, mitochondrial dynamics, microbiota-gut-brain axis and their interactions. *Clin. Nutr.* **2025**, *52*, 27–45. [CrossRef]
175. Frintrop, L.; Trinh, S.; Seitz, J.; Kipp, M. The Role of Glial Cells in Regulating Feeding Behavior: Potential Relevance to Anorexia Nervosa. *J. Clin. Med.* **2021**, *11*, 186. [CrossRef] [PubMed]
176. Frank, M.G.; Nguyen, K.H.; Ball, J.B.; Hopkins, S.; Kelley, T.; Baratta, M.V.; Fleshner, M.; Maier, S.F. SARS-CoV-2 spike S1 subunit induces neuroinflammatory, microglial and behavioral sickness responses: Evidence of PAMP-like properties. *Brain Behav. Immun.* **2022**, *100*, 267–277. [CrossRef] [PubMed]
177. Nunez, S.G.; Rabelo, S.P.; Subotic, N.; Caruso, J.W.; Knezevic, N.N. Chronic Stress and Autoimmunity: The Role of HPA Axis and Cortisol Dysregulation. *Int. J. Mol. Sci.* **2025**, *26*, 9994. [CrossRef]
178. Wei, Y.; Peng, S.; Lian, C.; Kang, Q.; Chen, J. Anorexia nervosa and gut microbiome: Implications for weight change and novel treatments. *Expert Rev. Gastroenterol. Hepatol.* **2022**, *16*, 321–332. [CrossRef] [PubMed]
179. Shobeiri, P.; Bagherieh, S.; Mirzayi, P.; Kalantari, A.; Mirmosayyeb, O.; Teixeira, A.L.; Rezaei, N. Serum and plasma levels of brain-derived neurotrophic factor in individuals with eating disorders (EDs): A systematic review and meta-analysis. *J. Eat. Disord.* **2022**, *10*, 105. [CrossRef]
180. Chen, H.; Martins, I.; Kroemer, G. The autophagy-repressive tissue hormone DBI/ACBP (diazepam binding inhibitor, acyl-CoA binding protein) suppresses anorexia. *Autophagy* **2024**, *20*, 2827–2829. [CrossRef]
181. Haleem, D.J. Improving therapeutics in anorexia nervosa with tryptophan. *Life Sci.* **2017**, *178*, 87–93. [CrossRef]
182. Jenkins, Z.M.; Eikelis, N.; Phillipou, A.; Castle, D.J.; Wilding, H.E.; Lambert, E.A. Autonomic Nervous System Function in Anorexia Nervosa: A Systematic Review. *Front. Neurosci.* **2021**, *15*, 682208. [CrossRef]
183. Clemente-Suárez, V.J.; Ramírez-Goerke, M.I.; Redondo-Flórez, L.; Beltrán-Velasco, A.I.; Martín-Rodríguez, A.; Ramos-Campo, D.J.; Navarro-Jiménez, E.; Yáñez-Sepúlveda, R.; Tornero-Aguilera, J.F. The Impact of Anorexia Nervosa and the Basis for Non-Pharmacological Interventions. *Nutrients* **2023**, *15*, 2594. [CrossRef]
184. Brennan, C.; Green, G.; Morgan, A.; Baudinet, J. The Role of the Dietitian Within a Day Programme for Adolescent Anorexia Nervosa: A Reflexive Thematic Analysis of Child and Adolescent Eating Disorder Clinician Perspectives. *J. Hum. Nutr. Diet.* **2025**, *38*, e70070. [CrossRef] [PubMed]
185. Bohon, C.; Le Grange, D.; Attia, E.; Golden, N.H.; Steinberg, D. United States-based practice guidelines for children and adolescents with eating disorders: Synthesis of clinical practice guidelines. *J. Eat. Disord.* **2025**, *13*, 66. [CrossRef] [PubMed]
186. Dozio, E.; Alonge, M.; Tori, G.; Caumo, A.; Russo, R.G.; Scuttari, E.; Fringuelli, L.; Terruzzi, I. Dynamic Nutrition Strategies for Anorexia Nervosa: Marker-Based Integration of Calories and Proteins. *Nutrients* **2025**, *17*, 560. [CrossRef]
187. Brennan, C.; Felemban, L.; McAdams, E.; Walsh, K.; Baudinet, J. An Exploration of Clinical Characteristics and Treatment Outcomes Associated with Dietetic Intervention in Adolescent Anorexia Nervosa. *Nutrients* **2024**, *16*, 4117. [CrossRef]
188. Ayrolles, A.; Clarke, J.; Godart, N.; André-Carletti, C.; Barbe, C.; Bargiacchi, A.; Blanchet, C.; Bergametti, F.; Bertrand, V.; Caldagues, E.; et al. Early-onset anorexia nervosa: A scoping review and management guidelines. *J. Eat. Disord.* **2024**, *12*, 182. [CrossRef] [PubMed]
189. Teo, D.E.S.Y.; Teong, V.W.L.; Ramachandran, R.; Lim, S.L.; Lin, C.X. Characteristics and outcome of patients with anorexia nervosa on medical nutritional therapy: An institutional study with review of literature. *Singap. Med. J.* **2024**, *65*, 564–570. [CrossRef]

190. Del Portillo, R.C.; Milla, S.P.; Martín, P.M.; Loria-Kohen, V.; Olmos, M.Á.M.; Álvarez, M.T.M.; Alija, M.J.C.; Palmero, M.Á.M.; Lozano, E.C.; Valero-Pérez, M.; et al. A consensus report by the Working Group on Eating Disorders of Sociedad Española de Nutrición Clínica y Metabolismo (GTTCA-SENPE). Evaluation, medical and nutritional management of anorexia nervosa. Update 2023. *Nutr. Hosp.* **2024**, *41*, 1–60. [CrossRef]
191. Hunter, R.; Platygeni, M.; Moore, E. Plant-based recovery from restrictive eating disorder: A qualitative enquiry. *Appetite* **2024**, *194*, 107137. [CrossRef] [PubMed]
192. Cinelli, G.; Croci, I.; De Santis, G.L.; Chianello, I.; Miller, K.P.; Gualtieri, P.; Di Renzo, L.; De Lorenzo, A.; Tozzi, A.E.; Zanna, V. Adherence to a Mediterranean Diet, Body Composition and Energy Expenditure in Outpatients Adolescents Diagnosed with Anorexia Nervosa: A Pilot Study. *Nutrients* **2023**, *15*, 3223. [CrossRef] [PubMed]
193. Stheneur, C.; Hanachi, M. Somatic Outcomes and Nutritional Management of Anorexia Nervosa. *Nutrients* **2023**, *15*, 2541. [CrossRef]
194. van den Hoek Ostende, M.M.; Brizzi, G.; Meregalli, V.; Schroeder, P.A.; Collantoni, E. Psychological distance affects real movements in virtual reality: Distance to food in anorexia nervosa. *J. Eat. Disord.* **2025**, *13*, 169. [CrossRef] [PubMed]
195. Himmerich, H.; Herpertz-Dahlmann, B.; Mörkl, S. Biological Therapies and Eating Disorders. *Eur. Eat. Disord. Rev.* **2025**, *33*, 619–625. [CrossRef] [PubMed]
196. Thorey, S.; Blanchet, C.; Guessoum, S.B.; Moro, M.R.; Ludot, M.; Carretier, E. Efficacy and tolerance of second-generation antipsychotics in anorexia nervosa: A systematic scoping review. *PLoS ONE* **2023**, *18*, e0278189. [CrossRef]
197. Peck, S.K.; Shao, S.; Gruen, T.; Yang, K.; Babakanian, A.; Trim, J.; Finn, D.M.; Kaye, W.H. Psilocybin therapy for females with anorexia nervosa: A phase 1, open-label feasibility study. *Nat. Med.* **2023**, *29*, 1947–1953. [CrossRef]
198. Rodan, S.C.; Maguire, S.; Meez, N.; Greenstien, K.; Zartarian, G.; Mills, K.L.; Suraev, A.; Bedoya-Pérez, M.A.; McGregor, I.S. Prescription and Nonprescription Drug Use Among People with Eating Disorders. *JAMA Netw. Open* **2025**, *8*, e2522406. [CrossRef]
199. Andries, A.; Frystyk, J.; Flyvbjerg, A.; Stoning, R.K. Dronabinol in severe, enduring anorexia nervosa: A randomized controlled trial. *Int. J. Eat. Dis.* **2014**, *47*, 18–23. [CrossRef]
200. Ragnhildstveit, A.; Slayton, M.; Jackson, L.K.; Brendle, M.; Ahuja, S.; Holle, W.; Moore, C.; Sollars, K.; Seli, P.; Robison, R. Ketamine as a Novel Psychopharmacotherapy for Eating Disorders: Evidence and Future Directions. *Brain Sci.* **2022**, *12*, 382. [CrossRef] [PubMed]
201. Xu, J.; Zhang, R.; Millischer, V.; Stiernborg, M.; Tume, C.E.; Mehdinia, S.; Barker, P.; Yilmaz, Z.; Gonçalves, V.F.; Lavebratt, C.; et al. Elevated plasma GDF15 combined with FGF21 suggests mitochondrial dysfunction in a subgroup of anorexia nervosa patients. *Transl. Psychiatry* **2025**, *15*, 215. [CrossRef]
202. Borner, T.; Shaulson, E.D.; Ghidewon, M.Y.; Barnett, A.B.; Horn, C.C.; Doyle, R.P.; Grill, H.J.; Hayes, M.R.; De Jonghe, B.C. GDF15 Induces Anorexia through Nausea and Emesis. *Cell Metab.* **2020**, *31*, 351–362. [CrossRef] [PubMed]
203. Breen, D.M.; Jagarlapudi, S.; Patel, A.; Zou, C.; Joaquim, S.; Li, X.; Kang, L.; Pang, J.; Hales, K.; Ziso-Qejvanaj, E.; et al. Growth differentiation factor 15 neutralization does not impact anorexia or survival in lipopolysaccharide-induced inflammation. *Iscience* **2021**, *24*, 102554. [CrossRef]
204. Hebebrand, J.; Milos, G.; Seitz, J.; Matthews, A. Recombinant human leptin for the treatment of anorexia nervosa. *Lancet* **2024**, *404*, 1639–1640. [CrossRef]
205. Brown, R.J.; Oral, E.A.; Cochran, E.; Araújo-Vilar, D.; Savage, D.B.; Long, A.; Fine, G.; Salinardi, T.; Gorden, P. Long-term effectiveness and safety of metreleptin in the treatment of patients with generalized lipodystrophy. *Endocrine* **2018**, *60*, 479–489. [CrossRef]
206. Welt, C.K.; Chan, J.L.; Bullen, J.; Murphy, R.; Smith, P.; DePaoli, A.M.; Karalis, A.; Mantzoros, C.S. Recombinant human leptin in women with hypothalamic amenorrhea. *N. Engl. J. Med.* **2004**, *351*, 987–997. [CrossRef]
207. Hebebrand, J.; Antel, J.; Peters, T. Case report: Clinical improvements observed in first off-label metreleptin treatment of a patient with atypical anorexia nervosa. *Eur. Child Adolesc. Psychiatry* **2024**, *33*, 2267–2272. [CrossRef]
208. Kounatidis, D.; Vallianou, N.G.; Karampela, I.; Grivakou, E.; Dalamaga, M. The intricate role of adipokines in cancer-related signaling and the tumor microenvironment: Insights for future research. *Semin. Cancer Biol.* **2025**, *113*, 130–150. [CrossRef]
209. Gradl-Dietsch, G.; Milos, G.; Wabitsch, M.; Bell, R.; Tschöpe, F.; Antel, J.; Hebebrand, J. Rapid Emergence of Appetite and Hunger Resulting in Weight Gain and Improvement of Eating Disorder Symptomatology during and after Short-Term Off-Label Metreleptin Treatment of a Patient with Anorexia Nervosa. *Obes. Facts* **2023**, *16*, 99–107. [CrossRef]
210. Vallianou, N.G.; Kounatidis, D.; Tsilingiris, D.; Panagopoulos, F.; Christodoulatos, G.S.; Evangelopoulos, A.; Karampela, I.; Dalamaga, M. The Role of Next-Generation Probiotics in Obesity and Obesity-Associated Disorders: Current Knowledge and Future Perspectives. *Int. J. Mol. Sci.* **2023**, *24*, 6755. [CrossRef]
211. Dhopatkar, N.; Keeler, J.L.; Mutwalli, H.; Whelan, K.; Treasure, J.; Himmerich, H. Gastrointestinal symptoms, gut microbiome, probiotics and prebiotics in anorexia nervosa: A review of mechanistic rationale and clinical evidence. *Psychoneuroendocrinology* **2023**, *147*, 105959. [CrossRef]

212. Žaja, O.; Fiolić, M.; Ćuk, M.C.; Tiljak, M.K. The role of L. reuteri DSM17938 in nutritional recovery and treatment of constipation in children and adolescents with anorexia nervosa—A randomized, double blind, placebo controlled study. *Clin. Nutr. ESPEN* **2021**, *46*, 47–53. [CrossRef]
213. Maschek, S.; Østergaard, T.H.; Krych, L.; Zachariassen, L.F.; Sørensen, D.B.; Junker Mentzel, C.M.; Hansen, A.K.; Sjögren, J.M.; Barfod, K.K. Investigating fecal microbiota transplants from individuals with anorexia nervosa in antibiotic-treated mice using a cross-over study design. *J. Eat. Disord.* **2025**, *13*, 82. [CrossRef] [PubMed]
214. Kooij, K.L.; Andreani, N.A.; van der Gun, L.L.; Keller, L.; Trinh, S.; van der Vijgh, B.; Luijendijk, M.; Dempfle, A.; Herpertz-Dahlmann, B.; Seitz, J.; et al. Fecal microbiota transplantation of patients with anorexia nervosa did not alter flexible behavior in rats. *Int. J. Eat. Disord.* **2024**, *57*, 1868–1881. [CrossRef] [PubMed]
215. Quagebeur, R.; Dalile, B.; Raes, J.; Van Oudenhove, L.; Verbeke, K.; Vrieze, E. The role of short-chain fatty acids (SCFAs) in regulating stress responses, eating behavior, and nutritional state in anorexia nervosa: Protocol for a randomized controlled trial. *J. Eat. Disord.* **2023**, *11*, 191. [CrossRef]
216. Hemmings, A.; Gallop, L.; İnce, B.; Cutinha, D.; Kan, C.; Simic, M.; Zadeh, E.; Malvisi, I.; McKenzie, K.; Zocek, L.; et al. A randomised controlled feasibility trial of intermittent theta burst stimulation with an open longer-term follow-up for young people with persistent anorexia nervosa (RaISE): Study protocol. *Eur. Eat. Disord. Rev.* **2024**, *32*, 575–588. [CrossRef]
217. Himmerich, H.; Lewis, Y.D.; Conti, C.; Mutwalli, H.; Karwautz, A.; Sjögren, J.M.; Uribe Isaza, M.M.; Tyszkiewicz-Nwafor, M.; Aigner, M.; McElroy, S.L.; et al. World Federation of Societies of Biological Psychiatry (WFSBP) guidelines update 2023 on the pharmacological treatment of eating disorders. *World J. Biol. Psychiatry* **2023**, *24*, 643–706. [CrossRef]

biomolecules

MDPI

Review

The Impact of Glucagon-like Peptide-1 Receptor Agonists on Erectile Function: Friend or Foe?

Dimitris Kounatidis [1,*], Natalia G. Vallianou [2], Eleni Rebelos [1,3,4], Kalliopi Vallianou [5], Evanthia Diakoumopoulou [1], Konstantinos Makrilakis [1] and Nikolaos Tentolouris [1]

[1] Diabetes Center, First Department of Propaedeutic Internal Medicine, Medical School, National and Kapodistrian University of Athens, Laiko General Hospital, 11527 Athens, Greece; elenirebelos@gmail.com (E.R.); evitadiak@gmail.com (E.D.); kmakrila@med.uoa.gr (K.M.); ntentol@med.uoa.gr (N.T.)

[2] First Department of Internal Medicine, Sismanogleio General Hospital, 15126 Athens, Greece; natalia.vallianou@gmail.com

[3] Turku PET Centre, University of Turku, 20014 Turku, Finland

[4] Department of Clinical and Experimental Medicine, University of Pisa, 56126 Pisa, Italy

[5] Clinic of Nephrology and Renal Transplantation, Medical School, National and Kapodistrian University of Athens, Laiko General Hospital, 11527 Athens, Greece; kallia_harry@hotmail.com

* Correspondence: dimitriskounatidis82@outlook.com

Abstract: Erectile dysfunction (ED) is a common yet frequently underrecognized microvascular complication of diabetes, affecting up to three out of four individuals. Key contributing factors include advancing age, long-standing disease duration, and suboptimal glycemic control, as well as insulin resistance and androgen deficiency—the latter being particularly common in men with type 2 diabetes (T2D) and obesity. While numerous studies have investigated the effects of various antidiabetic therapies on diabetes-related ED, the results remain inconsistent, limiting definitive conclusions. In recent years, increasing attention has focused on a novel class of antidiabetic medications, namely glucagon-like peptide-1 receptor agonists (GLP-1 RAs). These agents have become central to the treatment of T2D due to their potent glucose-lowering properties and well-documented benefits on cardiovascular outcomes, and weight loss. Given these pleiotropic effects, GLP-1 RAs have been presumed to positively influence erectile function—a hypothesis supported by a growing body of experimental and clinical research. However, preliminary reports have also raised concerns about a possible association between GLP-1 RA use and ED. This narrative review aims to synthesize current evidence regarding the impact of GLP-1 RAs on erectile function, providing a platform for future research in this evolving field.

Keywords: diabetes; dulaglutide; erectile dysfunction; GIP; GLP-1 receptor agonists; hypogonadism; liraglutide; semaglutide; testosterone; tirzepatide

1. Introduction

Erectile function is a cornerstone of male sexual health, and its impairment can negatively affect overall quality of life (QoL), emotional intimacy, and relationship satisfaction. Erectile dysfunction (ED) is defined as the persistent inability to achieve or maintain an erection sufficient for satisfactory sexual performance and can significantly diminish sexual well-being [1,2]. The diagnosis of ED relies on a comprehensive clinical approach, including detailed medical and sexual history, physical examination, laboratory evaluations, and standardized assessment questionnaires [1,3]. Among the most widely used instruments is the 15-item International Index of Erectile Function (IIEF), while its abridged 5-item version

(IIEF-5) provides a practical alternative for clinical settings. In addition to facilitating diagnosis, the IIEF-5 enables classification of ED severity, with scores ranging from severe ED (5–7) to no ED (22–25) [4]. First-line treatment for ED typically involves oral phosphodiesterase type 5 (PDE5) inhibitors, such as sildenafil, tadalafil, vardenafil, and avanafil. These agents act by inhibiting the PDE5 enzyme, thereby increasing cyclic guanosine monophosphate (cGMP) levels. Elevated cGMP promotes smooth muscle relaxation and enhances blood flow to the corpus cavernosum during sexual arousal, thereby facilitating penile erection [5].

ED is particularly prevalent among men with diabetes and has also been reported in individuals with prediabetes [6]. Hyperglycemia, insulin resistance (IR) and hypogonadism, particularly in the presence of obesity, constitutes the core pathophysiological triad underlying diabetes-induced ED [7]. Men with diabetes are approximately 3.5 times more likely to experience ED compared to their non-diabetic counterparts [8]. Although numerous epidemiological studies have sought to quantify the prevalence of diabetes-related ED, reported rates vary considerably, ranging from 35% to 75% [9]. A recent umbrella review encompassing data from 108,030 men with diabetes estimated a global pooled prevalence of 65.8% [10]. Notably, prevalence appears to be lower among men with type 1 diabetes (T1D), with reported rates up to 42.5% [11,12].

Several risk factors have been implicated in the development of diabetes-associated ED. These factors encompass advancing age, long-standing disease duration, and poor glycemic control as captured by high glycated hemoglobin (HbA1c) levels, and glycemic variability detected via continuous glucose monitoring (CGM). Additional risk factors include cardiometabolic comorbidities, including hypercholesterolemia, hypertension, cardiovascular disease (CVD), and obesity, as well as diabetes-related microvascular complications [11–14]. Notably, a recent study introduced a predictive model with strong performance (concordance index [C-index] = 0.827, 95% confidence interval [CI] = 0.772–0.882) for early detection of ED in subjects with type 2 diabetes (T2D), incorporating variables such as diabetes duration, carotid intima-media thickness (CIMT), and low-density lipoprotein cholesterol (LDL-C) levels [15].

Research has also explored the impact of various antidiabetic therapies on erectile function; however, the current evidence remains limited and ambiguous. In this context, increasing attention has been directed toward the potential effect of newer incretin-based therapies on erectile health, particularly glucagon-like peptide-1 receptor agonists (GLP-1 RAs) [16]. GLP-1 RAs have transformed the landscape of diabetes care by providing not only effective glycemic control but also substantial cardiovascular (CV) protection and significant weight loss [17,18]. Given that endothelial integrity and body weight are critical determinants of erectile function, it is plausible to hypothesize that these therapeutic advantages might extend to this domain as well. But does the available evidence truly support this assumption?

This narrative review examines the current evidence regarding the influence of GLP-1 RAs on erectile function. It begins with a discussion of the physiology of penile erection, and the pathophysiological mechanisms underlying diabetes-related ED, followed by a summary of the pharmacological and clinical characteristics of GLP-1 RAs. Subsequently, it presents the available preclinical and clinical data regarding their role in modulating erectile physiology. Finally, the review considers the impact of other antidiabetic drug classes on erectile function, offering a comparative perspective with GLP-1 RAs. Limitations of the existing literature are discussed, along with current challenges and future directions for research aimed at elucidating the therapeutic potential of GLP-1 RAs in this context.

2. Physiological Mechanisms Underlying Male Sexual Arousal

Sexual desire may be elicited by a variety of stimuli, including visual, auditory, tactile, or cognitive cues, which activate neural pathways that facilitate the relaxation of vascular and cavernosal smooth muscle. This relaxation allows for increased arterial inflow into the corpora cavernosa, a process governed by neurovascular mechanisms originating from hypothalamic centers and the sacral spinal cord (SSC) [19].

Central to this process is the neurotransmitter nitric oxide (NO), which is synthesized and released by two main categories of neurons. Non-adrenergic, non-cholinergic (NANC) neurons located in the pelvic plexus and pelvic nerves express neuronal nitric oxide synthase (nNOS) and release NO directly into adjacent vascular smooth muscle cells (VSMCs). In parallel, cholinergic parasympathetic neurons, whose cell bodies reside in the sacral parasympathetic nucleus (S2–S4) and project to the pelvic plexus and corpora cavernosa, release acetylcholine. Acetylcholine acts on M3 muscarinic receptors expressed on endothelial cells, activating endothelial nitric oxide synthase (eNOS) and promoting NO production [20,21]. Once produced, NO diffuses into neighboring VSMCs, where it stimulates soluble guanylyl cyclase (sGC), catalyzing the conversion of guanosine triphosphate (GTP) into cGMP. Elevated intracellular cGMP levels activate protein kinase G type 1 (PKG-1), which lowers cytosolic calcium concentrations and activates calcium-dependent potassium (KCa) channels, favoring VSMC hyperpolarization and smooth muscle relaxation [22,23].

Testosterone secreted by the testes exerts both genomic and non-genomic effects [24]. Through its interaction with nuclear androgen receptors (ARs) in endothelial and VSMCs, the hormone modulates gene expression to support vascular integrity. In addition, testosterone exerts non-genomic actions via membrane-bound androgen receptors (mARs) that function independently of classical AR activation [25]. These mechanisms collectively enhance endothelial NO production and stimulate the release of endothelium-derived hyperpolarizing factors (EDHFs), such as epoxyeicosatrienoic acids, hydrogen peroxide, and potassium ions, which further contribute to vasodilation within the erectile tissue [26].

As the corpora cavernosa fill with blood and expand, the resulting increase in intracavernosal pressure compresses the subtunical venous plexus beneath the tunica albuginea, restricting venous outflow and helping to maintain the erection. Additionally, contraction of pelvic floor muscles, such as the bulbospongiosus and ischiocavernosus further compresses the corpora cavernosa, enhancing penile rigidity and erection stability [27]. Figure 1 illustrates the main pathophysiological mechanisms driving male erection.

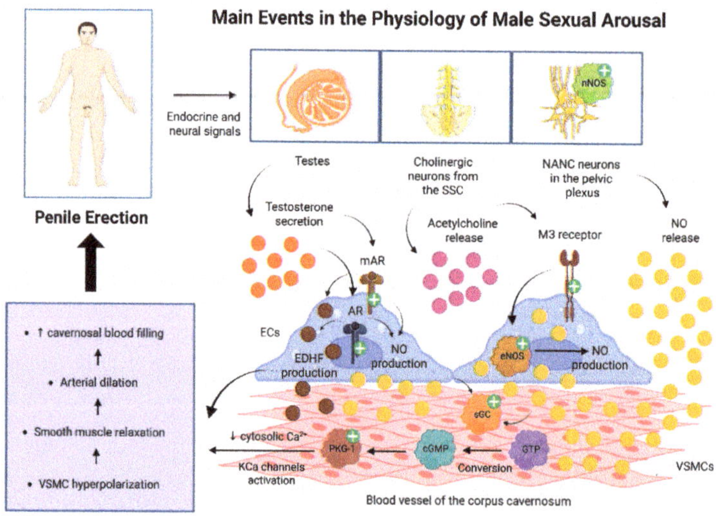

Figure 1. Key pathophysiological mechanisms underlying penile erection. NO plays a central role in the regulation of male erection, with two primary sources: (1) activation of nNOS in NANC neurons of

the pelvic plexus, leading to direct NO release into VSMCs, and (2) activation of eNOS in endothelial cells upon stimulation of muscarinic M3 receptors by Ach released from cholinergic neurons in the SSC, with NO subsequently diffusing to VSMCs. In both cases, NO activates sGC in VSMCs, catalyzing the conversion of GTP to cGMP. Elevated cGMP levels activate PKG-1, which in turn decreases cytosolic calcium concentrations and activates KCa channels, resulting in VSMC hyperpolarization. Testosterone also enhances NO production via activation of androgen receptors (ARs and mARs), contributing to both nNOS and eNOS pathways. Additionally, testosterone stimulates the production of EDHFs, which further support VSMC hyperpolarization. The resulting hyperpolarization induces smooth muscle relaxation and vasodilation, leading to increased cavernosal blood flow and penile erection [20–26]. Abbreviations: Ach: acetylcholine; AR: androgen receptor; cGMP: cyclic guanosine monophosphate; ECs: endothelial cells; EDHF: endothelium-derived hyperpolarizing factor; eNOS: endothelial nitric oxide synthase; GTP: guanosine triphosphate; KCa: calcium-dependent potassium; M3 receptor: muscarinic M3 receptor; mAR: membrane-bound androgen receptor; NANC: non-adrenergic, non-cholinergic; nNOS: neuronal nitric oxide synthase; NO: nitric oxide; PKG-1: protein kinase G type 1; sGC: soluble guanylyl cyclase; SSC: sacral spinal cord; VSMC: vascular smooth muscle cell; +: activation; ↑: increase; ↓: decrease. Created in BioRender. Kounatidis, D. (2025) https://BioRender.com/rykt1qj (assessed on 27 August 2025).

3. Pathophysiology of Erectile Dysfunction in Patients with Diabetes

3.1. Hyperglycemia

Poor glycemic control is a major contributing factor to the development of ED, regardless of diabetes type. Persistent hyperglycemia contributes to ED through several interconnected mechanisms, primarily driven by the generation of advanced glycation end products (AGEs) and reactive oxygen species (ROS) [28]. AGEs constitute a heterogeneous group of compounds formed via non-enzymatic reactions between reducing sugars (e.g., glucose) and proteins, lipids, or nucleic acids. Prolonged hyperglycemia accelerates this glycation process, resulting in the accumulation of AGEs both endogenously through metabolic pathways and exogenously from dietary sources. Over time, AGEs undergo further irreversible modifications that disrupt cellular and tissue function. These effects are mainly mediated through their interaction with the receptor for advanced glycation end products (RAGE) [29].

AGEs exacerbate ED by promoting oxidative stress and attenuating NO bioavailability, thereby impairing the relaxation of cavernosal smooth muscle [30]. This dysfunction is further exacerbated by the decreased expression and activity of both nNOS and eNOS, leading to impaired neurovascular regulation of penile erection [31–33]. Additionally, in diabetes, a state of NO resistance emerges, characterized by diminished responsiveness of cavernosal smooth muscle cells to NO signaling. This resistance is primarily driven by increased oxidative stress, and desensitization of the sGC–cGMP–PKG signaling cascade [34]. Moreover, hyperglycemia impairs maximal cavernosal smooth muscle relaxation in response to both acetylcholine and nitrergic nerve stimulation. This dysfunction is linked to enhanced activation of key signaling pathways and molecules, including Rho-associated kinase 2 (ROCK2), p38 mitogen-activated protein kinase (p38 MAPK), and arginase II [35]. ROCK2 signaling is critical in regulating vascular tone, cytoskeletal dynamics, intercellular adhesion, and endothelial barrier function [36]. Research has shown that in diabetic mice with heterozygous deficiency of ROCK2, these pathological changes are significantly reduced or even absent [35].

Beyond impairments in cavernosal relaxation, ROS accumulation and oxidative stress linked to AGE formation promote reductions in antioxidant responses, lipid peroxidation, DNA damage, and oxidative protein modifications [37,38]. Nitration of tyrosine residues to form 3-nitrotyrosine (3-NT) alters protein structure and function, disrupting enzymatic activity and intracellular signaling. These reactions, mediated by reactive nitrogen species

(RNS), further reduce NO availability, ultimately worsening endothelial dysfunction and impairing vasorelaxation [39]. Interestingly, recent evidence supports that the combined accumulation of AGEs and ROS triggers various forms of programmed cell death, including apoptosis, pyroptosis, and ferroptosis, resulting in ED aggravation [40].

Low-intensity pulsed ultrasound (LIPA) may exert protective effects against AGEs-induced endothelial injury in the corpus cavernosum by promoting mitophagy, thereby preserving mitochondrial function and cellular homeostasis [41]. Furthermore, sustained high glucose levels might facilitate corpus cavernosum fibrosis via the transforming growth factor beta 1/SMAD family proteins/connective tissue growth factor (TGF-β1/Smad/CTGF) signaling pathway [42]. Epidemiological evidence indicates an association between diabetes and Peyronie's disease, with diabetes recognized as a risk factor for its development [43].

The frequent coexistence of comorbidities in diabetes can intensify ED by amplifying the aforementioned pathophysiological processes. For instance, the coexistence of diabetes and hypertension might exhibit a synergistic negative effect on erectile function by enhancing oxidative stress, inflammation, and apoptosis [44]. Notably, ED may persist or even worsen despite appropriate antihypertensive treatment, as certain medications, especially thiazide diuretics, beta-adrenergic blockers, and aldosterone antagonists, have been linked to detrimental effects [45]. Moreover, hypercholesterolemia may further impair NO signaling by promoting the overactivation of nicotinamide adenine dinucleotide phosphate hydrogen (NADPH) oxidase subunits, leading to elevated superoxide production and eNOS uncoupling [46].

3.2. Insulin Resistance

IR may contribute to ED development, primarily through its noxious impact on endothelial function. By impairing endothelial NO synthesis and attenuating insulin-mediated vasodilation, IR disrupts the vascular mechanisms essential for erection [47]. In a model of insulin-resistant, obese mice, short-term pharmacological improvement of IR was sufficient to restore erectile function. This restoration was attributed to the normalization of both endothelium-dependent and nerve-mediated corpus cavernosum relaxation, along with reactivation of the cGMP signaling pathway and attenuation of cavernosal hypercontractility [48]. Clinical research further supports this association. In a prospective study involving 283 men with ED lasting at least six months, IR was identified in nearly half of the participants. Notably, IR was independently associated not only with the presence of ED but also with its severity [49]. Several studies, including analyses from large-scale datasets such as the National Health and Nutrition Examination Survey (NHANES), have identified surrogate markers related to IR that could potentially facilitate early risk assessment for ED. These indices include the homeostasis model assessment of insulin resistance (HOMA-IR), the triglyceride-glucose (TyG) index, as well as the metabolic score for insulin resistance (METS-IR). However, considerable heterogeneity among studies and variability in the diagnostic performance of these indices currently limit their clinical utility as predictive tools for ED [50–52].

3.3. Hypogonadism

Hypogonadism, defined by impaired testosterone production due to the dysfunction of the testes, hypothalamus, or pituitary gland, serves as a key metabolic disturbance linking diabetes and ED [53]. Both reduced circulating testosterone levels and diminished concentrations of its primary plasma binding protein, sex hormone-binding globulin (SHBG), have been identified as independent predictors of T2D development [54,55]. When present alongside obesity, androgen deficiency augments IR, creating a bidirectional loop that worsens both T2D and ED [56]. Although obesity is commonly associated with T2D, it is increasingly observed in T1D, giving rise to the concept of "double diabetes" [57]. Moreover, hypertension and other comorbidities may further compound testosterone-driven impairments in erectile function [58].

Androgen deprivation disrupts vascular homeostasis by promoting vasoconstriction, VSMC apoptosis, adipocyte infiltration, and fibrosis of the tunica albuginea [59–62]. Within the corpora cavernosa, the absence of adequate androgen signaling leads to reduced expression of eNOS and PDE5 mRNA, with the severity of the functional consequences correlating with the extent of enzymatic downregulation [63]. Persistent high glucose levels have been linked to decreased testosterone levels, in part due to the downregulation of vascular endothelial growth factor (VEGF), which compromises testicular microcirculation, leading to Leydig cell dysfunction and blunted testosterone synthesis [64]. On the other hand, disrupted insulin signaling mitigates hypothalamic kisspeptin expression, perturbing the hypothalamic–pituitary–gonadal (HPG) axis via altered mammalian target of rapamycin (mTOR)–kisspeptin pathway activity and impaired gonadotropin-releasing hormone (GnRH) secretion [65,66].

The HPG axis is further suppressed in obesity, through increased aromatase activity in adipose depots, driving peripheral conversion of androgens to estrogens. Elevated estradiol activates hypothalamic estrogen receptors, triggering a negative feedback on gonadotropin release and further lowering testosterone output [67]. This androgen deficiency skews stromal progenitor cells (SPCs) toward adipogenic differentiation, causing ectopic lipid accumulation within erectile tissue and worsening structural dysfunction [68]. Interestingly, a recent transcriptomic analysis has identified aberrant expression patterns of adiponectin receptors in men with ED [69].

Collectively, the combined effects of prolonged hyperglycemia, obesity-driven hormonal imbalance, and androgen deprivation promote the development of a chronic pro-inflammatory milieu that favors progressive endothelial dysfunction [70]. Figure 2 presents key pathophysiological events and mechanisms underlying ED in individuals with diabetes.

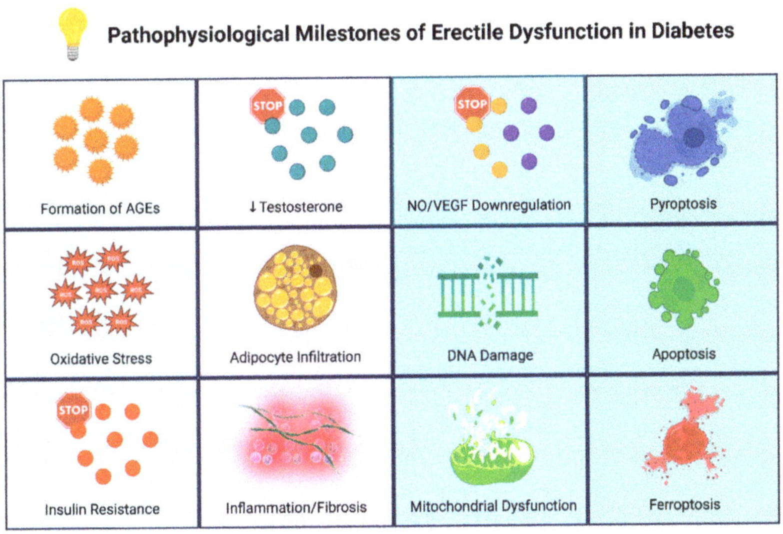

Figure 2. Pathophysiological milestones of erectile dysfunction in diabetes. Abbreviations: AGEs: advanced glycation end products; NO: nitric oxide; ROS: reactive oxygen species; VEGF: vascular endothelial growth factor [28–69]. Created in BioRender. Kounatidis, D. (2025) https://BioRender.com/ebz5ezk (assessed on 28 July 2025).

4. GLP-1 Receptor Agonists: Therapeutic Impact and Safety Considerations

4.1. Pharmacological Profile and Clinical Benefits of GLP-1 Receptor Agonists

GLP-1 RAs constitute a novel class of antidiabetic medications that have reshaped the management of T2D. These agents function by mimicking the actions of endogenous GLP-1, binding to its receptor to enhance glucose-dependent insulin secretion from pancreatic β-cells while suppressing glucagon release from α-cells. This glucose-dependent mechanism offers the benefit of avoiding hypoglycemia, while simultaneously contributing to effective glycemic control [71]. GLP-1 RAs are classified into short-acting and long-acting agents based on their pharmacokinetic properties. Short-acting agents, such as exenatide twice-daily and lixisenatide, primarily target postprandial glucose excursions by delaying gastric emptying. In contrast, long-acting GLP-1 RAs, including exenatide long-acting release (exenatide LAR), liraglutide, dulaglutide, and semaglutide, provide more sustained reductions in fasting plasma glucose (FPG) by continuously enhancing insulin secretion and suppressing glucagon release [72]. In contemporary clinical practice, once-weekly injectable formulations, particularly semaglutide are preferred, due to their superior efficacy and ease of use [71].

Beyond glucose control, GLP-1 RAs have gained prominence for their CV benefits, especially in individuals with established CVD. These benefits are not limited to patients with T2D but also extend to those with obesity, even in the absence of T2D [73]. The landmark LEADER (Liraglutide Effect and Action in Diabetes: Evaluation of Cardiovascular Outcome Results) trial, which enrolled 9340 individuals with T2D at high CV risk, demonstrated that liraglutide significantly reduced the incidence of major adverse cardiovascular events (MACE) compared with placebo over a median follow-up of 3.8 years (Hazard Ratio [HR]: 0.87; 95% CI: 0.78–0.97; $p = 0.01$ for superiority; $p < 0.001$ for noninferiority) [74]. Similar CV benefits were observed in the SUSTAIN-6 trial for semaglutide. This study included 3297 patients with T2D at high CV risk, 83% of whom had established CVD, chronic kidney disease (CKD), or both. Once-weekly semaglutide significantly reduced the incidence of the primary composite outcome (first occurrence of CV death, nonfatal myocardial infarction, or nonfatal stroke) compared with placebo (6.6% vs. 8.9%; HR: 0.74; 95% CI: 0.58–0.95; $p < 0.001$ for noninferiority [75]. Prescription of semaglutide may also provide renal protective benefits in patients with T2D and CKD. Recent evidence indicated that its use reduced the risk of a composite kidney-specific outcome, including progression to kidney failure or a sustained ≥50% decline in estimated glomerular filtration rate (eGFR), by 21% (HR: 0.79; 95% CI: 0.66–0.94) and attenuated the annual decline in eGFR by 1.16 mL/min/1.73 m^2 ($p < 0.001$) [76].

One of the most impactful therapeutic benefits of GLP-1 RAs is their capacity to induce substantial and sustained weight loss, particularly for liraglutide and semaglutide. Although the precise mechanisms underlying this effect are not fully elucidated, it is thought to involve central appetite suppression, enhanced satiety, and delayed gastric emptying [77]. High-dose formulations of these agents, specifically liraglutide at 3.0 mg and semaglutide at 2.4 mg, have been approved for the treatment of obesity, independent of glycemic status [78]. More recently, tirzepatide, a dual glucose-dependent insulinotropic polypeptide (GIP)/GLP-1 receptor agonist, has demonstrated superior weight-reducing effects compared with both placebo and semaglutide, particularly in the SURMOUNT trials, with these benefits sustained for up to 72 weeks [79–81].

GLP-1 RAs have also shown efficacy in metabolic dysfunction-associated steatotic liver disease (MASLD) and are now recommended as a first-line antidiabetic treatment for patients with metabolic dysfunction-associated steatohepatitis (MASH) and diabetes [82]. In a 24-week trial, dulaglutide administration produced a 26.4% relative decrease in liver fat content alongside significant reductions in gamma-glutamyl transpeptidase (GGT) levels [83]. Semaglutide has yielded even greater benefits, with its maximal dosage over 72 weeks increasing the rate of MASH resolution without fibrosis progression (59% vs. 17% with placebo; $p < 0.001$) [84]. The wide-ranging biological effects of GLP-1 RAs have further prompted investigation into other conditions, including Alzheimer's disease (AD), where their impact appears favorable, although the current evidence remains preliminary [85].

4.2. Adverse Effects and Safety Profile of GLP-1 Receptor Agonists

The most frequently reported adverse effects of GLP-1 RAs are gastrointestinal, including nausea, vomiting, diarrhea, and constipation. These events are typically dose-dependent and occur most often during treatment initiation or dose escalation. They are attributed to mechanisms overlapping with the weight-reducing properties of GLP-1 RAs, particularly delayed gastric emptying, which is considered the predominant factor [86,87]. A modest increase in heart rate has also been observed, although current evidence does not support an increased risk of arrhythmia [86].

GLP-1 RAs have also been associated with pancreatitis, which has predominantly been linked to gallbladder disease as a secondary effect of weight loss [88]. In addition, evidence suggests that GLP-1 RAs may increase pancreatic enzyme secretion, while the expression and subsequent stimulation of GLP-1 receptors in pancreatic islet and exocrine duct cells have been correlated with overgrowth of the ductal epithelial cells, resulting in hyperplasia and chronic low-grade or acute inflammation, potentially leading to pancreatitis [89]. Nevertheless, the true incidence appears to be low, and several studies have questioned the existence of a causal relationship [90–92]. Earlier experimental data are consistent with this view, showing that although GLP-1 receptor activation can increase pancreatic mass and alter gene expression, it does not confer susceptibility to pancreatitis [93]. GLP-1 RAs have also been associated with an increased risk of pancreatic cancer, given the established link between pancreatitis and pancreatic malignancy. Nevertheless, recent evidence largely supports the safety profile of GLP-1 RAs. Regarding other malignancies, preclinical and clinical studies suggest potential anti-cancer properties through suppression of tumor initiation and progression. An important exception includes individuals with a personal or family history of medullary thyroid carcinoma (MTC), where GLP-1 RAs are contraindicated. This recommendation is primarily based on rodent studies demonstrating increased calcitonin secretion and C-cell hyperplasia, although only minimal GLP-1 receptor expression has been reported in human C-cells [94].

With the increasing adoption of GLP-1 RAs, less common adverse effects are receiving increased attention. Recent pharmacovigilance data have implicated GLP-1 RAs, particularly semaglutide, in the development of androgenetic alopecia [95,96]. The precise mechanisms remain insufficiently defined, with available data relying largely on speculations. The leading hypothesis posits that rapid weight loss may induce metabolic and nutrient alterations that trigger telogen effluvium [97]. Ocular safety has also emerged as a key consideration in GLP-1 RA therapy, encompassing worsening of diabetic retinopathy (DR), particularly in semaglutide users with pre-existing DR at baseline, as well as ischemic optic neuropathy (ION) [75,98,99]. Current evidence suggests that these effects are more likely attributable to rapid improvements in glycemic control rather than a direct pharmacological action of GLP-1 RAs. In contrast, other investigations have not confirmed this association, while preclinical research has suggested a potential protective role through modulation of the ROCK2 pathway within the retinal microvasculature [100–103]. Of note, a very recent in vitro study demonstrated that semaglutide exerted potent antioxidative and cytoprotective effects in retinal endothelial cells, thereby facilitating wound healing [104]. Figure 3 summarizes the main advantages and disadvantages of GLP-1 RA prescription.

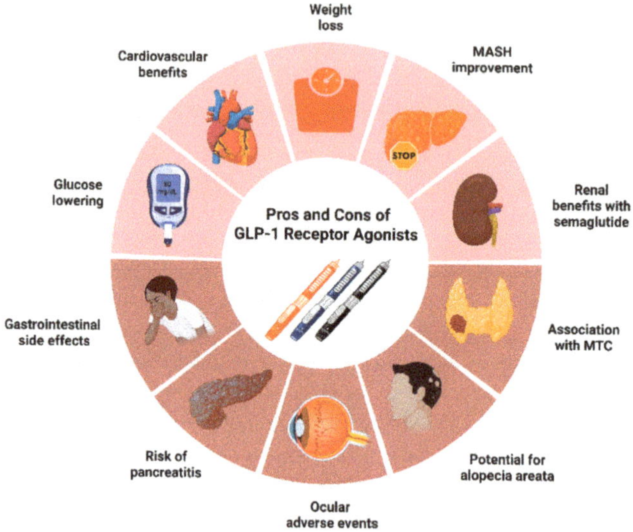

Figure 3. Benefits and drawbacks of GLP-1 receptor agonist use. Abbreviations: GLP-1: glucagon-like peptide-1; MASH: metabolic dysfunction-associated steatohepatitis; MTC: medullary thyroid carcinoma [71,73,76,77,82,86,90,94–96,98,99]. Created in BioRender. Kounatidis, D. (2025) https://BioRender.com/ez8dhoa (assessed on 28 July 2025).

The broader implications of GLP-1 RAs on diabetic microvascular disease, including ED, remain only partially understood. Both animal and human studies increasingly support the beneficial impact of GLP-1 RAs on erectile performance, primarily by improving penile vascular health and endothelial function, as well as through their weight-reducing properties [105,106]. Nevertheless, a limited number of studies have reported potential deterioration in erectile function following GLP-1 RA initiation [107]. Given these contradictory findings, the subsequent sections will shed light on the existing literature to determine whether current evidence permits any definitive clinical conclusions regarding the impact of GLP-1 RAs on erectile function.

5. Cardiovascular and Potential Erectile Implications of GLP-1 Receptor Agonists: Insights from Preclinical Research

5.1. Cardiovascular Effects of GLP-1 Receptor Agonists

A significant body of preclinical research has illuminated the beneficial impact of GLP-1 RAs on endothelial function, largely attributed to the widespread distribution of GLP-1 receptors on endothelial cells throughout different anatomical regions [108,109]. For instance, GLP-1 receptor expression has been identified in human umbilical vein endothelial cells (HUVECs), where GLP-1 appears to exert direct anti-inflammatory effects. These actions are primarily mediated via suppression of RAGE expression, through activation of the cyclic adenosine monophosphate (cAMP) signaling pathway [110]. Additional studies have indicated that GLP-1 enhances endothelial function in diabetic vasculature by reducing endothelial permeability and contractility, while also offering protection against oxidative stress and autophagic processes [111,112].

Endothelium-protective effects have been observed with both short- and long-acting GLP-1 RAs. For instance, exenatide has been shown to ameliorate endothelial dysfunction induced by hyperglycemia and hyperlipidemia through direct activation of eNOS via the AMP-activated protein kinase (AMPK) pathway [113]. Lixisenatide not only upregulates eNOS but may also promote VEGF expression, with evidence suggesting that some of its vascular effects may occur independently of GLP-1 receptor activation [114]. In the case of liraglutide, research demonstrated enhanced endothelial function via stimulation of the mammalian target of rapamycin complex 2/protein kinase B (mTORC2)/Akt pathway, resulting in increased NO production, elevated telomerase activity, and anti-apoptotic effects [115]. Liraglutide has also been reported to attenuate VSMC proliferation and atherosclerotic progression through AMPK activation and cell cycle arrest mechanisms [116]. Interestingly, in endothelial cells derived from individuals with diabetes, liraglutide alleviated endoplasmic reticulum stress, reduced c-Jun N-terminal kinase (JNK) activity, and restored insulin-mediated eNOS activation [117].

Once-weekly formulations have also produced favorable outcomes. Dulaglutide has been shown to reduce hyperglycemia-induced endothelial injury, potentially by preserving sirtuin 1 (SIRT1) levels and curbing mitochondrial fission [118], as well as via the down-regulation of the NLR family pyrin domain containing 3 (NLRP3) inflammasome [119]. On the other hand, semaglutide has exhibited vascular protective effects by mitigating inflammatory damage to endothelial progenitor cells (EPCs), likely through suppression of microRNA-155 (MiR-155) in macrophage-derived exosomes [120]. It may also contribute to the maintenance of endothelial barrier integrity by limiting extracellular matrix components associated with heightened permeability [121]. Further findings revealed that semaglutide reversed cerebral microvascular rarefaction induced by high-fat diet and supported the structural coherence of the neurovascular unit [122]. Notably, the drug appears to outperform dietary interventions in improving endothelial and cardiac function in obesity-associated heart failure, possibly via modulation of inflammatory pathways, immune cell responses, and adipose tissue remodeling [123].

GLP-1 RAs may also confer indirect benefits on endothelial function through their weight-reducing properties. Weight loss not only improves glycemic control but also reduces ectopic adipose tissue depots, such as perivascular and epicardial fat, that exert direct deleterious effects on the vascular endothelium [124]. Weight loss also contributes to improved insulin sensitivity. Beyond these indirect effects, GLP-1 RAs have been shown to exert direct actions on insulin signaling, including enhanced glucose uptake, reduced oxidative stress, and improved lipid metabolism [125]. These mechanisms collectively underscore the central role of GLP-1 RAs in promoting CV health—mechanisms that may likewise translate to improved erectile function, given the shared pathophysiology between CVD and ED [126]. However, dedicated studies specifically evaluating the effects of GLP-1 RAs on erectile function remain scarce.

5.2. Potential Impact of GLP-1 Receptor Agonists on Erectile Function

Existing data point to a positive role of exenatide in facilitating NO–dependent smooth muscle relaxation within the corpus cavernosum, likely mediated by reduced NADPH oxidase activity. In a rat model of cavernosal dysfunction, Dalaklioglu et al. reported that exenatide significantly improved erectile function, correlating with restoration of endothelium-dependent and neurogenic relaxation, diminished oxidative stress and apoptosis, and normalization of signaling within the eNOS and ROCK2 pathways [127]. Similar effects have been reported for liraglutide, which targets overlapping molecular pathways involved in smooth muscle tone regulation. Beyond its vasodilatory properties, liraglutide has been shown to promote structural remodeling of penile tissue by enhancing autophagic activity, thereby facilitating tissue regeneration independently of its glucose-lowering or weight-reducing effects [128]. Furthermore, liraglutide may enhance the Akt/eNOS signaling pathway by promoting Akt-dependent phosphorylation of eNOS, a crucial process for maintaining NO bioavailability and supporting endothelial function in penile erection [129,130]. Indirect mechanisms have also been proposed, particularly in the context of androgen deficiency. In an orchiectomized rat model, liraglutide was found to improve metabolic disturbances, such as increased adiposity and impaired glucose regulation [131]. By contrast, to the best of our knowledge, there is currently no direct experimental evidence exploring the effects of dulaglutide or semaglutide on erectile function.

Emerging evidence indicates that GLP-1 RAs may have therapeutic potential in diabetes-related ED through a mitochondria-targeted, piezoelectric nanosystem. In a diabetic mouse model of ED, ultrasound-activated barium titanate ($BaTiO_3$)-based nanosystem (BaTCG) generated localized piezoelectric currents that consumed protons (H^+) on the mitochondrial outer membrane. This process reduced mitochondrial matrix H^+ availability, collapsed the mitochondrial membrane potential, and triggered mitophagy, thereby attenuating oxidative stress, inflammation, and apoptosis. Simultaneously, the nanosystem enabled sustained release of long-acting GLP-1 RAs, which activated pancreatic β-cell receptors to enhance insulin secretion and improve glycemic control. The combined effects of restored metabolic balance, reduced mitochondrial injury, and suppressed inflammation promoted tissue regeneration within the corpus cavernosum, ultimately improving erectile function. Figure 4 provides a schematic representation of this nanosystem, highlighting the role of GLP-1 RA-mediated glycemic regulation in mitochondrial recovery and erectile tissue repair [132]. Several similar nanoplatforms have demonstrated promise in optimizing antidiabetic therapy; however, their clinical application remains investigational [133].

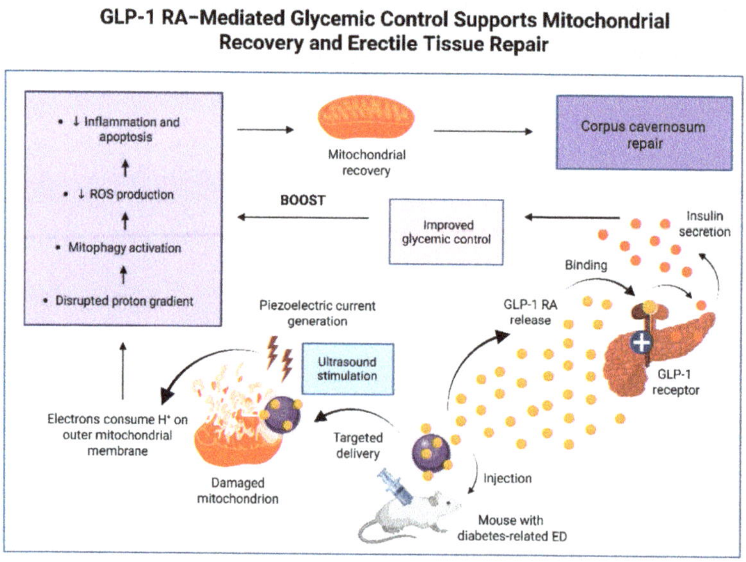

Figure 4. Nanotechnology-based GLP-1 RA delivery system supports mitochondrial recovery and erectile tissue repair. This figure illustrates how the improvement in glycemic control mediated by the action of GLP-1 RAs enhanced the anti-oxidant, anti-inflammatory, and anti-apoptotic effects of a mitochondria-targeted piezoelectric nanosystem, providing an additional benefit to the overall

amelioration of erectile dysfunction [132]. Abbreviations: ED: erectile dysfunction; GLP-1: glucagon-like peptide 1; GLP-1 RA: glucagon-like peptide 1 receptor agonist; H^+: protons; ROS: reactive oxygen species; +: activation. ↓: reduction. Created in BioRender. Kounatidis, D. (2025) https://BioRender.com/agueqv0 (assessed on 20 July 2025).

6. GLP-1 Receptor Agonists and Erectile Function: Insights from Clinical Research

6.1. Evidence Supporting a Beneficial Effect of GLP-1 Receptor Agonists on Erectile Function

While high-quality clinical data remain limited, accumulating evidence increasingly supports a potential therapeutic role for GLP-1 RAs in the management of ED. A small prospective, randomized, open-label study involving men with obesity-related functional hypogonadism unresponsive to lifestyle modification examined the effects of liraglutide (3 mg daily) and testosterone replacement therapy (TRT). Both interventions led to significant increases in total testosterone (TT) (+2.6 ± 3.5 nmol/L with liraglutide vs. +5.9 ± 7.2 nmol/L with TRT), with comparable improvements in sexual function, including libido and erectile performance. Furthermore, 16 weeks of liraglutide treatment outperformed TRT in promoting weight loss (−7.9 ± 3.8 kg vs. −0.9 ± 4.5 kg with TRT, $p < 0.001$), while a marked differential effect on the hypothalamic–pituitary–testicular (HPT) axis emerged, as TRT induced suppression of gonadotropins, whereas liraglutide produced a significant increase ($p < 0.001$) [134]. On the other hand, a subsequent prospective comparative study involving young men (aged 18–35) with metabolic hypogonadism, randomized to gonadotropins (Group A), liraglutide (Group B), or TRT (Group C), demonstrated the superiority of liraglutide in improving sexual function. After 4 months, Group B demonstrated significantly greater improvements in erectile function, with IIEF-5 rising from 4 ± 2 to 21 ± 4 ($p < 0.05$), while PDE5 inhibitor use declined most notably in Group B (Group A: 48%, Group B: 31%, Group C: 45%). Liraglutide treatment was associated with improved sperm quality, accompanied by a significant increase in TT of 192.9% (1.4 ± 0.6 vs. 4.1 ± 0.5 ng/mL, $p < 0.05$) and an elevation in SHBG of 157.1% (14.0 ± 3.0 vs. 36.0 ± 4.0 nmol/L, $p < 0.05$), both relative to baseline and in comparison with the other two groups at the end of treatment. In addition, patients receiving liraglutide exhibited significantly higher gonadotropin concentrations compared with the other groups, with FSH (Group A: 0.9 ± 0.2 IU/L vs. Group B: 2.6 ± 0.2 IU/L vs. Group C: 0.2 ± 0.1 IU/L, $p < 0.05$) and LH (Group A: 1.0 ± 0.3 IU/L vs. Group B: 3.2 ± 0.2 IU/L vs. Group C: 0.3 ± 0.1 IU/L, $p < 0.05$) [135].

The therapeutic benefits of GLP-1 RAs have also been reported when used as an adjunct to metformin. In a retrospective one-year cohort study involving male outpatients with T2D, Lisco et al. found that the combination of GLP-1 RAs with metformin, compared with metformin monotherapy, produced marked increases in TT (+41.41 ± 6.11 ng/dL, $p < 0.0001$) and free testosterone (fT) (+0.44 ± 0.09 ng/dL, $p < 0.0001$) levels. These hormonal changes were accompanied by a significant rise in IIEF-5 score (+2.26 ± 0.26, $p < 0.0001$). Interestingly, the magnitude of benefit was greater in men with higher baseline IIEF-5 score ($p = 0.045$), those presenting with carotid artery stenosis ($p = 0.045$), and those who experienced weight loss during treatment ($p = 0.013$) [106]. These results corroborate an earlier observational study by the same research group, which further demonstrated that the addition of liraglutide or dulaglutide to metformin enhanced erectile function, irrespective of baseline gonadal status ($p < 0.01$) [136].

Liraglutide also appears to provide additional benefits when used alongside TRT and metformin in men with T2D, obesity, hypogonadism, and recent-onset ED. After one year of combined TRT and metformin, patients experienced significant improvements in erectile function ($p < 0.01$), although none reached the glycemic target (HbA1c < 7.5%). Among the 43 participants, 26 showed a limited glycemic response and were considered poor responders. These individuals received liraglutide at 1.2 mg daily during the second year, while the remaining 17 continued on TRT and metformin combination. At the end of the second year, patients who remained on TRT and metformin exhibited no further gains in IIEF scores and experienced significant increases in HbA1c ($p < 0.05$) and body weight ($p < 0.04$). In contrast, subjects receiving liraglutide not only reached glycemic targets and lost weight ($p < 0.01$) but also showed additional increases in SHBG ($p < 0.05$) and TT ($p < 0.01$), with concurrent marked improvements in erectile function [137]. Genetic data provide further support for a causal link between GLP-1 RA use and reduced ED risk. A Mendelian randomization study found genetically proxied GLP-1 RA exposure to be inversely associated with ED (OR: 0.493; 95% CI: 0.430–0.565; $p < 0.001$), with mediation analysis indicating that conventional risk factors such as T2D, obesity, hypertension, and CVD explained only a small portion of the observed effect [138].

To date, the main randomized control trial (RCT) addressing erectile outcomes with GLP-1 RA therapy is an exploratory analysis of the REWIND (Researching Cardiovascular Events With a Weekly INcretin in Diabetes) trial. Among 5312 men with T2D, 3725 completed baseline and follow-up IIEF assessments. Dulaglutide treatment was associated with a modest but significant reduction in moderate-to-severe ED compared with placebo (21.3 vs. 22.0 per 100 person-years; HR: 0.92, 95% CI: 0.85–0.99; $p = 0.021$). Dulaglutide also attenuated erectile function decline over time, with a least squares (LS) mean difference in erectile function subscore of 0.61 (95% CI: 0.18–1.05; $p = 0.006$) [139].

6.2. Evidence Suggesting a Potential Adverse Impact of GLP-1 Receptor Agonists on Erectile Function

Despite increasing evidence supporting the beneficial effects of GLP-1 RAs on erectile function, a limited number of studies have raised concerns regarding potential adverse sexual outcomes. A retrospective cohort study using the TriNetX Research database evaluated the incidence of ED in 3094 non-diabetic men with obesity (aged 18–50) prescribed semaglutide for weight management, matched to an equal number of semaglutide-naïve men with obesity. After excluding patients with prior ED, T2D, or confounding pharmacologic treatments, semaglutide use was associated with a higher incidence of newly diagnosed ED or initiation of PDE5 inhibitor therapy (1.47% vs. 0.32%; RR: 4.5). Reported rates of testosterone deficiency were also higher among semaglutide users (1.53% vs. 0.80%; RR: 1.9). While ED and testosterone deficiency were uncommon in younger men, non-diabetic men with obesity on semaglutide showed a higher risk. Thus, the authors emphasized the importance of informing patients, particularly those at elevated risk, about this potential adverse effect, while also highlighting the need for further research to elucidate underlying mechanisms and optimize therapeutic benefits [140].

Additional signals have emerged from a cross-sectional analysis of the FDA Adverse Event Reporting System (FAERS) database (2003–2024). This pharmacovigilance study detected 182 GLP-1 RA-related sexual adverse events, extending beyond ED, to include reduced libido and orgasmic disorders. Most reports involved men aged 40–60 with T2D being the main indication of use (43.9%). Exenatide (24.2%) and semaglutide (21.4%) were the most frequently implicated agents, followed by liraglutide (20.9%) and dulaglutide (18.7%). Among the different types of sexual dysfunction identified, ED accounted for the majority, representing 71.4% of cases. Although the analysis revealed a statistically significant association ($p < 0.0001$), the study had important limitations, including those inherent to the FAERS database. Furthermore, the signal strength was weak, insufficient to indicate a significant clinical risk, highlighting the need for cautious interpretation of the study results [107]. Table 1 summarizes the main characteristics and findings of clinical studies evaluating the impact of GLP-1 RAs on erectile function.

Table 1. Main clinical studies assessing the impact of GLP-1 receptor agonists on erectile function.

Author, Year	Study Design	Results	Conclusions
Giagulli et al., 2015 [137]	✓ Retrospective observational study ✓ Objective: To evaluate whether adding liraglutide to lifestyle changes, metformin, and TRT improves erectile function ✓ Population: 43 men aged 45–59 with obesity, T2D, overt hypogonadism, and recent-onset ED ✓ Groups (post-pubertal hypogonadism $n = 30$, pre-pubertal hypogonadism $n = 13$): ▪ Year 1 (all): TRT 1 g/12 weeks plus metformin 2–3 g/day ▪ Year 2: good responders continued TRT plus metformin; poor responders added liraglutide 1.2 mg/day ✓ Duration: 2 years	✓ After 1 year: ▪ All patients showed improved erectile function and metabolic parameters ▪ No patients achieved HbA1c < 7.5% ▪ TT normalized to young adult range ✓ Good responders after 2 years: ▪ No further improvement in IIEF score ▪ ↑ HbA1c ($p < 0.05$) and ↑ body weight ($p < 0.04$) ✓ Liraglutide group after 2 years: ▪ ↓ HbA1c < 7.5% ▪ ↓ in body weight ($p < 0.01$) ▪ ↑ SHBG ($p < 0.05$) ▪ ↑ TT ($p < 0.01$) ▪ ↑ in IIEF score	✓ TRT combined with metformin initially improves erectile and metabolic parameters, regardless of hypogonadism onset ✓ In patients with poor metabolic control, adding liraglutide restores testosterone, achieves glycemic targets, reduces body weight, and enhances erectile performance
Jensterle et al., 2019 [134]	✓ Prospective, randomized, open-label study ✓ Objective: To compare the effects of liraglutide versus TRT in men with obesity-related functional hypogonadism, who were poor responders to lifestyle modification ✓ Population: 30 men with obesity (BMI 41.2 ± 8.4 kg/m^2) and functional hypogonadism, aged 46.5 ± 10.9 years ✓ Groups: ▪ 50 mg 1% TRT gel daily ▪ 3.0 mg liraglutide daily ✓ Duration: 16 weeks	✓ ↑ TT in both groups: ▪ TRT: +5.9 ± 7.2 nmol/L ▪ Liraglutide: +2.6 ± 3.5 nmol/L ✓ Both arms improved sexual function ✓ Liraglutide group showed ↑ LH and FSH ($p < 0.001$ for between-treatment effect) ✓ Weight loss: ▪ TRT: 0.9 ± 4.5 kg ▪ Liraglutide: 7.9 ± 3.8 kg ✓ MetS resolved in 2 liraglutide patients vs. 0 in TRT	✓ When lifestyle modification fails and MBS is not indicated, liraglutide should be preferred over TRT for overall health improvement in men with obesity-related functional hypogonadism
Bajaj et al., 2021 [139]	✓ Exploratory analysis of the double-blind, randomized, placebo-controlled REWIND trial ✓ Objective: To evaluate the incidence, prevalence, and progression of ED in men with T2D receiving dulaglutide vs. placebo, and to examine whether the drug's impact on erectile function aligns with its broader effects on diabetes-related outcomes ✓ 3725 men with T2D: ▪ Mean age 65.5 years ▪ 39.9% with established CVD ▪ 56.5% presenting with moderate to severe ED at baseline ✓ Duration: approximately 5 years	✓ Dulaglutide was associated with a lower incidence of moderate/severe ED compared with placebo (21.3 vs. 22.0 per 100 person-years; HR 0.92, 95% CI 0.85–0.99; $p = 0.021$) ✓ Dulaglutide group showed a smaller decline in IIEF scores compared with placebo (LS mean difference 0.61; 95% CI 0.18–1.05; $p = 0.006$).	✓ Long-term dulaglutide treatment may lower the risk of developing moderate or severe ED in men with T2D

Table 1. *Cont.*

Author, Year	Study Design	Results	Conclusions
Lisco et al., 2022 [136]	✓ Retrospective observational study ✓ Objective: To explore the impact of GLP-1 RAs (liraglutide and dulaglutide) on erectile function in men with T2D, both with and without baseline hypogonadism, when used as an add-on to metformin ✓ Population: 110 men with T2D and ED: ▪ Men with hypogonadism (HP): $n = 48$ and eugonadal patients (EP): $n = 62$ ▪ Mean age 51–64 years ▪ Diabetes duration 5–10 years ▪ 6% with established CVD ▪ Patients with eGFR < 60 mL/min per 1.73 m^2 were excluded ✓ Study protocol: ▪ Men with HbA1c < 7.2% received metformin (2 g/day) ▪ Men with HbA1c > 7.2% received a GLP-1RA (52% liraglutide, 1.2 mg/day; 48% dulaglutide, 1.5 mg/week) plus metformin ✓ Groups: ▪ 28 HP were treated with GLP-1RA plus metformin (HPs) ▪ 20 HP were treated with metformin alone (HPc) ▪ 38 EP were treated with GLP-1RA plus metformin (EPs) ▪ 30 EP were treated with metformin alone (EPc) ✓ Duration: 1 year	✓ HPs and EPs: Significant ↓ in HbA1c (−0.7 ± 0.3%; $p < 0.001$) ✓ HPc and EPc: Slight ↑ in HbA1c (+0.4 ± 0.2) ✓ HPs and EPs: Significant improvement in erectile function (↑ IIEF-5 score, all $p < 0.01$)	✓ Liraglutide and dulaglutide appear to improve erectile function in men with T2D, regardless of baseline hypogonadism status ✓ Additional controlled studies are required to confirm these preliminary findings
La Vignera et al., 2023 [135]	✓ Prospective interventional comparative study ✓ Objective: To investigate the effects of liraglutide on reproductive and sexual function in men of childbearing age with metabolic hypogonadism ✓ Population: 110 men aged 18–35 with metabolic hypogonadism, divided into three groups based on fertility intentions ✓ Groups: ▪ Group A ($n = 35$): Desiring fatherhood, treated with gonadotropins (urofollitropin 150 IU three times a week, plus hCG 2000 IU twice a week) ▪ Group B ($n = 35$): not seeking fatherhood, treated with liraglutide 3 mg/day ▪ Group C ($n = 40$): already fathered a child, treated with TRT 60 mg/day ✓ Duration: 4 months	✓ Group B (liraglutide) showed: ▪ Significant improvement in conventional sperm parameters compared with baseline and Group A ▪ Improved erectile function compared with baseline and both Groups A and C ▪ Significantly ↑ levels of TT and SHBG both relative to baseline and in comparison with the other two Groups: 1. TT: 192.9% (1.4 ± 0.6 vs. 4.1 ± 0.5 ng/mL, $p < 0.05$) 2. SHBG: 157.1% (14.0 ± 3.0 vs. 36.0 ± 4.0 nmol/L, $p < 0.05$) ▪ Significantly ↑ gonadotropin levels compared with both Groups A and C: 1. FSH: Group A: 0.9 ± 0.2 IU/L vs. Group B: 2.6 ± 0.2 IU/L vs. Group C: 0.2 ± 0.1 IU/L, $p < 0.05$ 2. LH: Group A: 1.0 ± 0.3 IU/L vs. Group B: 3.2 ± 0.2 IU/L vs. Group C: 0.3 ± 0.1 IU/L, $p < 0.05$	✓ Liraglutide is a promising and safe pharmacological option for men with obesity and metabolic hypogonadism, showing advantages over traditional therapies in improving both reproductive and sexual function

Table 1. *Cont.*

Author, Year	Study Design	Results	Conclusions
Lisco et al., 2024 [106]	✓ Retrospective cohort study ✓ Objective: To assess the effects of GLP-1 RAs combined with metformin vs. metformin alone on erectile function and metabolic parameters in men with T2D ✓ Population: 108 male outpatients with T2D and ED (median age 60 years) ✓ Groups: ▪ Metformin alone ($n = 45$) ▪ GLP-1RAs plus metformin ($n = 63$) ✓ Duration: 12 months	✓ After 12 months, GLP-1RAs plus metformin improved glucose control more than metformin alone: HbA1c ↓ from 8.3 ± 0.2% to 7.0 ± 0.3% ($p < 0.0001$) vs. 7.0 ± 0.5% to 7.3 ± 0.4% ($p = 0.0007$) ✓ GLP-1RAs plus metformin vs. metformin alone: ▪ Significant weight loss: −5.82 ± 0.69 kg ($p < 0.0001$) ▪ ↓ in WC: −4.99 ± 0.6 cm ($p < 0.0001$) ▪ Improvement in HbA1c: −0.56% ± 0.13% ($p < 0.0001$) ▪ ↓ in FPG: −25.54 ± 3.09 mg/dL ($p < 0.0001$) ▪ ↑ in TT: +41.41 ± 6.11 ng/dL ($p < 0.0001$) ▪ ↑ in fT: +0.44 ± 0.09 ng/dL ($p < 0.0001$) ▪ Improvement in self-reported erectile function: IIEF5 score +2.26 ± 0.26 ($p < 0.0001$) ✓ Predictors of greater IIEF5 improvement: ▪ Higher baseline IIEF-5 score ($p = 0.045$) ▪ Presence of carotid stenosis ($p = 0.045$) ▪ Greater weight loss from baseline ($p = 0.013$) ✓ 1-year GLP-1RAs plus metformin treatment was the main determinant of improved IIEF5 score (2.74 ± 0.53, $p < 0.0001$)	✓ GLP-1RAs combined with metformin improved erectile function in men with T2D compared with metformin alone, independent of baseline characteristics ✓ Improvements may be due to positive vascular effects ✓ Due to the retrospective nature of the study, a definitive cause-effect relationship cannot be established; RCTs are needed to confirm the efficacy of GLP-1RAs in treating ED in T2D
Able et al., 2025 [140]	✓ Retrospective cohort study ✓ Objective: To evaluate the risk of newly diagnosed ED or PDE5 inhibitor use, and testosterone deficiency, following at least one month of semaglutide prescription in non-diabetic men using it for weight loss ✓ Population: Population: 3094 non-diabetic men with obesity, aged 18–50, who received semaglutide, matched 1:1 to non-diabetic men with obesity without semaglutide exposure	✓ ED and/or PDE5 inhibitor use: ↑ in semaglutide users vs. non-users (1.47% vs. 0.32%; RR 4.5, 95% CI 2.3–9.0) ✓ Testosterone deficiency: ↑ in semaglutide users vs. non-users (1.53% vs. 0.80%; RR 1.9, 95% CI 1.2–3.1)	✓ ED, PDE5 inhibitor prescription, and testosterone deficiency were more frequent in non-diabetic men with obesity on semaglutide, despite weight loss benefits ✓ Patients should be informed of potential sexual health risks, particularly those at higher risk ✓ Further research is needed to clarify underlying mechanisms and optimize treatment outcomes-
Pourabhari Langroudi et al., 2025 [107]	✓ Pharmacovigilance study using the FAERS database (2003–2024) ✓ Objective: To investigate the association between GLP-1 RAs and male sexual dysfunction, including: ▪ ED ▪ Orgasmic dysfunction ▪ ↓ libido	✓ Total cases: 182 ✓ Most common indication: T2D (43.9%), followed by weight loss (4.9%) ✓ Patient age: Predominantly 40–60 years ✓ Trends over time: Reports ↑ in recent years, especially 2023; semaglutide reports rising, tirzepatide surpassed semaglutide in 2023 and Q1 2024 ✓ Subcutaneous route of administration most common ✓ Drug distribution: Exenatide 24.2%, lixisenatide 0.5%, liraglutide 20.9%, dulaglutide 18.7%, semaglutide 21.4%, tirzepatide 14.3% ✓ Type of sexual dysfunction: ED: 71.4%; ↓ libido: 15.1%; orgasmic abnormalities: 1.6%; other sexual dysfunctions: 11.0%	✓ The link between GLP-1 RAs and male sexual dysfunction appears weak, with current evidence not indicating a significant clinical risk ✓ Nevertheless, ongoing monitoring is advised as GLP-1 use grows, while further research is needed to clarify their impact on sexual function

Abbreviations: BMI: body mass index; CI: 95% confidence interval; CVD: cardiovascular disease; ED: erectile dysfunction; eGFR: estimated glomerular filtration rate; FAERS: FDA Adverse Event Reporting System; FPG: fasting plasma glucose; FSH: follicle-stimulating hormone; fT: free testosterone; GLP-1 RA: glucagon-like peptide-1 receptor agonist; HbA1c: glycated hemoglobin A1c; HPT: hypothalamic–pituitary–testicular; HR: hazard ratio; hCG: human chorionic gonadotropin; IIEF: international index of erectile function; LH: luteinizing hormone; LS: least squares; MBS: metabolic bariatric surgery; MetS: metabolic syndrome; PDE5: phosphodiesterase type 5; Q1: first quartile; RCT: randomized controlled trial; REWIND: Researching Cardiovascular Events With a Weekly INcretin in Diabetes; RR: relative risk; SHBG: sex hormone-binding globulin; T2D: type 2 diabetes; TRT: testosterone replacement therapy; TT: total testosterone; WC: waist circumference; ↑: increase; ↓: decrease.

7. The Influence of Other Antidiabetic Drugs on Erectile Function: Comparative Evidence with GLP-1 Receptor Agonists

7.1. Erectile Effects of Other Antidiabetic Therapies: Mechanistic Insights and Clinical Evidence

Multiple studies have examined the effects of antidiabetic therapies on erectile function, but findings remain inconsistent and at times contradictory. Metformin, for example, has demonstrated both beneficial and adverse effects. In a preclinical model of angiotensin II–induced ED, metformin normalized corpus cavernosum contractility and enhanced smooth muscle relaxation through increased eNOS phosphorylation [141]. When combined with antioxidants, it improved intracavernosal pressure and eNOS activity, while reducing mitochondrial autophagy, AGE generation, and the collagen-to-smooth muscle ratio in erectile tissue [142]. Conversely, short-term metformin use has been linked to reduced testosterone and libido, potentially blunting the testosterone rise typically associated with improved glycemic control [143,144].

Pioglitazone, an insulin sensitizer that functions as a peroxisome proliferator-activated receptor gamma (PPAR-γ) agonist, may exert protective effects against ED [145,146]. PPAR-γ is expressed in VSMCs, where its activation reduces oxidative stress and suppresses pro-inflammatory signaling [147]. Pioglitazone also attenuated endothelin-1–induced vasoconstriction by downregulating cyclooxygenase-2 (COX-2) via activator protein-1 (AP-1) and nuclear factor kappa-light-chain-enhancer of activated B cells (NF-κB), while enhancing NO bioavailability [148]. In a rat model of bilateral cavernous nerve injury, it restored erectile function via insulin-like growth factor-1 (IGF-1) signaling activation [149]. Clinically, pioglitazone has been shown to improve erectile response in men with sildenafil-resistant ED [150].

Insulin therapy, particularly when combined with antioxidants, has been shown to mitigate ED by inhibiting the AGEs–RAGE–oxidative stress axis [151]. It may also promote recovery by modulating apoptotic protein expression and by restoring androgen receptor expression in penile tissue [152,153]. Evidence on insulin secretagogues such as SUs indicates mixed effects. Genetic studies suggest an increased ED risk with gliclazide [154], possibly due to SU-mediated inhibition of ATP-sensitive potassium channels essential for prostaglandin E1 (PGE1)- and cAMP-dependent vasodilation [155]. Conversely, SUs have been associated with increased testosterone, enhanced libido, and improved erectile function [156]. Notably, adulterated sexual enhancement products containing glibenclamide or glyburide have caused outbreaks of severe hypoglycemia [157,158].

The role of dipeptidyl peptidase-4 inhibitors (DPP-4is) in erectile performance remains uncertain, although preliminary evidence suggests vascular benefits, including endothelial repair via endothelial progenitor cells (EPCs) mobilization [159]. Saxagliptin has been demonstrated to improve ED by upregulating stromal cell-derived factor 1 (SDF-1), which enhances EPC homing, and by activating phosphoinositide 3-kinase/protein kinase B (PI3K)/Akt signaling, thereby diminishing oxidative stress and apoptosis [160]. Additional research has revealed that toleragliptin improved erectile responses in a vincristine-induced ED model via increased nNOS expression and endothelial signaling [161]. Conversely, the VILDA study reported a 22.8% incidence of ED over two years with vildagliptin use [162].

Sodium-glucose cotransporter-2 inhibitors (SGLT-2is) are well recognized for their CV benefits, including improved arterial stiffness and endothelial function [163]; however, their effects on sexual health remain ambiguous. Dapagliflozin and empagliflozin have been shown to restore NO availability under tumor necrosis factor-alpha (TNF-α)-induced endothelial inflammation through ROS reduction. Interestingly, these effects appear independent of eNOS activity or barrier integrity [164,165]. Data on canagliflozin are limited, though one study suggested potential interaction with tadalafil metabolism via cytochrome

P450 3A4 inhibition, warranting caution in coadministration [166]. Clinically, dapagliflozin improved erectile function and enhanced tadalafil responsiveness in men with T2D [167]. In contrast, a large observational study reported increased incident ED risk with SGLT-2i therapy (adjusted HR: 1.55; 95% CI: 1.40–1.72) [168].

7.2. Potential Comparative Advantage of GLP-1 Receptor Agonists in Erectile Function

Although limited, comparative evidence suggests that GLP-1 RAs may exert superior effects on erectile function relative to other antidiabetic agents. In a trial comparing exenatide plus metformin with glimepiride plus metformin, TT increases over 3 months were significantly greater with exenatide (121.72 ± 56.73 ng/dL vs. 34.67 ± 16.30 ng/dL) [169]. Another study reported higher IIEF-5 scores in men treated with GLP-1 RAs versus insulin (16.7 ± 4.7 vs. 12.9 ± 6.2; $p = 0.02$), a difference that remained significant after adjustment for age and diabetes duration ($p = 0.01$). No differences in erectile function were observed between metformin, SUs, DPP-4is, or SGLT-2is [170].

The comparative advantage of GLP-1 RAs has also been supported by recent systematic reviews and meta-analyses. A meta-analysis by Yang et al. demonstrated superiority of GLP-1 RAs over metformin in improving erectile function ($p = 0.02$), with greater benefits in men with BMI ≥ 30 kg/m^2 ($p = 0.02$) [171]. Similarly, a meta-analysis by Rastrelli et al., including 4112 patients, showed that observational data supported improvements in erectile health with GLP-1 RAs, SGLT-2is, and metformin, though not superior to lifestyle interventions. In contrast, RCTs demonstrated significant IIEF improvements with metformin, pioglitazone, and dulaglutide. The authors suggested that GLP-1 RAs and SGLT-2is may be particularly effective in advanced metabolic dysfunction, whereas metformin may remain preferable for less complex cases [172]. The same research group also found that GLP-1 RA therapy significantly increased TT, fT, SHBG, and gonadotropin levels, with concurrent improvements in erectile function. Meta-regression analysis demonstrated a strong correlation between GLP-1 RA–induced weight loss and TT elevation, implicating weight reduction as a principal mediator ($p < 0.0001$) [173].

These findings are further supported by two additional systematic reviews and meta-analyses, both highlighting the favorable impact of GLP-1 RAs on functional hypogonadism, largely attributable to weight loss [174,175]. Van Cauwenberghe et al. [174] reported minimal or no hormonal effects from other antidiabetic agents. Notably, an individual study within this review yielded insights such as the neutral impact of exenatide on serum testosterone in men with T2D [176]. In contrast, Salvio et al. found that while androgenic effects of GLP-1 RAs were comparable to those of other antidiabetic therapies, they offered superior benefits in terms of BMI reduction, increased gonadotropin levels, and improved erectile function scores [175]. Table 2 presents the data, findings, and conclusions of these meta-analyses.

Table 2. Evidence from recently published meta-analyses on the comparative role of GLP-1 receptor agonists in erectile dysfunction relative to other antidiabetic therapies.

Author, Year, Reference	Yang et al., 2025 [171]	Rastrelli, et al., 2025 [172]	Corona, et al., 2025 [173]	Salvio et al., 2025 [175]
Meta-analysis characteristics	✓ Objective: To evaluate the impact of different classes of antidiabetic drugs on ED in patients with diabetes ✓ 3 cohort studies included ✓ Number of participants: 314 (209 eligible) men ✓ Baseline characteristics: ▪ Mean age: 58.9 years ▪ Mean BMI 32.1 kg/m^2 ✓ Mean follow-up: 2.3 years	✓ Objective: To assess the impact of antidiabetic drugs on ED in individuals with diabetes or prediabetes ✓ 3 observational and 3 placebo-controlled RCTs included ✓ Number of participants: 4112 subjects ✓ Baseline characteristics: ▪ Mean age: 57.8 years ▪ Mean BMI: 31.4 kg/m^2 ▪ Mean testosterone: 13.1 nmol/L ✓ Mean follow-up: 87.4 weeks	✓ Objective: To explore the effects of GLP-1 RAS on the HPT axis and male sexual function ✓ 6 studies included (type not defined) ✓ Number of participants: 386 individuals ✓ Baseline characteristics: ▪ Mean age 47.1 ± 11.1 years ▪ Mean BMI 35.2 ± 3.9 kg/m^2 ✓ Mean follow-up: 31.4 ± 19.1 weeks	✓ Objective: To investigate the effects of GLP-1 RAs on hormone secretion in overweight and men with obesity, and to compare their impact on testicular and erectile function with other antidiabetic or weight-loss therapies ✓ 6 cohorts and 1 randomized open-label study included ✓ Number of participants: 680 overweight/men with obesity ✓ Baseline characteristics: ▪ Mean age: 47.65 years ▪ Mean BMI: 34.44 kg/m^2 ✓ Mean follow-up: 8.43 months
Results	✓ GLP-1RAs were more effective than metformin in improving ED in diabetic patients ($p = 0.02$) and the effect was more pronounced in patients with higher BMI ($p = 0.02$) ✓ TZDs may offer benefits in sexual function, though further safety and efficacy studies are needed ✓ Insulin and SGLT-2is showed potential benefit but lack strong supporting evidence ✓ The effect of metformin and SUs remains unclear due to mixed or inconclusive data	✓ Observational studies: ▪ Metformin, GLP1 RAs, and SGLT2is were associated with improved ED ▪ No significant benefit compared to diet (for all drugs including SUs and DPP-4is) ✓ RCTs showed significant improvement in IIEF score with metformin, pioglitazone, and dulaglutide	✓ GLP-1 RA use was associated with: ▪ Significant ↑ in TT, calculated fT, and SHBG ▪ ↑ gonadotropins (either LH or FSH) at the study endpoint ▪ Improved erectile function across studies ✓ Meta-regression analysis: ▪ Significant positive correlation between the degree of GLP-1RA–induced weight loss and testosterone ↑ ($p < 0.0001$) ▪ No significant heterogeneity ($p = 0.429$)	✓ GLP-1RAs significantly ↑ TT levels ($p < 0.0001$), with similar ↑ in FT, SHBG, LH, and FSH ✓ Weight, BMI, WC, and HbA1c were significantly ↓ following GLP-1RA treatment ✓ Meta-regression analysis showed a negative correlation between TT ↑ and ↓ in weight and BMI ✓ Compared with other treatments, GLP-1RAs had similar effects on serum androgens, but greater ↓ in BMI and greater ↑ in gonadotropins and erectile function indices
Conclusions	✓ GLP-1 RAs appear to outperform metformin in improving erectile function in diabetic men, especially those with obesity ✓ Other drug classes may also offer positive effects on ED, but more high-quality clinical trials are needed to confirm these findings and address safety considerations	✓ Given the link between ED and future CVD risk, GLP-1 RAs and SGLT-2is are considered the preferred options, due to their broad role in reducing long-term complications ✓ Metformin is a suitable alternative for less complex patients	✓ GLP-1 RAs may offer a viable alternative therapeutic option for men with severe obesity-related testosterone deficiency, showing both hormonal and sexual function benefits alongside weight loss	✓ GLP-1RAs may play a beneficial role in managing functional hypogonadism and improving erectile function in overweight and men with obesity, primarily through weight loss–related mechanisms ✓ Current evidence does not confirm a direct effect of GLP-1 RAs on testicular function due to limitations in study design and available data

Abbreviations: BMI: body mass index; CVD: cardiovascular disease; DPP-4i; dipeptidyl peptidase-4 inhibitor; ED: erectile dysfunction; fT: free testosterone; FSH: follicle-stimulating hormone; GLP-1 RA: glucagon-like peptide-1 receptor agonist; HbA1c: glycated hemoglobin A1c; HPT: hypothalamic–pituitary–testicular; IIEF: international index of erectile function; LH: luteinizing hormone; RCT: randomized controlled trial; SHBG: sex hormone-binding globulin; SGLT-2i: sodium–glucose cotransporter-2 inhibitor; SU: sulfonylurea; TT: total testosterone; TZD: thiazolidinedione; WC: waist circumference; ↑: increase; ↓: decrease.

8. Research Gaps, Challenges, Future Perspectives, and Conclusions

8.1. Evaluating Mechanistic and Clinical Evidence Supporting GLP-1 Receptor Agonists–Mediated Erectile Benefits

ED is recognized as an early and independent predictor of CVD, with evidence indicating a 45% increased risk among individuals with ED [177]. The two conditions share overlapping pathophysiological mechanisms, including systemic inflammation, endothelial dysfunction, and atherosclerosis [178]. Current experimental data on the role of GLP-1 RAs in ED remains limited, primarily focusing on exenatide and liraglutide, with an emphasis on vascular outcomes [127–130]. On the other hand, understanding the mechanisms through which GLP-1 RAs exert their beneficial effects on CVD provides a theoretical framework for their potential effects on erectile function.

Potential benefits on erectile performance may arise from glycemic control–related mechanisms, including inhibition of AGE generation and reduction of oxidative stress [179], as well as preservation of mitochondrial integrity via modulation of programmed cell death pathways, such as pyroptosis and ferroptosis [180,181]. Furthermore, GLP-1 RAs may exert synergistic anti-inflammatory and antioxidant effects through reduced neutrophil extracellular trap (NET) formation and inhibition of NLRP3 inflammasome activation [182]. In diabetic ED models, a transcriptomic analysis demonstrated downregulation of genes involved in smooth muscle contractility and mitochondrial function, alongside upregulation of Wnt signaling and extracellular matrix remodeling [183].

Additional mechanisms may involve the modulation of key signaling pathways. The protein kinase A (PKA) pathway supports NO–mediated smooth muscle relaxation via cAMP signaling [184], whereas the Hippo–YAP1 axis regulates endothelial integrity, glucose metabolism, and fibrosis [185]. GLP-1 RA–mediated modulation of these pathways has shown benefits in other models and may potentially extent to erectile health [186,187]. Beyond endothelial dysfunction, pericyte impairment has been identified as a contributor to diabetic ED [188]. Given that GLP-1 receptors are expressed on cerebral pericytes, where they mediate vascular stabilization, similar effects in penile pericytes are plausible [189]. Moreover, the detection of GLP-1 receptors in testicular tissue raises the possibility of effects on gonadal function, mediated not only through hormonal pathways [190]. Although preclinical models frequently employ dosing regimens that limit direct translational applicability, they remain essential for delineating pharmacodynamic mechanisms, including those underlying the well-established CV benefits of GLP-1 RAs, which could also have implications for ED [191].

On the clinical front, an increasing number of studies in recent years have suggested favorable effects of GLP-1 RAs on erectile and sexual function, as well as on reproductive outcomes in both men and women [192]. However, these observations are subject to significant methodological constraints. Clinical trials investigating the effects of GLP-1 RAs on erectile function are typically limited by small sample sizes, short follow-up durations, and lack of evaluation of dose–response and exposure–response relationships. Furthermore, most available studies are observational in design and display considerable heterogeneity. To date, no RCT has been conducted with erectile function as a predefined primary endpoint [106,134–139].

These limitations also apply to comparative analyses of GLP-1 RAs versus other antidiabetic agents, for which the impact on erectile function remains uncertain. A plausible explanation for the apparent advantage of GLP-1 RAs may involve their superior weight-loss efficacy relative to other antidiabetic drugs, particularly SGLT-2is, which themselves confer substantial CV benefits. Few studies have specifically evaluated symptoms of androgen deficiency, and reported changes in testosterone levels are generally modest and subject to variability in study design [169–175]. Finally, the reliability and validity of instruments used to assess ED, such as the IIEF, remain debated, with evidence suggesting important limitations [193].

8.2. Assessing Emerging Concerns and Theoretical Risks of GLP-1 Receptor Agonists on Erectile Function

While the preponderance of available evidence suggests a beneficial effect of GLP-1 RAs on erectile function, rare reports of potential harm warrant careful consideration, even though the underlying studies face significant constraints, restricting causal inference. For instance, Able et al. relied on claims-based International Classification of Diseases, Tenth Revision (ICD-10), without verification of medication adherence. Lack of confirmation of actual drug intake limits the ability to draw conclusions about long-term effects on sexual health. Additional confounders included baseline BMI differences and the inability to distinguish between specific forms of sexual dysfunction [140]. On the other hand, although the FAERS analysis identified a statistically significant association between GLP-1 RA use and sexual dysfunction, the signal was weak, and inherent database limitations, including reporting bias and residual confounding, were present [107]. Combining the results of these two studies suggests that, while the overall risk of ED appears low or even uncertain, the increasing use of GLP-1 RAs highlights the need for ongoing clinical vigilance [107,140,194]. Of note, a recent U.S. surveillance study of emergency department visits for semaglutide-related adverse effects reported no cases of ED, though such visits are inherently uncommon [195,196].

To date, no experimental studies have demonstrated a harmful effect of GLP-1 RAs on erectile function. Consequently, the mechanisms through which GLP-1 RAs might negatively influence sexual health remain uncertain. It has been proposed that GLP-1 receptors in brain regions regulating appetite and reward may modulate neurotransmitters, such as dopamine and serotonin, potentially influencing sexual arousal [107,197]. Additional hypotheses include the rapid correction of glycemia following GLP-1 RA administration, similar to the exacerbation of DR, as well as long-term effects of prior hyperglycemia via persistent changes in endothelial gene expression, often referred to as "glucose memory" [198,199]. Other potential mechanisms involve altered testosterone pulsatility and GLP-1–mediated smooth muscle relaxation in penile tissue [140,200]. The development of ED may also be linked to the weight-lowering features of GLP-1 RAs, which can cause significant lean body mass loss, occasionally progressing to sarcopenia, a factor associated with ED [201,202]. On the other hand, in normal-weight individuals, caloric restriction has been linked to reduced testosterone levels, suggesting that GLP-1 RA use in such T2D patients may contribute to ED [203,204].

Overall, although caution is warranted, the aforementioned data should not dissuade clinicians from prescribing GLP-1 RAs, as such an association is not firmly established, and even in cases with semaglutide, the incidence was particularly low, approximately 1.4% [140,205]. In instances where ED develops following GLP-1 RA therapy, discontinuation is indicated, accompanied by a thorough urological evaluation. According to the 2025 European Association of Urology (EAU) guidelines, this evaluation should include a comprehensive medical and sexual history as well as early-morning fasting TT measurement. PDE5 inhibitors constitute the first-line treatment, while adjunctive TRT may be considered for hypogonadal non-responders, with careful attention to potential side effects, such as gastrointestinal symptoms, increased blood pressure, and thrombosis [206]. From a diabetology perspective, a reasonable approach involves substituting GLP-1 RA therapy with an alternative antidiabetic agent. If erectile function and testosterone levels improve but GLP-1 RA therapy remains clinically necessary, re-initiation at a previously tolerated lower dose or switching to a different GLP-1 RA may be considered, with close monitoring and informed consent. Notably, the timeline for testosterone recovery remains uncertain and may be prolonged [207]. Among available GLP-1 RAs, dulaglutide may represent the safest option, with even favorable effects [139], further supported by a recent RCT showing that in eugonadal normal-weight men, dulaglutide did not adversely affect sexual desire, the HPG axis, or sperm parameters compared with placebo. However, the study included a small number of participants and had a limited follow-up of only one month [208]. Despite these various hypotheses and therapeutic considerations, it is clear that well-designed, long-term RCTs are needed to provide definitive answers.

8.3. Comparing GLP-1 Receptor Agonists with Other Weight Loss Interventions

Assessment of the impact of GLP-1 RAs on erectile function is further limited by the lack of comparative data with other weight loss interventions. Theoretically, tirzepatide may confer additional erectile health benefits through its GIP component. However, the vascular effects of GIP remain poorly defined due to inconsistent GIP receptor (GIPR) expression across different endothelial beds [209]. Furthermore, while both GIP and GLP-1 exert protective effects on the vasculature under normal glucose levels, only GLP-1 enhances endothelial function in states of hyperglycemia. This occurs via increased eNOS expression and subsequent NO production through cAMP signaling [210]. This divergence may have implications for diabetes-associated ED, where NO bioavailability is essential. Moreover, although GIPR is expressed in the testes in preclinical models, its role in human reproductive physiology appears limited based on gene expression data [211]. On the other hand, tirzepatide may offer superior benefits over GLP-1 RAs due to its greater efficacy in weight loss. In a pilot study involving men with obesity and metabolic hypogonadism, tirzepatide improved both erectile function and hormonal profiles ($p < 0.05$), outperforming lifestyle modification and TRT [212]. It is worth noting that pharmacovigilance data, which linked GLP-1 RA administration with ED development, reported tirzepatide in 14.3% of cases—a lower proportion compared to liraglutide, dulaglutide, and semaglutide. Nevertheless, the annual distribution of adverse event reports revealed that tirzepatide exceeded semaglutide in 2023 and early 2024, consistent with its rising clinical use [107].

Relevant insights also arise from non-pharmacological weight loss interventions, such as metabolic bariatric surgery (MBS). Evidence suggests that MBS improves erectile function through mechanisms related to weight loss and T2D remission. On the contrary, although postoperative testosterone levels increase, they do not seem to independently improve ED [213]. Notably, metabolomic studies indicate that GLP-1 RAs induce modest metabolic changes primarily related to glycemic control and appetite, whereas MBS triggers broader shifts in gut hormone signaling and nutrient sensing [214,215]. Despite emerging evidence, comparative data evaluating the effects of GLP-1 RAs, tirzepatide, and MBS on erectile function remain lacking.

In conclusion, ED represents a common yet frequently underrecognized microvascular complication of diabetes, influenced by a broad spectrum of risk factors. Emerging preclinical and clinical evidence suggests that GLP-1 RAs may exert beneficial effects on erectile function, potentially exceeding those of other antidiabetic therapies. Nevertheless, conclusions remain premature. Concurrently, recent reports have raised concerns regarding possible ED development among GLP-1 RA users. However, these observations are limited by methodological shortcomings, and the strength of the association appears weak. Further mechanistic investigations are warranted to delineate the underlying biological pathways, while rigorously designed, long-term RCTs with erectile function as a predefined primary endpoint are essential. Until such data become available, clinicians should remain vigilant for symptoms of ED following the initiation or dose titration of GLP-1 RAs and be prepared to implement appropriate therapeutic strategies. Figure 5 summarizes the key features of diabetes-related ED and outlines the current evidence regarding the effects of GLP-1 RAs on erectile function.

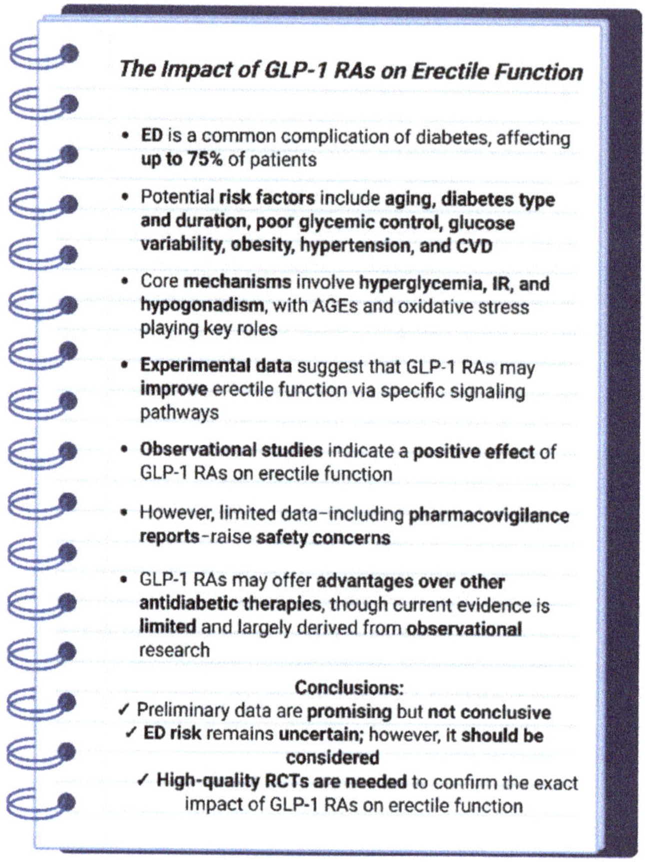

Figure 5. Diabetes-associated erectile dysfunction and the role of GLP-1 receptor agonists: An evidence-based snapshot. Abbreviations: AGEs: advanced glycation end products; CVD: cardiovascular disease; ED: erectile dysfunction; GLP-1 RA: glucagon-like peptide-1 receptor agonist; RCT: randomized controlled trial. Created in BioRender. Kounatidis, D. (2025) https://BioRender.com/21i8b6j (assessed on 27 August 2025).

Author Contributions: D.K. and N.G.V. conceived the idea of the manuscript and organized its plan. D.K. wrote the whole manuscript; E.R. and K.V. performed the literature search and were responsible for the references and the tables; D.K. was responsible for the figures; N.G.V., E.D., K.M. and N.T. reviewed, edited and supervised the manuscript. All authors have read and agreed to the published version of the manuscript.

Funding: This research received no external funding.

Institutional Review Board Statement: Not applicable.

Informed Consent Statement: Not applicable.

Data Availability Statement: Not applicable.

Conflicts of Interest: The authors declare no conflicts of interest.

Abbreviations

3-NT: 3-nitrotyrosine; Ach: acetylcholine; AD: Alzheimer's disease; AGEs: advanced glycation end products; Akt: protein kinase B; AMPK: AMP-activated protein kinase; AP-1: activator protein-1; AR: androgen receptor; BaTCG: barium titanate-based composite; BMI: body mass index; cAMP: cyclic adenosine monophosphate; CGM: continuous glucose monitoring; cGMP: cyclic guanosine monophosphate; CI: 95% confidence interval; C-index: concordance index; CIMT: carotid intima-media thickness; CKD: chronic kidney disease; COX-2: cyclooxygenase-2; CTGF: connective tissue growth factor; CV: cardiovascular; CVD: cardiovascular disease; DPP-4i: dipeptidyl peptidase-4 inhibitor; EAU: European Association of Urology; ED: erectile dysfunction; EDHF: endothelium-derived hyperpolarizing factor; eGFR: estimated glomerular filtration rate; eNOS: endothelial nitric oxide synthase; EPC: endothelial progenitor cell; Exenatide LAR: exenatide long-acting release; FAERS: FDA Adverse Event Reporting System; FPG: fasting plasma glucose; FSH: follicle-stimulating hormone; fT: free testosterone; GGT: gamma-glutamyl transferase; GIP: gastric inhibitory polypeptide; GIPR: GIP receptor; GLP-1 RA: glucagon-like peptide-1 receptor agonist; GnRH: gonadotropin-releasing hormone; GTP: guanosine triphosphate; HbA1c: glycated hemoglobin; hCG: human chorionic gonadotropin; HOMA-IR: homeostasis model assessment of insulin resistance; HPG: hypothalamic-pituitary-gonadal; HR: hazard ratio; HUVEC: human umbilical vein endothelial cell; ICD-10: International Classification of Diseases, Tenth Revision; IGF-1: insulin-like growth factor-1; IIEF: International Index of Erectile Function; ION: ischemic optic neuropathy; IR: insulin resistance; JNK: c-Jun N-terminal kinase; KCa: calcium-dependent potassium; LDL-C: low-density lipoprotein cholesterol; LEADER: Liraglutide Effect and Action in Diabetes: Evaluation of Cardiovascular Outcome Results; LH: luteinizing hormone; LIPA: low-intensity pulsed ultrasound; LS: least squares; M3 receptor: muscarinic M3 receptor; MACE: major cardiovascular events; mAR: membrane-bound androgen receptor; MASH: metabolic dysfunction-associated steatohepatitis; MBS: metabolic bariatric surgery; MetS: metabolic syndrome; METS-IR: metabolic score for insulin resistance; MiR-155: microRNA-155; mTORC2: mammalian target of rapamycin complex 2; MTC: medullary thyroid carcinoma; mTOR: mammalian target of rapamycin; NADPH: nicotinamide adenine dinucleotide phosphate; NANC: non-adrenergic, non-cholinergic; NET: neutrophil extracellular trap; NF-κB: nuclear factor kappa-light-chain-enhancer of activated B cells; NHANES: National Health and Nutrition Examination Survey; NLRP3: NLR family pyrin domain containing 3; nNOS: neuronal nitric oxide synthase; NO: nitric oxide; p38 MAPK: p38 mitogen-activated protein kinase; PDE5: phosphodiesterase type 5; PGE1: prostaglandin E1; PI3K: phosphoinositide 3-kinase; PKA: protein kinase A; PPAR-γ: peroxisome proliferator-activated receptor gamma; PKG-1: protein kinase G type 1; Q1: first quartile; QoL: quality of life; RAGE: receptor for advanced glycation end products; RCT: randomized controlled trial; REWIND: Researching Cardiovascular Events With a Weekly INcretin in Diabetes; RNS: reactive nitrogen species; ROCK2: Rho-associated kinase 2; ROS: reactive oxygen species; sGC: soluble guanylyl cyclase; SDF-1: stromal cell-derived factor 1; SGLT-2i: sodium–glucose cotransporter-2 inhibitor; SHBG: sex hormone-binding globulin; SIRT1: sirtuin 1; Smad: SMAD family proteins; SPC: stromal progenitor cell; SSC: sacral spinal cord; SU: sulfonylurea; T1D: type 1 diabetes; T2D: type 2 diabetes; TGF-β1; transforming growth factor beta 1; TNFα: tumor necrosis factor-alpha; TRT: testosterone replacement therapy; TT: total testosterone; TyG: triglyceride-glucose; TZD: thiazolidinedione; VEGF: vascular endothelial growth factor; VSMC: vascular smooth muscle cell; WC: waist circumference.

References

1. Yafi, F.A.; Jenkins, L.; Albersen, M.; Corona, G.; Isidori, A.M.; Goldfarb, S.; Maggi, M.; Nelson, C.J.; Parish, S.; Salonia, A.; et al. Erectile dysfunction. *Nat. Rev. Dis. Primers* **2016**, *2*, 16003. [CrossRef] [PubMed]
2. Obi, P.C.; Njoku, G.H.; Mbaike, A.C.; Ihim, A.C.; Nwako, O.F.; Anyanwu, A.C.; Ohiri, J.U.; Nwazor, E.; Ubani, B.C.; Ogbonna, S.U.; et al. Quality of Sexual Life and its Correlates among Men with Diabetes Mellitus and Erectile dysfunction attending a tertiary hospital in Owerri, Nigeria. *Niger. Med. J.* **2025**, *66*, 36–43. [CrossRef] [PubMed]

3. Pang, K.; Pan, D.; Xu, H.; Ma, Y.; Wang, J.; Xu, P.; Wang, H.; Zang, G. Advances in physical diagnosis and treatment of male erectile dysfunction. *Front. Physiol.* **2023**, *13*, 1096741. [CrossRef]
4. Rhoden, E.L.; Telöken, C.; Sogari, P.R.; Vargas Souto, C.A. The use of the simplified International Index of Erectile Function (IIEF-5) as a diagnostic tool to study the prevalence of erectile dysfunction. *Int. J. Impot. Res.* **2002**, *14*, 245–250. [CrossRef] [PubMed]
5. Pyrgidis, N.; Mykoniatis, I.; Haidich, A.B.; Tirta, M.; Talimtzi, P.; Kalyvianakis, D.; Ouranidis, A.; Hatzichristou, D. Effect of phosphodiesterase-type 5 inhibitors on erectile function: An overview of systematic reviews and meta-analyses. *BMJ Open* **2021**, *11*, e047396. [CrossRef] [PubMed]
6. Vishwakarma, R.K.; Patel, N.P.; Goutam, P. Evaluation of Erectile Dysfunction in Prediabetic Males at Tertiary Care Center in Central India. *J. Assoc. Physicians India* **2024**, *72*, e10–e13. [CrossRef]
7. Dandona, P.; Dhindsa, S. Update: Hypogonadotropic hypogonadism in type 2 diabetes and obesity. *J. Clin. Endocrinol. Metab.* **2011**, *96*, 2643–2651. [CrossRef]
8. Kouidrat, Y.; Pizzol, D.; Cosco, T.; Thompson, T.; Carnaghi, M.; Bertoldo, A.; Solmi, M.; Stubbs, B.; Veronese, N. High prevalence of erectile dysfunction in diabetes: A systematic review and meta-analysis of 145 studies. *Diabet. Med.* **2017**, *34*, 1185–1192. [CrossRef]
9. Chu, N.V.; Edelman, S.V. Erectile dysfunction and diabetes. *Curr. Diabetes Rep.* **2002**, *2*, 60–66. [CrossRef]
10. Kitaw, T.A.; Abate, B.B.; Tilahun, B.D.; Yilak, G.; Rede, M.B.; Getie, A.; Haile, R.N. The global burden of erectile dysfunction and its associated risk factors in diabetic patients: An umbrella reviews. *BMC Public Health* **2024**, *24*, 2816. [CrossRef]
11. Maiorino, M.I.; Bellastella, G.; Della Volpe, E.; Casciano, O.; Scappaticcio, L.; Cirillo, P.; Giugliano, D.; Esposito, K. Erectile dysfunction in young men with type 1 diabetes. *Int. J. Impot. Res.* **2017**, *29*, 17–22. [CrossRef]
12. Zhang, X.; Zhu, Z.; Tang, G. Global prevalence of erectile dysfunction and its associated risk factors among men with type 1 diabetes: A systematic review and meta-analysis. *Int. J. Impot. Res.* **2024**, *36*, 365–374. [CrossRef]
13. Yao, S.; Xing, S.; Shan, X.; Dai, W.; Cao, Y.; Hu, B. Importance of assessing erectile dysfunction in patients with type 2 diabetes mellitus based on glucose fluctuation: A Cross-Sectional study. *Endocrine* **2025**, *88*, 650–659. [CrossRef]
14. Dilixiati, D.; Waili, A.; Tuerxunmaimaiti, A.; Tao, L.; Zebibula, A.; Rexiati, M. Risk factors for erectile dysfunction in diabetes mellitus: A systematic review and meta-analysis. *Front. Endocrinol.* **2024**, *15*, 1368079. [CrossRef]
15. Lu, M.; Ni, D.; Wu, W.; Xu, C.; Zhang, Y. Analysis of Risk Factors Associated with Organic Erectile Dysfunction in Patients with Type 2 Diabetes Mellitus and Erectile Dysfunction. *Discov. Med.* **2025**, *37*, 141–151. [CrossRef]
16. Cignarelli, A.; Genchi, V.A.; D'Oria, R.; Giordano, F.; Caruso, I.; Perrini, S.; Natalicchio, A.; Laviola, L.; Giorgino, F. Role of Glucose-Lowering Medications in Erectile Dysfunction. *J. Clin. Med.* **2021**, *10*, 2501. [CrossRef]
17. Westermeier, F.; Fisman, E.Z. Glucagon like peptide-1 (GLP-1) agonists and cardiometabolic protection: Historical development and future challenges. *Cardiovasc. Diabetol.* **2025**, *24*, 44. [CrossRef]
18. Pantea, I.; Repanovici, A.; Andreescu, O. The Influence of GLP1 on Body Weight and Glycemic Management in Patients with Diabetes—A Scientometric Investigation and Visualization Study. *Medicina* **2024**, *60*, 1761. [CrossRef]
19. Calabrò, R.S.; Cacciola, A.; Bruschetta, D.; Milardi, D.; Quattrini, F.; Sciarrone, F.; la Rosa, G.; Bramanti, P.; Anastasi, G. Neuroanatomy and function of human sexual behavior: A neglected or unknown issue? *Brain Behav.* **2019**, *9*, e01389. [CrossRef]
20. Azadzoi, K.M.; Yang, J.; Siroky, M.B. Neural regulation of sexual function in men. *World J. Clin. Urol.* **2013**, *2*, 32–41. [CrossRef]
21. Liu, X.; Yu, Y.; Zhang, H.; Zhang, M.; Liu, Y. The Role of Muscarinic Acetylcholine Receptor M3 in Cardiovascular Diseases. *Int. J. Mol. Sci.* **2024**, *25*, 7560. [CrossRef]
22. Lincoln, T.M.; Dey, N.; Sellak, H. Invited review: cGMP-dependent protein kinase signaling mechanisms in smooth muscle: From the regulation of tone to gene expression. *J. Appl. Physiol.* **2001**, *91*, 1421–1430. [CrossRef]
23. Morgado, M.; Cairrão, E.; Santos-Silva, A.J.; Verde, I. Cyclic nucleotide-dependent relaxation pathways in vascular smooth muscle. *Cell. Mol. Life Sci.* **2012**, *69*, 247–266. [CrossRef]
24. Lucas-Herald, A.K.; Alves-Lopes, R.; Montezano, A.C.; Ahmed, S.F.; Touyz, R.M. Genomic and non-genomic effects of androgens in the cardiovascular system: Clinical implications. *Clin. Sci.* **2017**, *131*, 1405–1418. [CrossRef] [PubMed]
25. Foradori, C.D.; Weiser, M.J.; Handa, R.J. Non-genomic actions of androgens. *Front. Neuroendocrinol.* **2008**, *29*, 169–181. [CrossRef]
26. Goto, K.; Ohtsubo, T.; Kitazono, T. Endothelium-Dependent Hyperpolarization (EDH) in Hypertension: The Role of Endothelial Ion Channels. *Int. J. Mol. Sci.* **2018**, *19*, 315. [CrossRef]
27. Prieto, D. Physiological regulation of penile arteries and veins. *Int. J. Impot. Res.* **2008**, *20*, 17–29. [CrossRef]
28. Ma, J.; Chen, Y.; Si, Y.; Qian, J.; Wang, C.; Jin, J.; He, Q. The multifaceted nature of diabetic erectile dysfunction: Uncovering the intricate mechanisms and treatment strategies. *Front. Endocrinol.* **2024**, *15*, 1460033. [CrossRef]
29. Asadipooya, K.; Uy, E.M. Advanced Glycation End Products (AGEs), Receptor for AGEs, Diabetes, and Bone: Review of the Literature. *J. Endocr. Soc.* **2019**, *3*, 1799–1818. [CrossRef]

30. Castela, Â.; Costa, C. Molecular mechanisms associated with diabetic endothelial-erectile dysfunction. *Nat. Rev. Urol.* **2016**, *13*, 266–274. [CrossRef] [PubMed]
31. Cartledge, J.J.; Eardley, I.; Morrison, J.F. Advanced glycation end-products are responsible for the impairment of corpus cavernosal smooth muscle relaxation seen in diabetes. *BJU Int.* **2001**, *87*, 402–407. [CrossRef]
32. Kaltsas, A.; Zikopoulos, A.; Dimitriadis, F.; Sheshi, D.; Politis, M.; Moustakli, E.; Symeonidis, E.N.; Chrisofos, M.; Sofikitis, N.; Zachariou, A. Oxidative Stress and Erectile Dysfunction: Pathophysiology, Impacts, and Potential Treatments. *Curr. Issues Mol. Biol.* **2024**, *46*, 8807–8834. [CrossRef]
33. Cai, Z.; Yuan, S.; Zhong, Y.; Deng, L.; Li, J.; Tan, X.; Feng, J. Amelioration of Endothelial Dysfunction in Diabetes: Role of Takeda G Protein-Coupled Receptor 5. *Front. Pharmacol.* **2021**, *12*, 637051. [CrossRef]
34. Bahadoran, Z.; Mirmiran, P.; Kashfi, K.; Ghasemi, A. Vascular nitric oxide resistance in type 2 diabetes. *Cell Death Dis.* **2023**, *14*, 410. [CrossRef]
35. Toque, H.A.; Nunes, K.P.; Yao, L.; Liao, J.K.; Webb, R.C.; Caldwell, R.B.; Caldwell, R.W. Activated Rho kinase mediates diabetes-induced elevation of vascular arginase activation and contributes to impaired corpora cavernosa relaxation: Possible involvement of p38 MAPK activation. *J. Sex. Med.* **2013**, *10*, 1502–1515. [CrossRef]
36. Zhou, M.; Li, G.; Zhu, L.; Zhou, H.; Lu, L. Arctiin attenuates high glucose-induced human retinal capillary endothelial cell proliferation by regulating ROCK1/PTEN/PI3K/Akt/VEGF pathway in vitro. *J. Cell. Mol. Med.* **2020**, *24*, 5695–5706. [CrossRef]
37. Kilarkaje, N.; Yousif, M.H.; El-Hashim, A.Z.; Makki, B.; Akhtar, S.; Benter, I.F. Role of angiotensin II and angiotensin-(1-7) in diabetes-induced oxidative DNA damage in the corpus cavernosum. *Fertil. Steril.* **2013**, *100*, 226–233. [CrossRef]
38. Deng, W.; Bivalacqua, T.J.; Champion, H.C.; Hellstrom, W.J.; Murthy, S.N.; Kadowitz, P.J. Superoxide dismutase—A target for gene therapeutic approach to reduce oxidative stress in erectile dysfunction. *Methods Mol. Biol.* **2010**, *610*, 213–227. [CrossRef]
39. Radi, R. Protein tyrosine nitration: Biochemical mechanisms and structural basis of functional effects. *Acc. Chem. Res.* **2013**, *46*, 550–559. [CrossRef]
40. Zhang, J.; Xin, S.; Mao, J.; Liu, X.; Wang, T.; Liu, J.; Song, X.; Song, W. The role of pogrammed cell death in diabetes mellitus-induced erectile dysfunction: From mechanisms to targeted therapy. *Reprod. Biol. Endocrinol.* **2025**, *23*, 32. [CrossRef]
41. Chen, Y.; Xiao, M.; Zhao, L.; Huang, Y.; Lin, Y.; Xie, T.; Tian, J.; Wang, Q.; Tang, Y.; Su, Z. Low-Intensity Pulsed Ultrasound Counteracts Advanced Glycation End Products-Induced Corpus Cavernosal Endothelial Cell Dysfunction via Activating Mitophagy. *Int. J. Mol. Sci.* **2022**, *23*, 14887. [CrossRef]
42. Qabazard, B.; Yousif, M.; Mousa, A.; Phillips, O.A. GYY4137 attenuates functional impairment of corpus cavernosum and reduces fibrosis in rats with STZ-induced diabetes by inhibiting the TGF-β1/Smad/CTGF pathway. *Biomed. Pharmacother.* **2021**, *138*, 111486. [CrossRef]
43. Kadioglu, A.; Dincer, M.; Salabas, E.; Culha, M.G.; Akdere, H.; Cilesiz, N.C. A Population-Based Study of Peyronie's Disease in Turkey: Prevalence and Related Comorbidities. *Sex. Med.* **2020**, *8*, 679–685. [CrossRef]
44. Yang, C.C.; Liao, P.H.; Cheng, Y.H.; Chien, C.Y.; Cheng, K.H.; Chien, C.T. Diabetes associated with hypertension exacerbated oxidative stress-mediated inflammation, apoptosis and autophagy leading to erectile dysfunction in rats. *J. Chin. Med. Assoc.* **2022**, *85*, 346–357. [CrossRef]
45. Chrysant, S.G. Antihypertensive therapy causes erectile dysfunction. *Curr. Opin. Cardiol.* **2015**, *30*, 383–390. [CrossRef]
46. Musicki, B.; Liu, T.; Lagoda, G.A.; Strong, T.D.; Sezen, S.F.; Johnson, J.M.; Burnett, A.L. Hypercholesterolemia-induced erectile dysfunction: Endothelial nitric oxide synthase (eNOS) uncoupling in the mouse penis by NAD(P)H oxidase. *J. Sex. Med.* **2010**, *7*, 3023–3032. [CrossRef]
47. Potenza, M.A.; Addabbo, F.; Montagnani, M. Vascular actions of insulin with implications for endothelial dysfunction. *Am. J. Physiol. Endocrinol. Metab.* **2009**, *297*, 568–577. [CrossRef]
48. Silva, F.H.; Alexandre, E.C.; Calmasini, F.B.; Calixto, M.C.; Antunes, E. Treatment with Metformin Improves Erectile Dysfunction in a Murine Model of Obesity Associated with Insulin Resistance. *Urology* **2015**, *86*, 423.e1-6. [CrossRef]
49. Chen, S.; Wu, R.; Huang, Y.; Zheng, F.; Ou, Y.; Tu, X.; Zhang, Y.; Gao, Y.; Chen, X.; Zheng, T.; et al. Insulin resistance is an independent determinate of ED in young adult men. *PLoS ONE* **2013**, *8*, e83951. [CrossRef]
50. Jalali, S.; Zareshahi, N.; Behnoush, A.H.; Azarboo, A.; Shirinezhad, A.; Hosseini, S.Y.; Javidan, A.; Ghaseminejad-Raeini, A. Association of insulin resistance surrogate indices and erectile dysfunction: A systematic review and meta-analysis. *Reprod. Biol. Endocrinol.* **2024**, *22*, 148. [CrossRef]
51. Wu, S.; Xia, W. Association between different insulin resistance surrogates and erectile dysfunction in non-diabetic men: A large population-based study. *BMC Public Health* **2025**, *25*, 1949. [CrossRef]
52. Sun, C.; Gao, Y.; Liang, Z.; Liu, C.; Chen, M. Association of METS-IR index with prevalence of erectile dysfunction in US adults: A cross-sectional study. *Int. Urol. Nephrol.* **2024**, *56*, 2157–2164. [CrossRef]

53. De Silva, N.L.; Papanikolaou, N.; Grossmann, M.; Antonio, L.; Quinton, R.; Anawalt, B.D.; Jayasena, C.N. Male hypogonadism: Pathogenesis, diagnosis, and management. *Lancet Diabetes Endocrinol.* **2024**, *12*, 761–774. [CrossRef]
54. Ding, E.L.; Song, Y.; Manson, J.E.; Hunter, D.J.; Lee, C.C.; Rifai, N.; Buring, J.E.; Gaziano, J.M.; Liu, S. Sex hormone-binding globulin and risk of type 2 diabetes in women and men. *N. Engl. J. Med.* **2009**, *361*, 1152–1163. [CrossRef]
55. Khalil, S.H.A.; Dandona, P.; Osman, N.A.; Assaad, R.S.; Zaitoon, B.T.A.; Almas, A.A.; Amin, N.G. Diabetes surpasses obesity as a risk factor for low serum testosterone level. *Diabetol. Metab. Syndr.* **2024**, *16*, 143. [CrossRef]
56. Grossmann, M. Hypogonadism and male obesity: Focus on unresolved questions. *Clin. Endocrinol.* **2018**, *89*, 11–21. [CrossRef]
57. Infante, M.; Silvestri, F.; Padilla, N.; Pacifici, F.; Pastore, D.; Pinheiro, M.M.; Caprio, M.; Tesauro, M.; Fabbri, A.; Novelli, G.; et al. Unveiling the Therapeutic Potential of the Second-Generation Incretin Analogs Semaglutide and Tirzepatide in Type 1 Diabetes and Latent Autoimmune Diabetes in Adults. *J. Clin. Med.* **2025**, *14*, 1303. [CrossRef]
58. Mulligan, T.; Frick, M.F.; Zuraw, Q.C.; Stemhagen, A.; McWhirter, C. Prevalence of hypogonadism in males aged at least 45 years: The HIM study. *Int. J. Clin. Pract.* **2006**, *60*, 762–769. [CrossRef]
59. Traish, A.M. Androgens play a pivotal role in maintaining penile tissue architecture and erection: A review. *J. Androl.* **2009**, *30*, 363–369. [CrossRef]
60. Cai, J.J.; Wen, J.; Jiang, W.H.; Lin, J.; Hong, Y.; Zhu, Y.S. Androgen actions on endothelium functions and cardiovascular diseases. *J. Geriatr. Cardiol.* **2016**, *13*, 183–196. [CrossRef]
61. Traish, A.M.; Galoosian, A. Androgens modulate endothelial function and endothelial progenitor cells in erectile physiology. *Korean J. Urol.* **2013**, *54*, 721–731. [CrossRef]
62. Varlamov, O.; White, A.E.; Carroll, J.M.; Bethea, C.L.; Reddy, A.; Slayden, O.; O'Rourke, R.W.; Roberts, C.T., Jr. Androgen effects on adipose tissue architecture and function in nonhuman primates. *Endocrinology* **2012**, *153*, 3100–3110. [CrossRef] [PubMed]
63. Schardein, J.N.; Hotaling, J.M. The Impact of Testosterone on Erectile Function. *Androgens* **2022**, *3*, 113–124. [CrossRef]
64. Long, L.; Qiu, H.; Cai, B.; Chen, N.; Lu, X.; Zheng, S.; Ye, X.; Li, Y. Hyperglycemia induced testicular damage in type 2 diabetes mellitus rats exhibiting microcirculation impairments associated with vascular endothelial growth factor decreased via PI3K/Akt pathway. *Oncotarget* **2018**, *9*, 5321–5336. [CrossRef]
65. Schoeller, E.L.; Schon, S.; Moley, K.H. The effects of type 1 diabetes on the hypothalamic, pituitary and testes axis. *Cell Tissue Res.* **2012**, *349*, 839–847. [CrossRef]
66. Noori, T.; Sureda, A.; Shirooie, S. Role of natural mTOR inhibitors in treatment of diabetes mellitus. *Fundam. Clin. Pharmacol.* **2023**, *37*, 461–479. [CrossRef]
67. Kuryłowicz, A. Estrogens in Adipose Tissue Physiology and Obesity-Related Dysfunction. *Biomedicines* **2023**, *11*, 690. [CrossRef]
68. Traish, A.M.; Toselli, P.; Jeong, S.J.; Kim, N.N. Adipocyte accumulation in penile corpus cavernosum of the orchiectomized rabbit: A potential mechanism for veno-occlusive dysfunction in androgen deficiency. *J. Androl.* **2005**, *26*, 242–248. [CrossRef]
69. Begum, M.; Choubey, M.; Tirumalasetty, M.B.; Arbee, S.; Sadik, S.; Mohib, M.M.; Srivastava, S.; Minhaz, N.; Alam, R.; Mohiuddin, M.S. Exploring the Molecular Link Between Diabetes and Erectile Dysfunction Through Single-Cell Transcriptome Analysis. *Genes* **2024**, *15*, 1596. [CrossRef]
70. Kawai, T.; Autieri, M.V.; Scalia, R. Adipose tissue inflammation and metabolic dysfunction in obesity. *Am. J. Physiol. Cell Physiol.* **2021**, *320*, C375–C391. [CrossRef]
71. Müller, T.D.; Finan, B.; Bloom, S.R.; D'Alessio, D.; Drucker, D.J.; Flatt, P.R.; Fritsche, A.; Gribble, F.; Grill, H.J.; Habener, J.F.; et al. Glucagon-like peptide 1 (GLP-1). *Mol. Metab.* **2019**, *30*, 72–130. [CrossRef]
72. Yuan, X.; Gao, Z.; Hao, Z.; Ma, H.; Duan, K.; Yang, C. Effect of long-acting versus short-acting glucagon-like peptide-1 receptor agonists on improving body weight and related metabolic parameters in type 2 diabetes: A head-to-head meta-analysis. *Medicine* **2023**, *102*, e35739. [CrossRef] [PubMed]
73. Lincoff, A.M.; Brown-Frandsen, K.; Colhoun, H.M.; Deanfield, J.; Emerson, S.S.; Esbjerg, S.; Hardt-Lindberg, S.; Hovingh, G.K.; Kahn, S.E.; Kushner, R.F.; et al. Semaglutide and Cardiovascular Outcomes in Obesity without Diabetes. *N. Engl. J. Med.* **2023**, *389*, 2221–2232. [CrossRef] [PubMed]
74. Marso, S.P.; Daniels, G.H.; Brown-Frandsen, K.; Kristensen, P.; Mann, J.F.; Nauck, M.A.; Nissen, S.E.; Pocock, S.; Poulter, N.R.; Ravn, L.S.; et al. Liraglutide and Cardiovascular Outcomes in Type 2 Diabetes. *N. Engl. J. Med.* **2016**, *375*, 311–322. [CrossRef]
75. Marso, S.P.; Bain, S.C.; Consoli, A.; Eliaschewitz, F.G.; Jódar, E.; Leiter, L.A.; Lingvay, I.; Rosenstock, J.; Seufert, J.; Warren, M.L.; et al. Semaglutide and Cardiovascular Outcomes in Patients with Type 2 Diabetes. *N. Engl. J. Med.* **2016**, *375*, 1834–1844. [CrossRef]
76. Perkovic, V.; Tuttle, K.R.; Rossing, P.; Mahaffey, K.W.; Mann, J.F.E.; Bakris, G.; Baeres, F.M.M.; Idorn, T.; Bosch-Traberg, H.; Lausvig, N.L.; et al. Effects of Semaglutide on Chronic Kidney Disease in Patients with Type 2 Diabetes. *N. Engl. J. Med.* **2024**, *391*, 109–121. [CrossRef]
77. Moiz, A.; Filion, K.B.; Tsoukas, M.A.; Yu, O.H.; Peters, T.M.; Eisenberg, M.J. Mechanisms of GLP-1 Receptor Agonist-Induced Weight Loss: A Review of Central and Peripheral Pathways in Appetite and Energy Regulation. *Am. J. Med.* **2025**, *138*, 934–940. [CrossRef] [PubMed]

78. Rubio-Herrera, M.A.; Mera-Carreiro, S.; Sánchez-Pernaute, A.; Ramos-Levi, A.M. Impact of Treatment with GLP1 Receptor Agonists, Liraglutide 3.0 mg and Semaglutide 1.0 mg, While on a Waiting List for Bariatric Surgery. *Biomedicines* **2023**, *11*, 2785. [CrossRef]
79. Jastreboff, A.M.; Aronne, L.J.; Ahmad, N.N.; Wharton, S.; Connery, L.; Alves, B.; Kiyosue, A.; Zhang, S.; Liu, B.; Bunck, M.C.; et al. Tirzepatide Once Weekly for the Treatment of Obesity. *N. Engl. J. Med.* **2022**, *387*, 205–216. [CrossRef]
80. Garvey, W.T.; Frias, J.P.; Jastreboff, A.M.; le Roux, C.W.; Sattar, N.; Aizenberg, D.; Mao, H.; Zhang, S.; Ahmad, N.N.; Bunck, M.C.; et al. Tirzepatide once weekly for the treatment of obesity in people with type 2 diabetes (SURMOUNT-2): A double-blind, randomised, multicentre, placebo-controlled, phase 3 trial. *Lancet* **2023**, *402*, 613–626. [CrossRef]
81. Aronne, L.J.; Horn, D.B.; le Roux, C.W.; Ho, W.; Falcon, B.L.; Gomez Valderas, E.; Das, S.; Lee, C.J.; Glass, L.C.; Senyucel, C.; et al. Tirzepatide as Compared with Semaglutide for the Treatment of Obesity. *N. Engl. J. Med.* **2025**, *393*, 26–36. [CrossRef]
82. European Association for the Study of the Liver (EASL); European Association for the Study of Diabetes (EASD); European Association for the Study of Obesity (EASO). EASL-EASD-EASO Clinical Practice Guidelines on the management of metabolic dysfunction-associated steatotic liver disease (MASLD). *J. Hepatol.* **2024**, *81*, 492–542. [CrossRef]
83. Kuchay, M.S.; Krishan, S.; Mishra, S.K.; Choudhary, N.S.; Singh, M.K.; Wasir, J.S.; Kaur, P.; Gill, H.K.; Bano, T.; Farooqui, K.J.; et al. Effect of dulaglutide on liver fat in patients with type 2 diabetes and NAFLD: Randomised controlled trial (D-LIFT trial). *Diabetologia* **2020**, *63*, 2434–2445. [CrossRef]
84. Newsome, P.N.; Buchholtz, K.; Cusi, K.; Linder, M.; Okanoue, T.; Ratziu, V.; Sanyal, A.J.; Sejling, A.S.; Harrison, S.A.; NN9931-4296 Investigators. A Placebo-Controlled Trial of Subcutaneous Semaglutide in Nonalcoholic Steatohepatitis. *N. Engl. J. Med.* **2021**, *384*, 1113–1124. [CrossRef]
85. Fessel, J. All GLP-1 Agonists Should, Theoretically, Cure Alzheimer's Dementia but Dulaglutide Might Be More Effective Than the Others. *J. Clin. Med.* **2024**, *13*, 3729. [CrossRef] [PubMed]
86. Reiss, A.B.; Gulkarov, S.; Lau, R.; Klek, S.P.; Srivastava, A.; Renna, H.A.; De Leon, J. Weight Reduction with GLP-1 Agonists and Paths for Discontinuation While Maintaining Weight Loss. *Biomolecules* **2025**, *15*, 408. [CrossRef]
87. Wan, J.; Ferrari, C.; Tadros, M. GLP-1RA Essentials in Gastroenterology: Side Effect Management, Precautions for Endoscopy and Applications for Gastrointestinal Disease Treatment. *Gastroenterol. Insights* **2024**, *15*, 191–212. [CrossRef]
88. He, L.; Wang, J.; Ping, F.; Yang, N.; Huang, J.; Li, Y.; Xu, L.; Li, W.; Zhang, H. Association of Glucagon-Like Peptide-1 Receptor Agonist Use with Risk of Gallbladder and Biliary Diseases: A Systematic Review and Meta-analysis of Randomized Clinical Trials. *JAMA Intern. Med.* **2022**, *182*, 513–519. [CrossRef] [PubMed]
89. Hou, Y.; Ernst, S.A.; Heidenreich, K.; Williams, J.A. Glucagon-like peptide-1 receptor is present in pancreatic acinar cells and regulates amylase secretion through cAMP. *Am. J. Physiol. Gastrointest. Liver Physiol.* **2016**, *310*, G26–G33. [CrossRef]
90. Guo, H.; Guo, Q.; Li, Z.; Wang, Z. Association between different GLP-1 receptor agonists and acute pancreatitis: Case series and real-world pharmacovigilance analysis. *Front. Pharmacol.* **2024**, *15*, 1461398. [CrossRef]
91. Ayoub, M.; Chela, H.; Amin, N.; Hunter, R.; Anwar, J.; Tahan, V.; Daglilar, E. Pancreatitis Risk Associated with GLP-1 Receptor Agonists, Considered as a Single Class, in a Comorbidity-Free Subgroup of Type 2 Diabetes Patients in the United States: A Propensity Score-Matched Analysis. *J. Clin. Med.* **2025**, *14*, 944. [CrossRef] [PubMed]
92. Lomeli, L.D.; Kodali, A.M.; Tsushima, Y.; Mehta, A.E.; Pantalone, K.M. The incidence of acute pancreatitis with GLP-1 receptor agonist therapy in individuals with a known history of pancreatitis. *Diabetes Res. Clin. Pract.* **2024**, *215*, 111806. [CrossRef] [PubMed]
93. Koehler, J.A.; Baggio, L.L.; Lamont, B.J.; Ali, S.; Drucker, D.J. Glucagon-like peptide-1 receptor activation modulates pancreatitis-associated gene expression but does not modify the susceptibility to experimental pancreatitis in mice. *Diabetes* **2009**, *58*, 2148–2161. [CrossRef]
94. Kounatidis, D.; Vallianou, N.G.; Karampela, I.; Rebelos, E.; Kouveletsou, M.; Dalopoulos, V.; Koufopoulos, P.; Diakoumopoulou, E.; Tentolouris, N.; Dalamaga, M. Anti-Diabetic Therapies and Cancer: From Bench to Bedside. *Biomolecules* **2024**, *14*, 1479. [CrossRef]
95. Burke, O.; Sa, B.; Cespedes, D.A.; Sechi, A.; Tosti, A. Glucagon-like peptide-1 receptor agonist medications and hair loss: A retrospective cohort study. *J. Am. Acad. Dermatol.* **2025**, *92*, 1141–1143. [CrossRef]
96. Godfrey, H.; Leibovit-Reiben, Z.; Jedlowski, P.; Thiede, R. Alopecia associated with the use of semaglutide and tirzepatide: A disproportionality analysis using the FDA adverse event reporting system (FAERS) from 2022 to 2023. *J. Eur. Acad. Dermatol. Venereol.* **2025**, *39*, e153–e154. [CrossRef]
97. Buontempo, M.G.; Santos, B.T. Exploring the hair loss risk in glucagon-like peptide-1 agonists: Emerging concerns and clinical implications. *J. Eur. Acad. Dermatol. Venereol.* **2025**, *39*, 263–264. [CrossRef]
98. Sharma, A.; Parachuri, N.; Kumar, N.; Saboo, B.; Tripathi, H.N.; Kuppermann, B.D.; Bandello, F. Semaglutide and the risk of diabetic retinopathy-current perspective. *Eye* **2022**, *36*, 10–11. [CrossRef]

99. Lakhani, M.; Kwan, A.T.H.; Mihalache, A.; Popovic, M.M.; Nanji, K.; Xie, J.S.; Feo, A.; Rabinovitch, D.; Shor, R.; Sadda, S.; et al. Association of Glucagon-Like Peptide-1 Receptor Agonists with Optic Nerve and Retinal Adverse Events: A Population-Based Observational Study Across 180 Countries. *Am. J. Ophthalmol.* **2025**, *277*, 148–168. [CrossRef]
100. Cigrovski Berkovic, M.; Strollo, F. Semaglutide-eye-catching results. *World J. Diabetes.* **2023**, *14*, 424–434. [CrossRef] [PubMed]
101. Joo, J.H.; Sharma, N.; Shaia, J.; Wu, A.K.; Skugor, M.; Singh, R.P.; Rachitskaya, A.V. The Effect of Glucagon-Like Peptide-1 Receptor Agonists on Diabetic Retinopathy at a Tertiary Care Center. *Ophthalmol. Sci.* **2024**, *4*, 100547. [CrossRef]
102. Wang, F.; Mao, Y.; Wang, H.; Liu, Y.; Huang, P. Semaglutide and Diabetic Retinopathy Risk in Patients with Type 2 Diabetes Mellitus: A Meta-Analysis of Randomized Controlled Trials. *Clin. Drug Investig.* **2022**, *42*, 17–28. [CrossRef]
103. Wei, L.; Mo, W.; Lan, S.; Yang, H.; Huang, Z.; Liang, X.; Li, L.; Xian, J.; Xie, X.; Qin, Y.; et al. GLP-1 RA Improves Diabetic Retinopathy by Protecting the Blood-Retinal Barrier through GLP-1R-ROCK-p-MLC Signaling Pathway. *J. Diabetes Res.* **2022**, *2022*, 1861940. [CrossRef] [PubMed]
104. Anastasiou, I.A.; Tentolouris, A.; Sarantis, P.; Katsaouni, A.; Rebelos, E.; Mourouzis, I.; Pantos, C.; Tentolouris, N. Semaglutide Enhances Cellular Regeneration in Skin and Retinal Cells In Vitro. *Pharmaceutics* **2025**, *17*, 1115. [CrossRef]
105. Varnum, A.A.; Pozzi, E.; Deebel, N.A.; Evans, A.; Eid, N.; Sadeghi-Nejad, H.; Ramasamy, R. Impact of GLP-1 Agonists on Male Reproductive Health-A Narrative Review. *Medicina* **2023**, *60*, 50. [CrossRef]
106. Lisco, G.; Bartolomeo, N.; De Tullio, A.; De Pergola, G.; Guastamacchia, E.; Jirillo, E.; Piazzolla, G.; Triggiani, V.; Giagulli, V.A. Long-acting glucagon-like peptide 1 receptor agonists boost erectile function in men with type 2 diabetes mellitus complaining of erectile dysfunction: A retrospective cohort study. *Andrology* **2024**, *12*, 633–642. [CrossRef]
107. Pourabhari Langroudi, A.; Chen, A.L.; Basran, S.; Sommer, E.R.; Stinson, J.; Cheng, Y.S.; Del Giuduce, F.; Scott, M.; Eisenberg, M.L. Male sexual dysfunction associated with GLP-1 receptor agonists: A cross-sectional analysis of FAERS data. *Int. J. Impot. Res.* **2025**, *37*, 661–667. [CrossRef]
108. Helmstädter, J.; Frenis, K.; Filippou, K.; Grill, A.; Dib, M.; Kalinovic, S.; Pawelke, F.; Kus, K.; Kröller-Schön, S.; Oelze, M.; et al. Endothelial GLP-1 (Glucagon-Like Peptide-1) Receptor Mediates Cardiovascular Protection by Liraglutide In Mice with Experimental Arterial Hypertension. *Arterioscler. Thromb. Vasc. Biol.* **2020**, *40*, 145–158. [CrossRef]
109. Fu, Z.; Gong, L.; Liu, J.; Wu, J.; Barrett, E.J.; Aylor, K.W.; Liu, Z. Brain Endothelial Cells Regulate Glucagon-Like Peptide 1 Entry Into the Brain via a Receptor-Mediated Process. *Front. Physiol.* **2020**, *11*, 555. [CrossRef] [PubMed]
110. Ishibashi, Y.; Matsui, T.; Takeuchi, M.; Yamagishi, S. Glucagon-like peptide-1 (GLP-1) inhibits advanced glycation end product (AGE)-induced up-regulation of VCAM-1 mRNA levels in endothelial cells by suppressing AGE receptor (RAGE) expression. *Biochem. Biophys. Res. Commun.* **2010**, *391*, 1405–1408. [CrossRef] [PubMed]
111. Tang, S.T.; Tang, H.Q.; Su, H.; Wang, Y.; Zhou, Q.; Zhang, Q.; Wang, Y.; Zhu, H.Q. Glucagon-like peptide-1 attenuates endothelial barrier injury in diabetes via cAMP/PKA mediated down-regulation of MLC phosphorylation. *Biomed. Pharmacother.* **2019**, *113*, 108667. [CrossRef] [PubMed]
112. Cai, X.; She, M.; Xu, M.; Chen, H.; Li, J.; Chen, X.; Zheng, D.; Liu, J.; Chen, S.; Zhu, J.; et al. GLP-1 treatment protects endothelial cells from oxidative stress-induced autophagy and endothelial dysfunction. *Int. J. Biol. Sci.* **2018**, *14*, 1696–1708. [CrossRef]
113. Koska, J.; Sands, M.; Burciu, C.; D'Souza, K.M.; Raravikar, K.; Liu, J.; Truran, S.; Franco, D.A.; Schwartz, E.A.; Schwenke, D.C.; et al. Exenatide Protects Against Glucose- and Lipid-Induced Endothelial Dysfunction: Evidence for Direct Vasodilation Effect of GLP-1 Receptor Agonists in Humans. *Diabetes* **2015**, *64*, 2624–2635. [CrossRef]
114. Abdel-Latif, R.G.; Heeba, G.H.; Taye, A.; Khalifa, M.M.A. Lixisenatide ameliorates cerebral ischemia-reperfusion injury via GLP-1 receptor dependent/independent pathways. *Eur. J. Pharmacol.* **2018**, *833*, 145–154. [CrossRef]
115. Wu, H.; Xiao, C.; Zhao, Y.; Yin, H.; Yu, M. Liraglutide Improves Endothelial Function via the mTOR Signaling Pathway. *J. Diabetes Res.* **2021**, *2021*, 2936667. [CrossRef]
116. Jojima, T.; Uchida, K.; Akimoto, K.; Tomotsune, T.; Yanagi, K.; Iijima, T.; Suzuki, K.; Kasai, K.; Aso, Y. Liraglutide, a GLP-1 receptor agonist, inhibits vascular smooth muscle cell proliferation by enhancing AMP-activated protein kinase and cell cycle regulation, and delays atherosclerosis in ApoE deficient mice. *Atherosclerosis* **2017**, *261*, 44–51. [CrossRef]
117. Bretón-Romero, R.; Weisbrod, R.M.; Feng, B.; Holbrook, M.; Ko, D.; Stathos, M.M.; Zhang, J.Y.; Fetterman, J.L.; Hamburg, N.M. Liraglutide Treatment Reduces Endothelial Endoplasmic Reticulum Stress and Insulin Resistance in Patients with Diabetes Mellitus. *J. Am. Heart Assoc.* **2018**, *7*, e009379. [CrossRef]
118. Mei, X.; Li, Y.; Wu, J.; Liao, L.; Lu, D.; Qiu, P.; Yang, H.L.; Tang, M.W.; Liang, X.Y.; Liu, D. Dulaglutide restores endothelial progenitor cell levels in diabetic mice and mitigates high glucose-induced endothelial injury through SIRT1-mediated mitochondrial fission. *Biochem. Biophys. Res. Commun.* **2024**, *716*, 150002. [CrossRef] [PubMed]
119. Luo, X.; Hu, Y.; He, S.; Ye, Q.; Lv, Z.; Liu, J.; Chen, X. Dulaglutide inhibits high glucose- induced endothelial dysfunction and NLRP3 inflammasome activation. *Arch. Biochem. Biophys.* **2019**, *671*, 203–209. [CrossRef] [PubMed]
120. Pan, X.; Yang, L.; Wang, S.; Liu, Y.; Yue, L.; Chen, S. Semaglutide alleviates inflammation-Induced endothelial progenitor cells injury by inhibiting MiR-155 expression in macrophage exosomes. *Int. Immunopharmacol.* **2023**, *119*, 110196. [CrossRef]

121. Yue, L.; Chen, S.; Ren, Q.; Niu, S.; Pan, X.; Chen, X.; Li, Z.; Chen, X. Effects of semaglutide on vascular structure and proteomics in high-fat diet-induced obese mice. *Front. Endocrinol.* **2022**, *13*, 995007. [CrossRef] [PubMed]
122. Estato, V.; Obadia, N.; Chateaubriand, P.H.; Figueiredo, V.; Curty, M.; Costa Silva, M.; Ferreira, R.G.L.; Santa-Ritta, J.; Campos Baroni, M.; Aragão, A.; et al. Semaglutide restores astrocyte-vascular interactions and blood-brain barrier integrity in a model of diet-induced metabolic syndrome. *Diabetol. Metab. Syndr.* **2025**, *17*, 2. [CrossRef] [PubMed]
123. Withaar, C.; Meems, L.M.G.; Nollet, E.E.; Schouten, E.M.; Schroeder, M.A.; Knudsen, L.B.; Niss, K.; Madsen, C.T.; Hoegl, A.; Mazzoni, G.; et al. The Cardioprotective Effects of Semaglutide Exceed Those of Dietary Weight Loss in Mice with HFpEF. *JACC Basic Transl. Sci.* **2023**, *8*, 1298–1314. [CrossRef] [PubMed]
124. Berg, G.; Barchuk, M.; Lobo, M.; Nogueira, J.P. Effect of glucagon-like peptide-1 (GLP-1) analogues on epicardial adipose tissue: A meta-analysis. *Diabetes Metab. Syndr.* **2022**, *16*, 102562. [CrossRef]
125. Bednarz, K.; Kowalczyk, K.; Cwynar, M.; Czapla, D.; Czarkowski, W.; Kmita, D.; Nowak, A.; Madej, P. The Role of Glp-1 Receptor Agonists in Insulin Resistance with Concomitant Obesity Treatment in Polycystic Ovary Syndrome. *Int. J. Mol. Sci.* **2022**, *23*, 4334. [CrossRef]
126. Randrup, E.; Baum, N.; Feibus, A. Erectile dysfunction and cardiovascular disease. *Postgrad. Med.* **2015**, *127*, 166–172. [CrossRef]
127. Dalaklioglu, S.; Tasatargil, A.; Kuscu, N.; Celik, S.; Celik-Ozenci, C.; Ozdem, S.; Barutcigil, A.; Kucukcetin, I. Protective effect of exendin-4 treatment on erectile dysfunction induced by chronic methylglyoxal administration in rats. *Peptides* **2018**, *106*, 1–8. [CrossRef]
128. Yuan, P.; Ma, D.; Gao, X.; Wang, J.; Li, R.; Liu, Z.; Wang, T.; Wang, S.; Liu, J.; Liu, X. Liraglutide Ameliorates Erectile Dysfunction via Regulating Oxidative Stress, the RhoA/ROCK Pathway and Autophagy in Diabetes Mellitus. *Front. Pharmacol.* **2020**, *11*, 1257. [CrossRef]
129. Yue, L.; Xu, J.L.; Dong, J.; Xjang, G.D.; Xiang, L.; Zhao, L.S.; Zhang, J.X.; Zhai, Z.Y.; Zhu, G.P.; Liu, M.; et al. Regulatory effect of liraglutide on the expression of eNOS in the corpus cavernosum of diabetic rats. *Zhonghua Nan Ke Xue* **2016**, *22*, 212–218.
130. Hurt, K.J.; Musicki, B.; Palese, M.A.; Crone, J.K.; Becker, R.E.; Moriarity, J.L.; Snyder, S.H.; Burnett, A.L. Akt-dependent phosphorylation of endothelial nitric-oxide synthase mediates penile erection. *Proc. Natl. Acad. Sci. USA* **2002**, *99*, 4061–4066. [CrossRef]
131. Model, J.F.A.; Lima, M.V.; Ohlweiler, R.; Sarapio, E.; Vogt, É.L.; Rocha, D.S.; de Souza, S.K.; Vinagre, A.S. Liraglutide treatment counteracts alterations in adipose tissue metabolism induced by orchiectomy in rats. *Life Sci.* **2021**, *278*, 119586. [CrossRef]
132. Wang, S.; Wang, Z.; Zang, Z.; Liang, X.; Jia, B.; Ye, T.; Lan, Y.; Shi, X. A Mitochondrion-Targeting Piezoelectric Nanosystem for the Treatment of Erectile Dysfunction via Autophagy Regulation. *Adv. Mater.* **2025**, *37*, e2413287. [CrossRef] [PubMed]
133. Vallianou, N.G.; Dalamaga, M.; Pavlou, A.; Rebelos, E.; Karamanolis, N.N.; Papachristoforou, E.; Mavrothalassitis, E.; Eleftheriadou, I.; Tentolouris, N.; Kounatidis, D. The Transformative Role of Nanotechnology in the Management of Diabetes Mellitus: Insights from Current Research. *Biomolecules* **2025**, *15*, 653. [CrossRef]
134. Jensterle, M.; Podbregar, A.; Goricar, K.; Gregoric, N.; Janez, A. Effects of liraglutide on obesity-associated functional hypogonadism in men. *Endocr. Connect.* **2019**, *8*, 195–202. [CrossRef] [PubMed]
135. La Vignera, S.; Condorelli, R.A.; Calogero, A.E.; Cannarella, R.; Aversa, A. Sexual and Reproductive Outcomes in Obese Fertile Men with Functional Hypogonadism after Treatment with Liraglutide: Preliminary Results. *J. Clin. Med.* **2023**, *12*, 672. [CrossRef] [PubMed]
136. Lisco, G.; Ramunni, M.I.; De Pergola, G.; Jirillo, E.; Guastamacchia, E.; Triggiani, V.; Giagulli, V.A. GLP-1 Receptor Agonists and Erectile Dysfunction in Diabetic Men with and without Hypogonadism: A 1-Year Retrospective Observational Study. *Endocr. Abstr.* **2022**, *81*, P355. [CrossRef]
137. Giagulli, V.A.; Carbone, M.D.; Ramunni, M.I.; Licchelli, B.; De Pergola, G.; Sabbà, C.; Guastamacchia, E.; Triggiani, V. Adding liraglutide to lifestyle changes, metformin and testosterone therapy boosts erectile function in diabetic obese men with overt hypogonadism. *Andrology* **2015**, *3*, 1094–1103. [CrossRef]
138. An, H.; Xie, K.; Gan, H. Glucagon-like peptide-1 receptor agonists and the risk of erectile dysfunction: A drug target Mendelian randomization study. *Front. Endocrinol.* **2024**, *15*, 1448394. [CrossRef]
139. Bajaj, H.S.; Gerstein, H.C.; Rao-Melacini, P.; Basile, J.; Colhoun, H.; Conget, I.; Cushman, W.C.; Dagenais, G.R.; Franek, E.; Hanefeld, M.; et al. Erectile function in men with type 2 diabetes treated with dulaglutide: An exploratory analysis of the REWIND placebo-controlled randomised trial. *Lancet Diabetes Endocrinol.* **2021**, *9*, 484–490. [CrossRef]
140. Able, C.; Liao, B.; Saffati, G.; Maremanda, A.; Applewhite, J.; Nasrallah, A.A.; Sonstein, J.; Alzweri, L.; Kohn, T.P. Prescribing semaglutide for weight loss in non-diabetic, obese patients is associated with an increased risk of erectile dysfunction: A TriNetX database study. *Int. J. Impot. Res.* **2025**, *37*, 315–319. [CrossRef]
141. Labazi, H.; Wynne, B.M.; Tostes, R.; Webb, R.C. Metformin treatment improves erectile function in an angiotensin II model of erectile dysfunction. *J. Sex. Med.* **2013**, *10*, 2154–2164. [CrossRef]

142. Defeudis, G.; Mazzilli, R.; Tenuta, M.; Rossini, G.; Zamponi, V.; Olana, S.; Faggiano, A.; Pozzilli, P.; Isidori, A.M.; Gianfrilli, D. Erectile dysfunction and diabetes: A melting pot of circumstances and treatments. *Diabetes Metab. Res. Rev.* **2022**, *38*, e3494. [CrossRef]
143. Cai, T.; Hu, Y.; Ding, B.; Yan, R.; Liu, B.; Cai, L.; Jing, T.; Jiang, L.; Xie, X.; Wang, Y.; et al. Effect of Metformin on Testosterone Levels in Male Patients with Type 2 Diabetes Mellitus Treated with Insulin. *Front. Endocrinol.* **2021**, *12*, 813067. [CrossRef]
144. Hu, Y.; Ding, B.; Shen, Y.; Yan, R.N.; Li, F.F.; Sun, R.; Jing, T.; Lee, K.O.; Ma, J.H. Rapid Changes in Serum Testosterone in Men with Newly Diagnosed Type 2 Diabetes with Intensive Insulin and Metformin. *Diabetes Care* **2021**, *44*, 1059–1061. [CrossRef] [PubMed]
145. Kounatidis, D.; Vallianou, N.G.; Rebelos, E.; Kouveletsou, M.; Kontrafouri, P.; Eleftheriadou, I.; Diakoumopoulou, E.; Karampela, I.; Tentolouris, N.; Dalamaga, M. The Many Facets of PPAR-γ Agonism in Obesity and Associated Comorbidities: Benefits, Risks, Challenges, and Future Directions. *Curr. Obes. Rep.* **2025**, *14*, 19. [CrossRef]
146. Idris-Khodja, N.; Ouerd, S.; Trindade, M.; Gornitsky, J.; Rehman, A.; Barhoumi, T.; Offermanns, S.; Gonzalez, F.J.; Neves, M.F.; Paradis, P.; et al. Vascular smooth muscle cell peroxisome proliferator-activated receptor γ protects against endothelin-1-induced oxidative stress and inflammation. *J. Hypertens.* **2017**, *35*, 1390–1401. [CrossRef]
147. Hamblin, M.; Chang, L.; Fan, Y.; Zhang, J.; Chen, Y.E. PPARs and the cardiovascular system. *Antioxid. Redox Signal.* **2009**, *11*, 1415–1452. [CrossRef]
148. Palacios-Ramírez, R.; Hernanz, R.; Martín, A.; Pérez-Girón, J.V.; Barrús, M.T.; González-Carnicero, Z.; Aguado, A.; Jaisser, F.; Briones, A.M.; Salaices, M.; et al. Pioglitazone Modulates the Vascular Contractility in Hypertension by Interference with ET-1 Pathway. *Sci. Rep.* **2019**, *9*, 16461. [CrossRef]
149. Heidenberg, D.J.; Haney, N.M.; Rezk, B.M.; Talwar, S.; Okpechi, S.C.; Srivastav, S.K.; Honda, M.; Song, B.; Swan, K.; Awadallah, S.; et al. Pioglitazone's beneficial effects on erectile function preservation after cavernosal nerve injury in the rat are negated by inhibition of the insulin-like growth factor-1 receptor: A preclinical study. *Int. J. Impot. Res.* **2019**, *31*, 1–8. [CrossRef] [PubMed]
150. Gholamine, B.; Shafiei, M.; Motevallian, M.; Mahmoudian, M. Effects of pioglitazone on erectile dysfunction in sildenafil poor-responders: A randomized, controlled study. *J. Pharm. Pharm. Sci.* **2008**, *11*, 22–31. [CrossRef] [PubMed]
151. Wang, L.; Xu, Y.; Li, H.; Lei, H.; Guan, R.; Gao, Z.; Xin, Z. Antioxidant icariside II combined with insulin restores erectile function in streptozotocin-induced type 1 diabetic rats. *J. Cell. Mol. Med.* **2015**, *19*, 960–969. [CrossRef]
152. Yamanaka, M.; Shirai, M.; Shiina, H.; Tanaka, Y.; Tsujimura, A.; Matsumiya, K.; Okuyama, A.; Dahiya, R. Diabetes induced erectile dysfunction and apoptosis in penile crura are recovered by insulin treatment in rats. *J. Urol.* **2003**, *170*, 291–297. [CrossRef]
153. Shirai, M.; Yamanaka, M.; Shiina, H.; Igawa, M.; Ogishima, T.; Fujime, M.; Ishii, N.; Okuyama, A.; Lue, T.F.; Dahiya, R. Androgen, estrogen, and progesterone receptor gene regulation during diabetic erectile dysfunction and insulin treatment. *Urology* **2004**, *64*, 1244–1249. [CrossRef] [PubMed]
154. Feng, L.; Jinhua, W.; Shulin, G.; Jiangping, X.; Zhongxiang, L.; Xiaohong, L. Causal association between antidiabetic drugs and erectile dysfunction: Evidence from Mendelian randomization. *Front. Endocrinol.* **2024**, *15*, 1414958. [CrossRef] [PubMed]
155. Ruiz Rubio, J.L.; Hernández, M.; Rivera de los Arcos, L.; Benedito, S.; Recio, P.; García, P.; García-Sacristán, A.; Prieto, D. Role of ATP-sensitive K+ channels in relaxation of penile resistance arteries. *Urology* **2004**, *63*, 800–805. [CrossRef]
156. Al-Kuraishy, H.M.; Al-Gareeb, A.I. Erectile Dysfunction and Low Sex Drive in Men with Type 2 DM: The Potential Role of Diabetic Pharmacotherapy. *J. Clin. Diagn. Res.* **2016**, *10*, FC21–FC26. [CrossRef] [PubMed]
157. Chan, T.Y. Outbreaks of severe hypoglycaemia due to illegal sexual enhancement products containing undeclared glibenclamide. *Pharmacoepidemiol. Drug Saf.* **2009**, *18*, 1250–1251. [CrossRef]
158. Kao, S.L.; Chan, C.L.; Tan, B.; Lim, C.C.; Dalan, R.; Gardner, D.; Pratt, E.; Lee, M.; Lee, K.O. An unusual outbreak of hypoglycemia. *N. Engl. J. Med.* **2009**, *360*, 734–736. [CrossRef]
159. Altabas, V.; Altabas, K. DPP-4 inhibition improves a sexual condition? *Med. Hypotheses* **2015**, *85*, 124–126. [CrossRef]
160. Sun, T.; Xu, W.; Wang, J.; Wang, T.; Wang, S.; Liu, K.; Liu, J. Saxagliptin alleviates erectile dysfunction through increasing stromal cell-derived factor-1 in diabetes mellitus. *Andrology* **2023**, *11*, 295–306. [CrossRef]
161. Kataoka, T.; Sanagawa, A.; Hotta, Y.; Kimura, K. Effect of DPP-4 Inhibitors on Erectile Dysfunction Caused by Vincristine in Rats. *J. Sex. Med.* **2022**, *19*, S141. [CrossRef]
162. Simon, D.; Detournay, B.; Eschwege, E.; Bouée, S.; Bringer, J.; Attali, C.; Dejager, S. Use of Vildagliptin in Management of Type 2 Diabetes: Effectiveness, Treatment Persistence and Safety from the 2-Year Real-Life VILDA Study. *Diabetes Ther.* **2014**, *5*, 207–224. [CrossRef] [PubMed]
163. Lee, D.M.; Battson, M.L.; Jarrell, D.K.; Hou, S.; Ecton, K.E.; Weir, T.L.; Gentile, C.L. SGLT2 inhibition via dapagliflozin improves generalized vascular dysfunction and alters the gut microbiota in type 2 diabetic mice. *Cardiovasc. Diabetol.* **2018**, *17*, 62. [CrossRef]
164. Uthman, L.; Homayr, A.; Juni, R.P.; Spin, E.L.; Kerindongo, R.; Boomsma, M.; Hollmann, M.W.; Preckel, B.; Koolwijk, P.; van Hinsbergh, V.W.M.; et al. Empagliflozin and Dapagliflozin Reduce ROS Generation and Restore NO Bioavailability in Tumor Necrosis Factor α-Stimulated Human Coronary Arterial Endothelial Cells. *Cell. Physiol. Biochem.* **2019**, *53*, 865–886. [CrossRef]

165. Assaly, R.; Gorny, D.; Compagnie, S.; Mayoux, E.; Bernabe, J.; Alexandre, L.; Giuliano, F.; Behr-Roussel, D. The Favorable Effect of Empagliflozin on Erectile Function in an Experimental Model of Type 2 Diabetes. *J. Sex. Med.* **2018**, *15*, 1224–1234. [CrossRef]
166. Ali, A.B.H.; Abdel-Aal, F.A.M.; Rageh, A.H.; Mohamed, A.I. Cytochrome P450 3A4-mediated pharmacokinetic interaction study between tadalafil and canagliflozin using high-performance thin-layer chromatography. *J. Sep. Sci.* **2022**, *45*, 4187–4197. [CrossRef]
167. Cannarella, R.; Condorelli, R.A.; Leanza, C.; Garofalo, V.; Aversa, A.; Papa, G.; Calogero, A.E.; La Vignera, S. Dapagliflozin improves erectile dysfunction in patients with type 2 diabetes mellitus: An open-label, non-randomized pilot study. *Diabet. Med.* **2024**, *41*, e15217. [CrossRef]
168. Hu, W.S.; Lin, C.L. Erectile Dysfunction Risk Among Patients with Diabetes Mellitus Using Sodium-Glucose Cotransporter 2 Inhibitors. *J. Cardiovasc. Pharmacol.* **2024**, *84*, 545–549. [CrossRef] [PubMed]
169. Shao, N.; Yu, X.Y.; Yu, Y.M.; Li, B.W.; Pan, J.; Wu, W.H.; Zhang, H.J.; Ma, X.F.; Hao, M.; Kuang, H.Y. Short-term combined treatment with exenatide and metformin is superior to glimepiride combined metformin in improvement of serum testosterone levels in type 2 diabetic patients with obesity. *Andrologia* **2018**, *50*, e13039. [CrossRef] [PubMed]
170. Defeudis, G.; Di Tommaso, A.M.; Di Rosa, C.; Cimadomo, D.; Khazrai, Y.M.; Faggiano, A.; Cincione, R.I.; Napoli, N.; Mazzilli, R. The Role of Antihyperglycemic Drugs and Diet on Erectile Function: Results from a Perspective Study on a Population with Prediabetes and Diabetes. *J. Clin. Med.* **2022**, *11*, 3382. [CrossRef]
171. Yang, B.; Cheng, H.; Hu, Y.; Chen, Y.; Xu, Y.; Huang, W.; Long, Y.; Gao, C. Effects of Anti-Diabetic Drugs on Erectile Dysfunction: A Systematic Review and Meta-Analysis. *Diabetes Metab. Syndr. Obes.* **2025**, *18*, 467–478. [CrossRef] [PubMed]
172. Rastrelli, G.; Corona, G.; Sparano, C.; Vignozzi, L.; Sforza, A.; Maggi, M. Impact of Diabetic Medications on Erectile Function: A Meta-Analysis. *J. Sex. Med.* **2025**, *22* (Suppl. 2), qdaf077.056. [CrossRef]
173. Corona, G.; Rastrelli, G.; Sparano, C.; Vignozzi, L.; Sforza, A.; Maggi, M. Role of Glucagon-Like Peptide Type 1 Analogues on Hypothalamus-Pituitary-Testis Axis and Sexual Function in Men: A Systematic Review and Meta-Analysis. *J. Sex. Med.* **2025**, *22* (Suppl. 2), qdaf077.063. [CrossRef]
174. Van Cauwenberghe, J.; De Block, C.; Vanderschueren, D.; Antonio, L. Effects of treatment for diabetes mellitus on testosterone concentrations: A systematic review. *Andrology* **2023**, *11*, 225–233. [CrossRef]
175. Salvio, G.; Ciarloni, A.; Ambo, N.; Bordoni, M.; Perrone, M.; Rossi, S.; Balercia, G. Effects of glucagon-like peptide 1 receptor agonists on testicular dysfunction: A systematic review and meta-analysis. *Andrology*, 2025; *Ahead of print*. [CrossRef]
176. Graybill, S.; Hatfield, J.; Kravchenko, M.; Beckman, D.; Tate, J.; Beauvais, A.; Clerc, P.; Davila, D.; Forbes, W.; Wardian, J.; et al. Neutral effect of exenatide on serum testosterone in men with type 2 diabetes mellitus: A prospective cohort. *Andrology* **2021**, *9*, 792–800. [CrossRef]
177. Mostafaei, H.; Mori, K.; Hajebrahimi, S.; Abufaraj, M.; Karakiewicz, P.I.; Shariat, S.F. Association of erectile dysfunction and cardiovascular disease: An umbrella review of systematic reviews and meta-analyses. *BJU Int.* **2021**, *128*, 3–11. [CrossRef]
178. Nguyen Ngoc Dang, H.; Viet Luong, T.; Kiem Pham, A.; Trung Le, T.; Duc Le, N.; Minh Nguyen, H.; Anh Hoang, T.; Anh Ho, B. Exploring the bidirectional link between erectile dysfunction and 10-year cardiovascular risk in men with diabetes and hypertension. *Sci. Rep.* **2024**, *14*, 28816. [CrossRef]
179. Peng, W.; Zhou, R.; Sun, Z.F.; Long, J.W.; Gong, Y.Q. Novel Insights into the Roles and Mechanisms of GLP-1 Receptor Agonists against Aging-Related Diseases. *Aging Dis.* **2022**, *13*, 468–490. [CrossRef]
180. Lu, C.; Xu, C.; Li, S.; Ni, H.; Yang, J. Liraglutide and GLP-1(9-37) alleviated hepatic ischemia-reperfusion injury by inhibiting ferroptosis via GSK3β/Nrf2 pathway and SMAD159/Hepcidin/FTH pathway. *Redox Biol.* **2025**, *79*, 103468. [CrossRef]
181. Yang, F.; Wang, X.; Qi, J.; Zhang, K.; Jiang, Y.; Feng, B.; Lv, T.; Yang, L.; Yang, Q.; Zhao, M.; et al. Glucagon-like Peptide 1 Receptor Activation Inhibits Microglial Pyroptosis via Promoting Mitophagy to Alleviate Depression-like Behaviors in Diabetic Mice. *Nutrients* **2022**, *15*, 38. [CrossRef] [PubMed]
182. Rroji, M.; Spahia, N.; Figurek, A.; Spasovski, G. Targeting Diabetic Atherosclerosis: The Role of GLP-1 Receptor Agonists, SGLT2 Inhibitors, and Nonsteroidal Mineralocorticoid Receptor Antagonists in Vascular Protection and Disease Modulation. *Biomedicines* **2025**, *13*, 728. [CrossRef] [PubMed]
183. Luo, C.; Peng, Y.; Gu, J.; Li, T.; Wang, Q.; Qi, X.; Wei, A. Single-cell RNA sequencing reveals critical modulators of extracellular matrix of penile cavernous cells in erectile dysfunction. *Sci. Rep.* **2024**, *14*, 5886. [CrossRef]
184. Papachristoforou, E.; Lambadiari, V.; Maratou, E.; Makrilakis, K. Association of Glycemic Indices (Hyperglycemia, Glucose Variability, and Hypoglycemia) with Oxidative Stress and Diabetic Complications. *J. Diabetes Res.* **2020**, *2020*, 7489795. [CrossRef]
185. Wei, L.; Gao, J.; Wang, L.; Tao, Q.; Tu, C. Hippo/YAP signaling pathway: A new therapeutic target for diabetes mellitus and vascular complications. *Ther. Adv. Endocrinol. Metab.* **2023**, *14*, 20420188231220134. [CrossRef] [PubMed]
186. Zheng, Z.; Zong, Y.; Ma, Y.; Tian, Y.; Pang, Y.; Zhang, C.; Gao, J. Glucagon-like peptide-1 receptor: Mechanisms and advances in therapy. *Signal Transduct. Target Ther.* **2024**, *9*, 234. [CrossRef]

187. Li, Y.; Du, J.; Zhu, E.; Zhang, J.; Han, J.; Zhao, W.; Sun, B.; Tian, D. Liraglutide suppresses proliferation and induces adipogenic differentiation of 3T3-L1 cells via the Hippo-YAP signaling pathway. *Mol. Med. Rep.* **2018**, *17*, 4499–4507. [CrossRef]
188. Bae, S.G.; Yin, G.N.; Ock, J.; Suh, J.K.; Ryu, J.K.; Park, J. Single-cell transcriptome analysis of cavernous tissues reveals the key roles of pericytes in diabetic erectile dysfunction. *elife* **2024**, *12*, RP88942. [CrossRef] [PubMed]
189. Bailey, J.; Coucha, M.; Bolduc, D.R.; Burnett, F.N.; Barrett, A.C.; Ghaly, M.; Abdelsaid, M. GLP-1 receptor nitration contributes to loss of brain pericyte function in a mouse model of diabetes. *Diabetologia* **2022**, *65*, 1541–1554. [CrossRef]
190. Zhu, L.; Zhou, J.; Pan, Y.; Lv, J.; Liu, Y.; Yu, S.; Zhang, Y. Glucagon-like peptide-1 receptor expression and its functions are regulated by androgen. *Biomed. Pharmacother.* **2019**, *120*, 109555. [CrossRef]
191. Dudal, S.; Bissantz, C.; Caruso, A.; David-Pierson, P.; Driessen, W.; Koller, E.; Krippendorff, B.F.; Lechmann, M.; Olivares-Morales, A.; Paehler, A.; et al. Translating pharmacology models effectively to predict therapeutic benefit. *Drug Discov. Today* **2022**, *27*, 1604–1621. [CrossRef]
192. Jensterle, M.; Janez, A.; Fliers, E.; DeVries, J.H.; Vrtacnik-Bokal, E.; Siegelaar, S.E. The role of glucagon-like peptide-1 in reproduction: From physiology to therapeutic perspective. *Hum. Reprod. Update* **2019**, *25*, 504–517. [CrossRef]
193. Neijenhuijs, K.I.; Holtmaat, K.; Aaronson, N.K.; Holzner, B.; Terwee, C.B.; Cuijpers, P.; Verdonck-de Leeuw, I.M. The International Index of Erectile Function (IIEF)-A Systematic Review of Measurement Properties. *J. Sex. Med.* **2019**, *16*, 1078–1091. [CrossRef] [PubMed]
194. Han, W.; Morris, R.; Bu, K.; Zhu, T.; Huang, H.; Cheng, F. Analysis of literature-derived duplicate records in the FDA Adverse Event Reporting System (FAERS) database. *Can. J. Physiol. Pharmacol.* **2025**, *103*, 56–69. [CrossRef]
195. Lovegrove, M.C.; Stone, N.D.; Geller, A.I.; Weidle, N.J.; Lind, J.N.; Cohen, P.A. U.S. Emergency Department Visits Attributed by Clinicians to Semaglutide Adverse Events, 2022–2023. *Ann. Intern. Med.* **2025**, *178*, 898–900. [CrossRef]
196. Hollander, J.E. Emergency department considerations regarding erectile dysfunction. *Am. J. Cardiol.* **2005**, *96*, 73M–75M. [CrossRef] [PubMed]
197. Arillotta, D.; Floresta, G.; Papanti Pelletier, G.D.; Guirguis, A.; Corkery, J.M.; Martinotti, G.; Schifano, F. Exploring the Potential Impact of GLP-1 Receptor Agonists on Substance Use, Compulsive Behavior, and Libido: Insights from Social Media Using a Mixed-Methods Approach. *Brain Sci.* **2024**, *14*, 617. [CrossRef] [PubMed]
198. Bain, S.C.; Klufas, M.A.; Ho, A.; Matthews, D.R. Worsening of diabetic retinopathy with rapid improvement in systemic glucose control: A review. *Diabetes Obes. Metab.* **2019**, *21*, 454–466. [CrossRef]
199. Pitocco, D.; Zaccardi, F.; Di Stasio, E.; Romitelli, F.; Santini, S.A.; Zuppi, C.; Ghirlanda, G. Oxidative stress, nitric oxide, and diabetes. *Rev. Diabet. Stud.* **2010**, *7*, 15–25. [CrossRef]
200. Jeibmann, A.; Zahedi, S.; Simoni, M.; Nieschlag, E.; Byrne, M.M. Glucagon-like peptide-1 reduces the pulsatile component of testosterone secretion in healthy males. *Eur. J. Clin. Investig.* **2005**, *35*, 565–572. [CrossRef]
201. Memel, Z.; Gold, S.L.; Pearlman, M.; Muratore, A.; Martindale, R. Impact of GLP- 1 Receptor Agonist Therapy in Patients High Risk for Sarcopenia. *Curr. Nutr. Rep.* **2025**, *14*, 63. [CrossRef]
202. Viken, A.F.; Siiak, S.P.; Schlünssen, V.; Thorarinsdottir, E.H.; Skulstad, S.M.; Gyawali, S.; Bertelsen, R.J.; Real, F.G. Muscle Strength and Male Sexual Function. *J. Clin. Med.* **2024**, *13*, 426. [CrossRef]
203. Schulte, D.M.; Hahn, M.; Oberhäuser, F.; Malchau, G.; Schubert, M.; Heppner, C.; Müller, N.; Güdelhöfer, H.; Faust, M.; Krone, W.; et al. Caloric restriction increases serum testosterone concentrations in obese male subjects by two distinct mechanisms. *Horm. Metab. Res.* **2014**, *46*, 283–286. [CrossRef]
204. Smith, S.J.; Teo, S.Y.M.; Lopresti, A.L.; Heritage, B.; Fairchild, T.J. Examining the effects of calorie restriction on testosterone concentrations in men: A systematic review and meta-analysis. *Nutr. Rev.* **2022**, *80*, 1222–1236. [CrossRef] [PubMed]
205. Kohn, T.P.; Lipshultz, L.I. Comment on: Male sexual dysfunction associated with GLP-1 receptor agonists: A cross-sectional analysis of FAERS data. *Int. J. Impot. Res.* 2025; *Ahead of print*. [CrossRef]
206. Salonia, A.; Capogrosso, P.; Boeri, L.; Cocci, A.; Corona, G.; Dinkelman-Smit, M.; Falcone, M.; Jensen, C.F.; Gül, M.; Kalkanli, A.; et al. European Association of Urology Guidelines on Male Sexual and Reproductive Health: 2025 Update on Male Hypogonadism, Erectile Dysfunction, Premature Ejaculation, and Peyronie's Disease. *Eur. Urol.* **2025**, *88*, 76–102. [CrossRef] [PubMed]
207. Bong, G.W.; Clarke, H.S., Jr.; Hancock, W.C.; Keane, T.E. Serum testosterone recovery after cessation of long-term luteinizing hormone-releasing hormone agonist in patients with prostate cancer. *Urology* **2008**, *71*, 1177–1180. [CrossRef]
208. Lengsfeld, S.; Probst, L.; Emara, Y.; Werlen, L.; Vogt, D.R.; Bathelt, C.; Baur, F.; Caviezel, B.; Vukajlovic, T.; Fischer, M.; et al. Effects of the glucagon-like peptide-1 receptor agonist dulaglutide on sexuality in healthy men: A randomised, double-blind, placebo-controlled crossover study. *EBioMedicine* **2024**, *107*, 105284. [CrossRef]
209. Zhong, Q.; Bollag, R.J.; Dransfield, D.T.; Gasalla-Herraiz, J.; Ding, K.H.; Min, L.; Isales, C.M. Glucose-dependent insulinotropic peptide signaling pathways in endothelial cells. *Peptides* **2000**, *21*, 1427–1432. [CrossRef]
210. Lim, D.M.; Park, K.Y.; Hwang, W.M.; Kim, J.Y.; Kim, B.J. Difference in protective effects of GIP and GLP-1 on endothelial cells according to cyclic adenosine monophosphate response. *Exp. Ther. Med.* **2017**, *13*, 2558–2564. [CrossRef]

211. Killion, E.A.; Hussien, R.; Shkumatov, A.; Davies, R.; Lloyd, D.J.; Véniant, M.M.; Lebrec, H.; Fort, M.M. GIPR gene expression in testis is mouse specific and can impact male mouse fertility. *Andrology* **2022**, *10*, 789–799. [CrossRef] [PubMed]
212. La Vignera, S.; Cannarella, R.; Garofalo, V.; Crafa, A.; Barbagallo, F.; Condorelli, R.A.; Calogero, A.E. Short-term impact of tirzepatide on metabolic hypogonadism and body composition in patients with obesity: A controlled pilot study. *Reprod. Biol. Endocrinol.* **2025**, *23*, 92. [CrossRef]
213. Małczak, P.; Wysocki, M.; Kawa, I.; Wikar, T.; Pisarska-Adamczyk, M.; Pędziwiatr, M.; Major, P. Improved erectile function after bariatric surgery: Role of testosterone and other factors-a cohort prospective study. *Surg. Obes. Relat. Dis.* **2025**, *21*, 648–654. [CrossRef] [PubMed]
214. Anastasiou, I.A.; Kounatidis, D.; Honka, M.-J.; Vallianou, N.G.; Rebelos, E.; Karamanolis, N.N.; Dalamaga, M.; Pantos, C.; Mourouzis, I. Metabolomic Alterations in Patients with Obesity and the Impact of Metabolic Bariatric Surgery: Insights for Future Research. *Metabolites* **2025**, *15*, 434. [CrossRef] [PubMed]
215. Jones, B.; Sands, C.; Alexiadou, K.; Minnion, J.; Tharakan, G.; Behary, P.; Ahmed, A.R.; Purkayastha, S.; Lewis, M.R.; Bloom, S.; et al. The Metabolomic Effects of Tripeptide Gut Hormone Infusion Compared to Roux-en-Y Gastric Bypass and Caloric Restriction. *J. Clin. Endocrinol. Metab.* **2022**, *107*, e767–e782. [CrossRef] [PubMed]

biomolecules

MDPI

Review

Exploring the Revolutionary Impact of YAP Pathways on Physical and Rehabilitation Medicine

Carmelo Pirri

Department of Neuroscience, Institute of Human Anatomy, University of Padova, 35121 Padova, Italy; carmelop87@hotmail.it or carmelo.pirri@unipd.it

Abstract: Cellular behavior is strongly influenced by mechanical signals in the surrounding microenvironment, along with external factors such as temperature fluctuations, changes in blood flow, and muscle activity, etc. These factors are key in shaping cellular states and can contribute to the development of various diseases. In the realm of rehabilitation physical therapies, therapeutic exercise and manual treatments, etc., are frequently employed, not just for pain relief but also to support recovery from diverse health conditions. However, the detailed molecular pathways through which these therapies interact with tissues and influence gene expression are not yet fully understood. The identification of YAP has been instrumental in closing this knowledge gap. YAP is known for its capacity to perceive and translate mechanical signals into specific transcriptional programs within cells. This insight has opened up new perspectives on how physical and rehabilitation medicine may exert its beneficial effects. The review investigates the involvement of the Hippo/YAP signaling pathway in various diseases and considers how different rehabilitation techniques leverage this pathway to aid in healing. Additionally, it examines the therapeutic potential of modulating the Hippo/YAP pathway within the context of rehabilitation, while also addressing the challenges and controversies that surround its use in physical and rehabilitation medicine.

Keywords: YAP; Hippo; TAZ; exercise; sport; rehabilitation; physical therapy; muscle; bone; cartilage

1. Introduction

Physical and rehabilitation medicine, including physical therapies, exercise, ultrasound, vibration, cold therapy, hyperthermia, and manual therapy, etc., employs mechanical and thermal stimuli to achieve therapeutic effects, but the underlying mechanism of these interventions is incompletely understood. The Hippo/YAP signaling pathway provides a new perspective on how physical therapies contribute to tissue repair and regeneration. YAP, as a key effector of this pathway, acts as a sensor of mechanical cues such as extracellular matrix stiffness, fascial stiffness, blood flow, and muscle contractions, converting these inputs into cellular responses that affect proliferation, differentiation, and migration. The Hippo/YAP signaling pathway was initially identified in Drosophila, where it plays a fundamental role in controlling organ size by regulating cell proliferation and apoptosis [1]. In mammals, this pathway has been shown to be highly conserved, with Yes-associated protein (YAP) and transcriptional coactivator with PDZ-binding motif (TAZ) serving as the primary effectors that translocate to the nucleus in response to changes in mechanical cues and promote the transcription of targeted genes [2]. The Hippo/YAP pathway is regulated through a kinase cascade that includes mammalian STE20-like protein

https://doi.org/10.3390/biom15010096

kinases 1 and 2 (MST1/2) and large tumor suppressor kinases 1 and 2 (LATS1/2), which phosphorylate YAP/TAZ and prevent their nuclear localization [3].

The application of mechanical stimuli, such as those encountered during physical therapy, is known to influence cellular fate and tissue homeostasis [4,5]. These stimuli, including therapeutic exercise, ultrasound, manual therapy, physical therapies, and mechanical loading, have been shown to affect processes such as stem cell differentiation, tissue remodeling, and inflammatory response [6]. For instance, mechanical forces generated during exercise or ultrasound therapy have been demonstrated to influence YAP activity and promote tissue regeneration and repair [4,5,7]. Recent studies have further demonstrated that mechanical stimuli, including vibration therapy and mechanical stretching, can enhance YAP-mediated signaling, thereby promoting angiogenesis and muscle regeneration [8]. This mechano-transduction capability makes the Hippo/YAP pathway a key player in rehabilitation medicine, where mechanical signals are harnessed to stimulate tissue repair and remodeling, reduce inflammation, and improve functional recovery. Furthermore, physical therapies are particularly relevant in the context of musculoskeletal, cardiovascular, and metabolic diseases, where tissue regeneration and remodeling are crucial components of rehabilitation [4]. The integration of the Hippo/YAP pathway in these therapies provides a molecular understanding of how mechanical and thermal signals are translated into beneficial biological outcomes to enhance the effectiveness of rehabilitation strategies. Understanding the specific role of YAP in different therapeutic modalities will aid in optimizing treatment protocols and potentially combining various forms of physical therapies, exercise, and manual therapy, etc., for enhanced outcomes [9].

The aim of this narrative review is to explore the role of the Hippo/YAP pathway in physical and rehabilitation medicine, delving into how this mechanism can explain the therapeutic effects of the different therapies of physical and rehabilitation medicine. Specifically, we will focus on how the mechanical and thermal stimuli used in these therapies—such as physical exercise, ultrasound, vibration, manual therapy, and cryotherapy—influence YAP activation, and consequently the cellular processes involved in tissue repair and regeneration. A deeper understanding of how YAP responds to mechanical stimuli will allow for the identification of new strategies to optimize the effectiveness of rehabilitative therapies, potentially combining different modalities to maximize regenerative effects and reduce inflammation. This integrated approach could lead to significant improvements in the management of musculoskeletal, cardiovascular, and metabolic conditions, enhancing functional recovery and the patient's quality of life.

2. Generality on Yes-Associated Protein (YAP)

Yes-associated protein (YAP) is a pivotal effector of the Hippo signaling pathway, a highly conserved kinase cascade that plays a critical role in regulating fundamental cellular processes such as proliferation, differentiation, and apoptosis [1]. YAP1 was first identified and named by Sudol et al. in 1995, who discovered and cloned this novel 65 KD protein. It was found to bind to the SH3 domain of the non-receptor tyrosine kinase YES [10]. The gene product of YAP1 has two main subtypes, YAP1-1 and YAP1-2, which differ by an extra 38 amino acids that form the WW domain. The YAP1-1 variant, containing a single proline, is also known as Yes-associated protein (YAP65) due to its binding to the proto-oncogene YES. In mammals, the YAP1 ortholog has either one (YAP1-1) or two (YAP1-2) WW domains. Structurally, YAP1 includes a transcription activator-binding protein (TID), a WW domain, and a transcription activation domain (TAD). Some YAP1 variants also feature an SH3-binding motif (SH3-BM) and a leucine zinc finger. At its N-terminal, YAP1 has a proline-rich region, followed by the TAD and PDZ-binding motif (PDZ-BM), which are crucial for activating specific gene transcription. Due to regulatory variations in transcription, translation, and modification levels, YAP1 functions differently across cell types. The modular structure of the two YAP1s is illustrated in Figure 1.

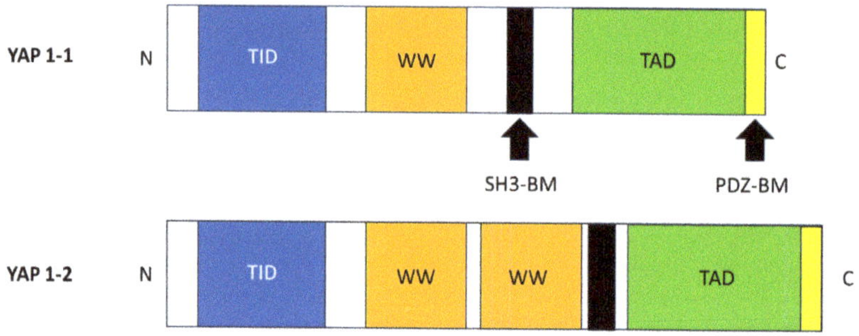

Figure 1. Modular structure of YAP.

The Hippo signaling pathway acts as a central regulator of organ size, tissue homeostasis, and regeneration, with YAP being one of the primary mediators through which these effects are exerted. Together with its homolog, transcriptional coactivator with PDZ-binding motif (TAZ), YAP functions as a mechanosensitive transcriptional regulator that integrates a variety of extracellular signals—including mechanical forces, oxidative stress, and biochemical cues—into specific cellular responses that determine cell fate [4,9]. The Hippo pathway itself consists of a series of kinases, including MST1/2 and LATS1/2, which phosphorylate and inactivate YAP/TAZ, thus retaining them in the cytoplasm [10]. When the Hippo pathway is inactive, YAP and TAZ translocate to the nucleus, where they interact with transcriptional enhancer activator domain (TEAD) family proteins, among others, to drive the transcription of genes involved in cell growth, survival, and tissue regeneration [11]. The activity of YAP is tightly regulated by upstream signals, such as cell–cell contact, cytoskeletal dynamics, and extracellular matrix (ECM) stiffness, allowing YAP to act as a key sensor of the mechanical and biochemical state of the tissue environment [3]. Mechanical stress is one of the most critical regulators of YAP activity. Cells residing within tissues are constantly exposed to mechanical forces, such as tension, compression, and shear stress, which alter the physical properties of the ECM. YAP/TAZ serve as mechanotransducers, responding to these forces by modulating gene expression programs that control cell proliferation and differentiation [4]. When cells experience high mechanical tension or increased ECM stiffness, YAP becomes activated and accumulates in the nucleus, promoting the expression of genes that drive cell growth and survival [5]. Conversely, under conditions of low mechanical tension, YAP is phosphorylated and sequestered in the cytoplasm, leading to decreased cell proliferation and increased differentiation and apoptosis. This mechanosensitive function of YAP is important in maintaining tissue homeostasis and orchestrating appropriate responses to injury and regeneration (Figure 2) [12].

YAP is also highly responsive to biochemical signals, including growth factors, hormones, and oxidative stress. Growth factors such as the epidermal growth factor (EGF) and transforming growth factor-β (TGF-β) can activate YAP through various signaling pathways, including the PI3K-Akt and Rho GTPase pathway, enhancing its nuclear localization and transcriptional activity [10,13]. YAP's interaction with these pathways allows it to act as a central integrator of diverse signaling inputs, thereby playing a key role in coordinating complex cellular responses, such as stem cell maintenance, tissue repair, and immune modulation [14]. Oxidative stress, which often occurs during inflammation or tissue damage, can also influence YAP activity, by altering its phosphorylation state or its interactions with other proteins in the cell [15]. In the nucleus, YAP functions as a transcriptional coactivator, lacking its own DNA-binding domain but instead partnering with TEAD transcription factors to regulate target gene expression [13]. The YAP-TEAD complex controls the expression of numerous genes involved in cell proliferation, such as CTGF (connective tissue growth factor) and CYR61 (cysteine-rich angiogenic inducer 61), which are crucial for tissue regeneration and wound healing [11]. Additionally, YAP interacts with other transcription factors, including SMADs, p73, and RUNX, enabling it to participate in multiple cellular pathways and processes. For example, YAP's interaction with SMADs allows it to modulate TGF-β signaling, which is essential for fibrosis, while its interaction with RUNX2 is important for osteoblast differentiation and bone formation [16].

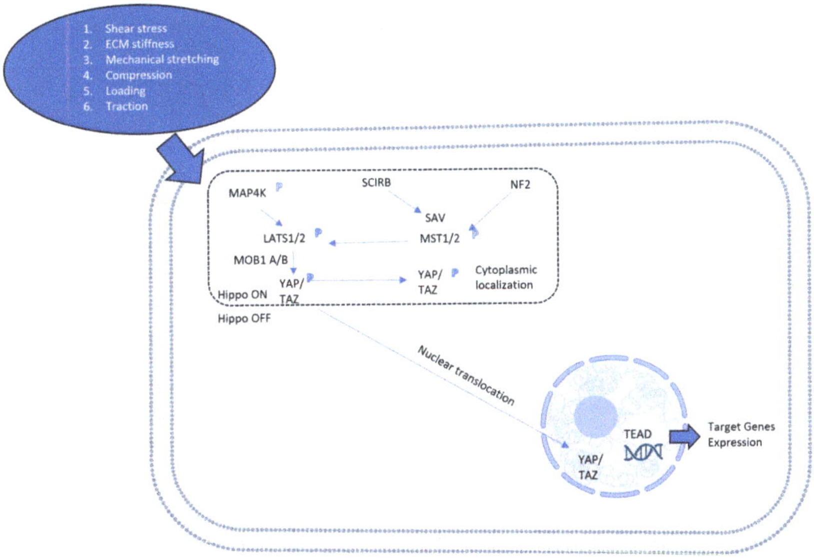

Figure 2. Mechano-transduction pathways regulating YAP/TAZ activity. This schematic shows how mechanical signals—such as ECM stiffness, mechanical stretch, and shear stress—modulate YAP/TAZ activity via Hippo-dependent and independent pathways. The Hippo pathway's core kinases, MST1/2 and LATS1/2, respond to mechanical cues to regulate YAP/TAZ phosphorylation and localization. Alternatively, these mechanical stimuli can bypass the kinase cascade, influencing YAP/TAZ localization through cytoplasmic and nuclear actin dynamics.

The regulation of YAP is also influenced by cell polarity and cell density. In conditions of high cell density, the Hippo pathway is activated, leading to YAP phosphorylation and cytoplasmic retention and inhibiting cell proliferation [13]. This contact inhibition mechanism prevents overgrowth and maintains tissue architecture. In contrast, in low-density environments or when cell polarity is disrupted, Hippo signaling is suppressed, allowing YAP to translocate to the nucleus and promote cell proliferation. This dynamic regulation underscores YAP's role in maintaining tissue homeostasis and its potential contribution to pathological conditions when dysregulated [17]. In summary, YAP is a versatile and dynamic effector of the Hippo signaling pathway, integrating mechanical, biochemical, and environmental signals to regulate cell behavior. Its ability to modulate gene expression in response to changes in the cellular microenvironment makes YAP a critical regulator of tissue homeostasis, regeneration, and development. Given its role in mediating key processes such as cell proliferation, differentiation and apoptosis, YAP has emerged as a promising therapeutic target for various conditions, including cancer, fibrosis, and degenerative diseases [9,18].

3. YAP and Diseases

The Hippo/YAP signaling pathway is an essential mechano-transducer that mediates cellular responses to mechanical cues, influencing disease progression across diverse contexts, including cancer, tissue regeneration, musculoskeletal integrity, and vascular health. In cancer, YAP responds to mechanical alterations in the tumor microenvironment, increasing ECM stiffness, which has been identified as a key factor in promoting tumor proliferation and invasiveness, notably in breast cancer [19,20]. YAP activation drives the transcription of genes like Skp2, which are critical for cell cycle progression, thereby enhancing tumorigenesis [19,20]. YAP inactivation leads to cell cycle arrest, effectively curbing tumor growth [19,20]. Additionally, the accumulation of advanced glycation end-products (AGEs) in the ECM can alter its viscoelasticity, promoting oncogenic β-catenin signaling in hepatocellular carcinoma (HCC). Mechanistically, AGEs increase ECM stiffness, activating the integrin-β1–tensin-1–YAP pathway, which in turn stimulates HCC cell proliferation and invasion, underscoring YAP's central role in converting ECM mechanical signals into tumor-promoting biochemical responses [21,22]. Beyond oncology, YAP is also essential in tissue regeneration, particularly in organs with a high regenerative capacity, like the liver. In response to mechanical cues such as shear stress, YAP facilitates hepatocyte proliferation by enabling them to re-enter the cell cycle in a process that is further amplified by integrin β1, a key shear sensor at the cell membrane [23,24]. Upon activation, YAP translocates to the nucleus, initiating transcriptional programs crucial for cell proliferation and tissue repair. Studies have shown that YAP activation following mechanical stress also supports skin regeneration by activating Engrailed-1, a protein essential for wound healing [25]. When YAP signaling is blocked, Engrailed-1 activation is inhibited, which compromises the regenerative response, highlighting YAP's critical function in tissue repair [25]. In the musculoskeletal system, YAP mediates responses to mechanical stress in cartilage and bone, which are highly responsive to mechanical signals due to their load-bearing roles. In cartilage, YAP activation promotes chondrocyte proliferation by upregulating cell cycle proteins like PCNA and cyclin D1 [26]. The nuclear translocation of YAP also depends on the substrate stiffness; soft substrates promote the expression of cartilage-specific markers like Col2α1, preserving the chondrocyte phenotype, while stiffer substrates drive YAP nuclear localization, leading to dedifferentiation. In bone tissue, YAP contributes to osteogenic differentiation in response to mechanical loading via the Piezo1 pathway, which regulates glutaminolysis through GLS1 activity. This pathway promotes bone cell differentiation and tissue integrity by enabling cells to adapt to mechanical stress, with YAP's role in skeletal development underscoring its importance for musculoskeletal health [27]. In vascular homeostasis, YAP's mechano-sensitivity also impacts endothelial cell behavior, particularly in the regulation of atherosclerosis [28]. Endothelial cells are continually exposed to shear stress from blood flow, with a stable, laminar flow inhibiting YAP's nuclear translocation, thereby reducing inflammation and oxidative stress. Disturbed or oscillatory flow, on the other hand, activates YAP, promoting atherogenic processes [29]. Mechanistically, YAP activation under a disturbed flow is regulated by integrin-Gα13 and JNK signaling pathways, while in vivo studies indicate that YAP knockdown in endothelial cells reduces plaque formation, highlighting its role in atherosclerosis progression [30]. Additionally, oscillatory shear stress induces YAP nuclear translocation, leading to endothelial activation through pathways dependent on integrin α5β1 and its downstream kinase c-Abl. Blocking these interactions reduces YAP activation and subsequently decreases early atherosclerotic changes [31]. Overall, YAP has an advanced mechanosensory function, converting physical stimuli into cellular responses that impact cell fate, growth, and disease progression. These pathways' ability to integrate and transduce complex mechanical signals positions YAP as a critical target for innovative treatments aimed at altering mechanical dynamics within tissues to modulate disease outcomes.

4. Hippo/YAP Signaling Pathway in Musculoskeletal Rehabilitation

In musculoskeletal rehabilitation, the mechanical stress generated during exercise has been shown to regulate, for example, chondrocyte proliferation and differentiation via YAP activation. Increased YAP expression promotes chondrocyte dedifferentiation under high mechanical loads, while its inhibition preserves the differentiated chondrocyte phenotype, which is essential for cartilage health. Additionally, mechanical loading through resistance exercises has been shown to enhance osteogenic differentiation by activating YAP, which in turn increases the expression of osteogenic markers such as RUNX2 and osteopontin [32]. This suggests that the Hippo/YAP pathway not only plays a crucial role in maintaining cartilage health, but also contributes to bone regeneration and strengthening during rehabilitation. Similarly, low-intensity pulsed ultrasound (LIPUS), a common therapeutic modality, enhances tissue regeneration by promoting YAP activation, which in turn regulates the transcription of genes involved in cell proliferation and tissue repair. LIPUS has also been reported to promote muscle regeneration by enhancing satellite cell activation through the Hippo/YAP pathway [33]. This effect is beneficial for muscle recovery following injury, where promoting the proliferation and differentiation of satellite cells is critical for restoring muscle function. Moreover, vibration therapy, which delivers mechanical stimulation at specific frequencies, has been demonstrated to enhance muscle hypertrophy and bone density by regulating YAP/TAZ signaling [34]. The vibration-induced activation of YAP leads to increased muscle protein synthesis and bone mineralization, which is essential for improving musculoskeletal health, especially in older adults or individuals with reduced mobility [34,35]. Recent findings also indicate that mechanical stretching applied to tendons and ligaments can modulate YAP activity, promoting tenocyte proliferation and enhancing tendon repair [36]. The Hippo/YAP pathway also plays a role in reducing inflammation during musculoskeletal rehabilitation. Mechanical loading has been shown to inhibit pro-inflammatory cytokine production by modulating YAP activity, thereby reducing tissue inflammation and promoting an environment conductive to healing [37]. This anti-inflammatory effect of YAP-mediated signaling is important in conditions such as osteoarthritis, where chronic inflammation impairs tissue regeneration and function. Recent research has also linked YAP to macrophage polarization, indicating that high YAP expression can promote macrophage polarization towards the pro-inflammatory M1 phenotype while inhibiting the anti-inflammatory M2 phenotype. This interplay can influence inflammation and tissue repair processes during musculoskeletal rehabilitation [37]. Overall, the Hippo/YAP signaling pathway is emerging as a key regulator of musculoskeletal health, responding to mechanical stimuli to drive tissue regeneration, reduce inflammation, and promote functional recovery. By understanding the mechanisms by which YAP modulates chondrocyte, muscle, bone and fascia cell behavior, rehabilitation protocols can be better tailored to enhance patient outcomes, in particular in individuals with musculoskeletal injuries or degenerative diseases (Figure 3 and Table 1).

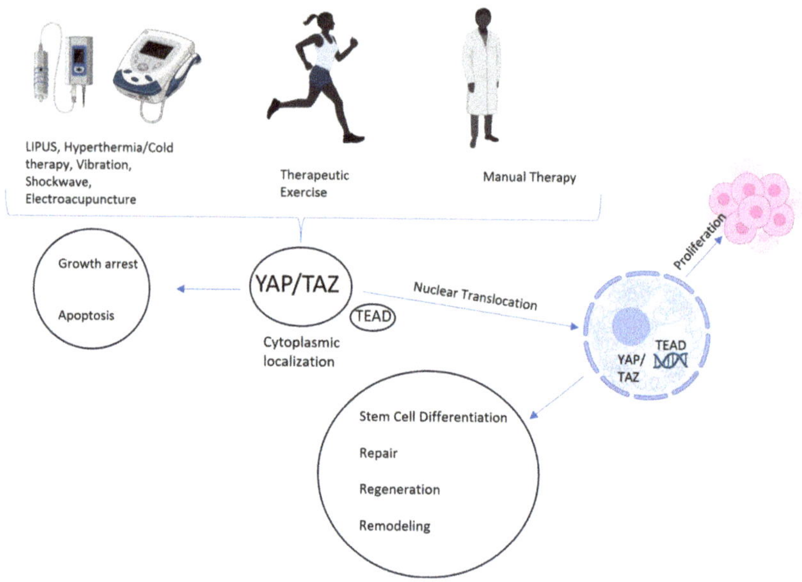

Figure 3. Hippo/YAP signaling pathway in physical and rehabilitation medicine. The Hippo/YAP pathway responds to mechanical stimuli, promoting tissue repair and regeneration in cartilage, bone, muscle, blood vessels, nerves, fascia, and tendons. YAP activation under different forms of mechanical loading—such as exercise, LIPUS, vibration therapy, etc.—reduces inflammation and supports cell proliferation and differentiation, as well as the regeneration and remodeling of various tissues. This pathway modulates key cellular processes, making it essential for tailoring rehabilitation protocols to enhance recovery in musculoskeletal injuries and degenerative conditions.

Table 1. The role of physical and rehabilitation medicine in modulating YAP. The table summarizes the intervention, the mechanism of YAP activation, YAP-mediated effects, and rehabilitative implications.

Intervention	Mechanism of YAP Activation	YAP-Mediated Effects	Rehabilitative Implications	References
Exercise (Including aerobic)	Mechanical stress and the ELABELA-APJ-Akt/YAP axis activate YAP through signaling pathway.	Enhances tissue repair by promoting cell migration and angiogenesis and reducing apoptosis.	Improves functional recovery, supports cardiovascular and musculoskeletal health, and prevents deconditioning.	[38–41]
LIPUS (Low-Intensity Pulsed Ultrasound)	Mechanical vibrations enhance YAP nuclear translocation by stabilizing the cytoskeleton and reducing phosphorylation.	Stimulates tissue regeneration, accelerates fracture healing, and inhibits adipocyte differentiation.	Optimizes recovery form fractures, supports weight management, and promotes healing in soft tissue injuries.	[38,42–45]
Cold therapy	Activates YAP via RBM3 or miR-429 pathways in response to low temperatures.	Promotes metabolic adaptation, browning of white adipose tissue, and neuroprotection.	Reduces inflammation, supports neural recovery, and aids in metabolic rehabilitation post injury or diseases.	[38,46,46–51]

Table 1. *Cont.*

Intervention	Mechanism of YAP Activation	YAP-Mediated Effects	Rehabilitative Implications	References
Hyperthermia	Enhances YAP nuclear activity through dephosphorylation of LATS1 under thermal stress.	Promotes osteogenesis and differentiation of ectomesenchymal stem cells for tissue repair.	Facilitates recovery from bone injuries, accelerates tissue repair, and relieves musculoskeletal pain.	[38,46,46–51]
Manual therapy	Modulates Hippo/YAP activity by altering cytoskeletal dynamics through mechano-transduction.	Enhances tissue flexibility, reduces local inflammation, and improves cellular recovery in affected regions.	Restores range of motion, alleviates pain, and promotes musculoskeletal and postural recovery.	[38,40]
Vibration (Low-intensity)	Mechanical oscillations activate YAP via RhoA signaling and cytoskeletal remodeling.	Enhances mesenchymal stem cell proliferation, supports osteogenic differentiation, and maintains bone density.	Prevents bone loss, supports muscle recovery, and improves tissue integrity in elderly or immobilized patients.	[35,39,52,53]
ESWT (Extracorporeal Shock Wave Therapy)	Stimulates YAP/TAZ via ECM-cytoskeleton interactions.	Facilitates connective tissue remodeling and nerve regeneration, promotes axonal repair, and enhances Schwann cell activity.	Improves peripheral nerve recovery, accelerates functional restoration, and mitigates neuropathic deficits.	[33,54–56]
Electroacupuncture	Electrical and mechanical stimuli activate YAP and mitochondrial fusion processes.	Reduces neuroinflammation, promotes neuroprotection, and enhances tissue repair.	Supports functional recovery post stroke, reduces neurological deficits, and aids in chronic pain management.	[37,38,57–61]

4.1. LIPUS and YAP

Low-intensity pulse ultrasound (LIPUS) is an increasingly popular therapeutic modality in regenerative medicine due to its ability to deliver mechanical stimulation in a non-invasive manner [42]. YAP is highly responsive to the mechanical cues generated by LIPUS, which can influence cellular processes [43]. Recent studies have shown that LIPUS enhances YAP nuclear translocation, thereby promoting the transcription of genes associated with cell proliferation and tissue regeneration, such as AREG, CYR61, and cyclin D1 [38,44]. In endothelial cells, LIPUS facilitates vascular remodeling by increasing YAP activity, leading to improved endothelial cell connections and angiogenesis, which is crucial for wound healing and tissue repair [45]. Additionally, LIPUS has been shown to inhibit adipocyte differentiation by enhancing YAP's nuclear localization, resulting in decreased expression of pro-adipogenic factors and promoting a phenotype conducive to tissue regeneration [45]. These findings underscore the role of LIPUS as a promising intervention that leverages YAP's mechanosensitive properties to enhance tissue healing and repair, making it a valuable tool in the field of regeneration medicine.

4.2. Vibration and YAP

Vibration therapy is another physical modality that exerts mechanical stimulation to promote tissue regeneration, with YAP playing a crucial role as a mechanosensitive regulator in this context. Vibration therapy applies cyclic mechanical forces to tissues, which enhances cellular mechano-transduction pathways, including the Hippo/YAP signaling pathway. YAP activation in response to vibration therapy has been demonstrated to promote muscle regeneration by stimulating satellite cell proliferation and differentiation, which is essential for repairing muscle damage and improving muscle strength [35,52]. Moreover, vibration-induced mechanical cues enhance osteogenic differentiation through YAP activation, promoting increased bone mineral density and bone strength, which is beneficial for individuals suffering from osteoporosis or reduced mobility [39]. Furthermore, studies have shown that vibration therapy also activates YAP/TAZ signaling in tenocytes, promoting tendon repair and improving the biomechanical properties of tendons, which is crucial for functional recovery after injury [53]. These findings highlight the potential of vibration therapy as a non-invasive intervention that harnesses YAP's mechanosensitive properties to enhance musculoskeletal health, making it an effective therapeutic option for improving tissue repair and functional outcomes in rehabilitation.

4.3. Electroacupuncture and YAP

Electroacupuncture (EA), a modality combining traditional acupuncture with electrical stimulation, has emerged as a promising technique in promoting tissue repair, partly by modulating the Hippo/YAP signaling pathway. Recent studies have demonstrated that EA promotes YAP activation and nuclear translocation, thereby enhancing the expression of genes related to tissue regeneration, including CTGF and ANKRD1 [37]. In musculoskeletal injuries, EA-induced YAP activation has been found to promote the proliferation and differentiation of mesenchymal stem cells (MSCs), thereby accelerating muscle repair and bone regeneration [57]. Additionally, EA has been shown to improve the repair of peripheral nerve injuries by activating YAP/TAZ signaling, which enhances Schwann cell proliferation and supports axon regrowth [58]. In ischemic stroke models, EA has been found to upregulate YAP expression in the penumbra area—the region surrounding the infarct. This upregulation of YAP reduces apoptosis and neuroinflammation, contributing to enhanced neurological recovery. By influencing the translocation of YAP into the nucleus, EA can modulate gene expression profiles that promote cell survival, reduce oxidative stress, and improve tissue repair mechanisms [59–61]. Specifically, EA treatment resulted in the increased expression of mitochondrial fusion proteins such as OPA1, MFN2, and MFN1, which played critical roles in maintaining mitochondrial integrity and reducing cortical damage and apoptosis in rat models of cerebral ischemia–reperfusion injury [59–61]. Moreover, YAP appears to function as a key regulatory hub in this process, acting through mechano-transduction pathways to mediate cellular responses to EA-induced mechanical stimuli. The interaction between connexin 43 and YAP has also been shown to promote the nuclear translocation of YAP, leading to the activation of astrocytes and mitigating the damage caused by ischemic brain injury [59–61]. These findings suggest that YAP not only aids in tissue recovery by reducing apoptosis, but also modulates the inflammatory response, thereby contributing to the overall neuroprotective effects of EA. The ability of EA to stimulate the Hippo/YAP pathway highlights its potential as a therapeutic intervention for enhancing neuroplasticity and promoting recovery in neurological diseases. By targeting key molecular players involved in cell survival and regeneration, EA offers a promising approach to improving clinical outcomes in patients suffering from ischemic stroke and potentially other neurodegenerative conditions [38].

4.4. Hyperthermia/Cold Therapy and YAP

Hyperthermia and cold therapy are common physical therapies used to treat various rehabilitative conditions, including musculoskeletal and neurological disorders. Recent research has highlighted the involvement of Hippo/YAP signaling pathways in mediating the effects of these thermal interventions on cellular processes [46,46]. Hyperthermia, which involves the application of heat to tissues, has been shown to activate YAP by promoting its nuclear translocation. During heat stress, YAP desphosphorylates and migrates to the nucleus, where it regulates the expression of genes involved in cell survival and tissue repair. This activation of YAP enhances the heat shock response, aiding in the recovery of damaged tissue and promoting cellular regeneration [38,46,46,47]. For instance, the heat-induced activation of YAP contributes to osteogenic differentiation in ectomesenchymal stem cells, enhancing bone regeneration and repair [38,48]. Cold therapy, on the other hand, has been found to modulate YAP activity in a different manner. Exposure to cold stress increases YAP expression in certain tissues, such as the brain and adipose tissue, where it plays a crucial role in maintaining cellular homeostasis. In the context of neural development, YAP has been implicated in the cold shock response by stabilizing its expression through interactions with RNA-binding proteins like RBM3 [49]. This upregulation promotes the proliferation and differentiation of neural stem cells [50], which is essential for maintaining brain health during cold stress. Furthermore, YAP collaborates with TEAD to regulate thermogenic activity in brown adipose tissue, contributing to adaptive responses to cold exposure by promoting the expression of thermogenic genes like UCP1 [38,50,51]. These findings illustrate the dual role of YAP in the Hippo/YAP signaling pathway in response to thermal therapies, highlighting its versatility in regulating cellular adaptations to environmental changes and its potential as a target for therapeutic interventions.

4.5. Shock Waves and YAP

Shock wave therapy, in particular low-intensity extracorporeal shock wave therapy (LIESWT), is an emerging physical therapy that has demonstrated potential in promoting tissue repair and regeneration. The mechanical stimuli generated by shock waves have been found to impact the Hippo/YAP signaling pathway. LIESWT induces mechanical stress in targeted tissues, which subsequently activates YAP by promoting its nuclear translocation and increasing its transcriptional activity [54]. This activation of YAP plays a key role in regulating cell proliferation, differentiation, and tissue repair mechanisms [54]. In models of sciatic nerve injury, LIESWT has been shown to upregulate YAP expression and enhance its nuclear translocation, thereby promoting Schwann cell activation and facilitating nerve regeneration. The activation of YAP by LIESWT acts as an integrated transcriptional complex that modifies gene expression to support cellular repair and regeneration. Additionally, YAP has been implicated in the enhancement of angiogenesis and the recovery of injured peripheral nerves, highlighting its importance in the context of neuro-regenerative therapy. The modulation of YAP activity through shock wave-induced mechanical stimuli can provide a promising approach for enhancing tissue repair and functional recovery in various clinical applications [33,54,55]. A recent study by Pirri et al. [56] has provided further insights into the mechanobiology of YAP in response to shock wave therapy. They investigated the presence and activation of YAP in deep fascia and demonstrated that YAP activation in fascial fibroblasts increased after focal extracorporeal shock waves (fESW) treatment. This activation was associated with an upregulation of ECM components, such as type I collagen (COL1A1) and hyaluronan-binding protein 2 (HABP2), indicating that YAP is integral to the remodeling and regenerative processes of the deep/muscular fascia [56]. These findings suggest that YAP activation through mechanical cues, such as shock wave therapy, can contribute to the repair and regeneration of musculoskeletal tissues by regulating ECM composition and fascial fibroblast activity [56].

4.6. Exercise and YAP

Therapeutic exercise is a cornerstone of rehabilitation, involving a range of activities designed to enhance muscle strength, flexibility, endurance, and overall functional capacity.

Recent research has demonstrated that during resistance training and aerobic exercises, mechanical signals generated by muscle contraction and stretching activate YAP, which translocates to the nucleus and promotes the expression of genes related to muscle growth, repair, and metabolism [40]. This process is essential for muscle hypertrophy, as well as for maintaining muscle mass during prolonged physical activity. YAP activation has also been implicated in improving mitochondrial function and metabolic regulation during endurance exercises. By enhancing mitochondrial biogenesis and promoting oxidative metabolism, YAP helps to improve muscular endurance and energy efficiency, which is beneficial in patients undergoing cardiovascular rehabilitation [41]. Moreover, studies suggest that therapeutic exercise that applies moderate to high levels of mechanical loading can enhance bone density by activating YAP in osteoblasts, promoting osteogenic differentiation and increasing bone mineral content [39]. In addition, therapeutic exercise has been shown to mitigate muscle wasting and sarcopenia through YAP-mediated pathways. In aging populations, resistance training increases YAP activity, which is associated with the activation of satellite cells and the prevention of muscle atrophy [38]. This highlights the importance of incorporating strength training exercises into rehabilitation programs for older adults, to maintain muscle mass and function [38,41]. Overall, the YAP signaling pathway is an important mediator of the adaptive responses to therapeutic exercise. By understanding how YAP contributes to muscle and bone health, as well as metabolic improvements, physiatrists can design more effective exercise programs that are tailored to the specific needs of patients, in order to optimize their recovery and functional outcomes.

4.7. Manual Therapy and YAP

Manual therapy, including joint mobilization, myofascial release, and spinal manipulation, etc., engages the Hippo/YAP signaling pathway to promote tissue healing and functional improvement. The mechanical forces applied during manual therapy activate YAP in a variety of cell types, including fibroblasts, chondrocytes, and intervertebral disc cells. This activation leads to several beneficial responses, such as enhanced ECM production, fibroblast proliferation, reduced apoptosis, and improved cellular resilience [40]. In particular, YAP activation has been shown to stimulate the production of key ECM components like collagen and hyaluronan, which are critical for maintaining the structural integrity of connective tissues and supporting tissue regeneration [38,40].

In intervertebral disc cells, manual therapy-induced YAP activation has been linked to reduced cell death and enhanced disc matrix synthesis, suggesting its role in mitigating degenerative disc disease. Moreover, YAP activation enhances the expression of fibrosis-related genes, including COL1A1 and HABP2 [40]. Furthermore, YAP signaling contributes to the modulation of inflammation by reducing pro-inflammatory cytokine production, thereby facilitating an environment conducive to tissue repair and pain reduction. This modulation occurs through pathways involving RhoA and ROCK, which are downstream effectors of YAP that influence cytoskeletal dynamics and cellular migration [38]. The role of YAP in mechano-transduction highlights how manual therapy can optimize tissue remodeling processes, enhance musculoskeletal function, and improve overall mobility. Understanding the molecular basis of these interventions provides valuable insight into how manual therapies can be tailored to maximize regenerative outcomes, offering a mechanistic foundation for the development of more targeted rehabilitation protocols.

5. Discussion

The Hippo/YAP signaling pathway plays a central role in mediating the effects of various rehabilitation interventions, spanning musculoskeletal, neurological, cardiovascular, and metabolic health. By understanding how YAP functions in response to different stimuli—including mechanical stress, thermal change, therapeutic exercise, and manual therapy, etc.—researchers and clinicians can create more targeted and effective rehabilitation protocols to promote tissue repair and functional recovery. In musculoskeletal rehabilitation, mechanical stimuli such as those generated through resistance exercise, ultrasound, vibration, and manual therapy, etc., activate YAP to regulate cell proliferation, differentiation, and tissue remodeling. The activation of YAP in chondrocytes, osteoblasts, fascial fibroblasts, and satellite cells has significant implications for cartilage health, bone regeneration, and muscle repair. For instance, YAP-mediated osteogenic differentiation in response to mechanical loading has the potential to accelerate bone healing, which is critical in patients recovering from fractures or osteoporosis. Similarly, the role of YAP in muscle hypertrophy and regeneration, in particular through satellite cell activation, underscores its importance in the rehabilitation of musculoskeletal injuries and preventing age-related sarcopenia [32–39,42–45,52,53].

Therapeutic exercise has emerged as a cornerstone of rehabilitation due to its ability to activate YAP and promote beneficial adaptations in muscle and bone tissues. By enhancing mitochondrial function, improving metabolic regulation, and driving anabolic processes, YAP activation supports the overall health and resilience of the musculoskeletal system [38,40,41]. These benefits are pronounced in older adults, where resistance training can mitigate muscle atrophy and improve functional capacity. The evidence suggests that tailoring exercise programs to optimize YAP activation can significantly enhance outcomes in patients undergoing rehabilitation for musculoskeletal disorders [38,40,41].

The role of YAP in vascular homeostasis further highlights its importance beyond musculoskeletal rehabilitation. YAP's response to shear stress and its involvement in endothelial cell proliferation, nitric oxide production, and angiogenesis is crucial in maintaining vascular health and promoting recovery form cardiovascular conditions. YAP's ability to regulate endothelial-to-mesenchymal transition (EndMT) also plays a pivotal role in preventing pathological fibrosis and maintaining vascular elasticity, which is essential for reducing cardiovascular risk factors and improving blood flow [38]. By modulating EndMT, YAP helps maintain endothelial cell integrity and prevents an excessive fibrotic response that can lead to vascular stiffening and reduced elasticity. Rehabilitation interventions that enhance YAP activity in endothelial cells may therefore have therapeutic potential in conditions such as atherosclerosis and peripheral artery disease. Moreover, YAP plays a role in promoting angiogenesis, which is essential for forming new blood vessels during tissue repair and in response to exercise [38]. Angiogenesis is a critical process, not only in wound healing but also in improving blood flow to ischemic tissue, which is relevant for patients with peripheral artery disease or other vascular impairments. YAP activation has been linked to the upregulation of vascular endothelial growth factor (VEGF) [38].

Thermal therapies, including hyperthermia and cold treatments, provide additional avenues for modulating YAP activity. The heat-induced activation of YAP can enhance tissue repair by promoting stem cell differentiation towards osteogenesis, while cold therapy's ability to reduce inflammation through YAP modulation highlights its role in creating a favorable environment for healing [46,46–51]. The dynamic application of heat and cold therapies, tailored to the specific needs of the patient, can provide synergistic benefits by leveraging the YAP pathway to optimize tissue repair, reduce pain, and improve overall recovery [46,46–51].

Manual therapy, including joint mobilization, fascial manipulation, spinal manipulation, and myofascial release, etc., also engages the Hippo/YAP signaling pathway to promote tissue healing and functional improvement. Mechanical forces applied during manual therapy can activate YAP in various cell types, leading to enhanced extracellular matrix production, reduced apoptosis, and tissue remodeling. Understanding the molecular basis of manual therapy provides valuable insights into how these interventions can be optimized to promote tissue regeneration, reduce pain, and improve mobility [40].

Finally, YAP's role in inflammation is complex, as it can act as both a pro- and anti-inflammatory mediator depending on the context [37]. In musculoskeletal rehabilitation, YAP's influence on macrophage polarization suggests that targeting YAP could help balance inflammation and tissue regeneration, providing new opportunities for improving rehabilitation outcomes [37]. Similarly, in vascular rehabilitation, YAP's interaction with NF-kb signaling may be leveraged to prevent vascular inflammation and enhance tissue repair in conditions such as atherosclerosis. Understanding these dual roles of YAP will be crucial for developing targeted interventions that can maximize its therapeutic benefits while minimizing potential adverse effects.

Future Perspectives in Rehabilitation

The rehabilitative potential of YAP-mediated mechanisms represents an exciting frontier in the field of rehabilitation. While existing research has laid a strong foundation, there remain several critical areas for investigation that could significantly enhance the rehabilitative applicability of these mechanisms: (1) personalized rehabilitation protocols; (2) the integration of innovative technologies; (3) rehabilitative validation and long-term safety; (4) collaborative and interdisciplinary research; and (5) exploring synergistic effects with other rehabilitation approaches.

1. Personalized rehabilitation protocols: rehabilitation outcomes often vary widely among individuals due to a range of factors, including age, underlying comorbidities, genetic predispositions, and even the biomechanical environment of the affected tissue. Developing personalized protocols that leverage YAP activation could transform rehabilitation practices by optimizing interventions for specific patient populations. For example, the mechanical parameters of LIPUS or vibration therapy—such as intensity, frequency, and duration—could be fine-tuned based on individual biological responses. Furthermore, understanding how systemic conditions (e.g., diabetes or osteoporosis, etc.) influence YAP activation may allow for the customization of therapeutic approaches, particularly for populations with impaired healing capabilities. This personalized approach could enhance efficacy, reduce recovery times, and minimize the risk of adverse effects.
2. Integration of innovative technologies: advances in wearable technology, machine learning, and bioengineering present unparalleled opportunities to enhance the monitoring and optimization of rehabilitation strategies. Wearable devices equipped with biomechanical sensors could track real-time data in the mechanical stimuli applied during therapy and correlate these with patient-specific YAP-related outcomes. Machine learning algorithms could analyze large datasets to identify the most effective rehabilitation strategies, offering predictive insights and enabling the continuous optimization of therapy. Additionally, technologies such as 3D bioprinting and tissue-engineered scaffolds could be designed to mimic the specific mechanical properties known to activate YAP, providing new avenues for tissue regeneration and repair in challenging cases, such as chronic injuries or post-surgical recovery.
3. Clinical validation and long-term safety: while preclinical studies have demonstrated the potential of YAP-targeted interventions, translating these findings into routine clinical practice remains a challenge. Future research should prioritize well-designed, multicenter clinical trials to validate the safety, efficacy, and scalability of therapies such as LIPUS, vibration therapy, and manual therapy, etc. Special attention should be paid to identifying optimal dosing regimens and understanding the limitation of these techniques in various patient populations. Additionally, longitudinal studies

are essential to investigate the long-term effects of sustained YAP activation. Questions regarding potential risks, such as fibrosis, hypertrophy, or other maladaptive responses must be addressed to ensure that these interventions are both effective and safe over extended periods.

4. Collaborative and interdisciplinary research: given the complexity of YAP-mediated mechanisms, interdisciplinary collaboration between researchers in biomechanics, molecular biology, bioengineering, and clinical rehabilitation is crucial. Such collaborations could drive the development of integrative models that account for the diverse roles of YAP in cellular adaptation, tissue repair, and functional recovery. Moreover, fostering partnerships between academia, industry, and healthcare providers may accelerate the translation of laboratory findings into practical, accessible solutions for patients.
5. Exploring the synergistic effects between different rehabilitation approaches: an exciting area of future research lies in investigating the potential synergy between YAP-targeted therapies and other established rehabilitation approaches, such as exercise-based therapy, pharmacological targets, or regenerative medicine techniques. For instance, combining mechanical therapies like LIPUS with growth factor delivery or stem cell therapies may enhance YAP activation and its downstream effects, leading to improved outcomes in tissue repair and regeneration. Similarly, understanding how YAP-mediated mechano-transduction interacts with systemic therapies for inflammation or oxidative stress could reveal new avenues for integrative treatment strategies.

By addressing these future directions, the field of rehabilitation stands poised to unlock the full potential of YAP-mediated mechanisms. The integration of personalized protocols, cutting-edge technologies, and rigorous clinical validation could not only improve patient outcomes, but also establish a new paradigm for evidence-based rehabilitation strategies.

6. Conclusions

YAP, with its signaling pathways, in particular Hippo/YAP, represents a pivotal mechanism through which various interventions used in physical and rehabilitation medicine exert their therapeutic effects. By acting as a mediator of mechanical stimuli, YAP plays a fundamental role in regulating tissue regeneration, reducing inflammation, and promoting functional recovery across different rehabilitation modalities. The insights gained from understanding YAP's role in musculoskeletal health, neurological health, and vascular homeostasis offer significant opportunities for enhancing rehabilitation outcomes. Future research should continue to explore how the targeted activation or inhibition of YAP can be leveraged to maximize the benefits of physical and rehabilitation medicine, leading to more effective and personalized rehabilitation strategies for patients with diverse medical conditions.

Funding: This research received no external funding.

Institutional Review Board Statement: Not applicable.

Informed Consent Statement: Not applicable.

Data Availability Statement: No new data were created or analyzed in this study.

Conflicts of Interest: The author declares no conflicts of interest.

References

1. Pan, D. The hippo signaling pathway in development and cancer. *Dev. Cell* **2010**, *19*, 491–505. [CrossRef] [PubMed] [PubMed Central]

2. Zhao, B.; Tumaneng, K.; Guan, K.L. The Hippo pathway in organ size control, tissue regeneration and stem cell self-renewal. *Nat. Cell Biol.* **2011**, *13*, 877–883. [CrossRef] [PubMed] [PubMed Central]
3. Meng, Z.; Moroishi, T.; Guan, K.L. Mechanisms of Hippo pathway regulation. *Genes Dev.* **2016**, *30*, 1–17. [CrossRef] [PubMed] [PubMed Central]
4. Dupont, S.; Morsut, L.; Aragona, M.; Enzo, E.; Giulitti, S.; Cordenonsi, M.; Zanconato, F.; Le Digabel, J.; Forcato, M.; Bicciato, S.; et al. Role of YAP/TAZ in mechanotransduction. *Nature* **2011**, *474*, 179–183. [CrossRef] [PubMed]
5. Aragona, M.; Panciera, T.; Manfrin, A.; Giulitti, S.; Michielin, F.; Elvassore, N.; Dupont, S.; Piccolo, S. A mechanical checkpoint controls multicellular growth through YAP/TAZ regulation by actin-processing factors. *Cell* **2013**, *154*, 1047–1059. [CrossRef] [PubMed]
6. Kim, E.; Riehl, B.D.; Bouzid, T.; Yang, R.; Duan, B.; Donahue, H.J.; Lim, J.Y. YAP mechanotransduction under cyclic mechanical stretch loading for mesenchymal stem cell osteogenesis is regulated by ROCK. *Front. Bioeng. Biotechnol.* **2024**, *11*, 1306002. [CrossRef] [PubMed] [PubMed Central]
7. Gabriel, B.M.; Hamilton, D.L.; Tremblay, A.M.; Wackerhage, H. The Hippo signal transduction network for exercise physiologists. *J. Appl. Physiol.* **2016**, *120*, 1105–1117. [CrossRef] [PubMed] [PubMed Central]
8. Ishikura, H. Effects of local vibration stimulation on muscle recovery and hypertrophy. *J. Phys. Ther. Sci.* **2024**, *36*, 441–446. [CrossRef] [PubMed] [PubMed Central]
9. Piccolo, S.; Dupont, S.; Cordenonsi, M. The biology of YAP/TAZ: Hippo signaling and beyond. *Physiol. Rev.* **2014**, *94*, 1287–1312. [CrossRef] [PubMed]
10. Yu, F.X.; Guan, K.L. The Hippo pathway: Regulators and regulations. *Genes Dev.* **2013**, *27*, 355–371. [CrossRef] [PubMed] [PubMed Central]
11. Liu, A.M.; Xu, Z.; Luk, J.M. An update on targeting Hippo-YAP signaling in liver cancer. *Expert Opin. Ther. Targets* **2012**, *16*, 243–247. [CrossRef] [PubMed]
12. Panciera, T.; Azzolin, L.; Cordenonsi, M.; Piccolo, S. Mechanobiology of YAP and TAZ in physiology and disease. *Nat. Rev. Mol. Cell Biol.* **2017**, *18*, 758–770. [CrossRef] [PubMed] [PubMed Central]
13. Zhao, B.; Li, L.; Lei, Q.; Guan, K.L. The Hippo-YAP pathway in organ size control and tumorigenesis: An updated version. *Genes Dev.* **2010**, *24*, 862–874. [CrossRef] [PubMed] [PubMed Central]
14. Totaro, A.; Panciera, T.; Piccolo, S. YAP/TAZ upstream signals and downstream responses. *Nat. Cell Biol.* **2018**, *20*, 888–899. [CrossRef] [PubMed] [PubMed Central]
15. Codelia, V.A.; Sun, G.; Irvine, K.D. Regulation of YAP by mechanical strain through Jnk and Hippo signaling. *Curr. Biol.* **2014**, *24*, 2012–2017. [CrossRef] [PubMed] [PubMed Central]
16. Varelas, X.; Sakuma, R.; Samavarchi-Tehrani, P.; Peerani, R.; Rao, B.M.; Dembowy, J.; Yaffe, M.B.; Zandstra, P.W.; Wrana, J.L. TAZ controls Smad nucleocytoplasmic shuttling and regulates human embryonic stem-cell self-renewal. *Nat. Cell Biol.* **2008**, *10*, 837–848. [CrossRef] [PubMed]
17. Yu, F.X.; Zhao, B.; Guan, K.L. Hippo Pathway in Organ Size Control, Tissue Homeostasis, and Cancer. *Cell* **2015**, *163*, 811–828. [CrossRef] [PubMed] [PubMed Central]
18. Zanconato, F.; Cordenonsi, M.; Piccolo, S. YAP/TAZ at the Roots of Cancer. *Cancer Cell* **2016**, *29*, 783–803. [CrossRef] [PubMed] [PubMed Central]
19. Franklin, J.M.; Wu, Z.; Guan, K.L. Insights into recent findings and clinical application of YAP and TAZ in cancer. *Nat. Rev. Cancer* **2023**, *23*, 512–525. [CrossRef] [PubMed]
20. Mao, W.; Mai, J.; Peng, H.; Wan, J.; Sun, T. YAP in pancreatic cancer: Oncogenic role and therapeutic strategy. *Theranostics* **2021**, *11*, 1753–1762. [CrossRef] [PubMed] [PubMed Central]
21. Fan, W.; Adebowale, K.; Váncza, L.; Li, Y.; Rabbi, M.F.; Kunimoto, K.; Chen, D.; Mozes, G.; Chiu, D.K.; Li, Y.; et al. Matrix viscoelasticity promotes liver cancer progression in the pre-cirrhotic liver. *Nature* **2024**, *626*, 635–642. [CrossRef] [PubMed] [PubMed Central]
22. Fan, S.; Gao, Y.; Qu, A.; Jiang, Y.; Li, H.; Xie, G.; Yao, X.; Yang, X.; Zhu, S.; Yagai, T.; et al. YAP-TEAD mediates PPAR α-induced hepatomegaly and liver regeneration in mice. *Hepatology* **2022**, *75*, 74–88. [CrossRef] [PubMed] [PubMed Central]
23. Iimuro, Y.; Kondo, Y.; Suzumura, K.; Uyama, N.; Asano, Y.; Hirano, T.; Yamanaka, J.; Iijima, H.; Nishiguchi, S.; Fujimoto, J. Regional hepatic regeneration after liver resection correlates well with preceding changes in the regional portal circulation in humans. *Dig. Dis. Sci.* **2013**, *58*, 3001–3009. [CrossRef] [PubMed]
24. Christ, B.; Collatz, M.; Dahmen, U.; Herrmann, K.H.; Höpfl, S.; König, M.; Lambers, L.; Marz, M.; Meyer, D.; Radde, N.; et al. Hepatectomy-Induced Alterations in Hepatic Perfusion and Function—Toward Multi-Scale Computational Modeling for a Better Prediction of Post-hepatectomy Liver Function. *Front. Physiol.* **2021**, *12*, 733868. [CrossRef] [PubMed] [PubMed Central]
25. Mascharak, S.; desJardins-Park, H.E.; Davitt, M.F.; Griffin, M.; Borrelli, M.R.; Moore, A.L.; Chen, K.; Duoto, B.; Chinta, M.; Foster, D.S.; et al. Preventing *Engrailed*-1 activation in fibroblasts yields wound regeneration without scarring. *Science* **2021**, *372*, eaba2374. [CrossRef] [PubMed] [PubMed Central]

26. Yang, K.; Wu, Y.; Cheng, P.; Zhang, J.; Yang, C.; Pi, B.; Ye, Y.; You, H.; Chen, A.; Xu, T.; et al. YAP and ERK mediated mechanical strain-induced cell cycle progression through RhoA and cytoskeletal dynamics in rat growth plate chondrocytes. *J. Orthop. Res.* **2016**, *34*, 1121–1129. [CrossRef] [PubMed]
27. Li, L.; Li, H.; He, Y.; Tang, H.; Dong, J.; Chen, X.; Lyu, F.; Dong, Y. Cyclic pulsation stress promotes bone formation of tissue engineered laminae through the F-actin/YAP-1/β-Catenin signaling axis. *NPJ Regen. Med.* **2021**, *6*, 51. [CrossRef] [PubMed] [PubMed Central]
28. Niu, N.; Xu, S.; Xu, Y.; Little, P.J.; Jin, Z.G. Targeting Mechanosensitive Transcription Factors in Atherosclerosis. *Trends Pharmacol. Sci.* **2019**, *40*, 253–266. [CrossRef] [PubMed] [PubMed Central]
29. Abe, J.; Berk, B.C. Novel mechanisms of endothelial mechanotransduction. *Arterioscler. Thromb. Vasc. Biol.* **2014**, *34*, 2378–2386. [CrossRef] [PubMed] [PubMed Central]
30. Wang, L.; Luo, J.Y.; Li, B.; Tian, X.Y.; Chen, L.J.; Huang, Y.; Liu, J.; Deng, D.; Lau, C.W.; Wan, S.; et al. Integrin-YAP/TAZ-JNK cascade mediates atheroprotective effect of unidirectional shear flow. *Nature* **2016**, *540*, 579–582. [CrossRef] [PubMed]
31. Li, B.; He, J.; Lv, H.; Liu, Y.; Lv, X.; Zhang, C.; Zhu, Y.; Ai, D. c-Abl regulates YAPY357 phosphorylation to activate endothelial atherogenic responses to disturbed flow. *J. Clin. Investig.* **2019**, *129*, 1167–1179. [CrossRef] [PubMed] [PubMed Central]
32. Zhong, W.; Li, Y.; Li, L.; Zhang, W.; Wang, S.; Zheng, X. YAP-mediated regulation of the chondrogenic phenotype in response to matrix elasticity. *J. Mol. Histol.* **2013**, *44*, 587–595. [CrossRef] [PubMed]
33. de Lucas, B.; Pérez, L.M.; Bernal, A.; Gálvez, B.G. Ultrasound Therapy: Experiences and Perspectives for Regenerative Medicine. *Genes* **2020**, *11*, 1086. [CrossRef] [PubMed] [PubMed Central]
34. Steppe, L.; Liedert, A.; Ignatius, A.; Haffner-Luntzer, M. Influence of Low-Magnitude High-Frequency Vibration on Bone Cells and Bone Regeneration. *Front. Bioeng. Biotechnol.* **2020**, *8*, 595139. [CrossRef] [PubMed] [PubMed Central]
35. McKeehen, J.N.; Novotny, S.A.; Baltgalvis, K.A.; Call, J.A.; Nuckley, D.J.; Lowe, D.A. Adaptations of mouse skeletal muscle to low-intensity vibration training. *Med. Sci. Sports Exerc.* **2013**, *45*, 1051–1059. [CrossRef] [PubMed] [PubMed Central]
36. Li, Y.; Xiao, C.; Li, R.; Zhong, W.; Xu, G.; Zhang, W. Role of Yes-associated protein (YAP) in regulation of mesenchymal stem cell tenogenic differentiation. *J. Mol. Histol.* **2022**, *53*, 273–283. [CrossRef] [PubMed]
37. Chen, L.; Jin, X.; Ma, J.; Xiang, B.; Li, X. YAP at the progression of inflammation. *Front. Cell Dev. Biol.* **2023**, *11*, 1204033. [CrossRef] [PubMed] [PubMed Central]
38. Pan, C.; Lu, F.; Hao, X.; Deng, X.; Liu, J.; Sun, K.; Hou, W.; Shang, X.; Chi, R.; Guo, F.; et al. Low-intensity pulsed ultrasound delays the progression of osteoarthritis by regulating the YAP-RIPK1-NF-κB axis and influencing autophagy. *J. Transl. Med.* **2024**, *22*, 286. [CrossRef] [PubMed] [PubMed Central]
39. Rubin, C.; Turner, A.S.; Müller, R.; Mittra, E.; McLeod, K.; Lin, W.; Qin, Y.X. Quantity and quality of trabecular bone in the femur are enhanced by a strongly anabolic, noninvasive mechanical intervention. *J. Bone Miner. Res.* **2002**, *17*, 349–357. [CrossRef] [PubMed]
40. Zhu, Z.; Tang, T.; He, Z.; Wang, F.; Chen, H.; Chen, G.; Zhou, J.; Liu, S.; Wang, J.; Tian, W.; et al. Uniaxial cyclic stretch enhances osteogenic differentiation of OPLL-derived primary cells via YAP-Wnt/β-catenin axis. *Eur. Cell Mater.* **2023**, *45*, 31–45. [CrossRef] [PubMed]
41. Xi, Y.; Li, Y.; Ren, W.; Bo, W.; Ma, Y.; Pan, S.; Gong, D.W.; Tian, Z. ELABELA-APJ-Akt/YAP Signaling Axis: A Novel Mechanism of Aerobic Exercise in Cardioprotection of Myocardial Infarction Rats. *Med. Sci. Sports Exerc.* **2023**, *55*, 1172–1183. [CrossRef] [PubMed]
42. Elvey, M.H.; Miller, R.; Khor, K.S.; Protopapa, E.; Horwitz, M.D.; Hunter, A.R. The use of low-intensity pulsed ultrasound in hand and wrist nonunions. *J. Plast. Surg. Hand Surg.* **2020**, *54*, 101–106. [CrossRef] [PubMed]
43. Lin, G.; Reed-Maldonado, A.B.; Lin, M.; Xin, Z.; Lue, T.F. Effects and Mechanisms of Low-Intensity Pulsed Ultrasound for Chronic Prostatitis and Chronic Pelvic Pain Syndrome. *Int. J. Mol. Sci.* **2016**, *17*, 1057. [CrossRef] [PubMed] [PubMed Central]
44. Pan, C.; Hao, X.; Deng, X.; Lu, F.; Liu, J.; Hou, W.; Xu, T. The roles of Hippo/YAP signaling pathway in physical therapy. *Cell Death Discov.* **2024**, *10*, 197. [CrossRef] [PubMed] [PubMed Central]
45. Wang, X.; Hu, G.; Gao, X.; Wang, Y.; Zhang, W.; Harmon, E.Y.; Zhi, X.; Xu, Z.; Lennartz, M.R.; Barroso, M.; et al. The induction of yes-associated protein expression after arterial injury is crucial for smooth muscle phenotypic modulation and neointima formation. *Arterioscler. Thromb. Vasc. Biol.* **2012**, *32*, 2662–2669. [CrossRef] [PubMed] [PubMed Central]
46. Luo, M.; Meng, Z.; Moroishi, T.; Lin, K.C.; Shen, G.; Mo, F.; Shao, B.; Wei, X.; Zhang, P.; Wei, Y.; et al. Publisher Correction: Heat stress activates YAP/TAZ to induce the heat shock transcriptome. *Nat. Cell Biol.* **2021**, *23*, 209, Erratum for: *Nat. Cell Biol.* **2020**, *22*, 1447–1459. [CrossRef] [PubMed]
47. Iwasa, H.; Maimaiti, S.; Kuroyanagi, H.; Kawano, S.; Inami, K.; Timalsina, S.; Ikeda, M.; Nakagawa, K.; Hata, Y. Yes-associated protein homolog, YAP-1, is involved in the thermotolerance and aging in the nematode Caenorhabditis elegans. *Exp. Cell Res.* **2013**, *319*, 931–945. [CrossRef] [PubMed]

48. Shi, W.; Wang, Z.; Bian, L.; Wu, Y.; HuiYa, M.; Zhou, Y.; Zhang, Z.; Wang, Q.; Zhao, P.; Lu, X. Periodic Heat Stress Licenses EMSC Differentiation into Osteoblasts via YAP Signaling Pathway Activation. *Stem Cells Int.* **2022**, *2022*, 3715471. [CrossRef] [PubMed] [PubMed Central]
49. Xia, W.; Su, L.; Jiao, J. Cold-induced protein RBM3 orchestrates neurogenesis via modulating Yap mRNA stability in cold stress. *J. Cell Biol.* **2018**, *217*, 3464–3479. [CrossRef] [PubMed] [PubMed Central]
50. Ye, C.; Duan, J.; Zhang, X.; Yao, L.; Song, Y.; Wang, G.; Li, Q.; Wang, B.; Ai, D.; Wang, C.; et al. Cold-induced Yes-associated-protein expression through miR-429 mediates the browning of white adipose tissue. *Sci. China Life Sci.* **2021**, *64*, 404–418. [CrossRef] [PubMed]
51. Tharp, K.M.; Kang, M.S.; Timblin, G.A.; Dempersmier, J.; Dempsey, G.E.; Zushin, P.H.; Benavides, J.; Choi, C.; Li, C.X.; Jha, A.K.; et al. Actomyosin-Mediated Tension Orchestrates Uncoupled Respiration in Adipose Tissues. *Cell Metab.* **2018**, *27*, 602–615.e4. [CrossRef] [PubMed] [PubMed Central]
52. Mettlach, G.; Polo-Parada, L.; Peca, L.; Rubin, C.T.; Plattner, F.; Bibb, J.A. Enhancement of neuromuscular dynamics and strength behavior using extremely low magnitude mechanical signals in mice. *J. Biomech.* **2014**, *47*, 162–167. [CrossRef] [PubMed] [PubMed Central]
53. Zhu, Z.; He, Z.; Tang, T.; Wang, F.; Chen, H.; Zhou, J.; Lin, C.; Chen, G.; Wang, J.; Li, J.; et al. Effect of mechanical stimulation on tissue heterotopic ossification: An in vivo experimental study. *Front. Physiol.* **2023**, *14*, 1225898. [CrossRef] [PubMed] [PubMed Central]
54. Xiang, J.; Wang, W.; Jiang, W.; Qian, Q. Effects of extracorporeal shock wave therapy on spasticity in post-stroke patients: A systematic review and meta-analysis of randomized controlled trials. *J. Rehabil. Med.* **2018**, *50*, 852–859. [CrossRef] [PubMed]
55. Yang, W.; He, Y.; Gan, L.; Zhang, F.; Hua, B.; Yang, P.; Liu, J.; Yang, L.; Guo, T. Cardiac shock wave therapy promotes arteriogenesis of coronary micrangium, and ILK is involved in the biomechanical effects by proteomic analysis. *Sci. Rep.* **2018**, *8*, 1814. [CrossRef] [PubMed] [PubMed Central]
56. Pirri, C.; Caroccia, B.; Angelini, A.; Piazza, M.; Petrelli, L.; Caputo, I.; Montemurro, C.; Ruggieri, P.; De Caro, R.; Stecco, C. A New Player in the Mechanobiology of Deep Fascia: Yes-Associated Protein (YAP). *Int. J. Mol. Sci.* **2023**, *24*, 15389. [CrossRef] [PubMed] [PubMed Central]
57. Han, R.; Chang, J.; Liu, Q.; Liu, H.; Li, J. Efficacy of electroacupunture at Zusanli (ST36) on jumping-injured muscle based on transcriptome sequencing and genes analysis. *J. Tradit. Chin. Med.* **2023**, *43*, 322–328. [CrossRef] [PubMed] [PubMed Central]
58. Poitelon, Y.; Lopez-Anido, C.; Catignas, K.; Berti, C.; Palmisano, M.; Williamson, C.; Ameroso, D.; Abiko, K.; Hwang, Y.; Gregorieff, A.; et al. YAP and TAZ control peripheral myelination and the expression of laminin receptors in Schwann cells. *Nat. Neurosci.* **2016**, *19*, 879–887. [CrossRef] [PubMed] [PubMed Central]
59. Liu, R.; Xu, N.G.; Yi, W.; Ji, C. Electroacupuncture Attenuates Inflammation after Ischemic Stroke by Inhibiting NF-κB-Mediated Activation of Microglia. *Evid. Based Complement. Altern. Med.* **2020**, *2020*, 8163052. [CrossRef] [PubMed] [PubMed Central]
60. Huang, J.; You, X.; Liu, W.; Song, C.; Lin, X.; Zhang, X.; Tao, J.; Chen, L. Electroacupuncture ameliorating post-stroke cognitive impairments via inhibition of peri-infarct astroglial and microglial/macrophage P2 purinoceptors-mediated neuroinflammation and hyperplasia. *BMC Complement. Altern. Med.* **2017**, *17*, 480. [CrossRef] [PubMed] [PubMed Central]
61. Huang, L.; Li, S.; Dai, Q.; Zhang, A.; Yu, Q.; Du, W.; Zhao, P.; Mo, Y.; Xu, K.; Chen, S.; et al. Astrocytic Yes-associated protein attenuates cerebral ischemia-induced brain injury by regulating signal transducer and activator of transcription 3 signaling. *Exp. Neurol.* **2020**, *333*, 113431. [CrossRef] [PubMed]

biomolecules

MDPI

Article

Mouse Exploratory Behaviour in the Open Field with and without NAT-1 EEG Device: Effects of MK801 and Scopolamine

Charmaine J. M. Lim [1], Jack Bray [1], Sanna K. Janhunen [2], Bettina Platt [1] and Gernot Riedel [1,*]

[1] Institute of Medical Sciences, University of Aberdeen, Aberdeen AB25 2ZD, UK; limcharmainejm@hotmail.com (C.J.M.L.); r05jb17@abdn.ac.uk (J.B.); b.platt@abdn.ac.uk (B.P.)

[2] Organon R&D Finland, Itäinen Pitkäkatu 4, 20520 Turku, Finland; sannajanhunen@hotmail.com

* Correspondence: g.riedel@abdn.ac.uk

Abstract: One aspect of reproducibility in preclinical research that is frequently overlooked is the physical condition in which physiological, pharmacological, or behavioural recordings are conducted. In this study, the physical conditions of mice were altered through the attachments of wireless electrophysiological recording devices (Neural Activity Tracker-1, NAT-1). NAT-1 devices are miniaturised multichannel devices with onboard memory for direct high-resolution recording of brain activity for >48 h. Such devices may limit the mobility of animals and affect their behavioural performance due to the added weight (total weight of approximately 3.4 g). The mice were additionally treated with saline (control), N-methyl-D-aspartate (NMDA) receptor antagonist MK801 (0.85 mg/kg), or the muscarinic acetylcholine receptor blocker scopolamine (0.65 mg/kg) to allow exploration of the effect of NAT-1 attachments in pharmacologically treated mice. We found only minimal differences in behavioural outcomes with NAT-1 attachments in standard parameters of locomotor activity widely reported for the open field test between the drug treatments. Hypoactivity was globally observed as a consistent outcome in the MK801-treated mice and hyperactivity in scopolamine groups regardless of NAT-1 attachments. These data collectively confirm the reproducibility for combined behavioural, pharmacological, and physiological endpoints even in the presence of lightweight wireless data loggers. The NAT-1 therefore constitutes a pertinent tool for investigating brain activity in, e.g., drug discovery and models of neuropsychiatric and/or neurodegenerative diseases with minimal effects on pharmacological and behavioural outcomes.

Keywords: reproducibility; open field test; electroencephalography; wireless; MK801; scopolamine

1. Introduction

With advancements made in preclinical research, laboratories often align the mechanistic and functional aspects of animal behavioural outcomes with those of dynamic neuronal events. This is often performed with tethered or, more recently, wireless miniature recording systems [1–8] to allow recording of electroencephalography (EEG) in freely moving animals with minimal discomfort. Other devices and their properties have been reviewed recently [9]; there are a limited number of data logging devices with onboard memory (Deuteron mouse-log16 and TSE Neurologger) that are similar in weight to the NAT-1 (~3 g) but differ for applications and sampling rates. Most devices are battery-powered high-weight transmitters (Neuralynx, Pinnacle; up to 23 g) with continuous data download but limited battery life. Most of these devices are not applicable to mice and are used for the recording of single units or single cells. In our laboratory, the use of an EEG wireless device includes the bilateral implantation of gold electrodes into the skull and the attachment of a lightweight neural activity tracker (NAT-1; Cybula, York, UK) device to this head stage prior to behavioural testing. These devices have been employed for sleep-staging and global activity analysis during home cage recordings [7,8]. However, the surgical procedure and additional weight a mouse must tolerate amounts to about 3 g (multifunctional

NAT-1 device including batteries), in addition to a small head stage modelled from dental cement and screw recording electrodes. The impact on the animal's head could therefore conceivably limit locomotor skills and the mobility of the animal and affect its natural explorative behaviour, and consequently the outcomes of behavioural findings, thereby altering the apparent effects of pharmacological agents.

Compounds commonly used in preclinical in vivo research such as N-methyl-D-aspartate receptor (NMDAR) antagonist (dizocilpine; [+]-MK801) and muscarinic antagonist, scopolamine, were inoculated in mice in this study. Both compounds induce schizophrenia- [10–13] or Alzheimer-related [14,15] cognitive decline respectively and are known to induce well measurable and robust behavioural outcomes in mice including dose-dependent hyperactivity/hypoactivity [13–19]. Inclusion of prespecified doses of MK801 and scopolamine (unpublished preliminary data) in this study will allow opposing nonsystematic movements in mice to accurately represent the effects of wireless recording device attachments in animals.

This exploratory study aims to: (1) evaluate potential differences in behavioural outcomes in the open field test in mice with and without surgically implanted head stages/EEG recording devices (NAT-1); and (2) explore the behavioural effects of such attachments in pharmacologically treated mice. Here, the open field test was the selected behavioural paradigm as it produces sufficiently reliable and repeatable measures under standardised conditions. All exploratory behaviour obtainable from video-observed tests are presented to avoid bias; EEG recordings obtained in this study were not deemed relevant to changes in behavioural outcomes in mice and are therefore omitted. Generally, analyses of activity in the open field test confirm that despite very similar performance with or without EEG devices in vehicle-treated control mice, subtle differences appeared in the drug groups that may inform of some behavioural alterations that affect outcomes other than activity.

2. Methods

2.1. Experimental Design

The study design, analysis, and reporting methods in this study were conducted in line with recommendations stated in the ARRIVE 2.0 guidelines [20] and are detailed in the relevant segments below. The study design comprised of two independent experiments conducted on separate occasions. The testing days of the week, number of animals tested per day, and testing seasons were, however, similar between the studies. To ensure that any differences between Experiments 1 (no NAT-1) and 2 (with NAT-1) were the result of the attachment of wireless data logging devices but not that of variations in procedural protocols or testing environment, special care was taken to ensure that the laboratory conditions, husbandry, avoidance of bias, and study designs were maintained between both experiments (detailed in Table 1). Variables which differed between Experiments 1 and 2 are marked with asterisks in Table 1 and mainly pertained to the surgical procedures and NAT-1-related handling requirements performed in Experiment 2.

Table 1. Details pertaining to the study designs, experimental conditions, and husbandry in Experiments 1 and 2.

	Experiment 1	Experiment 2
Study Outcome	Pharmacological studies without EEG attachments	Pharmacological studies in animals with EEG attachments
Persons involved (Blinder and experimenter)	2	3
Animals		
Strain	C57BL/6J (Wildtype; total n = 29)	C57BL/6J (Wildtype; total n = 37; lost 10, see Section 2.3)
Supplier/ Breeder	Charles River Laboratories, Kent, UK	Charles River Laboratories, Kent, UK

Table 1. *Cont.*

	Experiment 1	Experiment 2
Transport	By road	By road
Specific pathogen free conditions	Yes	Yes
Age at arrival and acclimatisation to facility	10 weeks; 2 weeks of acclimatisation	9 weeks; 2 weeks of acclimatisation
Housing/ husbandry; test room conditions		
Cage type	Conventional shoe box cages with wire top feeders; single housing	Conventional shoe box cages with wire top feeders (only 1 was moved to IVC cage); single housing
Enrichment; bedding	Cardboard rolls in all cages; corncob and aspen shavings (ratio 3:1)	No enrichment; corncob and aspen shavings (ratio 3:1)
Temperature, humidity	21 ± 2 °C; 50 ± 5%	21 ± 2 °C; 50 ± 5%
Exposure of animal to handling	Two/three times a day by facility staff	Two/three times a day by facility staff
Noise/background noise	Background noise (music) during light phase	Background noise (music) during light phase
Dark/light cycle	12:12 h; start of light phase at 07:00, no red light during dark phase	12:12 h; start of light phase at 07:00, no red light during dark phase
Diet	Normal chow pellets (SDS CRM) and room-temperature water ad libitum	Normal chow pellets (SDS CRM) and room-temperature water ad libitum
Handling of cages	Cleaning once every 2 weeks; water and food changed/topped up twice a week.	Cleaning once every 2 weeks; water and food changed/topped up twice a week.
Experimental: animals		
Age of animal at start to experiment	12–13 weeks (at the start of testing)	10–11 weeks (at surgery) and 12–13 weeks (at the start of testing)
Transport of animals in facility	Silent trolley type; cages were not blacked out or covered during transport	Silent trolley type; cages were not blacked out or covered during transport
Time of testing, day of week	Commencement at 10:00; testing was conducted from Monday to Thursday on the first week, and Monday to Wednesday on the second week.	Commencement at 10:00; testing was conducted from Monday to Thursday on the first week, and Monday to Wednesday on the second week.
Total duration of test phase per day	Approx. 5 h (4 animals tested am, 3 tested pm)	Approx. 5 h (3 animals tested am, 3 tested pm)
Animal under diet restrictions during experimental phase	No	No, except during NAT-1 application and drug administration: water bottles were removed, and cage lids were overturned to prevent NAT devices from accidentally wedging onto cage-lids and food-hoppers
*** Naivety to experimental paradigm and drugs**	Naïve to both experimental procedures and drugs	Home cage EEG recording and check, 24 h prior to open field test; naïve to drugs
Experimental: surgical protocols		
*** Time period between surgeries and start of experiment**	N.A.	2 weeks
*** Total duration of surgery per day**	N.A.	Approx. 7 h/day (commencement at 08.00; surgeries performed only on Mondays and Tuesdays

Table 1. *Cont.*

	Experiment 1	Experiment 2
* Anaesthetic	N.A.	Gas anaesthesia (isoflurane) 1.0 mL saline (intraperitoneal) and 0.2 mL of 0.1 mg/kg buprenorphine (subcutaneous; Vetergesic®, Cera Animal Health Ltd., High Wycombe, UK).
* Analgesia	N.A.	Drinking water was supplemented with Rimadyl® (Zoetis UK Ltd., Surrey, UK) for 2 days before and after the surgery
* Surgical procedure	N.A.	1 surgeon; 2 assistants
Experimental: design and methods/data analysis		
Drug, and concentration	0.9% saline; 0.65 mg/kg MK801; 0.85 mg/kg scopolamine. Intra-peritoneal administration of 10 mg/kg of treatment.	0.9% saline; 0.65 mg/kg MK801; 0.85 mg/kg scopolamine. Intra-peritoneal administration of 10 mg/kg of treatment.
* Experimenter	1 female	2 females and 1 male
* Frequency of handling	Handling during drug injections, immediately before and after test.	Handling during NAT-1 applications, drug injections, immediately before and after test.
Randomisation method for allocation of animals to test groups	random.org (https://www.random.org/sequences/, accessed 29 October 2018; Williams square design.	random.org (https://www.random.org/sequences/, accessed 19 February 2018); Williams square design.
Blinding of experimenter to group allocation	Yes.	Yes.
Standardisation of animal handling across experimenters (protocol and training)	Yes (1 in randomisation and blinding, 1 in drug administration and experiment)	Yes (2 involved in NAT-1 device application and drug administration and randomisation, 1 in experiment)
Software used in recording and data extraction	Automatic tracking software, ANY-Maze (v.5.1)	Automatic tracking software, ANY-Maze (v.5.1)
Inclusion and exclusion criteria	No inclusion criteria. Exclusion criteria based on analysis using Grubb's method before re-evaluation of animal behaviour (video).	No inclusion criteria. Exclusion criteria based on analysis using Grubb's method before re-evaluation of animal behaviour (video).
Statistical analysis	Bootstrapped one-way ANOVA and *t*-test for conventional analysis; repeated measures ANOVA for binned analysis; and estimation statistics to determine the magnitude of differences.	Bootstrapped one-way ANOVA and *t*-test for conventional analysis; repeated measures ANOVA for binned analysis; and estimation statistics to determine the magnitude of differences.
Experimental: complexity of test arena		
Test arena	Square; 50 × 50 × 50 cm; white Perspex	Square; 50 × 50 × 50 cm; white Perspex
Average light intensity in test arena	320 lux	320 lux
Environment in test room (temperature, humidity and decibels relative to the carrier)	20 ± 2 °C; 48 ± 5%; 30 ± 5 dBc; dedicted sound-attenuated room	20 ± 2 °C; 48 ± 5%; 30 ± 5 dBc; dedicted sound-attenuated room
Number of test arenas used at any given time	1	1
Cleaning agents	Non-fragrant and alcohol-free wet wipes	Non-fragrant and alcohol-free wet wipes

* Inconsistent experimental factors which differed between experiments.

2.2. Animals

Twenty-nine C57BL/6J (10–11-week-old male mice) from Charles River Laboratories (Kent, England) were obtained for Experiment 1 and 37 for Experiment 2. The mice were housed individually in standard Macrolon shoebox cages (Techniplast 1145) measuring at 369 × 165 × 132 mm, under a 12-h light/12-h dark cycle (lights on at 7 am). Temperature and humidity were set at 20 ± 5 °C and 48 ± 5% in both the housing and experimental rooms. The animals had free access to standard rodent food chow (SDS 801190 RM3, Special Diet Services (SDS)/Augy/France) and water ad libitum. All the animals were acclimatized to the facility environment and extensively handled for 2 weeks prior to the start of the surgeries and/or open field testing. The mice were tail-handled to avoid detachment of the head stages and management of severe stereotypic responses during transfer. All housing and handling of the animals were in accordance with the international standards on animal welfare (EU/63/2010) and Home Office (UK) regulations. At the end of the study, all the animals were euthanised and the brain tissue harvested.

2.3. Stereotactic Surgery and Home-Cage Recordings (Experiment 2 Only)

The animals had free access to water supplemented with carprofen Rimadyl® (5 mg/kg; Zoetis, UK) and were acclimatised to soft diet (Transbreed (E), SDS 801222, England; and Solid Drink® SDST-75, 3at-bio, Tiel, The Netherlands) 48 h prior to and after surgery. Stereotaxic surgeries were conducted as described previously [3,7,8,21]. Briefly, anaesthesia was induced with 3% isoflurane (IsoFlo®, Zoetis) in medical-grade oxygen and maintained on 1.5% isoflurane during surgery. The mice were held in a stereotaxic frame (Stoelting, Wood Dale, IL, USA). Epidural gold-plated screw electrodes were placed at the following stereotaxic locations [22] on exposed skulls from the prefrontal cortex (2 mm anterior to bregma/close to midline), left and right hippocampus (2 mm posterior to bregma, 1.5 mm lateral to midline). Reference and ground electrodes were placed in a neutral location above the occipital plate. The electrodes were soldered and assembled into a 6-pin adaptor and fixed on to the skull by a mixture of Durelon® dental cement and tissue adhesive. The pins measured 0.40 mm (square) with a spacing pitch of 1.27 mm. Once the dental cement dried, the animal was removed from the stereotaxic instrument and injected with 1.0 mL saline (intraperitoneal) and 0.2 mL of 0.1 mg/kg buprenorphine (subcutaneous; Vetergesic®, Cera Animal Health Ltd., High Wycombe, UK). The animals were placed on heating pads immediately after the procedure, and further analgesic treatment continued for 2–3 days as appropriate. Following surgery, the animals were weighed daily to monitor their recovery and were provided with a soft diet. A minimum of 7 days was allowed for recovery before the start of testing. Ten animals were not entered into this study due to dislodged head caps and were euthanised. From these, 6 head caps were lost during restraining for drug administration or NAT-1 attachment/removal, while an additional 4 lost their head stages in their home cages presumably while getting stuck in the wire tops. We have since then modified this procedure and hold animals in individually ventilated cages now. After the open field testing, all the operated mice went on to be tested for home cage activity and were finally euthanised to confirm electrode placement (not relevant here).

2.4. Drugs and Drug Formulation

The drug and dose selection were in line with the EQIPD project (see funding statement), in which between-laboratory robustness of the data was proposed and both drugs were utilized for drug comparison [23], but also for EEG recordings between laboratories [24].

(+)-MK801 maleate (CAS: 77086-22-7) and scopolamine hydrobromide (CAS: 114-49-8) were obtained from Tocris Bioscience (Bristol, UK) and diluted to 0.65 mg/kg MK801 and 0.85 mg/kg scopolamine (salt-corrected for both drugs) in 0.9% saline 24 h prior to intraperitoneal injections. Saline was administered as vehicle control. Reagents were administered (10 mL/kg) as a single dose, with a sterile injection needle, 30 min prior to the behavourial testing. The assignment of all the mice to different drugs was randomised

(Experiment 1: n = 10 for saline, n = 9 for MK801, and n = 10 for scopolamine; Experiment 2: n = 9 per drug group) using a Williams Square design [25].

2.5. Apparatus and Testing Protocols

Upon behavioural testing in the open field arena, the mice were aged 12–13 weeks old and weighed approximately 26.8 ± 2.0 g. Intraperitoneal administration of drugs and testing sequence were conducted in a blinded manner. Common to both experiments, the mice were habituated to the test room for 30 min, following drug inoculations. In Experiment 2, the mice were drug-treated immediately before the attachment of the NAT-1 devices. NAT-1 measures 18 × 22.2 × 10.3 mm and weighs approximately 2.2 g including a zinc battery (as). The total weight of the NAT-1 and head stage (ranging between 1.1 and 1.5 g) equals approximately 3.4 g (Figure S1). The attachment of the wireless NAT-1 device to the head stage involved temporarily restraining the mice between two pieces of faux fur fabric. This restraint is necessary to minimise movement and to ensure the secure placement of the device with minimal discomfort imposed on the mouse. The restraint was performed by the same trained experimenter throughout the study. A second trained experimenter carefully removed the pin protection and slid the NAT-1 onto the head stage. The entirety of the process generally took about 20 s.

To quantify the behavioural outcomes of using the NAT-1, we tested the activity of all the mice in the open field test [26–28]. The open field tests were conducted in a dedicated sound-attenuated room and in a square box (made up of white reflective Perspex) of 50 × 50 × 40 cm (length × width × height). Two white LED lights facing upwards from the arena and the main fluorescent room lights were switched on, giving rise to an average illumination of 320 lux. Temperature and humidity were maintained at 20 ± 5 °C and 48 ± 5%, respectively. An overhead camera (Imaging Source, DMK22AUCO3, Bremen, Germany), 125 cm from the floor of the arena, was connected to an automated recording system (ANY-Maze v. 5.1, Stoelting) for online tracking of the animal's centre of mass. Each mouse was placed into the centre of the open field and allowed to freely explore for 30 min, and the apparatus was thoroughly cleaned with odourless wet wipes between each mouse. The automatic tracking option was defaulted in ANY-Maze to allow the software to adjust tracking parameters by contrasting the apparatus background, illumination, and animal coat colour. Sampling rates were maintained at the programme's recommended 15 frames per second (fps) for small animals. All the analyses were conducted offline with the same software (ANY-Maze v. 6.1). Parameters which were widely recorded by different laboratories (see [29] for review) were explored. These included the following:

- *total distance* travelled in the arena;
- *line-crossings* in a 16-grid overlay;
- *meandering* (a measurement of an animal's *absolute turn angle* over the *total distance travelled*, which refers to the winding movement of animal);
- *frequency of rotations* (total of 360 degree turns to the left and 360 degree turns to the right);
- degree of thigmotaxis (a ratio of distance travelled in the periphery zone against the total distance travelled);
- *average distance from the centre point* (to provide meaningful indications of the general location of the mice without utilizing zonal outlines).

Thigmotaxis zone was defined as the zone 5 cm from the arena walls.

2.6. Exploratory Data Analyses

Representative track plots of the animals in each drug group of Experiments 1 and 2 were generated by extracting pixel values of the animal's centre point from ANY-Maze and converting them to millimetres using a MATLAB script. The track plots were further divided into two-time intervals (0–15 min and 15–30 min) for clarity and plotted using MATLAB (R2019a, Mathworks, Natick, MA, USA).

Additional exploratory analyses of all the outcomes obtainable from ANY-Maze were conducted to minimise and avoid hypothesis-driven selection of parameters in the open field test [30] by means of heat mapping [31,32]. A total of 68 analysis parameters were extracted by ANY-Maze and were listed for comparison. Heat maps were created to show all data points for the comparison of effects of NAT-1 on each drug condition. The heat maps were generated with *p*-values obtained from independent two-tailed *t*-tests with Satterthwaite correction for heteroscedasticity and bootstrapped 1000 times with replacement (seed set at 123456). The intensities of colours on the heat map were scaled according to the *p*-value threshold set at 95% confidence level. Parameters in the heat maps were clustered for clarity. *Apparatus measures* (#1–20) concerning *total distance* and other additional information regarding activity for the entirety of the test were categorized at the top of the map. *16-grid crossings* (#21), *centre-point* measures (#22–29), and *thigmotaxic zone* measures (#30–68) which are generally used as *standard parameters* were clustered next (Table 1). Heat maps represent global readouts over 30 min, so time-dependent differences and scoring by *segment of test* were not considered. *Immobility* and *freezing* parameters were defaulted in ANY-Maze as 2000 milliseconds (sensitivity level set at 65%) and 1000 milliseconds (threshold levels on at 30 and off at 40) as the minimum immobility and freeze durations, respectively, and no changes were made to either of the criteria. Parameters from head and tail tracking were not recorded.

2.7. Statistical Analysis

Parameters of interest obtained from the tracking software and calculated parameters of interest (e.g., total rotations, meandering, and thigmotaxis) were first tested for normality using the Lilliefors Kolmogorov-Smirnov test. The observed unpaired mean differences were reported and derived from the expected sampling error to illustrate the magnitude of differences using estimation analysis [33]. *t*-tests or analysis of variance (ANOVA; for total responses parameters) with post hoc Bonferroni's corrections were performed for all the conventional *p*-value analyses and bootstrapped 1000 times with replacement (seed set at 123456) to accommodate non-normally distributed outcomes [34]. Group differences were also considered negligible if the effect sizes (ηp^2) were less than 0.1. Only data for distance moved for the entirety of the test were further collapsed into 5 min bins; statistical differences were evaluated using two-way ANOVA for repeated measures, with experimental groups as between-subject factors and time as within-subject factors. Statistical significance was set at 95% confidence level for the conventional statistics. Outliers (with residuals more than two standard deviations away from the mean) were re-investigated for extreme experimental issues or recording errors. No convincing reasons to remove any data on this basis were found. All conventional analysis, analysis of estimates, and visualisation of outcomes were performed in R (v.4.3.0; [35]).

3. Results

The experimental and housing conditions for both independent experiments are detailed in Table 1; the surgical procedures and NAT-1 attachments were the only inconsistent variables between both experiments. Comparisons for each drug treatment were initially visualised in a heat map (Figure 1); significant differences where $p < 0.05$ thresholded are represented in blue. Head stage/NAT-1 attachments led to significant differences in 32 parameters for the saline-treated groups, 12 parameters for the MK801-treated groups, and 14 parameters for the scopolamine-treated groups. No parameter showed a statistical significance in all the comparisons between experiments for all the treatments (Figure 1 and Table S1).

The explorative strategies were visualised by track plots of representative animals in Experiments 1 and 2 (Figure 2a). Locomotor activity and spatial distribution in the open field were maintained between animals with or without NAT-1 of the same drug treatment in both the first and second 15 min interval of the experiment. For both experiments, the scopolamine-treated animals remained in the *periphery* whilst the MK801-treated animals remained in a section of the arena, generally localised away from the arena walls. The saline-treated animals traversed across the centre with a greater frequency than the two drug groups and hence did not show a clear preference for an exploration zone.

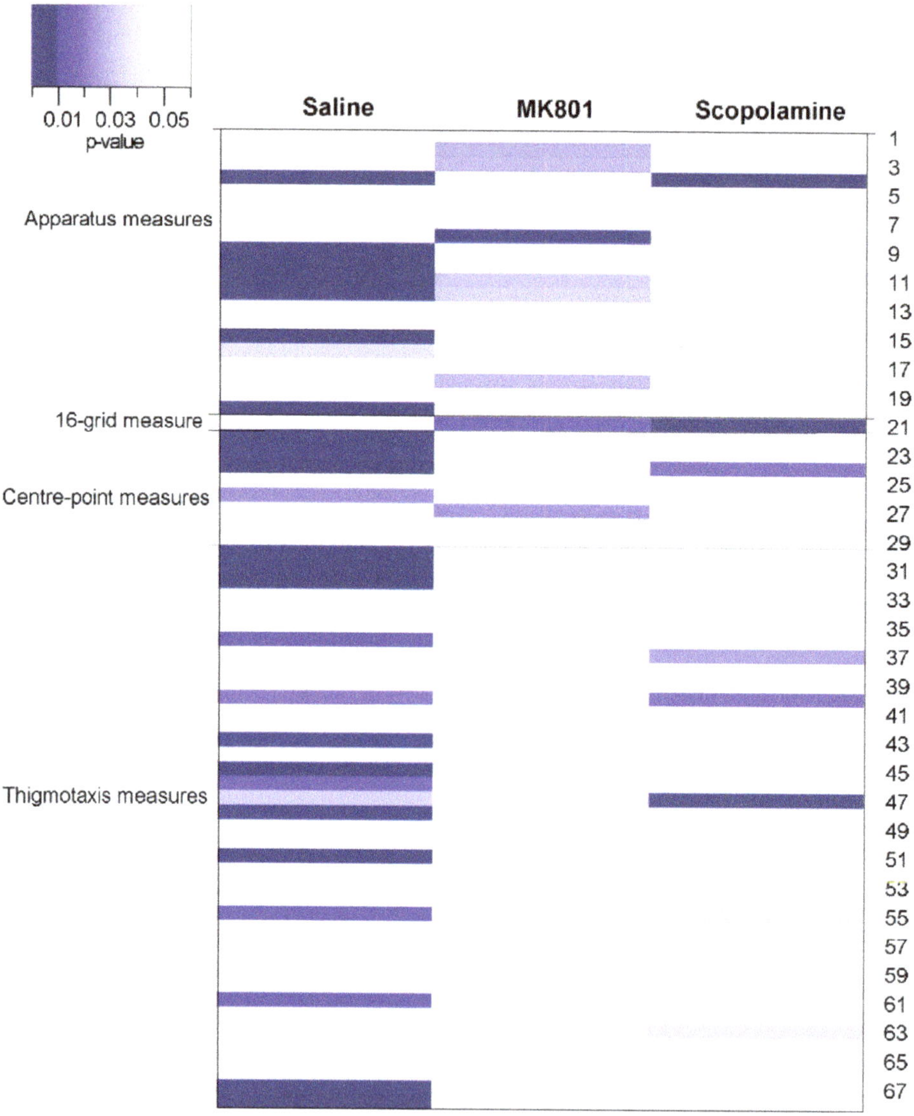

Figure 1. Heat map comparing experiment 1 (no head stage/NAT-1) and 2 (with head stage + NAT-1) for each drug treatment. Parameters are clustered into *apparatus measures* (#1–20; commonly used in the analysis of standard parameters); *16-grid measure* (#21); *centre-point measures* (#22–29); and *thigmotaxis measures* (#30–68). The intensity of colours on the heat map was scaled accordingly with the head stage/treatment cohorts, and the parameter presenting the greatest differences of p-values between head stage/treatment cohorts were isolated and analysed. Data presented in blue in the heat map represent a p-value with the threshold of 0.05 in the between-experiment comparisons. The heat map was visualized in R (v4.3.0).

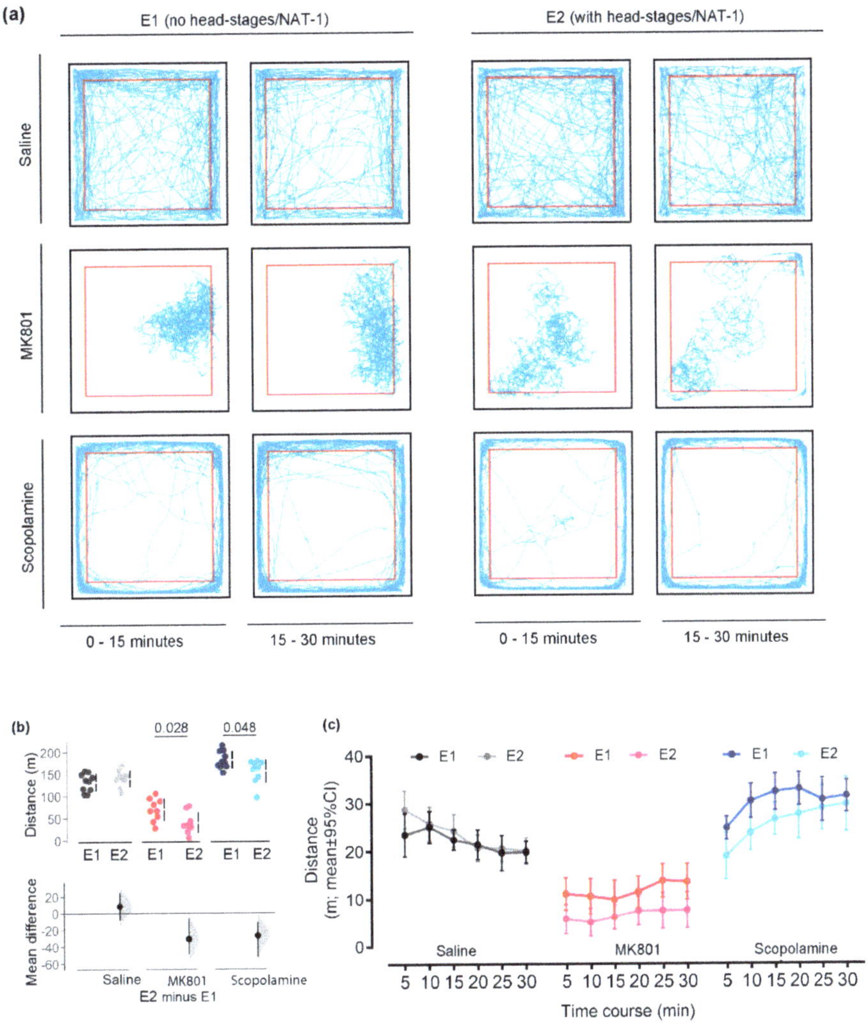

Figure 2. Behavioural responses for locomotor activity between Experiments 1 (E1) and 2 (E2). (**a**) Representative exploration paths of animals (with and without head stage + NAT-1) administered with vehicle, MK801 or scopolamine recorded in the open field test and visualised by track plots divided into time bins, 0–15 min and 15–30 min. The *periphery* (thigmotaxis zone) of the open field arena is demarcated by a red border, 5 cm away from the walls of the arena. Representative paths were illustrated using an in-house MATLAB (R2018a, Mathworks, USA) script. (**b**) Analysis of locomotor activity by *cumulative distance travelled* and (**c**) *total distance travelled* over 30 min of testing in the open field was visualised in 5 min time bins. Individuals are represented by single data points with their respective means (gap) ± standard deviation (vertical lines) in the top axis. The shaded curve and the error in the bottom axis demonstrate the distribution of sampling error and its respective 95% confidence interval for the difference between the means. Performance of animals in E1 is represented by the 0 line in the bottom axis. Conventional statistics (independent sample *t*-tests) revealing a significance set at 95% ($p < 0.05$) are annotated as text on (**b**). All analyses were bootstrapped at 1000 resamples, with seed set at 123456 and visualized in R (v.4.3.0). E1, with no head-stage/NAT-1; and E2with head-stage and NAT-1.

The head-stage/NAT-1 attachments decreased *total distance travelled* by 30.6 m and 26.9 m [95% CI −52.3, −10.4 m] in the MK801-and scopolamine- treated animals, respectively. The means of *total distance travelled* were outside of the interval estimate but were minimally increased by 8.37 m [95% CI −8.09, 24.6 m] in the saline-treated group (Figure 2b). The *Total distance travelled* analysed in 5 min time intervals are visualised in Figure 2c. No main effect of head stages was observed for the saline treatments, but significant reductions were yielded for the MK801- ($F(1,16) = 6.98$, $p = 0.018$, $\eta p^2 = 0.304$) and scopolamine- ($F(1,17) = 6.26$, $p = 0.023$, $\eta p^2 = 0.269$) treated cohorts.

Comparisons of animals with and without NAT-1 are statistically summarised in Table 2 and visualised in Figure 3. No notable differences due to the NAT-1 attachments were discerned in the *average distance from the centre-point* (m [95% CI]: saline, −0.020 [−0.026, −0.012]; MK801, 0.006 [−0.041, 0.036]; and scopolamine, −0.080 [−0.018, 0.030]); Figure 3a). *Line-crossings* differed between experimental groups for all treatments (Figure 3b). NAT-1 attachments induced the largest reductions in line-crossing counts in the MK801- (−409 counts [95% CI, −600, −158 counts]) and the scopolamine- (−362 counts [95% CI, −541, −242 counts]) treated animals, but minimally affected the saline-treated mice (−28.6 counts [95% CI, −148, 86.7 counts]).

Table 2. Summary of between-experiment comparisons for each drug treatment in pre-selected parameters. Data were obtained by bootstrapped independent *t*-tests (two-tailed), and mean difference was obtained using estimation analysis. All analyses were resampled 1000 times (seed starting at 123456) and obtained in R (v.4.3.0), and significance was considered where $p < 0.05$ (highlighted in red).

Parameters	Drug	Between-Experiment Comparisons			E2 minus E1
		E1	E2	*p*-Value	Mean Difference [95% CI]
		Mean ± SD	Mean ± SD		
Distance moved (m)	Saline	131.6 ± 20.2	139.9 ± 19.2	0.347	8.4 [−8.1; 24.6]
	MK801	70.0 ± 25.6	39.3 ± 23.5	0.028	−30.6 [−48.6; −6.0]
	Scopolamine	182.2 ± 20.0	155.2 ± 26.7	0.048	−26.9 [−52.3; −10.4]
Average distance from centre-point (cm)	Saline	0.218 ± 0.007	0.199 ± 0.010	0.001	−0.020 [−0.026; −0.012]
	MK801	0.155 ± 0.030	0.161 ± 0.050	0.782	0.006 [−0.041; 0.036]
	Scopolamine	0.237 ± 0.013	0.230 ± 0.012	0.275	−0.008 [−0.018; 0.003]
Thigmotaxis (ratio)	Saline	0.532 ± 0.052	0.463 ± 0.049	0.020	−0.068 [−0.11; −0.021]
	MK801	0.223 ± 0.157	0.285 ± 0.189	0.471	0.061 [−0.1; 0.207]
	Scopolamine	0.807 ± 0.147	0.777 ± 0.128	0.667	−0.030 [−0.131; 0.115]
Thigmotaxis (s)	Saline	1266.1 ± 96.3	1050.0 ± 105.7	0.002	−216 [−295; −123]
	MK801	545.1 ± 379.3	699.2 ± 445.6	0.435	154 [−199; 527]
	Scopolamine	1539.8 ± 184.5	1475.8 ± 163.8	0.437	−64 [−192; 107]
Line-crossings	Saline	1060.6 ± 145.3	1030.0 ± 132.49	0.442	−28.6 [−148; 86.7]
	MK801	866.9 ± 265.1	458.3 ± 238.5	0.014	−409 [−600; −158]
	Scopolamine	1374.2 ± 154.3	1012.0 ± 196.9	0.003	−362 [−541; −242]
Meandering (degrees/s)	Saline	912.5 ± 89.5	932.0 ± 107.4	0.668	19.5 [−63.6; 102]
	MK801	2716.4 ± 1008.4	3077 ± 1813.4	0.603	361 [−663.0; 2100]
	Scopolamine	705.6 ± 73.5	750.4 ± 89.2	0.325	44.7 [−25.6; 117]
Rotation	Saline	76.5 ± 10.5	98.8 ± 12.5	0.006	16.3 [5.6; 26]
	MK801	180.1 ± 142.7	158.2 ± 138.3	0.758	−21.9 [−156.0; 87.5]
	Scopolamine	99.8 ± 10.6	96.1 ± 18.2	0.623	−3.7 [−18.9; 7.16]

Thigmotaxis ratio was calculated by thigmotaxis distance divided by total distance moved; *SD*, standard deviation; CI, confidence intervals (upper, lower bounds); *E1*, Experiment 1 with no head-stage/NAT-1; and *E2*, Experiment 2 with head-stage/NAT-1.

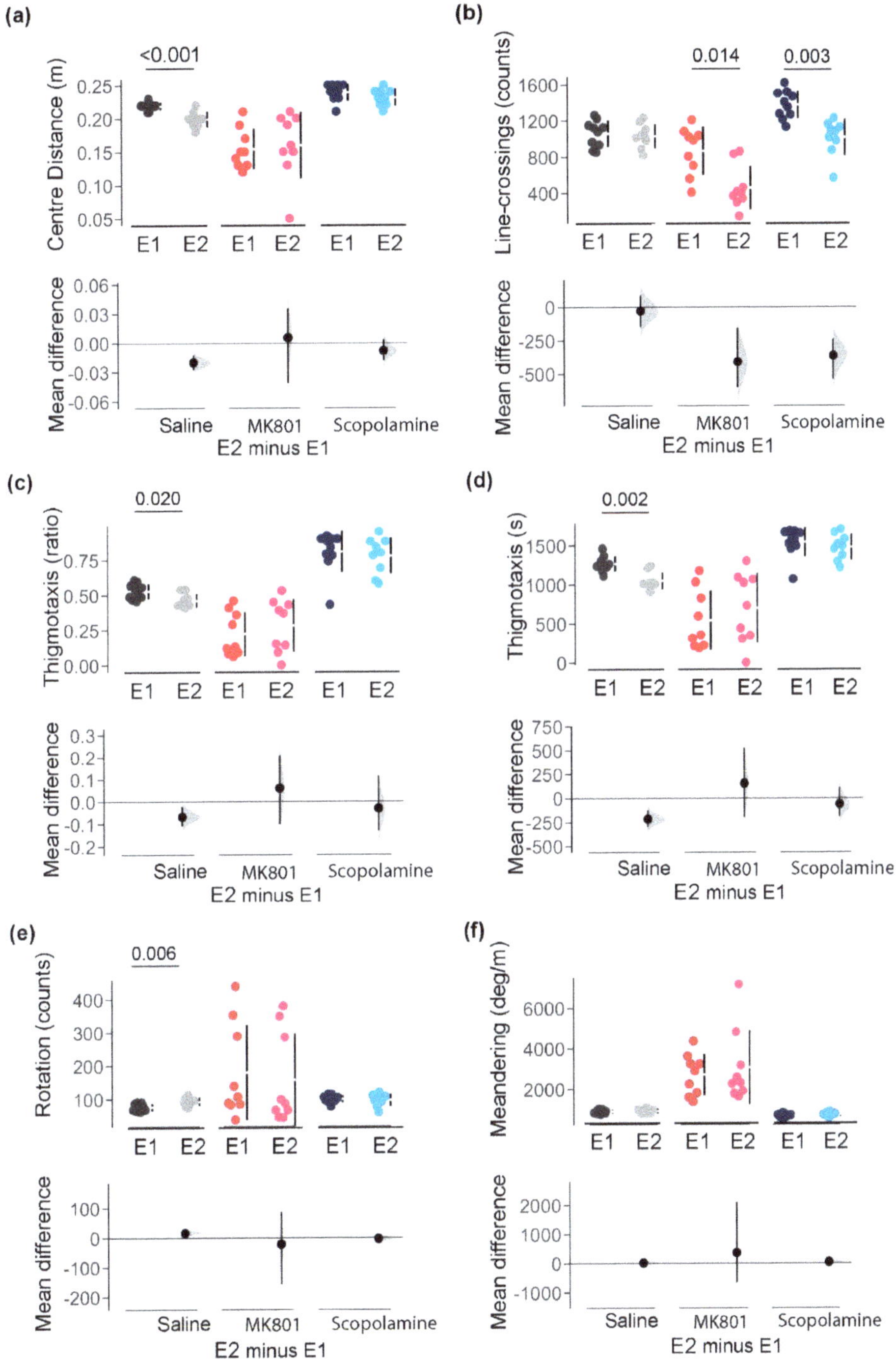

Figure 3. Comparison of behavioural responses for parameters of distance between Experiments 1 (E1) and 2 (E2). (**a**) *average distance from the centre-point;* (**b**) *line-crossings;* (**c**) *thigmotaxic ratio* (***d***) time;

spent in the thigmotaxic zone; (**e**) *total number of rotations*; and (**f**) *meandering*. Individuals are represented by single data points with their respective means (gap) ± standard deviation (vertical lines) in the top axis. The shaded curve and the error in the bottom axis demonstrate the distribution of sampling error and its respective 95% confidence interval for the difference between the means. Performance of animals in E1 is represented by the 0 line in the bottom axis. Conventional statistics (independent sample *t*-tests) revealing significance set at 95% ($p < 0.05$) are annotated as text on the graphs. All analyses were bootstrapped at 1000 resamples, with seed set at 123456, and visualized in R (v.4.3.0). E1 with no head-stage/NAT-1; and E2 with head-stage and NAT-1.

In parameters of localization, head-stage/NAT-1 attachment in animals was estimated to decrease *thigmotaxis ratio* and *thigmotaxis time* in saline-treated animals (−0.068 [95% CI, −0.11, −0.021] and −216 s [95% CI, −295, −123 s] respectively), but not under exposure to MK801 or scopolamine (Figure 3c,d). Comparisons between Experiments 1 and 2 also confirmed minimal differences in *rotations* (counts [95% CI]: saline, 16.3 [5.6, 26]; MK801, −21.9 [−156.0, 87.5]; and scopolamine, −3.7 [−18.9, 7.16]) or *meandering* (degrees/m [95% CI]: saline, 19.5 [−63.6, 102]; MK801, 361 [−663, 2100]; and scopolamine, 44.7 [−25.6, 117]) with NAT-1 attachments in all mice (Figure 3f).

The trend of activity in the MK801- and scopolamine-treated animals (relative to the saline-treated mice) is robust for both experimental groups. Within-experiment analyses also showed that the MK801-treated animals had decreased global activity in the open field (relative to the saline-treated animals; Table S2), independent of NAT-1 attachments. The contrary was observed for the scopolamine-treated animals.

4. Discussion

The primary objective of this study was to evaluate the robustness of behavioural findings between animals with and without head stage/NAT-1 attachments in the open field. This device differs in many respects from other implantable miniaturised telemetric EEG recording systems for mice [36–39] and is considered an advancement over many tethered recording devices (see [8] for discussion). While many of these systems are widely used, a comparison between unoperated and surgically operated and animals is not commonplace but important to define the robustness of behaviours under scrutiny. For the open field, the behavioural repertoire in saline-treated animals was shown to be generally maintained across experimental groups with minimal implications of the surgery, head-stage moulding, and NAT-1 attachments. Animals treated with MK801 or scopolamine on the other hand responded to these modifications with an overall lowering of locomotor activity, with less wall hugging and more frequent visits to the centre of the arena. This suggests that despite considerable weight, NAT-1 per se does not induce behavioural changes, but these can be visualised in the presence of drugs. Such behavioural anomalies may include stereotypy, but most obviously were reflected in hypoactivity in this study. Despite wider confidence intervals in the NAT-1 cohorts treated with either MK801 or scopolamine treatment, the magnitudes of the differences between the drug groups and the controls were in the same direction and numerically very similar, supporting the robustness of the data and the resilience of the mice against neurosurgical intervention and weight-bearing head-attached devices. We note that these experiments were performed with a single dose of each drug, and behavioural anomalies may differ if higher or lower doses are administered. Nevertheless, the current results provide confidence that the utility of different doses would also lead to identical outcomes in non-operated and operated animals alike.

We therefore confirm that lightweight untethered head-stage/NAT-1 attachments do not impose serious limitations on locomotor activity—the primary outcome commonly assessed in the open field. However, there were variations pertaining to parameters of movement types and location-specific proxies, particularly in the thigmotaxic zone. Differences in location-specific proxies can potentially be explained by the animals' somatosensory perception of the device on their head ('object permanence'; [40,41]. The device may present a steric hindrance for the mobility of the animals as indicated by reductions in thigmotaxis measures in the wall zone of the arena. Other non-prespecified endpoints in this study showed variable NAT-1 attachment-dependent differences and included parameters of ethological values such as *freezing* or *heading error, the zone which was first entered, path efficiency to zone,* and *corrected integrated path length.* Intriguingly, drug-dependent changes in these proxies were not influenced by NAT-1 attachments. A more high-resolution analysis looking at shorter fragments of behavioural bouts [42] may return some differences, and their behavioural meaning requires determination.

Surgical intervention in Experiment 2 is arguably the biggest differential that may contribute to irreproducibility in behavioural outcomes between groups. It is imperative to consider possible side effects of the implanted electrodes or surgical process which may provoke changes in sleep pattern [24]; induce meningeal lymphangiogenesis and affect brain homeostasis [43]; or increase microglial activation and seizure susceptibility [21]. Weight bearing and drug treatment may further exacerbate such side effects (see [44] for discussion). What appears crucial is that a direct comparison between unoperated and surgically operated specimens with telemetric devices should be performed for each behavioural test before animals with EEG recording devices can be classed as 'normal' controls. There is still abundant room for further assessments to determine if responses remain robust in more complex behavioural paradigms or in, e.g., other mouse lines or genetically modified strains. An aspect that may be addressed in the future is the effect of head-stage/NAT-1 attachments on more fine-grained behavioural fractionations using unsupervised, data-driven approaches [42,45] which require considerable hardware and software upgrades. This may be relevant to studies exploring severe behavioural anomalies in different etiological responses as indications of anxiety [46–48] or epilepsy [49].

Overall, our data provide compelling evidence for by and large undisturbed behaviour in terms of spatial exploration, as all the changes induced by the drugs in the reported parameters were similarly observed with and without NAT-1 attachment. This study confirms the robustness of behavioural responses in the context of using wireless recording devices in the open field. Data robustness per se is important, as many laboratories provide EEG findings aligned with behavioural responses, such as ours [3,7,50]. Normative behaviour is critical if EEG recordings are interpreted as representative for a 'normal' animal.

Supplementary Materials: The following supporting information can be downloaded at: https://www.mdpi.com/article/10.3390/biom14081008/s1, Figure S1: NAT-1 devise detailing dimensions of the microchip and positioning on a surgically operated animal. Table S1: List of obtainable parameters in ANY-Maze with their associated p-values from one-way analysis of variance and post-hoc analysis with Bonferroni's corrections (between-drug comparisons in each experiment) and independent T-test analysis (between-experiment comparisons; corresponding heat-map found in Figure 1). Significant associations were $p < 0.005$ are highlighted in red and significant associations common to both experiments are highlighted in red and bolded. Statistical analyses were obtained in R (v.4.3.0). Table S2: Statistical summary of comparisons between treatment groups in Experiments 1 and 2 in pre-selected parameters. Conventional statistical comparisons against saline which differ between experimental groups are highlighted in red- significance is considered where $p < 0.05$. All statistical analysis for comparisons to saline in each experimental group were resampled 1000 (seed starting at 123456) in R (v.4.3.0).

Author Contributions: C.J.M.L.: Conceptualization, Methodology, Investigation, Formal Analysis, Data archiving, Writing—original draft; J.B.: Methodology, Investigation; Review; S.K.J.: Conceptualization, Methodology, Review & Editing; B.P.: Conceptualization, Methodology, Supervision, Review and Editing; G.R.: Conceptualization, Project administration, Supervision, Methodology, Writing—Review & Editing, Funding. All authors have read and agreed to the published version of the manuscript.

Funding: This project included funding from the Innovative Medicines Initiative 2/EFPIA, European Quality in Preclinical Data (EQIPD) consortium under grant agreement number 777364.

Institutional Review Board Statement: The animal study protocol was approved by the Institutional Review Board (or Ethics Committee) of the University of Aberdeen and approved by a Home Office License (PPL70/8396, 04/09/2017).

Informed Consent Statement: Not applicable.

Data Availability Statement: All data are presented in this paper and numerically will be made available on request.

Acknowledgments: We would also like to acknowledge the staff of the Medical Research Facility, University of Aberdeen, for their support with animal care, handling, and behavioural experiments.

Conflicts of Interest: Author Sanna K. Janhunen was employed by the company Organon R&D Finland. The remaining authors declare that the research was conducted in the absence of any commercial or financial relationships that could be construed as a potential conflict of interest. The Organon R&D Finland had no role in the design of the study; in the collection, analyses, or interpretation of data; in the writing of the manuscript, or in the decision to publish the results.

References

1. Ahnaou, A.; Huysmans, H.; Jacobs, T.; Drinkenburg, W.H.I.M. Cortical EEG Oscillations and Network Connectivity as Efficacy Indices for Assessing Drugs with Cognition Enhancing Potential. *Neuropharmacology* **2014**, *86*, 362–377. [CrossRef] [PubMed]
2. Crispin-Bailey, C.; Moulds, A.; Platt, B.; Hollier, G.P.; Freeman, M.J.; Fargus, A.G. A Miniaturized 4-Channel, 2KSa/Sec Biosignal Data Recorder With 3-Axis Accelerometer and Infra-Red Timestamp Function. In Proceedings of the Seventh International Conference on Sensor Technologies and Applications, Porto, Portugal, 25–29 September 2013.
3. Crouch, B.; Yeap, J.M.; Pais, B.; Riedel, G.; Platt, B. Of Mice and Motion: Behavioural-EEG Phenotyping of Alzheimer's Disease Mouse Models. *J. Neurosci. Methods* **2019**, *319*, 89–98. [CrossRef] [PubMed]
4. Dhamne, S.C.; Silverman, J.L.; Super, C.E.; Lammers, S.H.T.; Hameed, M.Q.; Modi, M.E.; Copping, N.A.; Pride, M.C.; Smith, D.G.; Rotenberg, A.; et al. Replicable in Vivo Physiological and Behavioral Phenotypes of the Shank3B Null Mutant Mouse Model of Autism. *Mol. Autism* **2017**, *8*, 26. [CrossRef] [PubMed]
5. Drinkenburg, W.H.I.M.; Ruigt, G.S.F.; Ahnaou, A. Pharmaco-EEG Studies in Animals: An Overview of Contemporary Translational Applications. *Neuropsychobiology* **2015**, *72*, 151–164. [CrossRef] [PubMed]
6. Etholm, L.; Arabadzisz, D.; Lipp, H.P.; Heggelund, P. Seizure Logging: A New Approach to Synchronized Cable-Free EEG and Video Recordings of Seizure Activity in Mice. *J. Neurosci. Methods* **2010**, *192*, 254–260. [CrossRef]
7. Jyoti, A.; Plano, A.; Riedel, G.; Platt, B. EEG, Activity, and Sleep Architecture in a Transgenic AβPP Swe/PSEN1 A246E Alzheimer's Disease Mouse. *J. Alzheimers Dis.* **2010**, *22*, 873–887. [CrossRef] [PubMed]
8. Platt, B.; Drever, B.; Koss, D.; Stoppelkamp, S.; Jyoti, A.; Plano, A.; Utan, A.; Merrick, G.; Ryan, D.; Melis, V.; et al. Abnormal Cognition, Sleep, EEG and Brain Metabolism in a Novel Knock-in Alzheimer Mouse, PLB1. *PLoS ONE* **2011**, *6*, e27068. [CrossRef] [PubMed]
9. Jiang, Z.; Huxter, J.R.; Bowyer, S.A.; Blockeel, A.J.; Butler, J.; Imtiaz, S.A.; Wafford, K.A.; Phillips, K.G.; Tricklebank, M.D.; Marston, H.M.; et al. TaiNi: Maximizing Research Output Whilst Improving Animals' Welfare in Neurophysiology Experiments. *Sci. Rep.* **2017**, *7*, 8086. [CrossRef]
10. Rung, J.P.; Carlsson, A.; Rydén Markinhuhta, K.; Carlsson, M.L. (+)-MK-801 Induced Social Withdrawal in Rats; a Model for Negative Symptoms of Schizophrenia. *Prog. Neuropsychopharmacol. Biol. Psychiatry* **2005**, *29*, 827–832. [CrossRef]
11. Svoboda, J.; Stankova, A.; Entlerova, M.; Stuchlik, A. Acute Administration of MK-801 in an Animal Model of Psychosis in Rats Interferes with Cognitively Demanding Forms of Behavioral Flexibility on a Rotating Arena. *Front. Behav. Neurosci.* **2015**, *9*, 75. [CrossRef]
12. van der Staay, F.J.; Rutten, K.; Erb, C.; Blokland, A. Effects of the Cognition Impairer MK-801 on Learning and Memory in Mice and Rats. *Behav. Brain Res.* **2011**, *220*, 215–229. [CrossRef] [PubMed]
13. Wu, J.; Zou, H.; Strong, J.A.; Yu, J.; Zhou, X.; Xie, Q.; Zhao, G.; Jin, M.; Yu, L. Bimodal Effects of MK-801 on Locomotion and Stereotypy in C57BL/6 Mice. *Psychopharmacology* **2005**, *177*, 256–263. [CrossRef] [PubMed]
14. Cheon, S.Y.; Koo, B.-N.; Kim, S.Y.; Kam, E.H.; Nam, J.; Kim, E.J. Scopolamine Promotes Neuroinflammation and Delirium-like Neuropsychiatric Disorder in Mice. *Sci. Rep.* **2021**, *11*, 8376. [CrossRef] [PubMed]
15. Robinson, L.; Harbaran, D.; Riedel, G. Visual Acuity in the Water Maze: Sensitivity to Muscarinic Receptor Blockade in Rats and Mice. *Behav. Brain Res.* **2004**, *151*, 277–286. [CrossRef] [PubMed]
16. Mabunga, D.F.N.; Park, D.; Ryu, O.; Valencia, S.T.; Adil, K.J.L.; Kim, S.; Kwon, K.J.; Shin, C.Y.; Jeon, S.J. Recapitulation of Neuropsychiatric Behavioral Features in Mice Using Acute Low-Dose MK-801 Administration. *Exp. Neurobiol.* **2019**, *28*, 697–708. [CrossRef] [PubMed]
17. Qi, C.; Zou, H.; Zhang, R.; Zhao, G.; Jin, M.; Yu, L. Age-Related Differential Sensitivity to MK-801-Induced Locomotion and Stereotypy in C57BL/6 Mice. *Eur. J. Pharmacol.* **2008**, *580*, 161–168. [CrossRef] [PubMed]
18. Wegener, N.; Nagel, J.; Gross, R.; Chambon, C.; Greco, S.; Pietraszek, M.; Gravius, A.; Danysz, W. Evaluation of Brain Pharmacokinetics of (+)MK-801 in Relation to Behaviour. *Neurosci. Lett.* **2011**, *503*, 68–72. [CrossRef] [PubMed]

19. Zuo, D.Y.; Zhang, Y.H.; Cao, Y.; Wu, C.F.; Tanaka, M.; Wu, Y.L. Effect of Acute and Chronic MK-801 Administration on Extracellular Glutamate and Ascorbic Acid Release in the Prefrontal Cortex of Freely Moving Mice on Line with Open-Field Behavior. *Life Sci.* **2006**, *78*, 2172–2178. [CrossRef] [PubMed]
20. Percie du Sert, N.; Hurst, V.; Ahluwalia, A.; Alam, S.; Avey, M.T.; Baker, M.; Browne, W.J.; Clark, A.; Cuthill, I.C.; Dirnagl, U.; et al. The ARRIVE Guidelines 2.0: Updated Guidelines for Reporting Animal Research. *PLoS Biol.* **2020**, *18*, e3000410. [CrossRef]
21. Balzekas, I.; Hernandez, J.; White, J.; Koh, S. Confounding Effect of EEG Implantation Surgery: Inadequacy of Surgical Control in a Two Hit Model of Temporal Lobe Epilepsy. *Neurosci. Lett.* **2016**, *622*, 30–36. [CrossRef]
22. Paxinos, G.; Franklin, K.B.J. *The Mouse Brain in Stereotaxic Coordinates*, 2nd ed.; Academic Press: Cambridge, MA, USA, 2001; ISBN 978-0-12-547636-2.
23. Arroyo-Araujo, M.; Voelkl, B.; Laloux, C.; Novak, J.; Koopmans, B.; Waldron, A.-M.; Seiffert, I.; Stirling, H.; Aulehner, K.; Janhunen, S.K.; et al. Systematic Assessment of the Replicability and Generalizability of Preclinical Findings: Impact of Protocol Harmonization across Laboratory Sites. *PLoS Biol.* **2022**, *20*, e3001886. [CrossRef]
24. Ranjbaran, Z.; Keefer, L.; Stepanski, E.; Farhadi, A.; Keshavarzian, A. The Relevance of Sleep Abnormalities to Chronic Inflammatory Conditions. *Inflamm. Res.* **2007**, *56*, 51–57. [CrossRef]
25. Wang, B.-S.; Wang, X.-J.; Gong, L.-K. The Construction of a Williams Design and Randomization in Cross-Over Clinical Trials Using SAS. *J. Stat. Softw.* **2009**, *29*, 1–10. [CrossRef]
26. Robinson, L.; Spruijt, B.; Riedel, G. Between and within Laboratory Reliability of Mouse Behaviour Recorded in Home-Cage and Open-Field. *J. Neurosci. Methods* **2018**, *300*, 10–19. [CrossRef]
27. Robinson, L.; Riedel, G. Comparison of Automated Home-Cage Monitoring Systems: Emphasis on Feeding Behaviour, Activity and Spatial Learning Following Pharmacological Interventions. *J. Neurosci. Methods* **2014**, *234*, 13–25. [CrossRef]
28. Wahlsten, D. Standardizing Tests of Mouse Behavior: Reasons, Recommendations, and Reality. *Physiol. Behav.* **2001**, *73*, 695–704. [CrossRef]
29. Spruijt, B.M.; Peters, S.M.; de Heer, R.C.; Pothuizen, H.H.J.; van der Harst, J.E. Reproducibility and Relevance of Future Behavioral Sciences Should Benefit from a Cross Fertilization of Past Recommendations and Today's Technology: "Back to the Future". *J. Neurosci. Methods* **2014**, *234*, 2–12. [CrossRef]
30. Head, M.L.; Holman, L.; Lanfear, R.; Kahn, A.T.; Jennions, M.D. The Extent and Consequences of P-Hacking in Science. *PLoS Biol.* **2015**, *13*, e1002106. [CrossRef]
31. Gehlenborg, N.; Wong, B. Heat Maps. *Nat. Methods* **2012**, *9*, 213. [CrossRef]
32. Timotius, I.K.; Canneva, F.; Minakaki, G.; Moceri, S.; Plank, A.-C.C.; Casadei, N.; Riess, O.; Winkler, J.; Klucken, J.; Eskofier, B.; et al. Systematic Data Analysis and Data Mining in CatWalk Gait Analysis by Heat Mapping Exemplified in Rodent Models for Neurodegenerative Diseases. *J. Neurosci. Methods* **2019**, *326*, 108367. [CrossRef]
33. Ho, J.; Tumkaya, T.; Aryal, S.; Choi, H.; Claridge-Chang, A. Moving beyond P Values: Everyday Data Analysis with Estimation Plots. *Nat. Methods* **2019**, *16*, 565–566. [CrossRef]
34. Johnston, M.G.; Faulkner, C. A Bootstrap Approach Is a Superior Statistical Method for the Comparison of Non-Normal Data with Differing Variances. *New Phytol.* **2021**, *230*, 23–26. [CrossRef]
35. R Core Team. *R: A Language and Environment for Statistical Computing*; R Foundation for Statistical Computing: Vienna, Austria, 2023; Available online: https://www.R-Project.Org/ (accessed on 4 June 2020).
36. Lundt, A.; Wormuth, C.; Siwek, M.E.; Müller, R.; Ehninger, D.; Henseler, C.; Broich, K.; Papazoglou, A.; Weiergräber, M. EEG Radiotelemetry in Small Laboratory Rodents: A Powerful State-of-the Art Approach in Neuropsychiatric, Neurodegenerative, and Epilepsy Research. *Neural Plast.* **2016**, *2016*, 8213878. [CrossRef]
37. Mansouri, M.T.; Ahmed, M.T.; Cassim, T.Z.; Kreuzer, M.; Graves, M.C.; Fenzl, T.; García, P.S. Telemetric Electroencephalography Recording in Anesthetized Mice—A Novel System Using Minimally-Invasive Needle Electrodes with a Wireless OpenBCI^{TM} Cyton Biosensing Board. *MethodsX* **2023**, *10*, 102187. [CrossRef]
38. Papazoglou, A.; Lundt, A.; Wormuth, C.; Ehninger, D.; Henseler, C.; Soós, J.; Broich, K.; Weiergräber, M. Non-Restraining EEG Radiotelemetry: Epidural and Deep Intracerebral Stereotaxic EEG Electrode Placement. *J. Vis. Exp. JoVE* **2016**, *112*, 54216. [CrossRef]
39. Weiergräber, M.; Henry, M.; Hescheler, J.; Smyth, N.; Schneider, T. Electrocorticographic and Deep Intracerebral EEG Recording in Mice Using a Telemetry System. *Brain Res. Protoc.* **2005**, *14*, 154–164. [CrossRef]
40. Pennartz, C.M.A.; Farisco, M.; Evers, K. Indicators and Criteria of Consciousness in Animals and Intelligent Machines: An Inside-Out Approach. *Front. Syst. Neurosci.* **2019**, *13*, 25. [CrossRef]
41. Waldmann, M.R.; Schmid, M.; Wong, J.; Blaisdell, A.P. Rats Distinguish between Absence of Events and Lack of Evidence in Contingency Learning. *Anim. Cogn.* **2012**, *15*, 979–990. [CrossRef]
42. Wiltschko, A.B.; Tsukahara, T.; Zeine, A.; Anyoha, R.; Gillis, W.F.; Markowitz, J.E.; Peterson, R.E.; Katon, J.; Johnson, M.J.; Datta, S.R. Revealing the Structure of Pharmacobehavioral Space through Motion Sequencing. *Nat. Neurosci.* **2020**, *23*, 1433–1443. [CrossRef]
43. Hauglund, N.L.; Kusk, P.; Kornum, B.R.; Nedergaard, M. Meningeal Lymphangiogenesis and Enhanced Glymphatic Activity in Mice with Chronically Implanted EEG Electrodes. *J. Neurosci.* **2020**, *40*, 2371–2380. [CrossRef]

44. Ahuis, T.P.; Symk, M.K.; Laloux, C.; Jack, B.; Aulehner, K.; Waldron, A.-M.; Miljanovic, N.; Seiffert, I.; Dekun, S.; Boulanger, B.; et al. Evaluation of Variation in Preclinical EEG Spectral Power across Multiple Laboratories and Experiments: An EQIPD Study. *PLoS ONE* 2024, *in press*.
45. Egnor, S.E.R.; Branson, K. Computational Analysis of Behavior. *Annu. Rev. Neurosci.* **2016**, *39*, 217–236. [CrossRef] [PubMed]
46. Blanchard, D.C.; Summers, C.H.; Blanchard, R.J. The Role of Behavior in Translational Models for Psychopathology: Functionality and Dysfunctional Behaviors. *Neurosci. Biobehav. Rev.* **2013**, *37*, 1567–1577. [CrossRef] [PubMed]
47. Lister, R.G. Ethologically-Based Animal Models of Anxiety Disorders. *Pharmacol. Ther.* **1990**, *46*, 321–340. [CrossRef] [PubMed]
48. Peters, S.M.; Pothuizen, H.H.J.; Spruijt, B.M. Ethological Concepts Enhance the Translational Value of Animal Models. *Eur. J. Pharmacol.* **2015**, *759*, 42–50. [CrossRef] [PubMed]
49. Akman, O.; Raol, Y.H.; Auvin, S.; Cortez, M.A.; Kubova, H.; de Curtis, M.; Ikeda, A.; Dudek, F.E.; Galanopoulou, A.S. Methodologic Recommendations and Possible Interpretations of Video-EEG Recordings in Immature Rodents Used as Experimental Controls: A TASK1-WG2 Report of the ILAE/AES Joint Translational Task Force. *Epilepsia Open* **2018**, *3*, 437–459. [CrossRef]
50. Crouch, B.; Sommerlade, L.; Veselcic, P.; Riedel, G.; Schelter, B.; Platt, B. Detection of Time-, Frequency- and Direction-Resolved Communication within Brain Networks. *Sci. Rep.* **2018**, *8*, 1825. [CrossRef]

biomolecules

MDPI

Review

Nociceptor–Macrophage Interactions in Apical Periodontitis: How Biomolecules Link Inflammation with Pain

Nandita Menon and Anil Kishen *

Dental Research Institute, Faculty of Dentistry, University of Toronto, Toronto, ON M5G 1G6, Canada; nandita.menon@mail.utoronto.ca

* Correspondence: anil.kishen@dentistry.utoronto.ca

Abstract: Periradicular tissues have a rich supply of peripheral afferent neurons, also known as nociceptive neurons, originating from the trigeminal nerve. While their primary function is to relay pain signals to the brain, these are known to be involved in modulating innate and adaptive immunity by initiating neurogenic inflammation (NI). Studies have investigated neuroanatomy and measured the levels of biomolecules such as cytokines and neuropeptides in human saliva, gingival crevicular fluid, or blood/serum samples in apical periodontitis (AP) to validate the possible role of trigeminal nociceptors in inflammation and tissue regeneration. However, the contributions of nociceptors and the mechanisms involved in the neuro-immune interactions in AP are not fully understood. This narrative review addresses the complex biomolecular interactions of trigeminal nociceptors with macrophages, the effector cells of the innate immune system, in the clinical manifestations of AP.

Keywords: nociceptors; macrophages; neuropeptides; substance P; Calcitonin Gene-Related Peptide; pain; neurogenic inflammation; apical periodontitis

1. Introduction

Apical periodontitis (AP) is an inflammatory disease of periradicular tissue caused by oral opportunistic bacteria and their toxins retained in the apical regions of the periodontium [1,2]. Rich in blood vessels and nerve fibers, the periradicular tissue is composed of fibroblasts, epithelial cells, odontoblasts, cementoblasts, osteoblasts, osteoclasts, macrophages, and undifferentiated stem cells that form the periodontal ligament (PDL) and alveolar bone [3,4]. An infected root canal system in the apical periodontium causes all these immune cells and precursor cells to interact with one another, thereby eliciting an immune response against pathogens and pathogen-associated molecular patterns (PAMPs) [5]. However, AP can persist asymptomatically, without common clinical symptoms such as pain and swelling, while afferent sensory neurons that typically relay pain continue to engage in immune responses at the molecular level [6,7].

Neurogenic inflammation (NI) is set in motion when neuropeptides such as Substance P (SP) and Calcitonin Gene-Related Peptide (CGRP), among others, released by activated nociceptors act on resident immune cells and endothelial cells, causing vascular changes resulting in an influx of migrating immune cells to an inflamed tissue. To facilitate their function in NI, endothelial cells (vascular cells), mast cells, dendritic cells, T and B lymphocytes, and macrophages express cell-surface receptors for neuropeptides released by nociceptors [8,9]. Neuropeptides activate immune cells which release proinflammatory cytokines such as tumor necrosis factor (TNF)-α, interleukin (IL)-1β, IL-6, nerve growth factor (NGF), and prostaglandin E2 (PGE2) that are in turn known to activate sensory neurons [10,11]. Neuropeptides also play a role in regulating anti-inflammatory roles of immune cells. For example, CGRP has been shown to inhibit the production of pro-inflammatory cytokines and enhance the expression of a regulatory phenotype in activated macrophages [12]. This reflects on the pleiotropic nature of neuropeptides that are capable of modulating immune response.

It is important to note that neurogenic inflammation is caused by the activation of peripheral nociceptors, resulting in inflammatory events in tissues that involve host innate and adaptive immune systems. Neuroinflammation, however, describes a localized host immune response to inflammation specifically in the brain and spinal cord [13,14].

Of the myriad cellular interactions in periradicular tissues, the cross-talk between macrophages and sensory neurons stands out because macrophages constitute up to 46% of inflammatory cells in AP [7]. This review aims to provide a comprehensive understanding of the temporal interactions between sensory neurons and macrophages while highlighting the role of nociceptors in different clinical manifestations of chronic AP.

2. Overview of Macrophages in Apical Periodontitis

Inflammation is the physiological response to injury, stress, or infection. In terms of the immune cells involved, inflammation is the state in which innate and adaptive cells are activated, undergo differentiation, and migrate to resolve inflammation while restoring tissue homeostasis. A rapid host immune response is capable of resolving acute inflammation. On the other hand, prolonged exposure to triggers such as damage-associated molecular patterns (DAMPs), PAMPS, and pathogens results in a state of chronic inflammation that sees different immune and non-immune cells in action [15].

Acute AP is marked by the presence of periapical abscesses, while periapical granulomas are typically a sign of chronic AP, typically diagnosed via the radiographic visualization of rarefying osteitis in periapical tissues. AP is dynamic in nature because of the different immune cell types participating in different stages of AP. This dynamic nature can be observed via the progressive change in the distribution of immune cells in the periapical inflammatory infiltrate [16,17]. Immune cells can also undergo polarization to different phenotypes in order to facilitate the host immune response in disease or wound-healing [18,19].

Physiologically, the PDL has a key role in responding to microbes located within the pulp space. The PDL is a highly vascularized tissue, with up to ten times higher vascular volume than other fibrous connective tissues [20]. The blood vessels in PDL primarily participate in tooth support and shock absorption [21]. The activation of immune cells such as macrophages leads to the release of proinflammatory cytokines that increase the expression of cellular adhesion molecule (CAM-1) in endothelial cells, thereby promoting the local adhesion of monocytes and leukocytes [18,22]. This local response, combined with vasodilation caused by initial inflammation response, leads to the active infiltration of monocytes to the site of inflammation, where growth factors, pro-inflammatory cytokines, and microbial products induce cellular differentiation [23,24]. Once at the site of inflammation, these circulatory monocytes can be differentiated into macrophages or dendritic cells [25].

Toll-like receptor-4 (TLR4) is a sensory receptor predominantly expressed by human myeloid cells (monocytes and granulocytes) that recognize PAMPs such as lipopolysaccharides (LPS) expressed by Gram-negative bacteria [26]. Activated TLR4 recruits adaptor proteins that undergo phosphorylation. Myeloid differentiation primary response 88 (MyD88), Interleukin-1 receptor-associated kinase (IRAK), and tumor necrosis factor receptor-associated factor-6 (TRAF-6) are examples of such adaptor proteins. Phosphorylated adaptor proteins activate Nuclear Factor kappa B (NFκB) and mitogen-activated protein kinase (MAPK) signaling pathways, resulting in the expression of pro-inflammatory cytokines [27]. Accessory molecules such as myeloid differentiation 2 (MD2), cluster of differentiation (CD)36, and CD14 can expedite TLR4 signaling [28]. Macrophages are also antigen-presenting cells that activate members of the adaptive immunity, effectively mounting an adaptive immune response.

Macrophages are known to play an active role in resolving inflammation [29]. Activated macrophages can polarize into functional phenotypes of classically activated M1, or alternatively activated M2, to bring about pro- and anti-inflammatory effects, respectively. M1 macrophages produce pro-inflammatory cytokines TNF-α, IL-1β, and IL-6, which stimulate the production of tissue-damaging proteases and prostaglandins while mediating bone resorption. M1 macrophages also produce chemokines such as C-X-C motif chemokine ligand (CXCL)9 and CXCL10 that attract members of the adaptive immunity to the site of inflammation [30,31]. M2 macrophages are known to produce anti-inflammatory cytokines Transforming Growth Factor (TGF)-β and IL-10, which downregulate inflammation and immune response [32,33]. Their phagocytic capabilities help clear apoptotic cells and debris in healing tissues [5,34].

3. Role of Macrophages in Alveolar Bone Resorption

Resident macrophages in periodontal tissues secrete inflammatory cytokines and chemokines in response to pathogens. Qualitative and quantitative analyses have identified macrophages as the major constituent immune cell type in periapical granulomas [35]. This can be partly attributed to the vasodilation caused by the initial immune response, which in turn leads to the active infiltration of monocytes to the site of inflammation, wherein the growth factors, pro-inflammatory cytokines, and microbial products would induce differentiation to macrophages [24,36].

Alveolar bone resorption (loss) is a notable radiographic indication of chronic AP. The alveolar bone functions to hold teeth in the socket while absorbing any force applied to the teeth during biting and chewing. Subject to varying degrees of mechanical forces, the alveolar bone undergoes repeating cycles of bone formation and resorption. In a state of proinflammation, this equilibrium tends towards bone resorption, resulting in the loss of both mineral and organic components of the alveolar bone. The close association of macrophages with different stages of AP implicates these cells in the development of alveolar bone resorption. Macrophages were observed near osteoblasts in rat models of AP during the initial stages of bone resorption, while no macrophages were seen near the surfaces undergoing bone formation [19]. However, a study comparing data from immunohistochemical analyses of human specimens with clinical and radiological data reported no significant difference in the presence of macrophages between periapical granulomas and cysts [16]. Previous studies have highlighted disparities in the clinical presentation of AP [16,37]. Thus, a better understanding of the mechanisms underlying bone modeling and remodeling will help in identifying molecular pathways linking macrophages to alveolar bone resorption.

Osteoclastogenesis, or the development of mature osteoclasts (bone resorptive cells) from precursor cells, is mediated by receptor activator of nuclear factor kappa-β ligand (RANKL) [38]. RANKL is a cytokine produced by osteoblasts (which synthesize bone matrix and differentiate into osteocytes [39]) and stromal cells that binds to its receptor, RANK, to initiate the differentiation of osteoclast precursor cells [40]. This process is regulated by Osteoprotegerin (OPG), a decoy receptor of RANKL that is primarily produced by osteoblasts [41]. Hence, the RANKL-RANK-OPG trio are the main players in bone modeling and remodeling pathways [42]. GM-CSF (Granulocyte-Macrophage Colony-Stimulating Factor) is a cytokine that induces proliferation and maturation in macrophages, and it has been found to promote fusion of prefusion osteoclasts in the presence of RANKL, enabling bone resorption under pathologic and inflammatory conditions [43,44]. Besides GM-CSF, macrophages produce an array of inflammatory mediators in response to pathogens or microbial toxins that have been linked to bone resorption in AP [45]. IL-1 is one such proinflammatory cytokine that is produced by activated macrophages, which promote bone loss by enhancing RANKL expression by stimulating the synthesis of PGE2 in osteoblasts [46,47]. In addition, IL-1 is involved in the survival of osteoclasts and is known to induce multinucleation of osteoclasts, which is a hallmark of cell maturation [48]. Macrophages produce TNF-α, a proinflammatory cytokine that inhibits the differentiation and production of matrix proteins in osteoblasts [49,50]. TNF-α was found to play a role in osteoclastogenesis when increased RANKL mRNA expression was observed in osteocytes (which modulate bone remodeling [45]) that were cultured in vitro with exogenous TNF-α. This study observed an increased percentage of RANKL-positive osteocytes in mice injected with TNF-α [51].

4. Sensory Neuronal Activation in Apical Periodontitis

Periradicular tissues have an abundant supply of neurons that originate in the trigeminal ganglion (TG) [52]. Research groups in the 1970s identified nociceptive afferents with different activation thresholds and conduction velocities in the PDL [53]. Immunocytochemical analysis identified NGF-positive sensory nerves in periodontal tissues from rats [54], while ultrastructural analyses of the human PDL using electron microscopy identified Ruffini nerve endings, free nerve endings, and lamellated corpuscles [55]. Electrophysiological studies performed on cat models helped characterize trigeminal nerve endings in the PDL into low-threshold mechanoreceptors sensitive to mechanical load and vibration (A-β fibers) and nociceptors sensitive to pressure, temperature, and chemical irritants (A-δ, C-fibers) [56,57]. A-β fibers have large cell bodies, and these are myelinated [58,59]. A-δ fibers have medium-sized cell bodies, and C-fibers are considered the smallest [58–60]. Myelinated A-δ fibers have higher conduction velocities, resulting in a sharp pain sensation compared to the unmyelinated, slow C-fibers, which relay a delayed and dull, ache-like pain [58,61].

Studies implicating sensory neurons in inflammation have come a long way since the classic studies that concluded that neurogenic inflammation is based on axon reflexes [62,63]. While this theory did not pan out, some of the earlier works were successfully able to link activated sensory neurons to inflammatory phenomena such as vasodilation, setting stage for further studies [64,65]. As with the classical inflammatory response, neurogenic inflammation has also been associated with tissue regeneration [66]. It was reported that sensory denervation in rats, with pulp exposure to oral environment, showed the loss of pulpal tissue, while innervated teeth showed extensive neural sprouting and formation of reparative dentin and osteodentin [66].

Specialized surface receptors expressed by nociceptors play a crucial role in the cascade of events that led to NI (Figure 1). Transient receptor potential (TRP) channels are a class of integral membrane proteins that function as signal transducers in sensory neurons. These are polymodal in nature and can be activated by intracellular and/or extracellular mechanical, thermal, chemical, and osmotic pressure gradients. Based on sequence similarity, several classes of TRP channels exist. The TRPV subfamily (vanilloid subtype) has six members, of which TRPV1 has garnered attention for its role in the detection of noxious physical and chemical stimuli [67,68]. TRPA1 is the sole member of the TRPA subfamily (ankyrin subtype) [69,70]. The detailed 3D structure of the TRPA1 receptor was previously characterized by Cao et al. and Liao et al. [71,72]. Such studies made it possible to study the conformational changes that occur when TRPA1/TRPV1 receptors are activated. Their activation results in an active influx of extracellular calcium ions, which mediate action potentials in neurons and trigger exocytosis of CGRP and SP from large dense core vesicles (LDCVs) that involve members of the SNARE (soluble N-ethylmaleimide-sensitive factor attachment protein receptor) family of proteins [73,74]. Biologically active neuropeptides are derived from their precursor peptides that are cleaved in the endoplasmic reticulum, after which these pro-peptides undergo post-translational modifications to form active versions that are stored in LDCVs prior to release from the sensory nerve endings [75,76]. A-δ and C fibers express TRPV1 and TRPA1 channels, and hence, these channels are associated with chemical and thermal nociception [77]. Knockout models of TRPV1 and TRPA1 highlight the role of these receptors in eliciting host immune response besides nociception. Non-neuronal cells such as endothelial cells, fibroblast, odontoblasts, and immune cells, including mast cells, monocytes, macrophages, and dendritic cells, also express TRPV1 and TRPA1 [77–79]. In these cells too, TRPV1 and TRPA1 exhibit functional responses to similar triggers as seen in neurons; however, it is not clear yet if the sensitivity thresholds of these receptors in non-neuronal cells are similar to those in neurons [77,80].

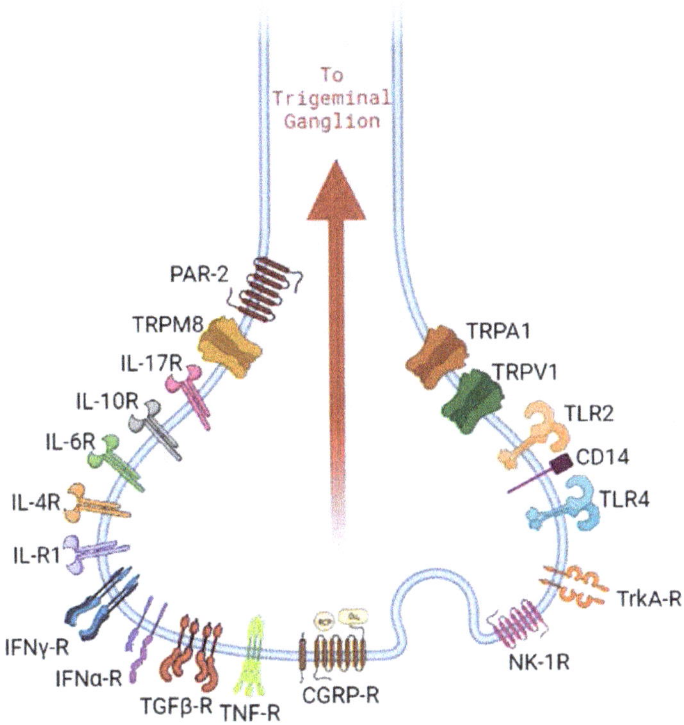

Figure 1. Nociceptor neurons express receptor molecules that recognize PAMPs, DAMPs, cytokines, and neuropeptides. This results in the activation of downstream signaling pathways that can contribute to neurogenic inflammation. Initialisms are listed in Appendix A.

Toll-like receptors (TLRs) are another class of surface receptors that have been identified on nociceptors. The nociceptor expression of TLR4 highlights its role in the host immune response to bacterial infections [81,82]. Peripheral sensory neurons showed the colocalized expression of TLR4 and CD14, indicating the capability of the direct detection of PAMPs [82]. Besides immunological implications, the LPS binding of TLR4 on trigeminal neurons was shown to increase sensitization of TRPV1 channels, leading to increased expression of CGRP, which suggests a possible mechanism for pain associated with bacterial infections [83–85]. Similar findings were reported in an independent study that demonstrated a concentration-dependent effect of LPS on TRPV1 in trigeminal neurons [83].

Different bacterial strains have been shown to evoke different degrees of nociceptor response [86]. Gram-positive bacteria operate through mechanisms that are distinct from those triggered by Gram-negative bacteria [87]. *S. aureus* utilize N-formylated peptides and pore-forming toxin α-hemolysin to instigate pain in mice [87]. This study also showed a strong correlation between the bacterial load of live *S. aureus* and pain expressed by mice. Such studies provide insight into the different mechanisms that may be involved in eliciting pain in bacteria-mediated diseases such as AP, where both Gram-negative and Gram-positive strains play active roles in disease pathogenesis [2,88].

5. Role of Sensory Neuropeptides in Pain and Inflammation

Neuropeptides are considered the primary mediators of neurogenic inflammation. A-δ and C-fibers are peptidergic in nature, meaning these release neuropeptides mediating NI in AP [89,90] (Table 1). CGRP and SP are the principal neuropeptides involved in AP.

Table 1. Neuropeptides (NPs) identified in apical periodontitis (AP), cells known to release NPs, cells shown to express receptors for NPs, and known functional effects of NPs in AP.

Neuropeptides in AP	Cells Releasing NPs	Cells Expressing NP Receptors	Functional Effects in AP
Calcitonin Gene-Related Peptide (CGRP)	Sensory neurons [9,75] T cells [91] Monocytes [91] Endothelial cells [75] Fibroblasts [92,93] Adipocytes [75]	Sensory neurons [90] Arterial vessels, Mononuclear immune cells, Schwann cells [91] Th1 cells [94] Langerhans cells [94] Monocytes [94] Macrophages [94] Dendritic cells [94] CD34^{+} hematopoietic progenitor cells [94] B & T lymphocytes [94] Epithelial cells [95]	Extensive degradation, resulting in reduced levels of CGRP observed in the GCF of diseased periodontal tissues [96,97] Vasodilation [98] Activation of innate and adaptive immune cells [99,100] Increase in blood flow [101] Prevention of monocyte/macrophage activation [102,103] Stimulation of osteogenesis [104,105] Endothelial cell proliferation [101]
Substance P (SP)	Neurons [106,107] Astrocytes [108] Microglia [107,108] Epithelial cells [108] Endothelial cells [108] T cells [108] Macrophages [108] Dendritic cells [106,108] Eosinophils [108] Mesenchymal stem cells [108] Leukocytes [106] Natural killer cells [106] Mast cells [106] Neutrophils [106]	Epithelial cells [108] Endothelial cells [106,108,109] White blood cells [109] Fibroblasts [108,109] Neurons [90,108,109] Smooth muscle cells [106,108] B & T lymphocytes [106,108] Natural killer cells [108] Dendritic cells [108,110] Monocytes [106,108] Macrophages [106,108] Microglia [106,108] Astrocytes [106,108] Eosinophils [106,108] Mast cells [106,108] Neutrophils [106]	Increased levels observed in GCF of diseased periodontal tissues [96,111] Desensitization of nociceptors [108] Neuronal activation [112] Onset of pain and inflammation [113] Inhibition of osteoblast differentiation [114] Induction of osteoclastogenesis [115] Increase in blood flow [113,116] Reduction in pro-inflammatory response of neutrophil, epithelial cells [117]
Vasoactive Intestinal Polypeptide (VIP)	Neurons [118] T cells [118] B cells [118] Mast cells [118] Eosinophils [118]	Epithelial cells [118] T cells [118] Macrophages [118] Dendritic cells [118] Mast cells [118] Neutrophils [118] Smooth muscle cells [118] Endothelial cells [118] Pancreatic B cells [118]	Found in increased levels in GCF of periodontitis patients; reduced levels as inflammation reduced [119] VIP reactivity observed in gingival biopsy, but no discernible difference compared to clinically healthy sites [120] Promotion of bone regeneration [105]
Neuropeptide Y (NPY)	Neurons [121] Pancreatic β cells [122] Vascular smooth muscle cells [123] Monocytes [123] Macrophages [123] Enterocytes [123] Endothelial cells [121] Langerhans cells [123] Microglia [123] B & T lymphocytes [123]	Neurons [121] Monocytes [123] Macrophages [122,123] Lymphocytes [123] Dendritic Cells [123] Neutrophils [123] B & T Lymphocytes [123] Leukocytes [123] Microglia [123] Osteoblasts [122] Endothelial Cells [122] PDL Stem Cells [105]	Proposed to be involved in initiation and regulation of periodontitis [105] Reduced levels observed in diseased sites of periodontitis patients [96] High levels observed in the saliva of periodontitis patients [124] Promotion of osteogenesis [105] Increase in PDL stem cells' osteogenic capacity in vitro [125]
Pituitary Adenylate Cyclase-Activating Peptide (PACAP)	Mast cells [126] Schwann cells [127]	Pancreatic cells [128] Sensory nerve terminals [128,129] Pituitary cells [127] Schwann cells [127] Satellite glial cells [129] Hepatocytes [129] Osteoblasts [129]	Promotion of bone regeneration [105] Dose-dependent inhibition of IL-12 in LPS-activated macrophages [130]
Neurokinin A (NKA)	Sensory neurons [131]	Sensory neurons [132] B & T cells [133] Dendritic cells [133] Macrophages [133]	Co-released with SP [131] Increased levels observed in the GCF of diseased periodontal tissues [134] Decreased levels of NK-A observed in diseased periodontal tissue [96]

CGRP was discovered in 1982 by Amara et al. [135]. This 37-residue peptide was found to be expressed in the central and peripheral nervous systems, promoting sensory, motor, and autonomic functions in the brain [136]. CGRP has two isoforms, α-CGRP and β-CGRP, which have over 90% similar homology and share similar biological effects in humans [75]. CALC I gene undergoes alternate splicing to generate α-CGRP, while β-CGRP is a product of CALC II gene transcription. While α-CGRP is predominant in the central and peripheral nervous systems, β-CGRP is found mainly in the enteric nervous system [75]. Vasodilatory effects of human and rat CGRP were discovered in 1985, effectively linking CGRP to hyperemia [137]. The development of CGRP antagonists helped understand the role of CGRP in migraines, linking neuropeptides with pain pathways in the orofacial region and in other tissues [138]. Recognizing the ability of sensory nerves to produce and release CGRP led to the discovery of different stimuli that can activate neural cells to release this neuropeptide and the pathways that lead to the production and release post-activation [139].

The CGRP receptor molecule is a G protein-coupled receptor and is composed of three components: receptor activity modifying protein 1 (RAMP1), Calcitonin-like receptor (CLR), and receptor component protein (RCP). All three components are required to form a fully functional CGRP receptor (CGRP-R). Upon activation by binding with CGRP, CGRP-R may couple with G protein α subunit, increasing the cyclical adenosine monophosphate (cAMP) levels, which activate protein kinase A (PKA). This can result in the phosphorylation of cAMP response-element-binding (CREB) protein leading to gene transcription or the activation of extracellular signal-regulated kinases (ERK) pathway within a neural cell [139]. This signaling pathway can also result in the opening of ATP-sensitive K^+ channels, which can lead to vasodilation [98].

Substance P (SP) is an undecapeptide that was first discovered in 1931 by Von Euler and Gaddum [140]. SP, neurokinin A (NKA), neuropeptide K (NPK), and neuropeptide γ (NPγ) are tachykinin neuropeptides encoded by the TAC1 gene; alternate RNA splicing and differential post-translational processing result in the expression of specific tachykinin peptides [141]. SP is best known for its role as a modulator of pain perception by altering cellular signaling pathways using G protein-coupled receptors called neurokinin 1 receptor (NK-1R) that can act through the cAMP secondary messenger system or inositol triphosphate (IP3)-diacylglycerol (DAG) system [108]. These downstream signaling pathways result in the expression of pro-inflammatory cytokines. SP can also exhibit autocrine effects on neural cells expressing NK-1R in a desensitizing process that results in the internal phosphorylation of NK-1R and recycling via endocytosis and acidification in endosomes [108].

CGRP and SP are known to colocalize, and these have been studied to assess the degree of pain experienced by patients with symptomatic AP [142]. Affirming their role in oral diseases, several studies have reported increased levels of CGRP and SP in the sites of inflammation in the dental pulp [143,144], periodontal ligament [145,146], and saliva and gingival crevicular fluid [111,147,148]. The levels of neuropeptides in these environments are correlated with the degrees of inflammation, confirming their role in inflammatory mechanisms [111,113,144].

CGRP and SP play vital roles in modulating bone resorption. In murine AP models, TRPV1-positive sensory neurons were shown to regulate osteoclastogenesis via CGRP signaling [104]. On the other hand, SP was found to up-regulate RANKL/OPG ratio in a rat model of periodontitis, promoting RANKL-induced osteoclast differentiation [115].

SP and CGRP have also been shown to mediate tissue repair. The sprouting of CGRP-positive fibers in odontoblast layers after injury induced in the cervical dentin of rats resulted in the formation of reparative dentin [149]. SP released by stimulated peripheral nerves increased blood flow to the dental pulp, thereby enabling tissue healing [113,116]. Hence, SP and CGRP can be considered immunomodulators that can have pro- and anti-inflammatory effects depending on molecular cues present in the tissue microenvironment.

6. Neuro-Immune Interactions Modulate Inflammation and Healing in AP

Evidence of a functional link between sensory neurons and immune cells comes from the proximity of sensory nerve endings and immune cells. This is corroborated by the expression of neuropeptide-specific receptors on immune cells and cytokine receptors and receptors capable of recognizing microbial pathogens on trigeminal nociceptor neurons [8,150] (Figures 1 and 2).

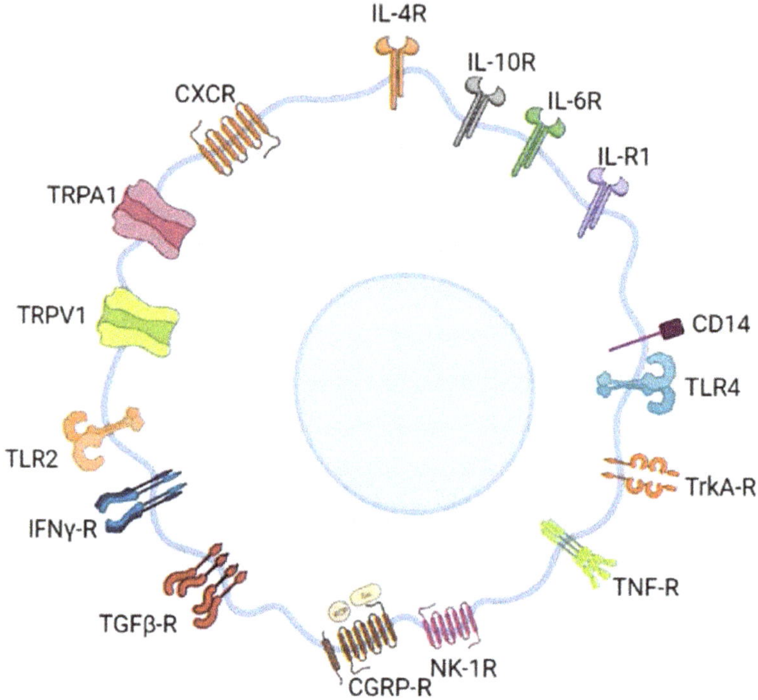

Figure 2. In addition to releasing neuropeptides, macrophages have been shown to express neuropeptide-specific receptors, indicating a significant role in neuro-immune cross-talk. Initialisms are listed in Appendix A.

In vitro studies have confirmed the expression of CGRP-R and NK-1R in murine and human macrophages, implying that neuropeptides CGRP and SP influence the activities of these immune cells [99,151]. Pro-inflammatory cytokines IL-1, IL-4, and interferon (IFN)-γ were also shown to induce NK-1R expression in macrophages [152,153]. Goldman et al. were among the first to conduct in vitro studies on peritoneal macrophages, showing enhanced phagocytosis when exogenous SP was applied to these cells [154]. The SP-mediated activation of macrophages was discovered when exposure to SP increased levels of eicosanoids in macrophages [155]. SP also acts as a chemoattractant for monocytes and macrophages, first demonstrated in vitro using human monocytes and peritoneal macrophages from guinea pigs [156,157]. Evidence of TNF-α, IL-1, and IL-6 release in macrophages in response to SP was also previously recorded [158]. In contrast, CGRP was found to actively prevent the activation of monocytes and macrophages [102,103]. Exogenous CGRP reduced proliferation in peripheral blood mononuclear cells and inhibited macrophage production of hydrogen peroxide, which is considered a classic response to inflammation [102,103].

Early identification of CGRP immunoreactive (CGRP-IR) nerve fibers in the PDL led to increased interest in studying their response to different stimuli. Dynamic changes were observed in CGRP-IR nerve densities surrounding blood vessels in PDL during tooth movement in rat molars [159]. Wakisaka et al. were among the first to study the distribution of SP-positive nerve fibers in rat PDL [160]. However, when immunoreactions for CGRP, SP, and neuropeptide Y (NPY) were evaluated in dental pulp, periodontal ligament, and gingiva in cats, denser innervation in the pulp was reported, and these nerves were found to be more immunoreactive to CGRP than to SP [161]. A similar trend was observed by the same group in a ferret model, affirming a similar abundance in sensory nerve innervation across different species [162]. Hence, several subsequent studies focused on studying CGRP-positive neurons. Kimberly and Byers induced pulpitis that led to periapical lesions in rat models and monitored changes in CGRP-IR nerves to inflammation for 35 days [163]. They observed that while inflammation initially caused nerve damage, increased neural sprouting was seen along the wound border. Even as inflammation advanced in severity and transitioned from the pulp to the periapical regions, CGRP-IR nerves sprouted extensively in both tissues. Independent studies reported similar patterns of neural sprouting in rat models, linking the response of pulpal innervation to inflammation and the induction of periapical lesions [164].

Dynamic changes in the distribution of sensory nerves during acute and chronic AP have drawn attention to understanding how these specialized neurons may modulate disease progression. Several studies measured increased levels of neuropeptides in saliva and gingival crevicular fluid (GCF) in animal models and in patients diagnosed with AP [165,166]. Similarly, investigations on the distribution patterns of neuropeptides in periodontitis suggested the potential of using neuropeptides as biomarkers for the early detection of periodontitis [167,168]. A significant increase in SP-like immunoreactivity was measured in GCF collected from periodontitis sites of volunteers [169]. Studies also found a significant correlation between SP and proinflammatory IL-1β, TNF-α, IL-8, and chemoattractant MCP-1 (monocyte chemoattractant protein-1) in GCF from patients with periodontitis [170]. CGRP exhibited a different profile in comparison; reduced CGRP immunoreactivity was observed in periodontitis sites compared to healthy periodontium [97]. Further analysis showed that carboxypeptidase activity in GCF resulted in a rapid breakdown of CGRP; however, this did not affect the levels of SP in GCF [171].

Macrophages express functional CGRP-R, i.e., receptors of CGRP. CGRP activation led to the proliferation and differentiation of murine macrophages into osteoclast-like cells via a cAMP/PKA-dependent signaling pathway [172]. Murine bone-marrow-derived macrophages (BMDMs) showed increased IL-6 expression as a response to CGRP activation, which led to speculations about the role of BMDMs in the negative inhibition of B cell development [173]. Similar findings were reported in rat models that demonstrated reduced IL-6 expression in invading macrophages upon perineural injection of a CGRP antagonist [100]. In vitro studies have shown that LPS-induced expression of pro-inflammatory cytokines, including IL-1β, IL-6, contributes to CGRP expression in macrophages [99].

Primarily produced by sensory neurons in the periodontal tissues, NGF is a neurotrophic factor that plays a critical role in neurogenesis, promoting the growth, differentiation, and survival of neurons [174,175]. Nerve Growth Factor-Tropomyosin receptor kinase A (NGF/TrkA) signaling suppresses CGRP production in monocytes as well as RAW264.7 macrophages [99]. In periodontal tissues that have a rich sensory nerve supply, NGF can induce increased neural sprouting and innervation under inflammation. NGF also controls the expression and release of neuropeptides in sensory neurons. NGF-mediated neural responses have direct implications on the host tissue response to inflammation [176,177]. NGF has been shown to induce monocyte differentiation in macrophages, thereby enhancing inflammatory effects [178]. Non-neural cells such as fibroblasts and osteoblasts can produce NGF in periodontal tissues, and several non-neural cell types express TrkA, which is a high-affinity NGF receptor molecule [179,180]. Hematopoietic stem cells express TrkA and produce their own NGF to support differentiation in these cells. This expression is maintained so that neutrophils, mast cells, and monocytes that arise from these stem cells also express TrkA. An interesting finding was that the expression of TrkA in monocytes and macrophages had functional implications that increased when these cells were stimulated with bacterial toxins [178,181]. For example, TrkA-stimulated human monocytes showed an upregulation of anti-apoptotic genes [182]. Another study showed that monocytes production of CGRP was reduced in response to LPS when NGF was neutralized [183]. TrkA-dependent ruffling of human-monocyte-derived macrophage cell membrane, induced by NGF, was found to increase phagocytic activity [184]. NGF also induced increased the macrophage expression of matrix metalloproteinase (MMP)-9 in this study. Although the functions and release mechanisms of neurotrophins in immune cells remain elusive, current knowledge substantiates the role of sensory-neuron-derived NGF in macrophage functions under inflammation.

SP was found to bind NK-1R receptors on murine macrophages and activate protein kinase C (PKC) or phosphatidylinositol 3-kinase/protein kinase B (PI3K/Akt), which triggers nuclear factor kappa-light-chain-enhancer of activated B cells transactivation and chemokine response via ERK1/2 and p38 MAPK signaling pathways [185,186]. Both resulted in the expression of the chemokines macrophage inflammatory protein-2 (MIP-2) and MCP-1. MIP-2 is a potent chemoattractant for neutrophils, and MCP-1 regulates monocyte migration and infiltration [187,188]. Macrophages produce SP in response to NF-κB activation, linking this phenomenon with the pro-inflammatory M1 macrophage phenotype. The administration of exogenous SP has been found to increase IL-10 expression in rat models, creating conducive environment for the activation of anti-inflammatory M2 macrophages, which can also produce SP [189,190]. SP was found to induce hemeoxygenase-1 (HO-1), an antioxidant, in LPS-stimulated murine macrophages in a dose-dependent manner [191]. HO-1 is expressed in response to oxidative stress and functions to suppress inflammation by inducing macrophage polarization in favor of an M2 anti-inflammatory phenotype [192]. SP also induced M2-like macrophages by activating PI3K/Akt in murine macrophages [193]. Hence, there is evidence of both pro- and anti-inflammatory effects of SP, which warrants detailed studies in AP.

7. Association of Pain with Nociceptor-Macrophage Cross-talk in AP

Pain is a natural response to tissue damage, and current investigations highlight a close association of this sensation with nociceptor–immune-cell interactions [194]. Several aspects can contribute to the presentation of pain in endodontic infections, including the volume of root canal space, the presence of pathogen diversity, the virulence of pathogenic load, the state of the immune system in an individual, and lesion dynamics, among others [195]. Despite standardized systems such as Numeric Rating Scale (NRS-11) and periapical index (PAI), which can correlate pain with clinical presentations in AP, these cannot be used to predict when and why pain is not experienced by patients with chronic AP [196,197]. Understanding pain perception from the perspectives of periapical inflammation on a molecular level may help improve diagnosis and develop better treatment plans and strategies.

Immune cells modulate nociceptor activity using cytokines and chemokines, and nociceptors in turn modulate functions of immune cells by expressing neuropeptides, resulting in a dynamic, bi-directional cross-talk between these two cell entities during tissue injury and inflammation [198] (Figure 3). Nociceptors express receptor molecules that are activated by PAMPs, DAMPs, and mediators released by immune cells. Neutrophils, mast cells, and macrophages have close encounters with nociceptive nerve terminals, and hence, mediators produced by these immune cells can induce pain sensitization and influence chronicity of pain. Macrophages produce TNF-α, IL-1β, IL-6, PGE2, and Leukotriene B4 (LTB4), covering a range of cytokines, growth factors, and lipids that bind to specific receptors on nociceptor neurons and result in direct pain sensitization or engage ion channels to increase neuronal cell excitation which then causes pain sensitization [194]. Such pain sensitization translates to hyperalgesia or allodynia in AP.

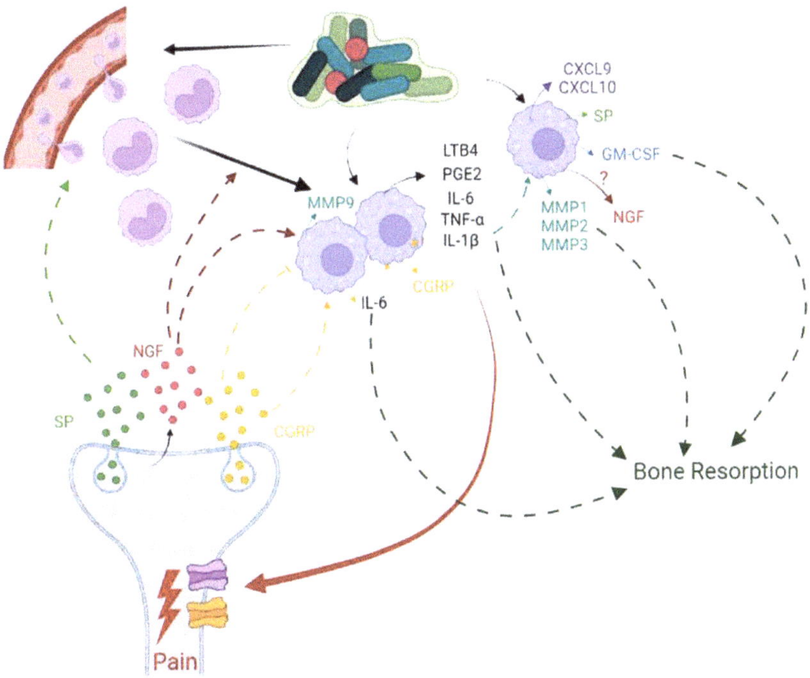

Figure 3. Bi-directional cross-talk between nociceptor neurons and macrophages has been implicated in inflammation, resulting in pain and bone resorption in apical periodontitis. Initialisms are listed in Appendix A.

Hyperalgesia and allodynia refer to increased nociceptive sensitivity in peripheral nerves. While both are caused by the reduced activation threshold of nociceptor neurons, hyperalgesia is caused by repeated or intense exposure to noxious mechanical or thermal stimuli, and allodynia is associated with non-toxic mechanical or thermal stimulation. A comparison of maximal bite force in contralateral, non-inflamed teeth in patients with acute periradicular periodontitis against teeth from healthy individuals reported allodynia in the contralateral teeth of acute AP patients. This was attributed to central sensitization, which refers to an increase in sensitivity in neurons in the central nervous system (CNS) [199,200]. Owatz et al. proposed that central sensitization may evoke mechanical allodynia in pulpal nociceptors independently of the local activation of the nociceptors in pulpal tissue [201]; this is yet to be confirmed. Studies have found correlations between nociceptive pain and neuropeptide CGRP [138,202]. Mechanical allodynia was reported when CGRP was injected into the hind paw skin of mice [203]. The intradermal injection of CGRP antagonist after 1 h of capsaicin injection resulted in a partially reversed hyperalgesia and mechanical allodynia in rats, confirming the role of CGRP in maintaining nociceptive sensitivity [204].

De França et al. evaluated M1 and M2 macrophage populations in human radicular cysts and periapical granulomas the presence of both are classic markers of apical periodontitis [205]. The authors also correlated pain with macrophage subpopulations in samples tested; a higher prevalence of macrophages was observed in areas with higher densities of microorganisms [205]. Based on sample collection, more cases with radicular cysts were symptomatic compared to periapical granulomas, meaning this trend could be attributed to the pro-inflammatory profile of cysts that were assumed to have increased levels of IL-1, IL-6, and PGE2, all of which are released by macrophages in response to microbial infection and have been identified in causing painful symptoms. However, periapical granulomas were characterized by pulsating pain in this study, and this was linked to higher inflammatory infiltrate that also had higher amounts of $CD68^+$ cell populations. A correlation between macrophages and CGRP, known to initiate and maintain pain, was reported when the increased presence of CGRP-immunoreactive fibers was observed along with an increased number of resident macrophages in rat models of AP [206]. These studies have directly or indirectly highlighted an important link between macrophages and pain sensations in AP. There is, however, a lack of consensus that successfully links macrophages and pain in AP progression from radicular cysts to periapical granulomas [205].

A recent study found correlations between the immunoexpression of MMPs and the presence of pain in patients with AP; periapical granulomas and periapical cysts in teeth extracted from patients who were evaluated for the presence of pain showed high expression levels of MMP-2 [207]. MMPs are a class of enzymes that participate in the degradation of matrix proteins and bone, contributing to the processes of tissue degradation and remodeling. Categorized into different classes such as collagenases, gelatinases, stromelysins, and others, MMPs are also capable of cleaving cytokines and chemokines, thereby modulating immune response. MMP-1, produced by activated macrophages, has been found in increased concentrations in rat and mouse models of periapical lesions [208,209]. MMP-2, MMP-3, MMP-8, MMP-9, and MMP-13 have also been identified in human periapical lesions; of these, macrophages have been shown to express all of these in AP [210–212]. Although their mechanism of action is not fully understood, it is known that, in response to different cues, MMPs are secreted by different cell types into the extracellular matrix, where they are activated proteolytically and their actions are regulated by endogenous inhibitors called tissue inhibitors of metalloproteinases (TIMPs) [213,214]. In the context of AP, pro-inflammatory cytokines TNF-α and IL-1β are known to activate MMP-1 and MMP-2; macrophages are a major source of these cytokines [213,214]. Stimulating pain in the upper central incisor of human volunteers resulted in elevated levels of MMP-8 in the gingival crevicular fluid [112]. This study observed an increase in SP levels in the GCF of stimulated teeth, thereby correlating neuronal activation associated with pain with immune activation.

8. Conclusions

The discovery of neurogenic inflammation highlighted the role of nociceptors in the immune response. The dental pulp and periapical tissues are rich in sensory nerves, and nociceptors are now known to activate key elements of the innate immune system besides sending pain signals to indicate infection. Neuropeptides and inflammatory mediators released by both nociceptors and immune cells regulate neuro-immune cross-talk. Animal models and in vitro studies have led to a deeper understanding of neuro-immune cross-talk; however, we are yet to understand how microbial pathogens, immune cells, and nociceptors interact with each other resulting in the modulation of pain and inflammation in AP. Chronic AP is especially challenging to treat or manage because it is usually asymptomatic, persisting in a state of continued inflammation with varying presenting symptoms. Macrophages play a critical role in sensitizing and desensitizing nociceptors, leading to hyperalgesia, allodynia, and central sensitization in patients with AP. A comprehensive insight into the molecular mechanisms that connect neuro-immune interactions, inflammation, and pain is essential to understand periapical host response in symptomatic and asymptomatic AP while offering significant translational potential.

Author Contributions: A.K. contributed to the conception of the review. N.M. and A.K. wrote the manuscript with supervision from A.K. Both authors provided critical feedback and helped to shape the manuscript. All authors have read and agreed to the published version of the manuscript.

Funding: The authors gratefully acknowledge the financial support from the Natural Sciences and Engineering Research Council of Canada (Discovery grant) (AK-RGPIN-2020-05844), Canada Research Chair program (AK), and the Dr. Lloyd and Mrs. Kay Chapman Chairship.

Institutional Review Board Statement: Not applicable.

Informed Consent Statement: Not applicable.

Data Availability Statement: No new data were created or analyzed in this study. Data sharing is not applicable to this article.

Conflicts of Interest: The authors declare no conflict of interest.

Appendix A. List of Acronyms

AP	Apical periodontitis
BMDMs	Bone-marrow-derived macrophages
CAM-1	Cellular adhesion molecule 1
cAMP	Cyclical adenosine monophosphate
CD	Cluster of differentiation
CGRP	Calcitonin Gene-Related Peptide
CGRP-IR	CGRP immunoreactive
CLR	Calcitonin-like receptor
CNS	Central nervous system
CREB	cAMP response element binding
CXC-L	Chemokine C-X-C motif chemokine ligand
DAG	Diacylglycerol
DAMPs	Damage-associated molecular patterns
ERKs	Extracellular signal-regulated kinases
GCF	Gingival crevicular fluid
GM-CSF	Granulocyte-Macrophage Colony-Stimulating Factor
HO-1	Hemeoxygenase-1
IFN-γ	Interferon-γ
IL-1	Interleukin-1
IP3	Inositol triphosphate
IRAK	Interleukin-1 receptor-associated kinase
LDCVs	Large dense core vesicles
LPS	Lipopolysaccharides
LTB4	Leukotriene B4
M1	Classically activated macrophages
M2	Alternatively activated macrophages
MAPK	Mitogen-activated protein kinase
MCP-1	Monocyte chemoattractant protein-1
MD2	Myeloid differentiation 2
MIP-2	Macrophage Inflammatory Protein-2
MMPs	Matrix metalloproteinases
MQ	Macrophage
MyD88	Myeloid differentiation primary response 88
NFκB	Nuclear Factor Kappa B
NGF	Nerve Growth Factor
NI	Neurogenic Inflammation
NKA	Neurokinin A
NPs	Neuropeptides
NP γ	Neuropeptide γ
NPK	Neuropeptide K
NPY	Neuropeptide Y
NRS-11	Numeric Rating Scale
OPG	Osteoprotegerin
PACAP	Pituitary adenylate cyclase-activating peptide
PAI	Periapical index
PAMPs	Pathogen-associated molecular patterns
PDL	Periodontal ligament
PGE2	Prostaglandin E2
PI3K/Akt	Phosphatidylinositol 3-kinase/protein kinase B
PKC	Protein Kinase C
RANKL	Receptor activator of nuclear factor kappa-β ligand
SNARE	Soluble N-ethylmaleimide-sensitive factor attachment protein receptor
SP	Substance P
TG	Trigeminal ganglion
TGF-β	Transforming growth factor-β
TIMPs	Tissue inhibitors of metalloproteinases
TLR	Toll-like receptor
TNF-α	Tumor necrosis factor-α
TRAF-6	Tumor necrosis factor Receptor Associated Factor-6
TrkA	Tropomyosin receptor kinase A
TRP	Transient receptor potential
VIP	Vasoactive intestinal peptide

References

1. Nair, P.N.R. On the causes of persistent apical periodontitis: A review. *Int. Endod. J.* **2006**, *39*, 249–281. [CrossRef] [PubMed]
2. Korona-Glowniak, I.; Piatek, D.; Fornal, E.; Lukowiak, A.; Gerasymchuk, Y.; Kedziora, A.; Bugla-Płoskonska, G.; Grywalska, E.; Bachanek, T.; Malm, A. Patterns of Oral Microbiota in Patients with Apical Periodontitis. *J. Clin. Med.* **2021**, *10*, 2707. [CrossRef] [PubMed]
3. Hassell, M.T. Tissues and cells of the periodontium. *Periodontology 2000* **1993**, *3*, 9–38. [CrossRef]
4. Nanci, A.; Bosshardt, D.D. Structure of periodontal tissues in health and disease*. *Periodontology 2000* **2006**, *40*, 11–28. [CrossRef] [PubMed]
5. Hussein, H.; Kishen, A. Local Immunomodulatory Effects of Intracanal Medications in Apical Periodontitis. *J. Endod.* **2022**, *48*, 430–456. [CrossRef] [PubMed]
6. Jonasson, P.; Kvist, T. Diagnosis of apical periodontitis in root-filled teeth. *Clin. Dent. Rev.* **2018**, *2*, 15. [CrossRef]
7. Li, F.-C.; Kishen, A. 3D Organoids for Regenerative Endodontics. *Biomolecules* **2023**, *13*, 900. [CrossRef]
8. Chiu, I.M.; von Hehn, C.A.; Woolf, C.J. Neurogenic inflammation and the peripheral nervous system in host defense and immunopathology. *Nat. Neurosci.* **2012**, *15*, 1063–1067. [CrossRef]
9. Assas, B.M.; Pennock, J.I.; Miyan, J.A. Calcitonin gene-related peptide is a key neurotransmitter in the neuro-immune axis. *Front. Neurosci.* **2014**, *8*, 23. [CrossRef]
10. Kabata, H.; Artis, D. Neuro-immune crosstalk and allergic inflammation. *J. Clin. Investig.* **2019**, *129*, 1475–1482. [CrossRef]
11. Crosson, T.; Roversi, K.; Balood, M.; Othman, R.; Ahmadi, M.; Wang, J.-C.; Seadi Pereira, P.J.; Tabatabaei, M.; Couture, R.; Eichwald, T.; et al. Profiling of how nociceptor neurons detect danger—New and old foes. *J. Intern. Med.* **2019**, *286*, 268–289. [CrossRef] [PubMed]
12. Baliu-Piqué, M.; Jusek, G.; Holzmann, B. Neuroimmunological communication via CGRP promotes the development of a regulatory phenotype in TLR4-stimulated macrophages. *Eur. J. Immunol.* **2014**, *44*, 3708–3716. [CrossRef] [PubMed]
13. Ji, R.-R.; Xu, Z.-Z.; Gao, Y.-J. Emerging targets in neuroinflammation-driven chronic pain. *Nat. Rev. Drug Discov.* **2014**, *13*, 533–548. [CrossRef] [PubMed]
14. Matsuda, M.; Huh, Y.; Ji, R.-R. Roles of inflammation, neurogenic inflammation, and neuroinflammation in pain. *J. Anesth.* **2019**, *33*, 131–139. [CrossRef] [PubMed]
15. Zehnder, M.; Belibasakis, G.N. On the dynamics of root canal infections—What we understand and what we don't. *Virulence* **2015**, *6*, 216–222. [CrossRef]
16. Azeredo, S.V.; Brasil, S.C.; Antunes, H.; Marques, F.V.; Pires, F.R.; Armada, L. Distribution of macrophages and plasma cells in apical periodontitis and their relationship with clinical and image data. *J. Clin.* **2017**, *9*, e1060–e1065. [CrossRef]
17. Márton, I.J.; Kiss, C. Overlapping Protective and Destructive Regulatory Pathways in Apical Periodontitis. *J. Endod.* **2014**, *40*, 155–163. [CrossRef]
18. Nair, P.N.R. Pathogenesis of Apical Periodontitis and the Causes of Endodontic Failures. *Crit. Rev. Oral Biol. Med.* **2004**, *15*, 348–381. [CrossRef]
19. Akamine, A.; Anan, H.; Hamachi, T.; Maeda, K. A histochemical study of the behavior of macrophages during experimental apical periodontitis in rats. *J. Endod.* **1994**, *20*, 474–478. [CrossRef]
20. Sims, M.R. A model of the anisotropic distribution of microvascular volume in the periodontal ligament of the mouse mandibular molar. *Aust. Orthod. J.* **1987**, *10*, 21–24.
21. Masset, A.; Staszyk, C.; Gasse, H. The blood vessel system in the periodontal ligament of the equine cheek teeth—Part II: The micro-architecture and its functional implications in a constantly remodelling system. *Ann. Anat. Anat. Anz.* **2006**, *188*, 535–539. [CrossRef] [PubMed]
22. Graunaite, I.; Lodiene, G.; Maciulskiene, V. Pathogenesis of apical periodontitis: A literature review. *J. Oral Maxillofac. Res.* **2012**, *2*, e1. [CrossRef] [PubMed]
23. Cekici, A.; Kantarci, A.; Hasturk, H.; Van Dyke, T.E. Inflammatory and immune pathways in the pathogenesis of periodontal disease. *Periodontology 2000* **2014**, *64*, 57–80. [CrossRef] [PubMed]
24. Davies, L.C.; Jenkins, S.J.; Allen, J.E.; Taylor, P.R. Tissue-resident macrophages. *Nat. Immunol.* **2013**, *14*, 986–995. [CrossRef]
25. Yang, J.; Zhang, L.; Yu, C.; Yang, X.-F.; Wang, H. Monocyte and macrophage differentiation: Circulation inflammatory monocyte as biomarker for inflammatory diseases. *Biomark. Res.* **2014**, *2*, 1. [CrossRef]
26. Molteni, M.; Gemma, S.; Rossetti, C. The Role of Toll-Like Receptor 4 in Infectious and Noninfectious Inflammation. *Mediat. Inflamm.* **2016**, *2016*, 6978936. [CrossRef]
27. Rayees, S.; Rochford, I.; Joshi, J.C.; Joshi, B.; Banerjee, S.; Mehta, D. Macrophage TLR4 and PAR2 Signaling: Role in Regulating Vascular Inflammatory Injury and Repair. *Front. Immunol.* **2020**, *11*, 2091. [CrossRef]
28. Perros, F.; Lambrecht, B.N.; Hammad, H. TLR4 signalling in pulmonary stromal cells is critical for inflammation and immunity in the airways. *Respir. Res.* **2011**, *12*, 125. [CrossRef]
29. Garlet, G.P.; Giannobile, W.V. Macrophages: The Bridge between Inflammation Resolution and Tissue Repair? *J. Dent. Res.* **2018**, *97*, 1079–1081. [CrossRef]
30. House, I.G.; Savas, P.; Lai, J.; Chen, A.X.Y.; Oliver, A.J.; Teo, Z.L.; Todd, K.L.; Henderson, M.A.; Giuffrida, L.; Petley, E.V.; et al. Macrophage-Derived CXCL9 and CXCL10 Are Required for Antitumor Immune Responses Following Immune Checkpoint Blockade. *Clin. Cancer Res.* **2020**, *26*, 487–504. [CrossRef]

31. Yao, Y.; Cai, X.; Ren, F.; Ye, Y.; Wang, F.; Zheng, C.; Qian, Y.; Zhang, M. The Macrophage-Osteoclast Axis in Osteoimmunity and Osteo-Related Diseases. *Front. Immunol.* **2021**, *12*, 664871. [CrossRef] [PubMed]
32. Italiani, P.; Boraschi, D. From Monocytes to M1/M2 Macrophages: Phenotypical vs. Functional Differentiation. *Front. Immunol.* **2014**, *5*, 514. [CrossRef] [PubMed]
33. Lappin, D.F.; Macleod, C.P.; Kerr, A.; Mitchell, T.J.; Kinane, D.F. Anti-inflammatory cytokine IL-10 and T cell cytokine profile in periodontitis granulation tissue. *Clin. Exp. Immunol.* **2001**, *123*, 294–300. [CrossRef] [PubMed]
34. Zhu, L.; Hua, F.; Ding, W.; Ding, K.; Zhang, Y.; Xu, C. The correlation between the Th17/Treg cell balance and bone health. *Immun. Ageing* **2020**, *17*, 30. [CrossRef]
35. Stern, M.H.; Dreizen, S.; Mackler, B.F.; Selbst, A.G.; Levy, B.M. Quantitative analysis of cellular composition of human periapical granuloma. *J. Endod.* **1981**, *7*, 117–122. [CrossRef]
36. Shapouri-Moghaddam, A.; Mohammadian, S.; Vazini, H.; Taghadosi, M.; Esmaeili, S.-A.; Mardani, F.; Seifi, B.; Mohammadi, A.; Afshari, J.T.; Sahebkar, A. Macrophage plasticity, polarization, and function in health and disease. *J. Cell. Physiol.* **2018**, *233*, 6425–6440. [CrossRef] [PubMed]
37. Lukić, A.; Danilovic, V.; Petrović, R. Comparative immunohistochemical and quantitative analysis of inflammatory cells in symptomatic and asymptomatic chronic periapical lesions. *Vojnosanit. Pregl.* **2008**, *65*, 435–440. [CrossRef]
38. Kawai, T.; Matsuyama, T.; Hosokawa, Y.; Makihira, S.; Seki, M.; Karimbux, N.Y.; Goncalves, R.B.; Valverde, P.; Dibart, S.; Li, Y.-P.; et al. B and T lymphocytes are the primary sources of RANKL in the bone resorptive lesion of periodontal disease. *Am. J. Pathol.* **2006**, *169*, 987–998. [CrossRef]
39. Dallas, S.L.; Bonewald, L.F. Dynamics of the transition from osteoblast to osteocyte. *Ann. N. Y. Acad. Sci.* **2010**, *1192*, 437–443. [CrossRef]
40. Pérez-Sayáns, M.; Somoza-Martín, J.M.; Barros-Angueira, F.; Rey, J.M.G.; García-García, A. RANK/RANKL/OPG role in distraction osteogenesis. *Oral Surg. Oral Med. Oral Pathol. Oral Radiol. Endod.* **2010**, *109*, 679–686. [CrossRef]
41. Wada, T.; Nakashima, T.; Hiroshi, N.; Penninger, J.M. RANKL–RANK signaling in osteoclastogenesis and bone disease. *Trends Mol. Med.* **2006**, *12*, 17–25. [CrossRef]
42. Boyce, B.F.; Xing, L. Functions of RANKL/RANK/OPG in bone modeling and remodeling. *Arch. Biochem. Biophys.* **2008**, *473*, 139–146. [CrossRef] [PubMed]
43. Ushach, I.; Zlotnik, A. Biological role of granulocyte macrophage colony-stimulating factor (GM-CSF) and macrophage colony-stimulating factor (M-CSF) on cells of the myeloid lineage. *J. Leukoc. Biol.* **2016**, *100*, 481–489. [CrossRef] [PubMed]
44. Lee, M.S.; Kim, H.S.; Yeon, J.-T.; Choi, S.-W.; Chun, C.H.; Kwak, H.B.; Oh, J. GM-CSF Regulates Fusion of Mononuclear Osteoclasts into Bone-Resorbing Osteoclasts by Activating the Ras/ERK Pathway1. *J. Immunol.* **2009**, *183*, 3390–3399. [CrossRef] [PubMed]
45. Hienz, S.A.; Paliwal, S.; Ivanovski, S. Mechanisms of Bone Resorption in Periodontitis. *J. Immunol. Res.* **2015**, *2015*, 615486. [CrossRef] [PubMed]
46. Gupta, G. Gingival crevicular fluid as a periodontal diagnostic indicator- II: Inflammatory mediators, host-response modifiers and chair side diagnostic aids. *J. Med. Life* **2013**, *6*, 7–13.
47. Lee, Y.-M.; Fujikado, N.; Manaka, H.; Yasuda, H.; Iwakura, Y. IL-1 plays an important role in the bone metabolism under physiological conditions. *Int. Immunol.* **2010**, *22*, 805–816. [CrossRef]
48. Jimi, E.; Shuto, T.; Koga, T. Macrophage colony-stimulating factor and interleukin-1 alpha maintain the survival of osteoclast-like cells. *Endocrinology* **1995**, *136*, 808–811. [CrossRef]
49. Gilbert, L.; He, X.; Farmer, P.; Boden, S.; Kozlowski, M.; Rubin, J.; Nanes, M.S. Inhibition of Osteoblast Differentiation by Tumor Necrosis Factor-α*. *Endocrinology* **2000**, *141*, 3956–3964. [CrossRef]
50. Lee, H.-L.; Yi, T.; Woo, K.M.; Ryoo, H.-M.; Kim, G.-S.; Baek, J.-H. Msx2 mediates the inhibitory action of TNF-α on osteoblast differentiation. *Exp. Mol. Med.* **2010**, *42*, 437–445. [CrossRef]
51. Marahleh, A.; Kitaura, H.; Ohori, F.; Kishikawa, A.; Ogawa, S.; Shen, W.-R.; Qi, J.; Noguchi, T.; Nara, Y.; Mizoguchi, I. TNF-α Directly Enhances Osteocyte RANKL Expression and Promotes Osteoclast Formation. *Front. Immunol.* **2019**, *10*, 2925. [CrossRef] [PubMed]
52. Willis, R.D.; DiCosimo, C.J. The absence of proprioceptive nerve endings in the human periodontal ligament: The role of periodontal mechanoreceptors in the reflex control of mastication. *Oral Surg. Oral Med. Oral Pathol.* **1979**, *48*, 108–115. [CrossRef] [PubMed]
53. Byers, M.R. Sensory innervation of periodontal ligament of rat molars consists of unencapsulated Ruffini-like mechanoreceptors and free nerve endings. *J. Comp. Neurol.* **1985**, *231*, 500–518. [CrossRef] [PubMed]
54. Byers, M.R. Segregation of NGF receptor in sensory receptors, nerves and local cells of teeth and periodontium demonstrated by EM immunocytochemistry. *J. Neurocytol.* **1990**, *19*, 765–775. [CrossRef] [PubMed]
55. Lambrichts, I.; Creemers, J.; van Steenberghe, D. Morphology of neural endings in the human periodontal ligament: An electron microscopic study. *J. Periodontal Res.* **1992**, *27*, 191–196. [CrossRef] [PubMed]
56. Tabata, T.; Karita, K. Response properties of periodontal mechanosensitive fibers in the superior dental nerve of the cat. *Exp. Neurol.* **1986**, *94*, 469–478. [CrossRef]
57. Karita, K.; Tabata, T. Response fields of the periodontal mechanosensitive units in the superior alveolar nerve of the cat. *Exp. Neurol.* **1985**, *90*, 558–565. [CrossRef]

58. Tender, G.C.; Li, Y.-Y.; Cui, J.-G. Vanilloid receptor 1-positive neurons mediate thermal hyperalgesia and tactile allodynia. *Spine J.* **2008**, *8*, 351–358. [CrossRef]
59. Christine, W.; Michael, G.H. *Neuroanatomy, Touch Receptor*; StatPearls Publishing: St. Petersburg, FL, USA, 2023.
60. Mengel, M.K.C.; Jyväsjärvi, E.; Kniffki, K.-D. Identification and characterization of afferent periodontal C fibres in the cat. *PAIN* **1992**, *48*, 413–420. [CrossRef] [PubMed]
61. Bender, I.B. Pulpal Pain Diagnosis—A Review. *J. Endod.* **2000**, *26*, 175–179. [CrossRef]
62. Bayliss, W.M. On the origin from the spinal cord of the vaso-dilator fibres of the hind-limb, and on the nature of these fibres1. *J. Physiol.* **1901**, *26*, 173–209. [CrossRef] [PubMed]
63. Lewis, T. The Blood Vessels of the Human Skin. *Br. Med. J.* **1926**, *2*, 61. [CrossRef] [PubMed]
64. Bayliss, W.M. *The Vaso-Motor System/by William M. Bayliss*; Longmans, Green: New York, NY, USA, 1923.
65. Langley, J.N. Antidromic action. *J. Physiol.* **1923**, *57*, 428–446. [CrossRef] [PubMed]
66. Byers, M.R.; Taylor, P.E. Effect of Sensory Denervation on the Response of Rat Molar Pulp to Exposure Injury. *J. Dent. Res.* **1993**, *72*, 613–618. [CrossRef] [PubMed]
67. Toledo Mauriño, J.J.; Fonseca-Camarillo, G.; Furuzawa-Carballeda, J.; Barreto-Zuñiga, R.; Martínez Benítez, B.; Granados, J.; Yamamoto-Furusho, J.K. TRPV Subfamily (TRPV2, TRPV3, TRPV4, TRPV5, and TRPV6) Gene and Protein Expression in Patients with Ulcerative Colitis. *J. Immunol. Res.* **2020**, *2020*, 2906845. [CrossRef]
68. Jara-Oseguera, A.; Simon, A.S.; Rosenbaum, T. TRPV1: On the Road to Pain Relief. *Curr. Mol. Pharmacol.* **2008**, *1*, 255–269. [CrossRef]
69. Samanta, A.; Hughes, T.E.T.; Moiseenkova-Bell, V.Y. Transient Receptor Potential (TRP) Channels. In *Membrane Protein Complexes: Structure and Function*; Harris, J.R., Boekema, E.J., Eds.; Springer: Singapore, 2018; pp. 141–165. [CrossRef]
70. Zygmunt, P.M.; Högestätt, E.D. TRPA1. In *Mammalian Transient Receptor Potential (TRP) Cation Channels: Volume I*; Nilius, B., Flockerzi, V., Eds.; Springer: Berlin/Heidelberg, Germany, 2014; pp. 583–630. [CrossRef]
71. Cao, E.; Liao, M.; Cheng, Y.; Julius, D. TRPV1 structures in distinct conformations reveal activation mechanisms. *Nature* **2013**, *504*, 113–118. [CrossRef]
72. Liao, M.; Cao, E.; Julius, D.; Cheng, Y. Structure of the TRPV1 ion channel determined by electron cryo-microscopy. *Nature* **2013**, *504*, 107–112. [CrossRef]
73. Sappington, R.M.; Sidorova, T.; Long, D.J.; Calkins, D.J. TRPV1: Contribution to Retinal Ganglion Cell Apoptosis and Increased Intracellular Ca2+ with Exposure to Hydrostatic Pressure. *Investig. Ophthalmol. Vis. Sci.* **2009**, *50*, 717–728. [CrossRef]
74. King, R.; Brain, S.D. Chapter 189—CGRP. In *Handbook of Biologically Active Peptides*, 2nd ed; Kastin, A.J., Ed.; Academic Press: Boston, MA, USA, 2013; pp. 1394–1401. [CrossRef]
75. Russell, F.A.; King, R.; Smillie, S.-J.; Kodji, X.; Brain, S.D. Calcitonin Gene-Related Peptide: Physiology and Pathophysiology. *Physiol. Rev.* **2014**, *94*, 1099–1142. [CrossRef]
76. Matteoli, M.; Haimann, C.; Torri-Tarelli, F.; Polak, J.M.; Ceccarelli, B.; De Camilli, P. Differential effect of alpha-latrotoxin on exocytosis from small synaptic vesicles and from large dense-core vesicles containing calcitonin gene-related peptide at the frog neuromuscular junction. *Proc. Natl. Acad. Sci. USA* **1988**, *85*, 7366–7370. [CrossRef] [PubMed]
77. Fernandes, E.; Fernandes, M.; Keeble, J. The functions of TRPA1 and TRPV1: Moving away from sensory nerves. *Br. J. Pharmacol.* **2012**, *166*, 510–521. [CrossRef] [PubMed]
78. Omari, S.A.; Adams, M.J.; Geraghty, D.P. Chapter Six—TRPV1 Channels in Immune Cells and Hematological Malignancies. In *Advances in Pharmacology*; Geraghty, D.P., Rash, L.D., Eds.; Academic Press: Cambridge, MA, USA, 2017; Volume 79, pp. 173–198.
79. Naert, R.; López-Requena, A.; Talavera, K. TRPA1 Expression and Pathophysiology in Immune Cells. *Int. J. Mol. Sci.* **2021**, *22*, 11460. [CrossRef] [PubMed]
80. Tamás, K.; Zsolt, B.; Erzsébet, L.; Enikő, T.P.; Nóra, E.; Ágnes, C.; Zoltán, P.; István, É.; Róbert, P.; Attila, T. Tissue-Specific Regulation of Microvascular Diameter: Opposite Functional Roles of Neuronal and Smooth Muscle Located Vanilloid Receptor-1. *Mol. Pharmacol.* **2008**, *73*, 1405. [CrossRef]
81. Helley, M.P.; Abate, W.; Jackson, S.K.; Bennett, J.H.; Thompson, S.W.N. The expression of Toll-like receptor 4, 7 and co-receptors in neurochemical sub-populations of rat trigeminal ganglion sensory neurons. *Neuroscience* **2015**, *310*, 686–698. [CrossRef]
82. Wadachi, R.; Hargreaves, K. Trigeminal nociceptors express TLR-4 and CD14: A mechanism for pain due to infection. *J. Dent. Res.* **2006**, *85*, 49–53. [CrossRef]
83. Diogenes, A.; Ferraz, C.C.R.; Akopian, A.N.; Henry, M.A.; Hargreaves, K.M. LPS Sensitizes TRPV1 via Activation of TLR4 in Trigeminal Sensory Neurons. *J. Dent. Res.* **2011**, *90*, 759–764. [CrossRef]
84. Meseguer, V.; Alpizar, Y.A.; Luis, E.; Tajada, S.; Denlinger, B.; Fajardo, O.; Manenschijn, J.-A.; Fernández-Peña, C.; Talavera, A.; Kichko, T.; et al. TRPA1 channels mediate acute neurogenic inflammation and pain produced by bacterial endotoxins. *Nat. Commun.* **2014**, *5*, 3125. [CrossRef]
85. Ferraz, C.C.R.; Henry, M.A.; Hargreaves, K.M.; Diogenes, A. Lipopolysaccharide From Porphyromonas gingivalis Sensitizes Capsaicin-Sensitive Nociceptors. *J. Endod.* **2011**, *37*, 45–48. [CrossRef]
86. Lagomarsino, V.N.; Kostic, A.D.; Chiu, I.M. Mechanisms of microbial–neuronal interactions in pain and nociception. *Neurobiol. Pain* **2021**, *9*, 100056. [CrossRef]
87. Chiu, I.M.; Heesters, B.A.; Ghasemlou, N.; Von Hehn, C.A.; Zhao, F.; Tran, J.; Wainger, B.; Strominger, A.; Muralidharan, S.; Horswill, A.R.; et al. Bacteria activate sensory neurons that modulate pain and inflammation. *Nature* **2013**, *501*, 52–57. [CrossRef]

88. Sun, X.; Yang, Z.; Nie, Y.; Hou, B. Microbial Communities in the Extraradicular and Intraradicular Infections Associated with Persistent Apical Periodontitis. *Front. Cell. Infect. Microbiol.* **2022**, *11*, 798367. [CrossRef]
89. Eftekhari, S.; Salvatore, C.A.; Calamari, A.; Kane, S.A.; Tajti, J.; Edvinsson, L. Differential distribution of calcitonin gene-related peptide and its receptor components in the human trigeminal ganglion. *Neuroscience* **2010**, *169*, 683–696. [CrossRef]
90. Messlinger, K.; Balcziak, L.K.; Russo, A.F. Cross-talk signaling in the trigeminal ganglion: Role of neuropeptides and other mediators. *J. Neural Transm.* **2020**, *127*, 431–444. [CrossRef] [PubMed]
91. Messlinger, K. The big CGRP flood—Sources, sinks and signalling sites in the trigeminovascular system. *J. Headache Pain* **2018**, *19*, 22. [CrossRef] [PubMed]
92. Li, W.; Zhang, Z.; Li, X.; Cai, J.; Li, D.; Du, J.; Zhang, B.; Xiang, D.; Li, N.; Li, Y. CGRP derived from cardiac fibroblasts is an endogenous suppressor of cardiac fibrosis. *Cardiovasc. Res.* **2020**, *116*, 1335–1348. [CrossRef]
93. Kim, Y.J.; Granstein, R.D. Roles of calcitonin gene-related peptide in the skin, and other physiological and pathophysiological functions. *Brain Behav. Immun. Health* **2021**, *18*, 100361. [CrossRef] [PubMed]
94. Holzmann, B. Modulation of immune responses by the neuropeptide CGRP. *Amino Acids* **2013**, *45*, 1–7. [CrossRef] [PubMed]
95. Yang, D.; Jacobson, A.; Meerschaert, K.A.; Sifakis, J.J.; Wu, M.; Chen, X.; Yang, T.; Zhou, Y.; Anekal, P.V.; Rucker, R.A.; et al. Nociceptor neurons direct goblet cells via a CGRP-RAMP1 axis to drive mucus production and gut barrier protection. *Cell* **2022**, *185*, 4190–4205.e4125. [CrossRef]
96. Sert, S.; Sakallioğlu, U.; Lütfioğlu, M.; Aydoğdu, A.; Acarel, E.; Günaydın, M. Neurogenic inflammation in periimplant and periodontal disease: A case–control split-mouth study. *Clin. Oral Implant. Res.* **2019**, *30*, 800–807. [CrossRef] [PubMed]
97. Lundy, F.T.; Shaw, C.; McKinnell, J.; Lamey, P.-J.; Linden, G.J. Calcitonin gene-related peptide in gingival crevicular fluid in periodontal health and disease. *J. Clin. Periodontol.* **1999**, *26*, 212–216. [CrossRef] [PubMed]
98. Nelson, M.T.; Huang, Y.; Brayden, J.E.; Hescheler, J.; Standen, N.B. Arterial dilations in response to calcitonin gene-related peptide involve activation of K+ channels. *Nature* **1990**, *344*, 770–773. [CrossRef]
99. Ma, W.; Dumont, Y.; Vercauteren, F.; Quirion, R. Lipopolysaccharide induces calcitonin gene-related peptide in the RAW264. 7 macrophage cell line. *Immunology* **2010**, *130*, 399–409. [CrossRef] [PubMed]
100. Ma, W.; Quirion, R. Increased calcitonin gene-related peptide in neuroma and invading macrophages is involved in the up-regulation of interleukin-6 and thermal hyperalgesia in a rat model of mononeuropathy. *J. Neurochem.* **2006**, *98*, 180–192. [CrossRef] [PubMed]
101. Sun, Y.; Tao, R.; Zhang, M.; Cao, X.; Wang, H.; Xue, L.; Wu, M. Expression of Calcitonin Gene-related Peptide in Rat Pulp and Periodontal Tissues by Indirect Immunofluorescence Method. *Monoclon. Antibodies Immunodiagn. Immunother.* **2013**, *32*, 404–408. [CrossRef]
102. Niedermühlbichler, M.; Wiedermann, C.J. Suppression of superoxide release from human monocytes by somatostatin-related peptides. *Regul. Pept.* **1992**, *41*, 39–47. [CrossRef]
103. Fox, F.E.; Kubin, M.; Cassin, M.; Niu, Z.; Hosoi, J.; Torii, H.; Granstein, R.D.; Trinchieri, G.; Rook, A.H. Calcitonin Gene-Related Peptide Inhibits Proliferation and Antigen Presentation by Human Peripheral Blood Mononuclear Cells: Effects on B7, Interleukin 10, and Interleukin12. *J. Investig. Dermatol.* **1997**, *108*, 43–48. [CrossRef]
104. Takahashi, N.; Matsuda, Y.; Sato, K.; de Jong, P.R.; Bertin, S.; Tabeta, K.; Yamazaki, K. Neuronal TRPV1 activation regulates alveolar bone resorption by suppressing osteoclastogenesis via CGRP. *Sci. Rep.* **2016**, *6*, 29294. [CrossRef]
105. Duan, Y.; Liang, Y.; Yang, F.; Ma, Y. Neural Regulations in Tooth Development and Tooth–Periodontium Complex Homeostasis: A Literature Review. *Int. J. Mol. Sci.* **2022**, *23*, 14150. [CrossRef]
106. Johnson, M.B.; Young, A.D.; Marriott, I. The Therapeutic Potential of Targeting Substance P/NK-1R Interactions in Inflammatory CNS Disorders. *Front. Cell. Neurosci.* **2017**, *10*, 296. [CrossRef]
107. Zieglgänsberger, W. Substance P and pain chronicity. *Cell Tissue Res.* **2019**, *375*, 227–241. [CrossRef] [PubMed]
108. Mashaghi, A.; Marmalidou, A.; Tehrani, M.; Grace, P.M.; Pothoulakis, C.; Dana, R. Neuropeptide substance P and the immune response. *Cell. Mol. Life Sci.* **2016**, *73*, 4249–4264. [CrossRef]
109. Graefe, S.B.; Mohiuddin, S.S. *Biochemistry, Substance P.*; StatPearls Publishing: St. Petersburg, FL, USA, 2022.
110. Perner, C.; Flayer, C.H.; Zhu, X.; Aderhold, P.A.; Dewan, Z.N.A.; Voisin, T.; Camire, R.B.; Chow, O.A.; Chiu, I.M.; Sokol, C.L. Substance P Release by Sensory Neurons Triggers Dendritic Cell Migration and Initiates the Type-2 Immune Response to Allergens. *Immunity* **2020**, *53*, 1063–1077.e1067. [CrossRef] [PubMed]
111. Pradeep, A.R.; Raj, S.; Aruna, G.; Chowdhry, S. Gingival crevicular fluid and plasma levels of neuropeptide Substance-P in periodontal health, disease and after nonsurgical therapy. *J. Periodontal Res.* **2009**, *44*, 232–237. [CrossRef] [PubMed]
112. Avellán, N.-L.; Sorsa, T.; Tervahartiala, T.; Forster, C.; Kemppainen, P. Experimental tooth pain elevates substance P and matrix metalloproteinase-8 levels in human gingival crevice fluid. *Acta Odontol. Scand.* **2008**, *66*, 18–22. [CrossRef]
113. Sacerdote, P.; Levrini, L. Peripheral Mechanisms of Dental Pain: The Role of Substance P. *Mediat. Inflamm.* **2012**, *2012*, 951920. [CrossRef]
114. Niedermair, T.; Schirner, S.; Seebröker, R.; Straub, R.H.; Grässel, S. Substance P modulates bone remodeling properties of murine osteoblasts and osteoclasts. *Sci. Rep.* **2018**, *8*, 9199. [CrossRef]
115. Yan, K.; Lin, Q.; Tang, K.; Liu, S.; Du, Y.; Yu, X.; Li, S. Substance P participates in periodontitis by upregulating HIF-1α and RANKL/OPG ratio. *BMC Oral Health* **2020**, *20*, 27. [CrossRef]

116. Kim, S.; Dorscher-Kim, J. Effects of Substance P on Blood Flow of the Dental Pulp in Dogs. In Proceedings of the Substance P and Neurokinins; Springer-Verlag New York Inc.: New York, NY, USA, 1987; pp. 197–199.
117. Augustyniak, D.; Roszkowiak, J.; Wiśniewska, I.; Skała, J.; Gorczyca, D.; Drulis-Kawa, Z. Neuropeptides SP and CGRP Diminish the *Moraxella catarrhalis* Outer Membrane Vesicle- (OMV-) Triggered Inflammatory Response of Human A549 Epithelial Cells and Neutrophils. *Mediat. Inflamm.* **2018**, *2018*, 4847205. [CrossRef]
118. Iwasaki, M.a.A.Y.; Kaunitz, J.D. Recent advances in vasoactive intestinal peptide physiology and pathophysiology: Focus on the gastrointestinal system [version 1; peer review: 4 approved]. *F1000Research* **2019**, *8*, 1629. [CrossRef]
119. Linden, G.J.; Mullally, B.H.; Burden, D.J.; Lamey, P.-J.; Shaw, C.; Ardill, J.; Lundy, F.T. Changes in vasoactive intestinal peptide in gingival crevicular fluid in response to periodontal treatment. *J. Clin. Periodontol.* **2002**, *29*, 484–489. [CrossRef] [PubMed]
120. Luthman, J.; Friskopp, J.; Dahliöf, G.; Ahlström, U.; Sjöström, L.; Johansson, O. Immunohistochemical study of neurochemical markers in gingiva obtained from periodontitis-affected sites. *J. Periodontal Res.* **1989**, *24*, 267–278. [CrossRef] [PubMed]
121. Abualsaud, N.; Caprio, L.; Galli, S.; Krawczyk, E.; Alamri, L.; Zhu, S.; Gallicano, G.I.; Kitlinska, J. Neuropeptide Y/Y5 Receptor Pathway Stimulates Neuroblastoma Cell Motility Through RhoA Activation. *Front. Cell Dev. Biol.* **2021**, *8*, 627090. [CrossRef] [PubMed]
122. Zhang, Y.; Liu, C.-Y.; Chen, W.-C.; Shi, Y.-C.; Wang, C.-M.; Lin, S.; He, H.-F. Regulation of neuropeptide Y in body microenvironments and its potential application in therapies: A review. *Cell Biosci.* **2021**, *11*, 151. [CrossRef]
123. Chen, W.-c.; Liu, Y.-b.; Liu, W.-f.; Zhou, Y.-y.; He, H.-f.; Lin, S. Neuropeptide Y Is an Immunomodulatory Factor: Direct and Indirect. *Front. Immunol.* **2020**, *11*, 580378. [CrossRef]
124. Haririan, H.; Andrukhov, O.; Böttcher, M.; Pablik, E.; Wimmer, G.; Moritz, A.; Rausch-Fan, X. Salivary neuropeptides, stress, and periodontitis. *J. Periodontol.* **2018**, *89*, 9–18. [CrossRef] [PubMed]
125. Winning, L.; El Karim, I.A.; Linden, G.J.; Irwin, C.R.; Killough, S.A.; Lundy, F.T. Differential regulation of NPY and SP receptor expression in STRO-1+ve PDLSCs by inflammatory cytokines. *J. Periodontal Res.* **2022**, *57*, 186–194. [CrossRef]
126. Okragly, A.J.; Morin, S.M.; DeRosa, D.; Martin, A.P.; Johnson, K.W.; Johnson, M.P.; Benschop, R.J. Human mast cells release the migraine-inducing factor pituitary adenylate cyclase-activating polypeptide (PACAP). *Cephalalgia* **2018**, *38*, 1564–1574. [CrossRef]
127. Maugeri, G.; D'Amico, A.G.; Musumeci, G.; Reglodi, D.; D'Agata, V. Effects of PACAP on Schwann Cells: Focus on Nerve Injury. *Int. J. Mol. Sci.* **2020**, *21*, 8233. [CrossRef]
128. Kuri, B.A.; Chan, S.-A.; Smith, C.B. PACAP regulates immediate catecholamine release from adrenal chromaffin cells in an activity-dependent manner through a protein kinase C-dependent pathway. *J. Neurochem.* **2009**, *110*, 1214–1225. [CrossRef]
129. Edvinsson, J.C.A.; Grell, A.-S.; Warfvinge, K.; Sheykhzade, M.; Edvinsson, L.; Haanes, K.A. Differences in pituitary adenylate cyclase-activating peptide and calcitonin gene-related peptide release in the trigeminovascular system. *Cephalalgia* **2020**, *40*, 1296–1309. [CrossRef] [PubMed]
130. Delgado, M.; Munoz-Elias, E.J.; Gomariz, R.P.; Ganea, D. VIP and PACAP inhibit IL-12 production in LPS-stimulated macrophages. Subsequent effect on IFNγ synthesis by T cells. *J. Neuroimmunol.* **1999**, *96*, 167–181. [CrossRef]
131. Trafton, J.A.; Abbadie, C.; Basbaum, A.I. Differential Contribution of Substance P and Neurokinin A to Spinal Cord Neurokinin-1 Receptor Signaling in the Rat. *J. Neurosci.* **2001**, *21*, 3656–3664. [CrossRef]
132. Onaga, T. Tachykinin: Recent developments and novel roles in health and disease. *Biomol. Concepts* **2014**, *5*, 225–243. [CrossRef] [PubMed]
133. Mishra, A.; Lal, G. Neurokinin receptors and their implications in various autoimmune diseases. *Curr. Res. Immunol.* **2021**, *2*, 66–78. [CrossRef] [PubMed]
134. Lundy, F.T.; El Karim, I.A.; Linden, G.J. Neuropeptide Y (NPY) and NPY Y1 receptor in periodontal health and disease. *Arch. Oral Biol.* **2009**, *54*, 258–262. [CrossRef] [PubMed]
135. Amara, S.G.; Jonas, V.; Rosenfeld, M.G.; Ong, E.S.; Evans, R.M. Alternative RNA processing in calcitonin gene expression generates mRNAs encoding different polypeptide products. *Nature* **1982**, *298*, 240–244. [CrossRef]
136. Fisher, L.A.; Kikkawa, D.O.; Rivier, J.E.; Amara, S.G.; Evans, R.M.; Rosenfeld, M.G.; Vale, W.W.; Brown, M.R. Stimulation of noradrenergic sympathetic outflow by calcitonin gene-related peptide. *Nature* **1983**, *305*, 534–536. [CrossRef]
137. Brain, S.D.; Williams, T.J.; Tippins, J.R.; Morris, H.R.; MacIntyre, I. Calcitonin gene-related peptide is a potent vasodilator. *Nature* **1985**, *313*, 54–56. [CrossRef]
138. Iyengar, S.; Ossipov, M.H.; Johnson, K.W. The role of calcitonin gene–related peptide in peripheral and central pain mechanisms including migraine. *Pain* **2017**, *158*, 543. [CrossRef]
139. Hay, D.L.; Garelja, M.L.; Poyner, D.R.; Walker, C.S. Update on the pharmacology of calcitonin/CGRP family of peptides: IUPHAR Review 25. *Br. J. Pharmacol.* **2018**, *175*, 3–17. [CrossRef]
140. v. Euler, U.S.; Gaddum, J.H. An unidentified depressor substance in certain tissue extracts. *J. Physiol.* **1931**, *72*, 74–87. [CrossRef]
141. Krause, J.E.; Takeda, Y.; Hershey, A.D. Structure, Functions, and Mechanisms of Substance P Receptor Action. *J. Investig. Dermatol.* **1992**, *98*, S2–S7. [CrossRef]
142. Arslan, H.; Yıldız, E.D.; Köseoğlu, S. Effects of endodontic treatment on salivary levels of CGRP and substance P: A pilot study. *Restor. Dent. Endod.* **2020**, *45*, e40. [CrossRef]
143. Wakisaka, S. Neuropeptides in the dental pulp: Distribution, origins, and correlation. *J. Endod.* **1990**, *16*, 67–69. [CrossRef] [PubMed]

144. Caviedes-Bucheli, J.; Muñoz, H.R.; Azuero-Holguín, M.M.; Ulate, E. Neuropeptides in Dental Pulp: The Silent Protagonists. *J. Endod.* **2008**, *34*, 773–788. [CrossRef]
145. Kato, J.; Wakisaka, S.; Kurisu, K. Immunohistochemical Changes in the Distribution of Nerve Fibers in the Periodontal Ligament during an Experimental Tooth Movement of the Rat Molar. *Acta Anat.* **2008**, *157*, 53–62. [CrossRef]
146. Caviedes-Bucheli, J.; Castellanos, F.; Vasquez, N.; Ulate, E.; Munoz, H.R. The influence of two reciprocating single-file and two rotary-file systems on the apical extrusion of debris and its biological relationship with symptomatic apical periodontitis. A systematic review and meta-analysis. *Int. Endod. J.* **2016**, *49*, 255–270. [CrossRef] [PubMed]
147. Lundy, F.T.; Mullally, B.H.; Burden, D.J.; Lamey, P.-J.; Shaw, C.; Linden, G.J. Changes in substance P and neurokinin A in gingival crevicular fluid in response to periodontal treatment. *J. Clin. Periodontol.* **2000**, *27*, 526–530. [CrossRef] [PubMed]
148. Burgener, B.; Ford, A.R.; Situ, H.; Fayad, M.I.; Hao, J.J.; Wenckus, C.S.; Johnson, B.R.; BeGole, E.A.; George, A. Biologic Markers for Odontogenic Periradicular Periodontitis. *J. Endod.* **2010**, *36*, 1307–1310. [CrossRef] [PubMed]
149. Stashenko, P.; Teles, R.; d'Souza, R. Periapical inflammatory responses and their modulation. *Crit. Rev. Oral Biol. Med.* **1998**, *9*, 498–521. [CrossRef] [PubMed]
150. Takeda, K.; Akira, S. Toll-Like Receptors. *Curr. Protoc. Immunol.* **2007**, *77*, 14.12.11–14.12.13. [CrossRef] [PubMed]
151. Lai, J.-P.; Ho, W.Z.; Kilpatrick, L.E.; Wang, X.; Tuluc, F.; Korchak, H.M.; Douglas, S.D. Full-length and truncated neurokinin-1 receptor expression and function during monocyte/macrophage differentiation. *Proc. Natl. Acad. Sci. USA* **2006**, *103*, 7771–7776. [CrossRef] [PubMed]
152. Marriott, I.; Bost, K.L. IL-4 and IFN-γ Up-Regulate Substance P Receptor Expression in Murine Peritoneal Macrophages1. *J. Immunol.* **2000**, *165*, 182–191. [CrossRef]
153. Simeonidis, S.; Castagliuolo, I.; Pan, A.; Liu, J.; Wang, C.-C.; Mykoniatis, A.; Pasha, A.; Valenick, L.; Sougioultzis, S.; Zhao, D.; et al. Regulation of the NK-1 receptor gene expression in human macrophage cells via an NF-κB site on its promoter. *Proc. Natl. Acad. Sci. USA* **2003**, *100*, 2957–2962. [CrossRef]
154. Goldman, R.; Bar-Shavit, Z.; Shezen, E.; Terry, S.; Blumberg, S. Enhancement of Phagocytosis by Neurotensin, a Newly Found Biological Activity of the Neuropeptide. In *Macrophages and Natural Killer Cells: Regulation and Function*; Normann, S.J., Sorkin, E., Eds.; Springer US: Boston, MA, USA, 1982; pp. 133–141. [CrossRef]
155. Hartung, H.-P. Activation of macrophages by neuropeptides. *Brain Behav. Immun.* **1988**, *2*, 275–281. [CrossRef]
156. Ruff, M.R.; Wahl, S.M.; Pert, C.B. Substance P receptor-mediated chemotaxis of human monocytes. *Peptides* **1985**, *6*, 107–111. [CrossRef]
157. Hartung, H.-P.; Toyka, K.V. Substance P, the Immune System and Inflammation. *Int. Rev. Immunol.* **1989**, *4*, 229–249. [CrossRef]
158. Cozens, P.J.; Rowe, F.M. Substance-P is a potent inducer of tnf and IL-1 secretion by macrophages—A potential role for tnf in the pathogenesis of asthma. *Immunobiology* **1987**, *175*, 7.
159. Saito, I.; Ishii, K.; Hanada, K.; Sato, O.; Maeda, T. Responses of calcitonin gene-related peptide-immunopositive nerve fibres in the periodontal ligament of rat molars to experimental tooth movement. *Arch. Oral Biol.* **1991**, *36*, 689–692. [CrossRef]
160. Wakisaka, S.; Nishikawa, S.; Ichikawa, H.; Matsuo, S.; Takano, Y.; Akai, M. The distribution and origin of substance P-like immunoreactivity in the rat molar pulp and periodontal tissues. *Arch. Oral Biol.* **1985**, *30*, 813–818. [CrossRef]
161. Heyeraas, K.J.; Kvinnsland, I.; Byers, M.R.; Jacobsen, E.B. Nerve fibers immunoreactive to protein gene product 9.5, calcitonin gene-related peptide, substance P, and neuropeptide Y in the dental pulp, periodontal ligament, and gingiva in cats. *Acta Odontol. Scand.* **1993**, *51*, 207–221. [CrossRef] [PubMed]
162. Berggreen Jacobsen, E.; Fristad, I.; Heyeraas, K.J. Nerve fibers immunoreactive to calcitonin gene-related peptide, substance P, neuropeptide Y, and dopamine ß-hydroxylase in innervated and denervated oral tissues in ferrets. *Acta Odontol. Scand.* **1998**, *56*, 220–228. [CrossRef] [PubMed]
163. Kimberly, C.L.; Byers, M.R. Inflammation of rat molar pulp and periodontium causes increased calcitonin gene-related peptide and axonal sprouting. *Anat. Rec.* **1988**, *222*, 289–300. [CrossRef]
164. Khayat, B.G.; Byers, M.R.; Taylor, P.E.; Mecifi, K.; Kimberly, C.L. Responses of nerve fibers to pulpal inflammation and periapical lesions in rat molars demonstrated by calcitonin gene-related peptide immunocytochemistry. *J. Endod.* **1988**, *14*, 577–587. [CrossRef]
165. Gašperšič, R.; Kovačič, U.; Cör, A.; Skalerič, U. Unilateral ligature-induced periodontitis influences the expression of neuropeptides in the ipsilateral and contralateral trigeminal ganglion in rats. *Arch. Oral Biol.* **2008**, *53*, 659–665. [CrossRef]
166. Kabashima, H.; Nagata, K.; Maeda, K.; Iijima, T. Involvement of substance P, mast cells, TNF-α and ICAM-1 in the infiltration of inflammatory cells in human periapical granulomas. *J. Oral Pathol. Med.* **2002**, *31*, 175–180. [CrossRef]
167. de Avila, E.D.; de Molon, R.S.; de Godoi Gonçalves, D.A.; Camparis, C.M. Relationship between levels of neuropeptide Substance P in periodontal disease and chronic pain: A literature review. *J. Investig. Clin. Dent.* **2014**, *5*, 91–97. [CrossRef]
168. Ferrin, J.; Kirakodu, S.; Jensen, D.; Al-Attar, A.; Peyyala, R.; Novak, M.J.; Dawson, D.; Al-Sabbagh, M.; Stromberg, A.J.; Orraca, L.; et al. Gene expression analysis of neuropeptides in oral mucosa during periodontal disease in non-human primates. *J. Periodontol.* **2018**, *89*, 858–866. [CrossRef] [PubMed]
169. Liinden, G.J.; McKinneil, J.; Shaw, C.; Lundy, F.T. Substance P and neurokinin A in gingival crevicular fluid in periodontal health and disease. *J. Clin. Periodontol.* **1997**, *24*, 799–803. [CrossRef]
170. Hanioka, T.; Takaya, K.; Matsumori, Y.; Matsuse, R.; Shizukuishi, S. Relationship of the substance P to indicators of host response in human gingival crevicular fluid. *J. Clin. Periodontol.* **2000**, *27*, 262–266. [CrossRef] [PubMed]

171. Lundy, F.T.; Salmon, A.L.; Lamey, P.-J.; Shaw, C.; Linden, G.J. Carboxypeptidase-mediated metabolism of calcitonin gene-related peptide in human gingival crevicular fluid—A rôle in periodontal inflammation? *J. Clin. Periodontol.* **2000**, *27*, 499–505. [CrossRef]
172. Owan, I.; Ibaraki, K. The role of calcitonin gene-related peptide (CGRP) in macrophages: The presence of functional receptors and effects on proliferation and differentiation into osteoclast-like cells. *Bone Miner.* **1994**, *24*, 151–164. [CrossRef]
173. Fernandez, S.; Knopf, M.A.; Bjork, S.K.; McGillis, J.P. Bone Marrow-Derived Macrophages Express Functional CGRP Receptors and Respond to CGRP by Increasing Transcription of c-fos and IL-6 mRNA. *Cell. Immunol.* **2001**, *209*, 140–148. [CrossRef] [PubMed]
174. Frielingsdorf, H.; Simpson, D.R.; Thal, L.J.; Pizzo, D.P. Nerve growth factor promotes survival of new neurons in the adult hippocampus. *Neurobiol. Dis.* **2007**, *26*, 47–55. [CrossRef] [PubMed]
175. Ulupinar, E.; Jacquin, M.F.; Erzurumlu, R.S. Differential effects of NGF and NT-3 on embryonic trigeminal axon growth patterns. *J. Comp. Neurol.* **2000**, *425*, 202–218. [CrossRef]
176. Minnone, G.; De Benedetti, F.; Bracci-Laudiero, L. NGF and Its Receptors in the Regulation of Inflammatory Response. *Int. J. Mol. Sci.* **2017**, *18*, 1028. [CrossRef]
177. Salio, C.; Averill, S.; Priestley, J.V.; Merighi, A. Costorage of BDNF and neuropeptides within individual dense-core vesicles in central and peripheral neurons. *Dev. Neurobiol.* **2007**, *67*, 326–338. [CrossRef]
178. Ehrhard, P.B.; Ganter, U.; Stalder, A.; Bauer, J.; Otten, U. Expression of functional trk protooncogene in human monocytes. *Proc. Natl. Acad. Sci. USA* **1993**, *90*, 5423–5427. [CrossRef]
179. Kurihara, H.; Shinohara, H.; Yoshino, H.; Takeda, K.; Shiba, H. Neurotrophins in Cultured Cells from Periodontal Tissues. *J. Periodontol.* **2003**, *74*, 76–84. [CrossRef]
180. Ochi, K.; Saito, I.; Hanada, K.; Maeda, T. Expression of TrkB-like immunoreactivity in non-neural cells of rat periodontal ligament. *Arch. Oral Biol.* **1997**, *42*, 455–464. [CrossRef] [PubMed]
181. Caroleo, M.C.; Costa, N.; Bracci-Laudiero, L.; Aloe, L. Human monocyte/macrophages activate by exposure to LPS overexpress NGF and NGF receptors. *J. Neuroimmunol.* **2001**, *113*, 193–201. [CrossRef] [PubMed]
182. la Sala, A.; Corinti, S.; Federici, M.; Saragovi, H.U.; Girolomoni, G. Ligand activation of nerve growth factor receptor TrkA protects monocytes from apoptosis. *J. Leukoc. Biol.* **2000**, *68*, 104–110. [CrossRef] [PubMed]
183. Bracci-Laudiero, L.; Aloe, L.; Caroleo, M.C.; Buanne, P.; Costa, N.; Starace, G.; Lundeberg, T. Endogenous NGF regulates CGRP expression in human monocytes, and affects HLA-DR and CD86 expression and IL-10 production. *Blood* **2005**, *106*, 3507–3514. [CrossRef]
184. Williams, K.S.; Killebrew, D.A.; Clary, G.P.; Seawell, J.A.; Meeker, R.B. Differential regulation of macrophage phenotype by mature and pro-nerve growth factor. *J. Neuroimmunol.* **2015**, *285*, 76–93. [CrossRef] [PubMed]
185. Sun, J.; Ramnath, R.D.; Zhi, L.; Tamizhselvi, R.; Bhatia, M. Substance P enhances NF-κB transactivation and chemokine response in murine macrophages via ERK1/2 and p38 MAPK signaling pathways. *Am. J. Physiol. Cell Physiol.* **2008**, *294*, C1586–C1596. [CrossRef]
186. Lee, S.-K.; Pi, S.-H.; Kim, S.-H.; Min, K.-S.; Lee, H.-J.; Chang, H.-S.; Kang, K.-H.; Kim, H.-R.; Shin, H.-I.; Kim, E.-C. Substance P regulates macrophage inflammatory protein 3α/chemokine C-C ligand 20 (CCL20) with heme oxygenase-1 in human periodontal ligament cells. *Clin. Exp. Immunol.* **2007**, *150*, 567–575. [CrossRef]
187. Keane, M.P.; Strieter, R.M.; Belperio, J.A. Macrophage inflammatory protein. In *Encyclopedia of Respiratory Medicine*; Laurent, G.J., Shapiro, S.D., Eds.; Academic Press: Oxford, UK, 2006; pp. 1–5. [CrossRef]
188. Deshmane, S.L.; Kremlev, S.; Amini, S.; Sawaya, B.E. Monocyte Chemoattractant Protein-1 (MCP-1): An Overview. *J. Interferon Cytokine Res.* **2009**, *29*, 313–326. [CrossRef]
189. Jiang, M.H.; Chung, E.; Chi, G.F.; Ahn, W.; Lim, J.E.; Hong, H.S.; Kim, D.W.; Choi, H.; Kim, J.; Son, Y. Substance P induces M2-type macrophages after spinal cord injury. *Neuroreport* **2012**, *23*, 786–792. [CrossRef]
190. Hong, H.S.; Son, Y. Substance P ameliorates collagen II-induced arthritis in mice via suppression of the inflammatory response. *Biochem. Biophys. Res. Commun.* **2014**, *453*, 179–184. [CrossRef]
191. Montana, G.; Lampiasi, N. Substance P Induces HO-1 Expression in RAW 264.7 Cells Promoting Switch towards M2-Like Macrophages. *PLoS ONE* **2016**, *11*, e0167420. [CrossRef] [PubMed]
192. Zhang, M.; Nakamura, K.; Kageyama, S.; Lawal, A.O.; Gong, K.W.; Bhetraratana, M.; Fujii, T.; Sulaiman, D.; Hirao, H.; Bolisetty, S.; et al. Myeloid HO-1 modulates macrophage polarization and protects against ischemia-reperfusion injury. *JCI Insight* **2019**, *3*, e120596. [CrossRef] [PubMed]
193. Lim, J.E.; Chung, E.; Son, Y. A neuropeptide, Substance-P, directly induces tissue-repairing M2 like macrophages by activating the PI3K/Akt/mTOR pathway even in the presence of IFNγ. *Sci. Rep.* **2017**, *7*, 9417. [CrossRef] [PubMed]
194. Pinho-Ribeiro, F.A.; Verri, W.A.; Chiu, I.M. Nociceptor Sensory Neuron–Immune Interactions in Pain and Inflammation. *Trends Immunol.* **2017**, *38*, 5–19. [CrossRef]
195. Rechenberg, D.K.; Munir, A.; Zehnder, M. Correlation between the clinically diagnosed inflammatory process and periapical index scores in severely painful endodontically involved teeth. *Int. Endod. J.* **2021**, *54*, 172–180. [CrossRef] [PubMed]
196. Wideman, T.H.; Edwards, R.R.; Walton, D.M.; Martel, M.O.; Hudon, A.; Seminowicz, D.A. The Multimodal Assessment Model of Pain: A Novel Framework for Further Integrating the Subjective Pain Experience Within Research and Practice. *Clin. J. Pain* **2019**, *35*, 212. [CrossRef]

197. Wilcox, C.E.; Mayer, A.R.; Teshiba, T.M.; Ling, J.; Smith, B.W.; Wilcox, G.L.; Mullins, P.G. The Subjective Experience of Pain: An FMRI Study of Percept-Related Models and Functional Connectivity. *Pain Med.* **2015**, *16*, 2121–2133. [CrossRef]
198. Tarnawski, L.; Olofsson, P.S. Inflammation neuroscience: Neuro-immune crosstalk and interfaces. *Clin. Transl. Immunol.* **2021**, *10*, e1352. [CrossRef]
199. Takeda, M.; Matsumoto, S.; Sessle, B.J.; Shinoda, M.; Iwata, K. Peripheral and Central Mechanisms of Trigeminal Neuropathic and Inflammatory Pain. *J. Oral Biosci.* **2011**, *53*, 318–329. [CrossRef]
200. Khan, A.A.; Owatz, C.B.; Schindler, W.G.; Schwartz, S.A.; Keiser, K.; Hargreaves, K.M. Measurement of Mechanical Allodynia and Local Anesthetic Efficacy in Patients with Irreversible Pulpitis and Acute Periradicular Periodontitis. *J. Endod.* **2007**, *33*, 796–799. [CrossRef]
201. Owatz, C.B.; Khan, A.A.; Schindler, W.G.; Schwartz, S.A.; Keiser, K.; Hargreaves, K.M. The Incidence of Mechanical Allodynia in Patients with Irreversible Pulpitis. *J. Endod.* **2007**, *33*, 552–556. [CrossRef] [PubMed]
202. Schou, W.S.; Ashina, S.; Amin, F.M.; Goadsby, P.J.; Ashina, M. Calcitonin gene-related peptide and pain: A systematic review. *J. Headache Pain* **2017**, *18*, 34. [CrossRef] [PubMed]
203. Shi, X.; Wang, L.; Li, X.; Sahbaie, P.; Kingery, W.S.; Clark, J.D. Neuropeptides Contribute to Peripheral Nociceptive Sensitization by Regulating Interleukin-1β Production in Keratinocytes. *Anesth. Analg.* **2011**, *113*, 175. [CrossRef]
204. Sun, R.-Q.; Lawand, N.B.; Willis, W.D. The role of calcitonin gene-related peptide (CGRP) in the generation and maintenance of mechanical allodynia and hyperalgesia in rats after intradermal injection of capsaicin. *PAIN* **2003**, *104*, 201–208. [CrossRef]
205. França, G.M.D.; Carmo, A.F.d.; Costa Neto, H.; Andrade, A.L.D.L.d.; Lima, K.C.d.; Galvão, H.C. Macrophages subpopulations in chronic periapical lesions according to clinical and morphological aspects. *Braz. Oral Res.* **2019**, *33*, e047. [CrossRef]
206. Toriya, Y.; Hashiguchi, I.; Maeda, K. Immunohistochemical examination of the distribution of macrophages and CGRP-immunoreactive nerve fibers in induced rat periapical lesions. *Dent. Traumatol.* **1997**, *13*, 6–12. [CrossRef] [PubMed]
207. Pereira Faustino, I.S.; Azevedo, R.S.; Takahama, A. Metalloproteinases 2 and 9 Immunoexpression in Periapical Lesions from Primary Endodontic Infection: Possible Relationship with the Histopathological Diagnosis and the Presence of Pain. *J. Endod.* **2016**, *42*, 547–551. [CrossRef]
208. Hong, C.-Y.; Lin, S.-K.; Kok, S.-H.; Cheng, S.-J.; Lee, M.-S.; Wang, T.-M.; Chen, C.-S.; Lin, L.-D.; Wang, J.-S. The role of lipopolysaccharide in infectious bone resorption of periapical lesion. *J. Oral Pathol. Med.* **2004**, *33*, 162–169. [CrossRef]
209. Lin, S.-K.; Kok, S.-H.; Kuo, M.Y.-P.; Wang, T.-J.; Wang, J.-T.; Yeh, F.T.-C.; Hsiao, M.; Lan, W.-H.; Hong, C.-Y. Sequential expressions of MMP-1, TIMP-1, IL-6, and COX-2 genes in induced periapical lesions in rats. *Eur. J. Oral Sci.* **2002**, *110*, 246–253. [CrossRef]
210. Jain, A.; Bahuguna, R. Role of matrix metalloproteinases in dental caries, pulp and periapical inflammation: An overview. *J. Oral Biol. Craniofacial Res.* **2015**, *5*, 212–218. [CrossRef]
211. Barreiros, D.; Nelson Filho, P.; Paula-Silva, F.W.G.; Oliveira, K.M.H.d.; Lucisano, M.P.; Rossi, A.D.; Silva, L.A.B.; Küchler, E.C.; Silva, R.A.B. MMP2 and MMP9 are Associated with Apical Periodontitis Progression and Might be Modulated by TLR2 and MyD88. *Braz. Dent. J.* **2018**, *29*, 43–47. [CrossRef] [PubMed]
212. Luchian, I.; Goriuc, A.; Sandu, D.; Covasa, M. The Role of Matrix Metalloproteinases (MMP-8, MMP-9, MMP-13) in Periodontal and Peri-Implant Pathological Processes. *Int. J. Mol. Sci.* **2022**, *23*, 1806. [CrossRef] [PubMed]
213. Laronha, H.; Caldeira, J. Structure and Function of Human Matrix Metalloproteinases. *Cells* **2020**, *9*, 1076. [CrossRef]
214. Elgezawi, M.; Haridy, R.; Almas, K.; Abdalla, M.A.; Omar, O.; Abuohashish, H.; Elembaby, A.; Christine Wölfle, U.; Siddiqui, Y.; Kaisarly, D. Matrix Metalloproteinases in Dental and Periodontal Tissues and Their Current Inhibitors: Developmental, Degradational and Pathological Aspects. *Int. J. Mol. Sci.* **2022**, *23*, 8929. [CrossRef] [PubMed]

biomolecules

MDPI

Article

HIV-1 Transcriptional Activator Tat Inhibits *IL2* Expression by Preventing the Presence of Pol II on the *IL2* Promoter

Spyridoula Anastasopoulou †, Tassos Georgakopoulos † and Athanasia Mouzaki *

Laboratory of Immunohematology, Division of Hematology, Department of Internal Medicine, Medical School, University of Patras, GR-26500 Patras, Greece; s_anastasopoulou@upatras.gr (S.A.); georgako@upatras.gr (T.G.)
* Correspondence: mouzaki@upatras.gr
† These authors contributed equally to this work.

Abstract: HIV-1 infection leads to a gradual loss of T helper cells, chronic immune activation, and eventual immune system breakdown. HIV-1 causes deregulation of the expression of IL-2, a cytokine important for T helper cell growth and survival, which is downregulated in HIV-1 patients. The present study addresses the regulation of *IL2* expression via HIV-1 Tat transcriptional activator. We used J-LAT cells, a T cell line that serves as a latency model for studies of HIV-1 expression in T cells, and as controls a T cell line lacking HIV-1 elements and a T cell line with a stably integrated copy of the HIV-1-LTR promoter. We show that endogenously expressed Tat inhibits *IL2* transcription in J-Lat cells via its presence in the ARRE-1/2 elements of the *IL2* promoter and that the inhibition of *IL2* expression is mediated by Tat inhibiting Pol II activity at the *IL2* promoter, which is mediated by preventing the presence of Pol II at the ARRE-1/2 elements. Overall, Tat is present at the *IL2* promoter, apart from its cognate HIV-1 LTR target. This supports our current knowledge of how HIV-1 affects the host transcriptional machinery and reflects the potential of Tat to disrupt transcriptional regulation of host genes to manipulate cell responses.

Keywords: HIV-1; *IL2*; Tat; Pol II; NFAT2; J-LAT; transcription factors

1. Introduction

HIV infection is characterized by chronic activation of the immune system, uncontrolled viral replication, and a gradual decline in CD4$^+$ T helper (Th) cells. The introduction of highly active antirctroviral therapy (HAART) immediately after diagnosis has resulted in chronic rather than fatal infection. Although the life expectancy of HIV-1-infected individuals who respond well to the drugs has improved dramatically with the long-term use of HAART [1], studies show that the virus is not completely eliminated in the plasma of patients and viremia is observed [2]. This inability to eliminate the virus is attributed to cells that serve as viral reservoirs and are responsible for the latency of infection [3,4]. The exact molecular mechanisms underlying viral latency are still unknown [5–7]. The viral reservoir consists of cells harboring transcriptionally integrated silent proviruses capable of producing infectious virions after treatment interruption. Once HIV is integrated into the host genome, its transcription is regulated by a complex of cis and trans factors [8]. Many studies suggest that repressive chromatin states, DNA methylation, and post-transcriptional changes in histones and non-histone proteins, as well as host transcription factors and the HIV-1 Tat protein, determine the transition to stable latency [7,8]. The presence of this latent reservoir is associated with chronic immune activation and persistent inflammation [9].

One of the most important effects of chronic HIV-1 infection is the dysregulation of the expression of a number of cytokines such as IL-2, IL-17, IFN-γ, and others [10]. The expression of IL-2 plays a critical role in infection, as it is the first cytokine secreted by naive Th cells after their activation by antigens [11], and it has been found to be downregulated in HIV-1 patients [12,13]. This was confirmed by in vitro studies using Jurkat T cells, which can express the stably integrated HIV-1 *Tat* [14,15] or the entire stably integrated virus [16] in their genome and, in this respect, behave exactly like $CD4^{+}CD25^{-}$ primary Th cells latently infected with HIV-1. Production of IL-2 is a measure of immunological activation of T cells via the downstream T cell receptor (TCR) pathway. Therefore, impaired IL-2 expression is associated with deficient development and proliferation of lymphocytes, monocytes, and macrophages, as well as inefficient viral replication and spread. For this reason, IL-2 is also important for latency. Its expression in acutely or latently infected T cells is impaired after in vitro stimulation with mitogens at the transcriptional level, albeit in a completely opposite manner. *IL2* mRNA levels are increased in acutely infected stimulated cells, whereas they are decreased in latently infected stimulated cells compared with stimulated uninfected cells [16]. Exogenous addition of IL-2 to patient T cells promotes their proliferation and protects them from apoptosis [17]. In addition, immunotherapy with IL-2 has been shown to increase Th cell numbers [18], reduce the pool of latently infected resting Th cells [19], and elicit HIV-1-specific T cell responses [20]. For these reasons, production of IL-2 by patients' T cells is essential for proper immune function.

IL2 gene activation occurs after binding of specific transcription factors to its promoter [21,22]. The transcription factors NF-κB and NFAT recognize purine-rich binding sites in the *IL2* promoter and play a central role in the signaling pathway that regulates *IL2* expression [21,23]. HIV-1 Tat protein is another transcription factor that affects *IL2* transcription [14,15,24].

The promoter region of *IL2*, which spans nucleotides −292 to −273 (ARRE-2 sequence) [22], is similar to the region of HIV-1-LTR, which spans nucleotides −279 to −250 (RATS sequence) [25]. The elements ARRE-2 and RATS are involved in suppressing the expression of the *IL2* gene and HIV-1, respectively [25–27]. Several transcription factors associated with *IL2* expression, including NF-κB and NFAT, are also involved in transcriptional activation of the HIV-1 genome via binding to the LTR viral sequence [24,28,29]. Binding of NF-κB to the LTR of HIV-1 stimulates its transcription, whereas NFAT appears to bind to the same NF-κB core element and not to the putative NFAT site (−255 to −217) in the LTR [28,29]. Many studies suggest that there is an interplay between host factors and the HIV-1 Tat protein in the regulation of HIV-1 [8,24,29,30].

Tat, a small RNA-binding protein, is the first protein expressed after HIV-1 infection and is required for efficient transcription and viral replication [24]. It is involved in apoptotic and survival mechanisms, alteration of T-cell proliferation, and expression of various cytokine genes [24]. The mechanism by which Tat regulates most cellular genes remains to be elucidated. Since Tat does not have a DNA-binding domain, many studies suggest that its activity is exerted on the expression of host genes either by modulating transcription through binding to TAR-like sequences in the 5′-untranslated region of nascent RNA in the promoter regions of genes or by interacting with transcription-regulating host proteins that form a Tat complex [24,31]. Tat has been reported to interact with various host proteins such as transcription factor Sp1, cyclin T, nuclear RNA polymerase II, and others [29,32,33]. The transcription factor NFAT1 also interacts with the HIV-1 Tat protein, and the two factors modulate each other's activity [34]. Tat also activates NF-κB through its interaction with IκB-α and p65 [35].

Previous studies have shown that transient expression of HIV-1 Tat acts mainly by enhancing *IL2* promoter activity [36–38]. However, cells stably expressing Tat show a decrease in *IL2* expression [14–16]. From this point of view, transiently expressed Tat behaves analogously to cells acutely or transiently infected with the whole HIV-1 genome, whereas stably expressed Tat behaves similarly to a stably integrated virus [14–16]. These contrasting results in the regulation of *IL2* may reflect differences in gene regulation between acutely HIV-1 infected and latently infected cells [16].

The present study focuses on the impairment of *IL2* expression by HIV-1 Tat transcriptional activator expressed in J-Lat cells, a $CD4^{+}CD25^{-}$ Th cell line that serves as a latency model for studies of HIV-1 expression in Th cells. Because HIV-1 Nef protein has also been reported to affect *IL2* expression, J-Lat cells, which have the complete HIV-1 virus integrated into their genome but without the Nef gene, are an ideal model to study the effect of Tat on *IL2* expression in the absence of Nef [39] and avoid the artificial results caused by transient overexpression of Tat. We show that endogenously expressed Tat in induced latently infected J-Lat cells exerts its inhibitory effect on *IL2* transcription through its presence in the ARRE-1 and ARRE-2 elements of the *IL2* promoter and by preventing the presence of Pol II in the same elements.

2. Materials and Methods

2.1. Cells and Cultures

Jurkat (American type Culture Collection), Jurkat-Lat 6.3 (J-Lat) (NIH AIDS Reagent Program catalog number #9849), and Jurkat-LTRG (J-LTRG) (NIH AIDS Reagent Program catalog number #11587) T cell lines were used. Jurkat cells are derived from human T-cell acute lymphoblastic leukemia. J-Lat cells are Jurkat-derived cells latently infected with the packaged retroviral construct HIV-R7/E-/GFP. This is a full-length HIV-1 genome with a non-functional Env gene due to a frame shift and the green fluorescent protein GFP gene in place of the Nef gene [40]. J-LTRG cells, also derived from Jurkat cells, contain a stably integrated copy of the HIV-1 LTR promoter linked to the GFP reporter as a reporter for HIV-1 LTR expression [41,42]. In the absence of HIV-1 Tat expression, the cells exhibit extremely low eGFP expression.

All cell lines were cultured in RPMI-1640 medium supplemented with 10% heat-inactivated fetal bovine serum (FBS), 2 mM L-glutamine, penicillin (100 U/mL), and 50 µM 2-mercaptoethanol (CM) at a concentration of 10^6 cells/mL in a humidified incubator at 37 °C and 5% CO_2. When required, cells were stimulated with the mitogens ionomycin (2 µM) and phorbol myristate acetate (20 ng/mL) (P/I) for 6 h (peak *IL2* expression [26]).

2.2. Quantitative Real Time PCR (qPCR)

Total RNA was isolated using the standard method of Trizol according to the manufacturer's instructions (Gibco). RNA yield and purity were determined by measuring absorbance at 260/280 nm on a Quawell microvolume spectrophotometer Q3000 (Quawell Technology). Total RNA (100 ng per reaction) was used for cDNA synthesis using M-MLV reverse transcriptase (200 U/µL) (Sigma-Aldrich, St. Louis, MI, USA) in 10× M-MLV reverse transcriptase buffer, 40 U/µL RNase inhibitor (Thermo Fisher Scientific-Invitrogen), 1 mM each dNTPs (Sigma-Aldrich), and 2.5 µM random hexanucleotide primers (Sigma-Aldrich). qPCR mRNA analysis was performed on an Mx3000P$_{TM}$ Quantitative PCR System Thermocycler (Stratagene, San Diego, MA, USA) using SYBR-green fluorescence quantification technology (KAPA SYBR FAST qPCR Kit, Kapa Biosystems, Wilmington, MA, USA). PCR conditions were 95 °C for 15 min, followed by 36 cycles of 95 °C for 30 s to denature cDNA, 58 °C for 30 s for annealing, and 72 °C for 30 s for extension. The results were analyzed using MxPro^{TM} software (Stratagene). Expression of the β-actin gene served as a normalizer. All measurements were performed in triplicate. The primers used for qPCR (Eurofins Genomics/Oligo Analysis Tool) were:

Tat—5′-AGGAAGTCAGCCTAAAACTGC-3′ and 5′-GCTCTTCGTCGCTGTCTCC-3′, yielding a PCR product of 132 bp;

GFP—5′-GGTGATACCCTTGTTAATAG-3′ and 5′-AGGTAATGGTTGTCTGGTAAA-3, yielding a product of 176 bp;

IL2—5′-TCACCAGGATGCTCACATTTAAGT-3′ and 5′-GAGGTTTGAGTTCTTCTTC TTCTACAC-3′, yielding a PCR product of 127 bp;

β-actin—5′-TTGGCAATGAGCGGTTCC-3′ and 5′-AGCACTGTGTTGGCGTAC-3′, yielding a PCR product of 137 bp.

2.3. HIV-1 Tat siRNA Transfection

J-Lat cells were transfected with siRNA oligonucleotides using Lipofectamine LTX DNA transfection reagents (Invitrogen, Life Technologies) according to the manufacturer's instructions. Cells were cultured in CM for 24 h (10^6 cells per experimental point). The culture medium was changed 30 min before addition of the transfection reagent. Cells were incubated for 18–24 h in the presence of 10 nM of:

RNA oligo A [5′-GGAGCCAGUAGAUCCUAGA)TT-3′] or

RNA oligo B [5′-GCUUGUACCAAUUGCUAUU)TT-3′] (Eurofins Genomics/siRNA Design Tool).

After 24 h, the culture medium was changed, and 6 h later, the cells were cultured with P/I for an additional 6 h. They were then subjected to qPCR analysis and Western immunoblotting, as indicated in the results.

2.4. Western Immunoblotting

Whole-cell extracts (10^6 cells per experimental point) were prepared using RIPA buffer (50 mM Tris-HCl pH 7.4, 150 mM NaCl, 1% Triton X-100, 1% sodium deoxycholate, 0.1% SDS, and 1 mM EDTA) supplemented with protease inhibitors (Sigma-Aldrich). Proteins were separated using SDS-PAGE and electrophoretically transferred to nitrocellulose membranes. Membranes were incubated with a rabbit polyclonal antibody to HIV-1 Tat (ab43014, Abcam, Cambridge, UK) and a rabbit monoclonal antibody to β-actin (Cell Signaling Technology #4970, Danvers, MA, USA), which served as a loading control. HRP-conjugated goat anti-rabbit IgG (sc-2004, Santa Cruz, CA, USA) was used as a secondary antibody. Protein levels were visualized using the ECL LumiGLO detection kit (Upstate, Biotechnology UBI).

2.5. Co-Immunoprecipitation

J-Lat cells were cultured in the presence (P/I) or absence (CM) of mitogens for 6 h. Whole protein extracts were then digested with RIPA buffer (50 mM Tris-HCl pH 7.4, 150 mM NaCl, 1% Triton X-100, 1% sodium deoxycholate, 0.1% SDS, and 1 mM EDTA) supplemented with protease inhibitors (Sigma-Aldrich). Extracts before immunoprecipitation were used as controls to ensure that all samples contained the same starting material (input). Extracts were subjected to immunoprecipitation with a rabbit polyclonal anti-Tat antibody (ab43014, Abcam) (0.1 μg/mL) using magnetic protein G beads (Dynabeads, Invitrogen). Protein–protein interactions between the transcription factors NFAT2 and Tip60 and HIV-1 Tat were examined using Western immunoblotting with a mouse monoclonal antibody to NFAT2 (sc-7294, Santa Cruz) and a mouse monoclonal antibody to Tip60 (sc-166323, Santa Cruz). An extract of J-Lat cells expressing NFAT and Tip60 without antibody (-Ab) was used as a negative control. A goat anti-rabbit IgG HRP-conjugated antibody (sc-2004, Santa Cruz) and a goat anti-mouse IgG HRP-conjugated antibody (sc-2005, Santa Cruz) were used as secondary antibodies.

2.6. Chromatin Immunoprecipitation (ChIP) Assays

For ChIP assays, Jurkat, Jurkat-Lat, and Jurkat-LTRG cells (10^7 cells per experimental point) were cultured in CM or P/I for 6 h. Cells were fixed in 1.1% formaldehyde for 10 min and then quenched with 125 mM glycine for 5 min. They were then lysed and sonicated to generate 200–500 bp DNA fragments. ChIP assays were performed as described [26]. All reactions were performed in 1 mL sample tubes using 10 μg of isolated chromatin with 40 μL of protein G beads (Dynabeads, Invitrogen) and 5 μg of the corresponding antibody for each ChIP reaction. The antibodies used were anti-HIV-1 Tat (ab43014, Abcam) and mouse monoclonal anti-Pol II (sc-9001, Santa Cruz). Immunoprecipitated DNA from each cell line was analyzed for detection of specific sites with qPCR using the KAPA SYBR FAST qPCR kit (Kapa Biosystems) and the Mx3000P$_{TM}$ Quantitative PCR System Thermocycler (Stratagene). The sequences of the ChIP primers used for the different genomic regions in qPCR (Eurofins Genomics/Oligo Analysis Tool) were:

ARRE-1/TATA—5′-TCTTTGGGGGTTTAAAGAAATTC-3′ and 5′-AGGAGTTGAGGT TACTGTGAG-3′, yielding a PCR product of 217 bp;

ARRE-2—5′-CTTGCTGTTGTCCACCAC-3′ and 5′-TGGATGTAGGTGAAATCCC-3′, yielding a PCR product of 201 bp;

RATS region of HIV-1-LTR—5′-CCTTTGGATGGTGCTACAAGC-3′ and 5′-GATGCAG CTCTCGGGCCA-3′, yielding a PCR product of 137 bp;

TATA region of the β2m promoter—5′-CGCCGATGTACAGACAGCAAA-3′ and 5′-TGCTGTCAGCTTCAGGAATG-3′, yielding a PCR product of 230 bp;

TATA region of the β-actin gene promoter—5′-CGGCGAAGCCGGTGAG-3′ and 5′-CTGGCGGGGGCTACGC-3′, yielding a PCR product of 175 bp.

The optimized PCR conditions were 95 °C for 10 min, followed by 40 cycles of 95 °C for 30 s and 60 °C for 30 s. The results represent the ChIP signal as a fold increase in signal compared with the background signal. In this normalization method, the fold enrichment is calculated (2^{-DDCt}) and divided by the non-specific adjustment (Ct IP)—(Ct mock). Ct is the cycle at which the threshold line is crossed.

2.7. Statistical Analysis

Data are given as means (SD or SE) of three independent experiments. Statistical probabilities were evaluated using Student's *t* test. The statistical significance level was set at $p < 0.05$. Data analysis and graphical representation were performed using GraphPad Prism v.9.0 software.

3. Results

3.1. Decreased IL2 mRNA Expression in Induced J-Lat Cells Due to Tat Inhibition

To investigate the role of Tat-HIV-1 in *IL2* transcription and to decipher the underlying mechanism, we determined *IL2* mRNA levels in J-Lat cells using qPCR. J-Lat cells were cultured under non-induced conditions (CM) and then induced with P/I for 6 h. The genetic background of J-Lat cells allowed us to avoid the effects of exogenous expression of Tat protein [36–38] (Supplementary Figure S1) and the effects of Nef protein on *IL2* mRNA expression [39].

Jurkat and J-LTRG cells were cultured under the same conditions (CM and P/I) as controls. The mRNA levels of *IL2*, GFP, and Tat were measured using qPCR in all cell lines under both conditions. As expected, *IL2* mRNA was upregulated in all cell lines after induction (Figure 1A). Consistent with our current knowledge of the function of Tat [24] and LTR-HIV-1 expression controlled by host transcription factors such as Sp1 and NF-κB [43] in the presence of mitogens, induction of J-Lat and J-LTRG cells with P/I resulted in GFP mRNA production in both cell lines (Figure 1B) and increased levels of endogenously produced *Tat* mRNA in induced J-Lat cells compared with non-induced cells (Figure 1C) [40,44,45]. The presence of Tat protein in J-Lat cells was detected using Western immunoblotting (Figure 1D,E). The discrepancy between the increased mRNA levels of Tat and the lower protein levels in induced J-Lat cells could be due to an asymmetry between the viral mRNAs produced and the actual functional viral mRNAs leading to the production of Tat protein [45], and/or to the repressive effect that Rev [46] and Gag gene products of HIV-1 [47] exert on the expression of Tat protein without affecting its mRNA expression.

When we compared the *IL2* mRNA levels produced after induction between cell lines, we found that they were significantly lower in induced J-Lat cells than in induced Jurkat or J-LTRG cells (Figure 1A). The difference between J-Lat, J-LTRG, and Jurkat cells under induced conditions is the complete expression of the integrated HIV-1 genome in J-Lat cells without the Nef and Env genes, suggesting that the Tat protein is likely a repressor of *IL2* expression. This observation is consistent with previous studies showing the same effect on *IL2* mRNA levels in cells with HIV-1 Tat or the entire HIV-1 virus integrated into their genome [14–16] and with the known role of Tat as a transcriptional repressor [15,48].

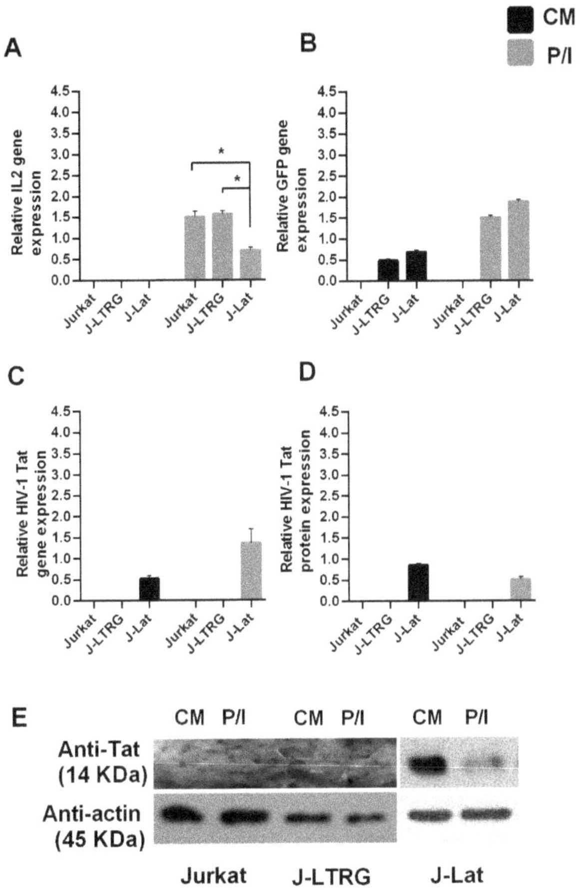

Figure 1. Induced J-Lat cells express lower levels of *IL2* mRNA compared with induced Jurkat and J-LTRG cells. (**A**) Relative *IL2* mRNA levels measured using qPCR in the T cell lines Jurkat, J-LTRG (carries HIV-1 LTR), and J-Lat (carries a noninfectious copy of HIV-1) cultured in CM or after induction with P/I for 6 h. (**B**) Relative mRNA synthesis of the GFP reporter gene. (**C**) Relative mRNA expression of the HIV-1 *Tat* gene. (**D**) Relative HIV-1 Tat protein expression in Jurkat, J-LTRG, and J-Lat cells measured using Western immunoblotting with specific antibodies to HIV-1 Tat. (**E**) Tat expression in Jurkat, J-LTRG, and J-Lat cells using Western immunoblotting with specific anti-Tat antibodies. Results are shown as means of three independent experiments; error bars represent SE (* $p < 0.05$; Student's t test). Actin was used as a protein loading control.

To investigate whether the HIV-1 Tat transcription factor per se was responsible for the low *IL2* mRNA levels in stimulated J-Lat cells, we silenced its endogenous expression with two different *Tat*-specific siRNA oligonucleotides, oligo A and B (Figure 2, see also M&M). After transfection, the cells were cultured ±P/I for 6 h. Tat protein reduction was detected using qPCR (Figure 2A) and Western blot analysis (Figure 2C,D). *IL2* mRNA expression was determined using qPCR. The results show that when Tat expression was inhibited in stimulated J-Lat cells, the expression of IL-2 mRNA was higher than that in induced J-Lat cells in which *Tat* was not silenced (Figure 2B), confirming that Tat can inhibit *IL2* expression independently of the presence of other viral genes. *IL2* mRNA in induced J-Lat cells in which *Tat* was silenced did not reach the level of mRNA expressed in induced Jurkat cells, because *Tat* was not completely silenced in these cells. Overall, these results demonstrate that endogenous expression of the HIV-1 Tat transcriptional activator in the context of a nearly intact viral genome, but in the absence of Nef and Env genes, does not allow complete transcriptional expression of the *IL2* gene after cell induction in T cells stably expressing Tat.

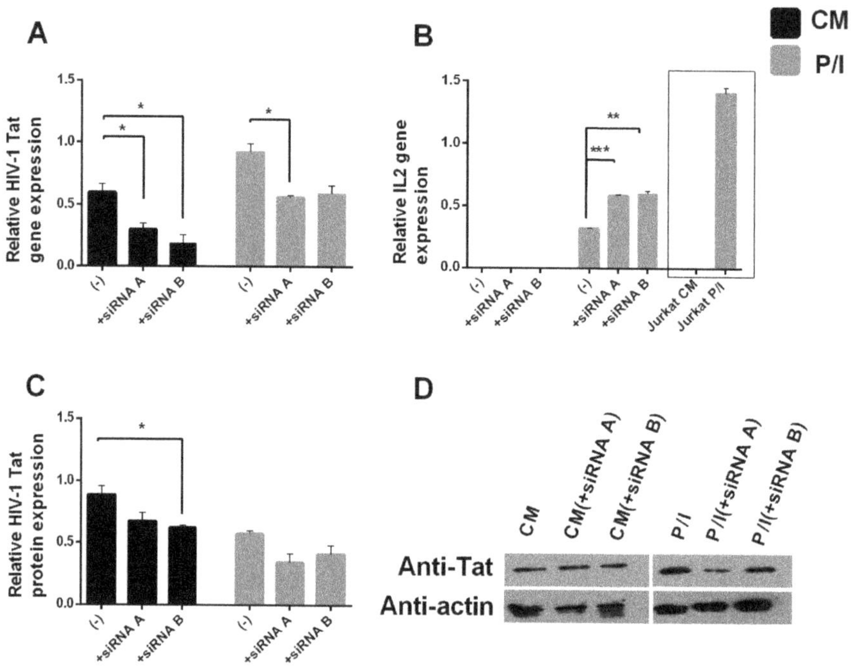

Figure 2. Silencing of endogenous HIV-1 *Tat* results in the relief of *IL2* mRNA expression in induced J-Lat cells. J-Lat cells were transfected with siRNA A and siRNA B and cultured in CM or with P/I for 6 h. Relative HIV-1 *Tat* (**A**) and *IL2* (**B**) mRNA levels in J-Lat cells. Additionally, shown in (**B**) for comparison are relative *IL2* mRNA levels in Jurkat cells cultured under the same conditions. (**C**) Relative HIV-1 Tat protein expression measured using Western immunoblotting with specific antibodies against HIV-1 Tat in J-Lat cells. (**D**) Representative Western blots of HIV-1 Tat protein expression in J-Lat cells. Results are presented as the result of the comparative quantification algorithm ΔΔCt. Results are presented as means of three independent experiments; error bars represent SE (* $p < 0.05$, ** $p < 0.01$, *** $p < 0.001$; Student's t test). Actin was used as a protein loading control.

3.2. Tat Inhibits IL2 Expression by Preventing the Presence of Pol II on the IL2 Promoter

Although it has been proposed that the impairment of *IL2* mRNA levels in HIV-1-infected cells is due to the impairment of the TCR pathway [16], we hypothesized that endogenously expressed Tat may play a more direct role in *IL2* transcription. *IL2* transcriptional activation is known to be under the control of several host transcription factors that either enhance or repress it. The ARRE-2 DNA element of the *IL2* promoter, which is also an NFAT2 binding site, plays a central role in the pro-inductive repression of *IL2* transcription in naive $CD4^{+}CD25^{-}$ Th cells and the Jurkat cell line [26]. Since Tat has been described elsewhere as a transcriptional repressor [15,48] and is present in a complex that can bind to the ARRE-2 element [34], we investigated whether the lower *IL2* transcription in induced J-Lat cells compared to *IL2* mRNA levels in induced Jurkat and J-LTRG cells was consistent with the presence of the Tat protein in the *IL2* promoter in J-Lat cells. To investigate the physiological interactions between Tat and the *IL2* promoter in vivo, we performed ChIP analysis using chromatin from untransfected Jurkat, J-LTRG, and J-Lat cells cultured in the presence or absence of P/I. Cross-linked chromatin was subjected to immunoprecipitation with antibodies to Tat and human Pol II or a negative control antibody to IgG. The association of Tat and Pol II with the *IL2* promoter and HIV-1 LTR was detected with qPCR using specific primer sets (Figure 3A). We examined the presence of Tat at ARRE-2 and ARRE-1, the region of the *IL2* core promoter that includes the TATA box (which also contains an NFAT-binding sequence) and the RATS element of LTR (Figure 3B). The relative positions of ARRE-2 and ARRE-1 are shown in Figure 3A. The chromatin regions encompassing the promoters of the human β2m and actin genes and containing the TATA element were selected as negative controls for Tat binding (Figure 3B).

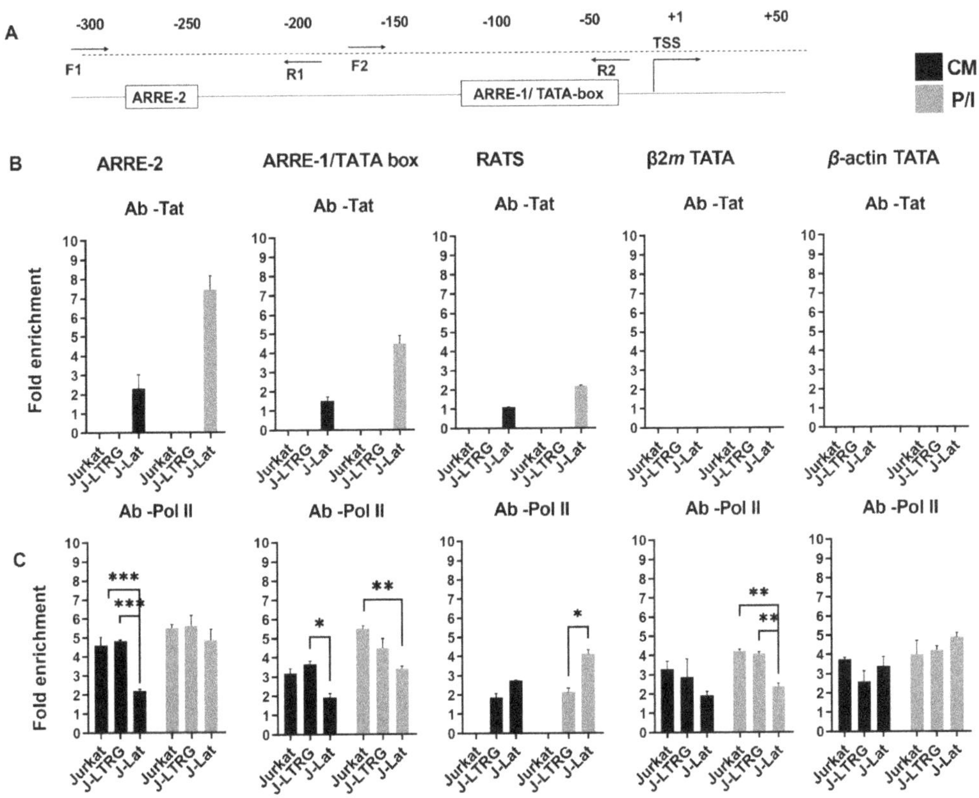

Figure 3. Binding affinity of endogenously expressed Tat to *IL2* loci. ChIP assays in Jurkat, J-LTRG, and J-Lat cells. (**A**) Schematic representation of the human *IL2* promoter showing the relative position of the ARRE-2 and ARRE-1/TATA box regulatory elements. The arrows represent the primer pairs used for each PCR amplification. (**B**,**C**) ChIP analysis in Jurkat, J-LTRG, and J-Lat cells cultured in CM or after induction with P/I for 6 h with qPCR. Binding to the *IL2* promoter regions ARRE-1/TATA and ARRE-2 and the RATS sequence of HIV-1 LTR. The binding affinity of the β2m TATA promoter region was used as a negative control. The β-actin TATA promoter was used as a control to check the specificity of the results. ChIP assays were performed with antibodies against Tat (**B**) and Pol II (**C**). Results represent ChIP signal as a fold increase in signal relative to background signal. Results are presented as means of three independent experiments; error bars represent SE (* $p < 0.05$, ** $p < 0.01$, *** $p < 0.001$; Student's t test).

Because Tat is not expressed in Jurkat and J-LTRG cells and Jurkat cells also do not have HIV-1 LTR in their genome, we were unable to detect Tat in ARRE-2, ARRE-1, and RATS in either cell line under either CM or P/I conditions (Figure 3B). In J-Lat cells expressing Tat under both CM and P/I conditions (Figure 1C,D), Tat protein was detected in ARRE-2, ARRE-1, and RATS, but at unequal levels. More specifically, Tat levels decreased from ARRE-2 to ARRE-1 to RATS under both CM and P/I conditions. At the same time, compared with CM, Tat binding increased three-fold after induction (P/I) in ARRE-2 and ARRE-1 elements and two-fold in RATS (Figure 3B Ab-Tat). The above results indicate that Tat shows a higher preference for the ARRE-2 and ARRE-1 elements of the *IL2* promoter than for the RATS element of HIV-1 LTR in latent J-Lat cells under both CM and P/I conditions, which may also mark its interference with *IL2* transcription. This preference and its three-fold higher presence in ARRE-2 and ARRE-1 after cell induction is accompanied by a shorter *IL2* mRNA induction compared with the other two cell lines that do not express Tat (Figure 1A). It is interesting to note that when we performed the same ChIP assays in previously transfected Jurkat and J-LTRG cells with a pcDNA3-based plasmid overexpressing Tat (1-101 a.a.), we failed to detect the presence of exogenously added Tat on the *IL2* promoter (Supplementary Figure S2), which may explain the discrepancy in *IL2* mRNA expression between endogenous and exogenously added Tat in the literature [14,15,36–38], (Supplementary Figure S1). Control ChIP assays with the Tat antibody showed no presence of Tat at the β2m or actin gene promoters under all conditions tested and in all cell types, a result confirming the specificity of our experiments.

Since the ARRE-2 element is known to be involved in the transcriptional repression of *IL2* [26], we examined the effects of endogenous Tat expression on the presence of Pol II in the *IL2* promoter before and after induction in all cell lines. In Jurkat and J-LTRG cells, Pol II was found to interact with the ARRE-2 element in CM and P/I to an almost similar extent between cell lines. In both cell lines, Pol II also interacted with ARRE-1 in CM, although to a lesser extent than with ARRE-2 under the same conditions. In both cell types, the presence of Pol II in ARRE-1 increased at P/I, consistent with transcriptional activation of the *IL2* gene after induction, and nearly reached the level of presence in ARRE-2 under the same conditions (Figure 3C). In contrast, the presence of Pol II in J-Lat cells was lower than in the other two cell lines in both ARRE-2 and ARRE-1 at CM. These differences between Jurkat, J-LTRG, and J-Lat cells in the presence of Pol II in ARRE-2 and ARRE-1 at CM may be attributed to the simultaneous presence of Tat in both elements at CM in J-Lat cells, leading to the exclusion of Pol II from the ARRE-2 and ARRE-1. The presence of Pol II at the ARRE-2 and ARRE-1 elements under CM conditions in all cell lines may reflect its association with one or more factors that bind to these elements in a ready-to-use state [26]. After induction (P/I) of J-Lat cells, the presence of Pol II in ARRE-2 reached almost the same levels as in the other two cell lines. This was not the case in ARRE-1, where the presence of Pol II in P/I was ~60% of that in Jurkat cells and ~75% of that in J-LTRG cells. This may be analogous to the exclusion of Pol II from Tat in ARRE-1 at CM in J-Lat, because the concomitant presence of Tat in ARRE-1 under P/I conditions results in impairment of *IL2* transcription during transcriptional activation, possibly by impeding the function of Pol II.

Because of the absence of an HIV-LTR sequence in Jurkat cells, we could not detect binding of Pol II with the specific primer set for the RATS element under both CM and P/I conditions (Figure 3C). In J-LTRG cells, the binding of Pol II to the RATS element was almost identical under CM and P/I conditions, suggesting that the underlying transcription of the HIV-1 LTR region in the absence of transcriptionally active virus, and particularly in the absence of Tat is due to the leakiness of the LTR in CM and during induction (P/I) to the action of host transcription factors such as NF-κB and NFAT [28,29] (Figure 3C). This result is consistent with GFP expression in J-LTRG cells grown under CM and/or P/I conditions (Figure 1B).

Similarly, Pol II is present in the genomic region surrounding the TATA box of the β2m promoter under both CM and P/I conditions in Jurkat and J-LTRG cells. In contrast, the presence of Pol II in the β2m TATA was reduced by almost half in induced J-Lat cells compared with induced Jurkat and J-LTRG cells, consistent with the fact that the previously reported [49] transcriptional repression of β2m mRNA expression in cells latently infected with HIV-1 is a completely different effect from that observed for *IL2* mRNA expression because there is no concomitant presence of Tat in the β2m promoter. In contrast, binding of Pol II to the TATA element of the β-actin promoter (where Tat is also absent) is not affected under either non-induced or induced conditions in all cell lines. Taken together, our results confirm that the suppressive effect of Tat on *IL2* mRNA expression in cells latently infected HIV-1 is Tat-specific and is related to its presence at the *IL2* promoter and, in particular, at ARRE1 by completely preventing the concomitant presence of Pol II.

Since Tat has no DNA-binding domain and its action on HIV-1 LTR requires the TAR element [24], its presence at the *IL2* promoter and subsequent hindrance of *IL2* transcriptional activation must be attributed to interactions of Tat with host transcription factors. Tat is known to interact with NFAT1 [34] and the p65 subunit of NF-κB [34], two factors involved in *IL2* transcriptional activation [21,22]. Since the interaction of Tat in both cases leads to an increase in the transcriptional activities of NFAT1 and NF-κB at many genes [34,35], we performed co-immunoprecipitation experiments to investigate whether Tat interacts with another NFAT member, NFAT2, to find a possible interaction to which we could attribute the inhibitory effect of endogenously expressed Tat on induced *IL2* transcription. We found that Tat did not interact with NFAT2 under either CM or P/I conditions. As a control for the accuracy of our assay, we demonstrated the known interaction between Tat and Tip60 protein [50] (Figure 4).

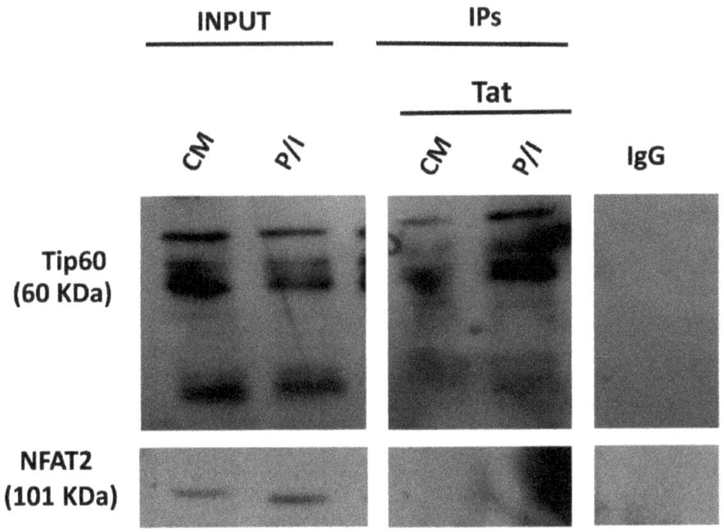

Figure 4. Tat and NFAT2 do not physically interact in J-Lat cells. J-Lat cells were cultured ±P/I for 6 h. Extracts before immunoprecipitation were used as controls to ensure that all samples contained the same starting material (input). Extracts were subjected to immunoprecipitation with an anti-Tat antibody. Immunoprecipitates were analyzed using Western immunoblotting with an anti-NFAT2 monoclonal antibody. An extract from J-Lat cells with IgG antibodies was used as a negative control. As a positive control, extracts from the same cells were immunoprecipitated with a human anti-Tat antibody to detect the interaction of Tat with Tip60. The bands shown in the gel represent the Tip60 isoforms. Tip60 isoforms were detected between 60 and 53 kDa.

4. Discussion

Overall, our results show that Tat, when endogenously expressed in J-Lat cells, has a high presence at the *IL2* promoter, apart from its cognate HIV-1-LTR target. This not only supports our current knowledge of HIV-1 hijacking the host transcriptional machinery, but also reflects the potential of the Tat activator to exert its multiple functions by disrupting transcriptional regulation of many genes, such as cytokine genes, to manipulate cell responses (Figure 5). Since Tat lacks a DNA binding domain, its presence in the *IL2* promoter in the two distinct elements ARRE-1 and ARRE-2, which are several base pairs apart, must be due to its interactions with host transcription factors involved in *IL2* transcriptional regulation. Because the inhibitory effect on *IL2* transcription in latently infected J-Lat cells occurs during cell induction, these interactions could involve NF-κB and/or NFAT1 and other transcription factors, but not NFAT2, as shown here. Future studies will elucidate the nature, type, and exact number of these Tat interactions.

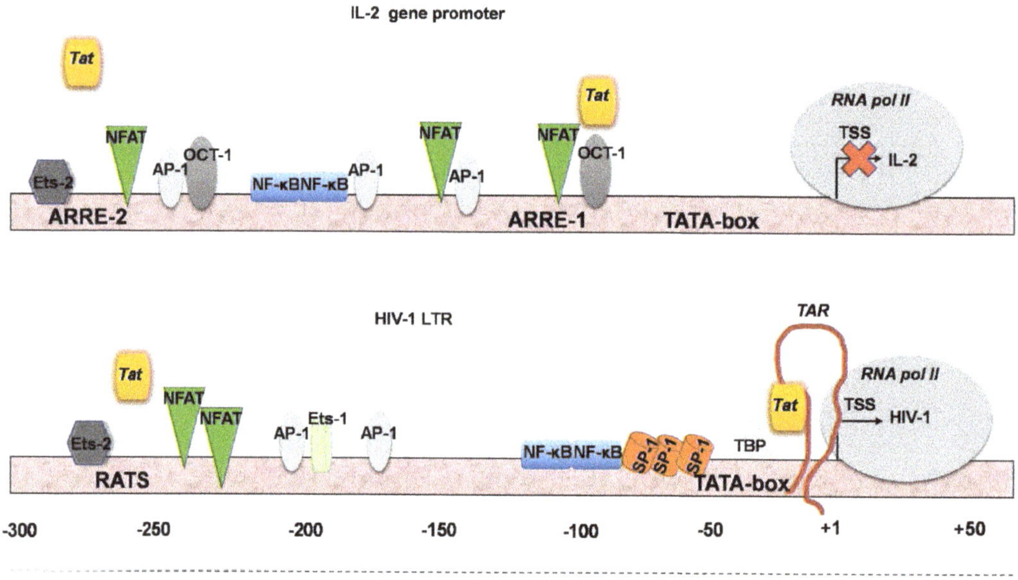

Figure 5. Model of Tat activity on the *IL2* promoter. Schematic representation of transcription factors binding to the regulatory elements of the *IL2* promoter ARRE-1/TATA and ARRE-2 and the RATS sequence of HIV-1 LTR.

Supplementary Materials: The following supporting information can be downloaded at: https://www.mdpi.com/article/10.3390/biom13060881/s1, Figure S1: Effects of exogenously added Tat on mRNA expression of IL-2 in Jurkat, J-LTRG, and J-Lat cells; Figure S2: Binding affinity of exogenously expressed Tat to the IL-2 promoter in transfected Jurkat, J-LTRG, and J-Lat cells. Reference [51] is cited in the supplementary materials.

Author Contributions: S.A. performed the experiments, analyzed the results, and wrote the manuscript; T.G. designed the study and methodology, analyzed and explained the results, and wrote the manuscript; A.M. designed the study, analyzed and explained the results, and wrote the manuscript. All authors have read and agreed to the published version of the manuscript.

Funding: This work was funded by grants from the Hellenic Institute of State Scholarships (IKY) (to S.A.) and GILEAD-ASKLEPIOS grants for 2018 and 2019 (to A.M.). The funders had no influence on the study design, data collection and analysis, decision to publish, or preparation of the manuscript.

Institutional Review Board Statement: Not applicable.

Informed Consent Statement: Not applicable.

Data Availability Statement: All raw data are available upon request from the corresponding author.

Conflicts of Interest: The authors declare no conflict of interest.

References

1. May, M.T.; Gompels, M.; Delpech, V.; Porter, K.; Orkin, C.; Kegg, S.; Hay, P.; Johnson, M.; Palfreeman, A.; Gilson, R.; et al. Impact on life expectancy of HIV-1 positive individuals of CD4+ cell count and viral load response to antiretroviral therapy. *AIDS* **2014**, *28*, 1193–1202. [CrossRef]
2. Palmer, S.; Maldarelli, F.; Wiegand, A.; Bernstein, B.; Hanna, G.J.; Brun, S.C.; Kempf, D.J.; Mellors, J.W.; Coffin, J.M.; King, M.S. Low-level viremia persists for at least 7 years in patients on suppressive antiretroviral therapy. *Proc. Natl. Acad. Sci. USA* **2008**, *105*, 3879–3884. [CrossRef]
3. Margolis, D.M.; Archin, N.M.; Cohen, M.S.; Eron, J.J.; Ferrari, G.; Garcia, J.V.; Gay, C.L.; Goonetilleke, N.; Joseph, S.B.; Swanstrom, R.; et al. Curing HIV: Seeking to Target and Clear Persistent Infection. *Cell* **2020**, *181*, 189–206. [CrossRef] [PubMed]
4. Schrager, L.K.; D'Souza, M.P. Cellular and anatomical reservoirs of HIV-1 in patients receiving potent antiretroviral combination therapy. *JAMA* **1998**, *280*, 67–71. [CrossRef]
5. Romani, B.; Allahbakhshi, E. Underlying mechanisms of HIV-1 latency. *Virus Genes* **2017**, *53*, 329–339. [CrossRef] [PubMed]
6. Castro-Gonzalez, S.; Colomer-Lluch, M.; Serra-Moreno, R. Barriers for HIV Cure: The Latent Reservoir. *AIDS Res. Hum. Retrovir.* **2018**, *34*, 739–759. [CrossRef] [PubMed]
7. Sengupta, S.; Siliciano, R.F. Targeting the Latent Reservoir for HIV-1. *Immunity* **2018**, *48*, 872–895. [CrossRef]
8. Mori, L.; Valente, S.T. Key Players in HIV-1 Transcriptional Regulation: Targets for a Functional Cure. *Viruses* **2020**, *12*, 529. [CrossRef]
9. Rojas, M.; Luz-Crawford, P.; Soto-Rifo, R.; Reyes-Cerpa, S.; Toro-Ascuy, D. The Landscape of IFN/ISG Signaling in HIV-1-Infected Macrophages and Its Possible Role in the HIV-1 Latency. *Cells* **2021**, *10*, 2378. [CrossRef] [PubMed]
10. Nakayama, K.; Nakamura, H.; Koga, M.; Koibuchi, T.; Fujii, T.; Miura, T.; Iwamoto, A.; Kawana-Tachikawa, A. Imbalanced production of cytokines by T cells associates with the activation/exhaustion status of memory T cells in chronic HIV type 1 infection. *AIDS Res. Hum. Retrovir.* **2012**, *28*, 702–714. [CrossRef]
11. Murphy, K.; Weaver, C. *Janeway's Immunobiology*, 9th ed.; Garland Science, Taylor & Francis Group, LLC: New York, NY, USA; London, UK, 2017.
12. Lane, H.C.; Fauci, A.S. Immunologic abnormalities in the acquired immunodeficiency syndrome. *Annu. Rev. Immunol.* **1985**, *3*, 477–500. [CrossRef]
13. Nakayama-Hosoya, K.; Ishida, T.; Youngblood, B.; Nakamura, H.; Hosoya, N.; Koga, M.; Koibuchi, T.; Iwamoto, A.; Kawana-Tachikawa, A. Epigenetic repression of interleukin 2 expression in senescent CD4+ T cells during chronic HIV type 1 infection. *J. Infect. Dis.* **2015**, *211*, 28–39. [CrossRef]
14. Purvis, S.F.; Georges, D.L.; Williams, T.M.; Lederman, M.M. Suppression of interleukin-2 and interleukin-2 receptor expression in Jurkat cells stably expressing the human immunodeficiency virus Tat protein. *Cell Immunol.* **1992**, *144*, 32–42. [CrossRef]
15. González, E.; Punzón, C.; González, M.; Fresno, M. HIV-1 Tat inhibits IL-2 gene transcription through qualitative and quantitative alterations of the cooperative Rel/AP1 complex bound to the CD28RE/AP1 composite element of the IL-2 promoter. *J. Immunol.* **2001**, *166*, 4560–4569. [CrossRef] [PubMed]
16. Shin, Y.; Yoon, C.-H.; Lim, H.; Park, J.; Roh, T.-Y.; Kang, C.; Choi, B.-S. Impaired IL-2 expression in latent HIV-1 infection. *Biochem. Biophys. Res. Commun.* **2015**, *463*, 1237–1242. [CrossRef] [PubMed]
17. Seder, R.A.; Grabstein, K.H.; Berzofsky, J.A.; McDyer, J.F. Cytokine interactions in human immunodeficiency virus-infected individuals: Roles of interleukin (IL)-2, IL-12, and IL-15. *J. Exp. Med.* **1995**, *182*, 1067–1077. [CrossRef]
18. Sellier, P.; Lafuente-Lafuente, C.; Bergmann, J.F. Interleukin-2 therapy in patients with HIV infection. *N. Engl. J. Med.* **2010**, *362*, 270–271. [PubMed]
19. Chun, T.-W.; Engel, D.; Mizell, S.B.; Hallahan, C.W.; Fischette, M.; Park, S.; Davey, R.T.; Dybul, M.; Kovacs, J.A.; Metcalf, J.A.; et al. Effect of interleukin-2 on the pool of latently infected, resting CD4+ T cells in HIV-1-infected patients receiving highly active anti-retroviral therapy. *Nat. Med.* **1999**, *5*, 651–655. [CrossRef]
20. Sullivan, A.K.; Hardy, G.A.; Nelson, M.R.; Gotch, F.; Gazzard, B.G.; Imami, N. Interleukin-2-associated viral breakthroughs induce HIV-1-specific CD4 T cell responses in patients on fully suppressive highly active antiretroviral therapy. *AIDS* **2003**, *17*, 628–629. [CrossRef] [PubMed]
21. Bunting, K.; Wang, J.; Shannon, M.F. Control of interleukin-2 gene transcription: A paradigm for inducible, tissue-specific gene expression. *Vitam Horm.* **2006**, *74*, 105–145.
22. Serfling, E.; Avots, A.; Neumann, M. The architecture of the interleukin-2 promoter: A reflection of T lymphocyte activation. *Biochim. Biophys. Acta.* **1995**, *1263*, 181–200. [CrossRef]
23. Chow, C.W.; Rincón, M.; Davis, R.J. Requirement for transcription factor NFAT in interleukin-2 expression. *Mol. Cell. Biol.* **1999**, *19*, 2300–2307. [CrossRef] [PubMed]

24. Clark, E.; Nava, B.; Caputi, M. Tat is a multifunctional viral protein that modulates cellular gene expression and functions. *Oncotarget* **2017**, *8*, 27569–27581. [CrossRef]
25. Mouzaki, A.; Doucet, A.; Mavroidis, E.; Muster, L.; Rungger, D. A repression-derepression mechanism regulating the transcription of human immunodeficiency virus type 1 in primary T cells. *Mol. Med.* **2000**, *6*, 377–390. [CrossRef]
26. Panagoulias, I.; Georgakopoulos, T.; Aggeletopoulou, I.; Agelopoulos, M.; Thanos, D.; Mouzaki, A. Transcription Factor Ets-2 Acts as a Preinduction Repressor of Interleukin-2 (IL-2) Transcription in Naive T Helper Lymphocytes. *J. Biol. Chem.* **2016**, *291*, 26707–26721. [CrossRef]
27. Panagoulias, I.; Karagiannis, F.; Aggeletopoulou, I.; Georgakopoulos, T.; Argyropoulos, C.P.; Akinosoglou, K.; Gogos, C.; Skoutelis, A.; Mouzaki, A. Ets-2 Acts as a Transcriptional Repressor of the Human Immunodeficiency Virus Type 1 through Binding to a Repressor-Activator Target Sequence of 5′-LTR. *Front. Immunol.* **2018**, *8*, 1924. [CrossRef]
28. Pessler, F.; Cron, R.Q. Reciprocal regulation of the nuclear factor of activated T cells and HIV-1. *Genes Immun.* **2004**, *5*, 158–167. [CrossRef]
29. Ne, E.; Palstra, R.J.; Mahmoudi, T. Transcription: Insights from the HIV-1 Promoter. *Int. Rev. Cell. Mol. Biol.* **2018**, *335*, 191–243.
30. El Kharroubi, A.; Piras, G.; Zensen, R.; Martin, M.A. Transcriptional activation of the integrated chromatin-associated human immunodeficiency virus type 1 promoter. *Mol. Cell. Biol.* **1998**, *18*, 2535–2544. [CrossRef]
31. Carvallo, L.; Lopez, L.; Fajardo, J.E.; Jaureguiberry-Bravo, M.; Fiser, A.; Berman, J.W. HIV-Tat regulates macrophage gene expression in the context of neuroAIDS. *PLoS ONE* **2017**, *12*, e0179882. [CrossRef]
32. Ott, M.; Geyer, M.; Zhou, Q. The control of HIV transcription: Keeping RNA polymerase II on track. *Cell. Host. Microbe* **2011**, *10*, 426–435. [CrossRef]
33. Mousseau, G.; Valente, S.T. Role of Host Factors on the Regulation of Tat-Mediated HIV-1 Transcription. *Curr. Pharm. Des.* **2017**, *23*, 4079–4090. [CrossRef]
34. Macián, F.; Rao, A. Reciprocal modulatory interaction between human immunodeficiency virus type 1 Tat and transcription factor NFAT1. *Mol. Cell. Biol.* **1999**, *19*, 3645–3653. [CrossRef]
35. Fiume, G.; Vecchio, E.; De Laurentiis, A.; Trimboli, F.; Palmieri, C.; Pisano, A.; Falcone, C.; Pontoriero, M.; Rossi, A.; Scialdone, A.; et al. Human immunodeficiency virus-1 Tat activates NF-κB via physical interaction with IκB-α and p65. *Nucleic Acids Res.* **2012**, *40*, 3548–3562. [CrossRef]
36. Westendorp, M.O.; Li-Weber, M.; Frank, R.W.; Krammer, P.H. Human immunodeficiency virus type 1 Tat upregulates interleukin-2 secretion in activated T cells. *J. Virol.* **1994**, *68*, 4177–4185. [CrossRef]
37. Vacca, A.; Farina, M.; Maroder, M.; Alesse, E.; Screpanti, I.; Frati, L.; Gulino, A. Human immunodeficiency virus type-1 tat enhances interleukin-2 promoter activity through synergism with phorbol ester and calcium-mediated activation of the NF-AT cis-regulatory motif. *Biochem. Biophys. Res. Commun.* **1994**, *205*, 467–474. [CrossRef]
38. Kwon, H.-S.; Brent, M.M.; Getachew, R.; Jayakumar, P.; Chen, L.-F.; Schnolzer, M.; McBurney, M.W.; Marmorstein, R.; Greene, W.C.; Ott, M. Human immunodeficiency virus type 1 Tat protein inhibits the SIRT1 deacetylase and induces T cell hyperactivation. *Cell. Host. Microbe* **2008**, *3*, 158–167. [CrossRef]
39. Markle, T.J.; Philip, M.; Brockman, M.A. HIV-1 Nef and T-cell activation: A history of contradictions. *Future Virol.* **2013**, *8*, 391–404. [CrossRef]
40. Jordan, A.; Bisgrove, D.; Verdin, E. HIV Reproducibly Establishes a Latent Infection after Acute Infection of T Cells in Vitro. *EMBO J.* **2003**, *22*, 1868–1877. [CrossRef]
41. Ochsenbauer-Jambor, C.; Jones, J.; Heil, M.; Zammit, K.P.; Kutsch, O. T-cell line for HIV drug screening using EGFP as a quantitative marker of HIV-1 replication. *Biotechniques* **2006**, *40*, 91–100. [CrossRef]
42. Kutsch, O.; Levy, D.N.; Bates, P.J.; Decker, J.; Kosloff, B.R.; Shaw, G.M.; Priebe, W.; Benveniste, E.N. Bis-anthracycline antibiotics inhibit human immunodeficiency virus type 1 transcription. *Antimicrob Agents Chemother.* **2004**, *48*, 1652–1663. [CrossRef]
43. Pereira, L.A.; Bentley, K.; Peeters, A.; Churchill, M.J.; Deacon, N.J. A compilation of cellular transcription factor interactions with the HIV-1 LTR promoter. *Nucleic Acids Res.* **2000**, *28*, 663–668. [CrossRef]
44. Spina, C.A.; Anderson, J.; Archin, N.M.; Bosque, A.; Chan, J.; Famiglietti, M.; Greene, W.C.; Kashuba, A.; Lewin, S.R.; Margolis, D.M.; et al. An in-depth comparison of latent HIV-1 reactivation in multiple cell model systems and resting CD4+ T cells from aviremic patients. *PLoS Pathog.* **2013**, *9*, e1003834. [CrossRef]
45. Telwatte, S.; Moron-Lopez, S.; Aran, D.; Kim, P.; Hsieh, C.; Joshi, S.; Montano, M.; Greene, W.C.; Butte, A.J.; Wong, J.K.; et al. Heterogeneity in HIV and cellular transcription profiles in cell line models of latent and productive infection: Implications for HIV latency. *Retrovirology* **2019**, *16*, 32. [CrossRef]
46. Lata, S.; Ali, A.; Sood, V.; Raja, R.; Banerjea, A.C. HIV-1 Rev downregulates Tat expression and viral replication via modulation of NAD(P)H:quinine oxidoreductase 1 (NQO1). *Nat. Commun.* **2015**, *6*, 7244. [CrossRef]
47. Hong, H.W.; Lee, S.W.; Myung, H. Induced degradation of Tat by nucleocapsid (NC) via the proteasome pathway and its effect on HIV transcription. *Viruses* **2013**, *5*, 1143–1152. [CrossRef]
48. Dhamija, N.; Choudhary, D.; Ladha, J.S.; Pillai, B.; Mitra, D. Tat predominantly associates with host promoter elements in HIV-1-infected T-cells-regulatory basis of transcriptional repression of c-Rel. *FEBS J.* **2015**, *282*, 595–610. [CrossRef]
49. Carroll, I.R.; Wang, J.; Howcroft, T.K.; Singer, D.S. HIV Tat represses transcription of the beta 2-microglobulin promoter. *Mol. Immunol.* **1998**, *35*, 1171–1178. [CrossRef]

50. Kamine, J.; Elangovan, B.; Subramanian, T.; Coleman, D.; Chinnadurai, G. Identification of a cellular protein that specifically interacts with the essential cysteine region of the HIV-1 Tat transactivator. *Virology* **1996**, *216*, 357–366. [CrossRef]
51. Frankel, A.D.; Biancalana, S.; Hudson, D. Activity of synthetic peptides from the Tat protein of human immunodeficiency virus type 1. *Proc. Natl. Acad. Sci. USA* **1989**, *86*, 7397–7401. [CrossRef]

biomolecules

MDPI

Review

Biological Catalysis and Information Storage Have Relied on *N*-Glycosyl Derivatives of β-D-Ribofuranose since the Origins of Life

Katarzyna Wozniak and Krzysztof Brzezinski *

Department of Structural Biology of Prokaryotic Organisms, Institute of Bioorganic Chemistry, Polish Academy of Sciences, Noskowskiego 12/14, 61-074 Poznan, Poland
* Correspondence: kbrzezinski@ibch.poznan.pl

Abstract: Most naturally occurring nucleotides and nucleosides are *N*-glycosyl derivatives of β-D-ribose. These *N*-ribosides are involved in most metabolic processes that occur in cells. They are essential components of nucleic acids, forming the basis for genetic information storage and flow. Moreover, these compounds are involved in numerous catalytic processes, including chemical energy production and storage, in which they serve as cofactors or coribozymes. From a chemical point of view, the overall structure of nucleotides and nucleosides is very similar and simple. However, their unique chemical and structural features render these compounds versatile building blocks that are crucial for life processes in all known organisms. Notably, the universal function of these compounds in encoding genetic information and cellular catalysis strongly suggests their essential role in the origins of life. In this review, we summarize major issues related to the role of *N*-ribosides in biological systems, especially in the context of the origin of life and its further evolution, through the RNA-based World(s), toward the life we observe today. We also discuss possible reasons why life has arisen from derivatives of β-D-ribofuranose instead of compounds based on other sugar moieties.

Keywords: cofactor; coenzyme; enzyme; *N*-glycosidic bond; nucleic acid; nucleoside; nucleotide; origin of life; ribozyme; Rossmann fold

1. Introduction

It is a challenge to define life based on the Natural Sciences in a simple yet exhaustive way [1,2]. Among the numerous attempts to describe life as a process, a definition coined by the US National Aeronautics and Space Administration (NASA), which refers to life as "a self-sustaining chemical system capable of Darwinian evolution," seems the most popular and the least controversial [2,3]. One of its many advantages is that it does not specify the chemical basis on which life would be based, which is essential, for example, from the perspective of questions regarding the origin of life on the Earth and whether it was based on the same molecular basis then as it is today.

The processes occurring in living organisms are highly complex and diverse. On the other hand, all currently known organisms function based on the central dogma of molecular biology, which determines the direction of information flow from the DNA through the RNA to proteins (excluding cases in which genetic information is encoded within an RNA molecule, e.g., an RNA template bound to active telomerase [4] or RNA virus genomes [5]). For this reason, the molecular basis of life is now the same in all living organisms and encompasses several basic aspects, that is: (i) genetic information carriers in the form of inherited DNA transcribed onto the RNA; (ii) catalytic processes mediated mainly by enzymes, and less often ribozymes, often supported by low-molecular cofactors; (iii) ions interacting with both nucleic acids and proteins; and (iv) lipids and (v) sugars performing a range of metabolic and structural functions, such as ensuring an intact environment for specific intracellular chemical processes. A careful look at these individual aspects can lead to two interesting observations. First of all, both DNA and RNA, alongside a number of low-molecular cofactors, are based on only a few structurally very similar building blocks, namely, *N*-glycosidic derivatives of β-D-ribofuranose. Secondly, elements based on these *N*-glycosides provide genetic information as well as enable catalytic processes to occur. Indeed, a number of those processes based on, among others, the *N*-ribosides are a driving force for the self-sustainability of living chemical systems. Additionally, the genetic information is not copied and transcribed perfectly, especially that encoded in RNA carriers, and, thus, is capable of Darwinian evolution. Interestingly, a hypothesis proposed by Yarus [6,7] suggests that the origin of Darwinian behavior at the molecular level in a primordial environment could be facilitated by small oligonucleotides. In particular, complementary 5′-5′-cofactor-like dimeric ribonucleotides could be involved in establishing an Initial Darwinian Ancestor through synthesizing and enriching molecules with eligible chemical properties. The above observations indicate that these compounds are crucial for life processes. Furthermore, the second observation provokes questions about the molecular basis of the origins of life and about the role of *N*-glycosidic derivatives of β-D-ribofuranoses in the origin of primary metabolic processes.

2. *N*-Glycosyl Derivatives of β-D-Ribofuranose: Toward the RNA-Based World(s) and Its Evolution

The Earth itself is estimated to be just over 4.5 billion years old [8], whereas the abiogenesis, i.e., the origin of life from inanimate matter, commenced about four billion years ago [9–11]. Due to the significant lapse of time, biogenesis theories are based on geochemistry-, geophysics-, geology- and biochemistry-based premises. Furthermore, bioinformatics analysis of the increasingly more available complete genomes of contemporary living organisms is an exceptionally useful tool for studying the history of life on Earth. Combining the results of these seemingly distant studies yields a coherent picture that can logically reflect the origins and subsequent evolution of life on Earth. It is now believed that the hydrothermal vents or geothermal fields on the sea or ocean floor provided an adequate environment for life on Earth to start, due to the availability of chemicals that would have been necessary to sustain the metabolic processes of the time [12–20]. Were these primary forms of life based on the transfer of information via the DNA –> RNA –> protein pathway, as life forms are today? It does not seem so.

2.1. *Why Has Life Arisen from N-Glycosyl Derivatives of β-D-Ribofuranose?*

Widely accepted hypotheses to explain that the origin of life includes the stage of the "RNA-based World" [21–27] that was based (entirely or partially) on β-D-ribonucleotides and sustained due to the ability of self-replication. However, the question that should be asked is why nature chose β-D-ribonucleotides as building blocks at the initial steps of abiogenesis. This question especially relates to their components: purine and pyrimidine bases, sugar moiety (and its chirality) and phosphate groups. An essential determinant was the presence of particular synthetic routes, which ensured access to specific chemical compounds in a prebiotic environment. However, conditions present on the Earth about four billion years ago presumably promoted numerous chemical reactions; thus, an environment contained various organic and inorganic compounds [28]. Therefore, the emergence of the chemical self-replication required a selection mechanism(s) of building blocks [29–33]. Nature chose four β-D-ribonucleotides based on adenine, cytosine, guanine and uracil, which ensured the success of the "RNA-based World". An interesting point is the homochirality of the sugar moiety in the "RNA-based World" [34], which contradicts the chemical paradigm of symmetry between a pair of enantiomers. It is of note that the geometry of synthetic nucleic acids that are based on β-L-ribofuranose derivatives agrees well with the stereochemistry of experimental models of naturally occurring nucleic acids [35]. Moreover, chirally inverted macromolecules are capable of performing their function in the mirror-image world (proteins and RNA built from D-amino acids and β-L-ribonucleotides, respectively), as revealed recently for T7 transcription [36]. It is unclear why biological molecules contain homochiral components, when this asymmetry appeared and, finally, why RNA is composed of β-D-ribonucleotides instead of their L-enantiomers. However, some structural studies of RNA enantiomers have revealed differences on the RNA–solvent interfaces [37,38]. However, the reason for such differences is not fully explained [38,39].

A definitive answer to why nature chose no other sugar moiety (aldose or ketose, tetrose or hexose, another pentose) but ribose as building blocks in the initial steps of abiogenesis and the origin of live is elusive. Herein, a severe obstacle is a lack of knowledge of the exact chemistry and physics in a prebiotic environment, which surely defined and promoted particular synthetic pathways as well as their stereoselectivity. However, some structural, chemical and biological premises suggest why *N*-glycosyl derivatives of β-D-ribofuranose could be preferred over other derivatives. In a water solution at standard conditions, D-ribose exists in an equilibrium of four cyclic (five- or six-membered) hemiacetal forms, namely, α- and β-D-ribofuranose and α- and β-D-ribopyranose, and the latter forms are more stable (a relative abundance of six-membered forms is approximately 80% and only approximately 13% for β-D-ribofuranose) [40]. On the other hand, at high temperatures, at a steady temperature gradient, the equilibrium is shifted, and the abundance of the ribofuranose ring that is present in *N*-ribosides is elevated [41]. Importantly, in this form, the sugar ring is relatively compact and very flexible, as the structure of oligonucleotides can be defined by its seven torsion angles, as well as the pseudorotation of the ring [42]. A polynucleotide based on a smaller, four-membered sugar ring (e.g., a tetrofuranose) would have a decreased structural flexibility due to a lower number of variable torsion angles. Furthermore, introducing a larger, six-membered sugar ring (e.g., hexofuranose) would have a similar effect due to the increased steric hindrances. The above observations indicate the necessity of a five-membered sugar moiety as a building block of flexible oligo- and polynucleotides. Analogously, the choice of ribose over the arabinose, xylose or lyxose may result from decreased structural flexibility, as the presence of C2′- and/or C3′-hydroxyl group(s) above the furanose ring in the latter three sugars may lead to steric clashes with the bulky moieties attached to the C1′ and C5′ positions [43,44]. Moreover, studies on non-enzymatic template-directed primer extension reactions with nucleotide substrates based on ribo-, threo-, arabino- and 2′-deoxynucleotides indicated significant preferences for ribonucleotides [32,33]. In particular, reactions conducted with mixtures of the nucleotides favored the incorporation of ribonucleotides. In turn, numerous cycles of replication enriched descendant strands in ribonucleotides and finally led to homogenous RNA strands. Thus, these results also revealed a possible mechanism for selecting ribonucleotides from a primordial soup. In addition, studies on enzymatic primer extension on DNA templates as well as thermal denaturation analysis also indicated preferences for β-D-ribonucleotides over nucleotides based on glycerol [29,30].

2.2. Origin of the RNA-Based World(s)

The "RNA World", the hypothesis formulated by Gilbert [26], which assumes that life evolved from self-replication and catalysis based mainly on RNA machinery, has been of great interest for many decades. Recently, this concept has been significantly modified, and new theories present the origin of life as RNA catalysis facilitated by amino acid- or peptide-based cofactors or the "RNA-peptide World". Nevertheless, each scenario had to be based on the prebiotic synthesis of activated nucleotides and RNA molecules, including ribozymes [45–55]. Following this remark, the very origin of any RNA-based World(s) from inanimate matter appears particularly interesting. There are two main theories which attempt to provide the answer. In line with them, the RNA developed either (i) through the chemical evolution of pre-RNA polymers (based, for example, on other sugar/amino acid residues and nitrogenous bases) [56], or (ii) it was directly synthesized from β-D-ribonucleotides that were available as a result of the geochemical processes of that time [57–62]. The second theory, which assumes that RNA is a direct "invention" of nature, based on the availability of β-D-ribonucleotides containing four nucleobases, namely, cytosine, uracil, adenine and guanine, which are able to combine and form single

RNA chains under appropriate conditions, is definitely more common nowadays [63–65]. Its widespread popularity was undoubtedly influenced by research on the possibility of β-D-ribonucleotide synthesis under the prebiotic conditions of the time [59,62,66]. These works solved two problems related to the origin of the "RNA World", namely, the so-called "prebiotic chemist's nightmare" [64] and the "problem of prebiotic β-D-nucleoside synthesis" [67]. The first problem relates to a situation in which a large number of products of prebiotic chemical reactions are synthesized at very low yields. Consequently, the molar concentration of these compounds was probably very low in a water environment. Thus, the formation of an RNA polymer composed of only four nucleotides that additionally occur at low concentrations in a complex mixture of other compounds was unlikely. The second issue relates to synthetic pathways of β-D-nucleotides in the prebiotic world. With the emergence of the "RNA World" theory, it was assumed the *N*-glycosidic bond was formed between the D-ribose and nitrogenous base molecules during the synthesis of β-D-ribonucleotides. However, this chemical pathway seems unlikely. The primary reason for that is the fact that, even if the D-ribose was produced in large quantities under prebiotic conditions, the chemical synthesis of β-D-nucleotides by using D-ribose and nitrogen heterocyclic bases as substrates is either very inefficient [67] (for purine bases) or does not occur at all or is strongly shifted toward hydrolysis to D-ribose and a free base [68] (for pyrimidine bases). Instead, Powner et al. [59,66] and Stairs et al. [62] have suggested different possible routes of synthesis of these *N*-glycosidic derivatives, in which both the sugar residue and the base are formed simultaneously on an activated precursor molecule containing a C-N bond of the emerging *N*-glycoside. This pathway of the synthesis of activated ribonucleotides, which are necessary for RNA synthesis, has another two advantages. The phosphate ions participating in the reaction are not only a substrate and a buffer but also serve as a catalyst. Additionally, such a pathway limits the variety of products at the initial stages of the reaction. A point of weakness of those chemical routes in the context of the "prebiotic chemist's nightmare" problem is that syntheses are not selective, and products are derived from four aldopentose isomers. Moreover, the synthesis efficiency of D-ribosides is lower than that of other pentosides. The above limitations were overcome by Kim and Benner [69], who described the stereoselective synthesis of purine and pyrimidine nucleotides by using phosphorylated carbohydrates and nucleobases as substrates, giving only the β-anomer of the nucleotides. Additionally, for adenine-based nucleotides, the reaction is *N*9-regioselective. Finally, Becker et al. [70] reported a unified synthesis of all four pyrimidine- and purine-based RNA nucleotides from ribose and small-molecule compounds that could potentially be present in the prebiotic environment. Herein, these reactions are driven by wet–dry cycles and occur in the presence of phosphate anions.

The fusion of the "RNA-based World" and the "lipid world", including the encapsulation of self-replicating RNA (ribozymes) into cellular-type lipid systems (e.g., liposomes), which took place either simultaneously or sometime later, could have marked the occurrence of primary life on Earth [15–17,71–73]. The compartmentalization of a so-called autocatalytic set containing catalytically active RNAs, possibly with other organic compounds (e.g., peptides, cofactors, etc.) ensured the self-replication process(es) to be sustainable and capable of evolution [17,74]. The diffusion processes, as well as micromolar concentrations of organic compounds in an oceanic environment, would otherwise prevent the self-replication of the system [15,28].

2.3. Evolution of the RNA-Based World toward More Complex Metabolism

2.3.1. Ribozyme-Based Catalysis

The success of the "RNA-based World(s)" most likely relied on an ability to self-replicate. Although a self-replicating single RNA molecule remains undiscovered [75], other resembling systems based on the recombination and ligation of RNA fragments, which could provide a basis for the simplest form of Gilbert's "RNA World", have been described [76]. These include an RNA polymerase ribozyme that catalyzes its own evolutionary ancestor, an RNA ligase ribozyme [77], or an autocatalytic set of two RNA ligase ribozymes that catalyze the synthesis of each other [78]. However, the latter system can utilize only RNA fragments, for example, trinucleotide triphosphates [79], instead of individual nucleotides. This requirement could be fulfilled by another catalytic feature of ribozymes, namely, the partial cleaving of RNA molecules. Indeed, numerous natural self-cleaving and cleaving ribozymes have been identified to date [24,80,81]. Recently, another self-replication system based on promoter recognition has been described [82], by which selective promotor-based replication could prevent the synthesis of others, e.g., parasitic RNA molecules. A phenomenon of self- or non-self-replication of the primordial encoding RNA seems to be a more complex problem and generates numerous questions. Replicated RNA strands must be long enough to encode a particular function but sufficiently short for efficient copying. Moreover, the synthesis of a descendant strand is inefficient with mononucleotides used as substrates. Additionally, a mechanism for separating both strands from the duplex product is unknown. The above issues were solved by works on non-enzymatic primer extension and ligation [83–85]. These works revealed that short fragments of RNAs could serve as a good template for replication and facilitate this process through their polymerization or ligation. Furthermore, short RNA molecules could interact with each other to form functional assemblies that are activated by other polymerization reactions [83,85]. Moreover, during the non-enzymatic primer extension process, synthetized duplex RNA could be unwound during so-called strand displacement replication, which facilitates the primer extension on templates that were previously occupied by a complementary strand [84].

It is of note that the aminoacylated RNA primer significantly enhances non-enzymatic primer extension and ligation reactions [51]. Moreover, descendant RNA strands could be linked with various amino acids with chemically diverse properties enabling new catalytic functions for amino acid-facilitated ribozyme catalysis. The above observation is simply an example of numerous discoveries that have significantly modified the simple "RNA World" theory. Indeed, new hypotheses assume the integration of RNA and amino acid/peptide chemistries at the initial stages of the origin of life. The inclusion of amino acids in metabolism may have been possible due to their common availability in the environment [86]. It had two very important consequences: (i) the development of a genetic code, as a result of which, information encoding in the RNA and its transcription into proteins became possible, and (ii) the shift to protein-mediated catalysis [87]. The origin of genetic code could evolve through molecular recognition between pairs: primitive tRNA-specific amino acids, giving the chemical basis for a selective formation of aminoacyl-tRNA sets [54]. Although a simple peptide bond formation between two amino acids can be catalyzed by the RNA only [88], the formation of the ribosome, a large peptide-RNA assembly that forms the peptide bond between two specific amino acids based on the information encoded in RNA, appeared to be an evolutionary milestone [89].

As the RNA molecules evolved over time, new features began to emerge, which enabled increasingly complex metabolism controlled by a ribozyme-based catalysis [1,17,90–93]. Owing to their capacity for the selective binding of nucleotide-derived cofactors (coribozymes), amino acids and peptides, substrates and environmental cations, in particular, metal cations such as Mg^{2+}, Mn^{2+}, Co^{2+} or Fe^{2+} [94–96], the range of reactions catalyzed by RNA molecules was broadened to include redox processes, as well as the transfer of electrons and simple functional groups. Interestingly, the vast majority of these primary substrates and coribozymes, such as adenosine-5′-triphosphate (ATP), oxidized/reduced forms of universal redox cofactors, including nicotinamide adenine dinucleotide (NAD^+/NADH and $NADP^+$/NADPH) and flavin adenine dinucleotide (FAD/$FADH_2$), adenosylcobalamin (AdoCbl), coenzyme A (CoA) or *S*-adenosyl-L-methionine (SAM), are C5′-substituted adenosine derivatives (Figure 1) [91,97–100].

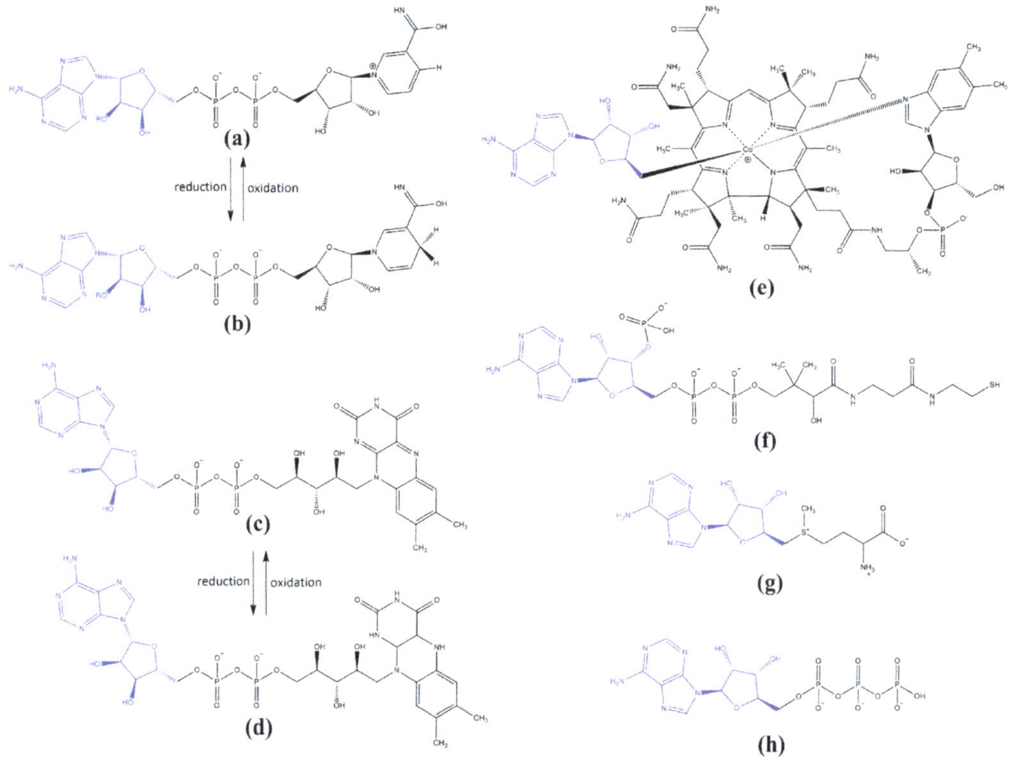

Figure 1. Common coribozymes and substrates in the "RNA-based World" were C5′-substituted adenosine derivatives: (**a**,**b**) nicotinamide adenine dinucleotide in its oxidized (NAD^+) and reduced (NADH) form, respectively, (**c**,**d**) and flavin adenine dinucleotide in its oxidized (FAD) and reduced ($FADH_2$) form; (**e**) adenosylcobalamin (AdoCbl); (**f**) coenzyme A (CoA); (**g**) *S*-adenosyl-L-methionine (SAM) and (**h**) adenosine-5′-triphosphate (ATP); C5′-substituted adenosine moiety is shown in blue.

It was probably due to the formation of aptamer, an RNA motif enabling selective binding of adenosine-based cofactors (recognition via adenosine or adenine moiety of a cofactor), so that the functional groups directly involved in the catalyzed reaction could be exposed on the surface of ribozymes [91,98,99]. Of note, the recognition of the cofactors mentioned above by an aptamer is not limited to an adenosine/adenine moiety of nucleotides. Recently, an RNA aptamer, which recognizes a redox-active compound, flavin (a part of the FAD cofactor), has been described [101]. Interestingly, the binding of flavin by aptamer shifts the reduction potential of this cofactor. This discovery may indicate that ribozymes could bind and utilize various cofactors and fine-tune their redox properties for particular chemical reactions.

2.3.2. Stabilization of Genetic Information

The increasing complexity of metabolism required more and more information to be encoded in RNA molecules, which, as genetic material, pose two important problems. The first of them is the speed and accuracy of ribozyme-catalyzed replication [102,103]. The second problem is the low stability of RNA molecules, which is linked mainly to the presence of the 2′-hydroxyl ribose group, which is essential for the ribozyme-catalyzed processes. The aforementioned group promotes the intramolecular hydrolysis of the 3′,5′-phosphodiester bond and, consequently, RNA molecule fragmentation (Figure 2a).

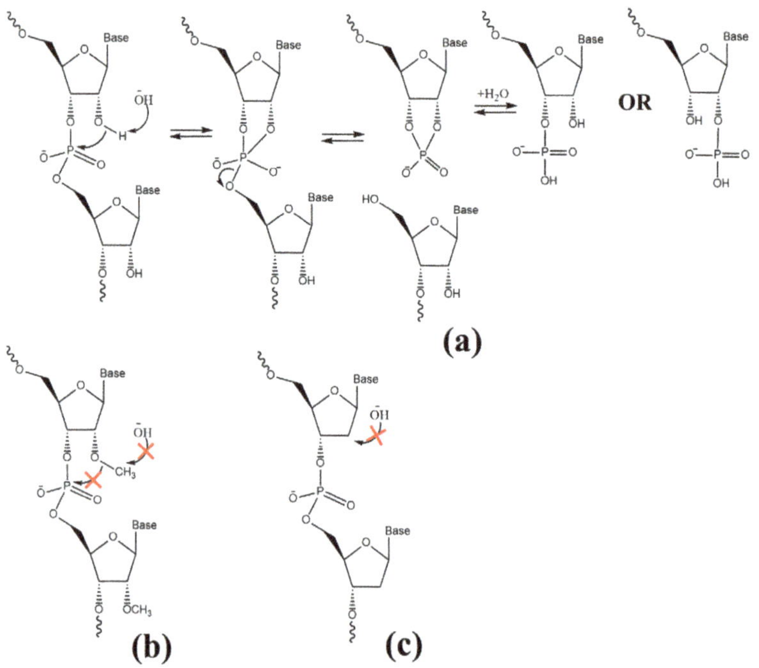

Figure 2. Stabilization of genetic information: (**a**) schematic of RNA degradation catalyzed in the presence of alkaline anions, (**b**) a methylation of 2′-OH group of ribose moiety abolishes its nucleophilic character and prevents the alkaline hydrolysis of RNA, (**c**) removal of 2′-OH group contributed to the change in genetic information carrier from less stable RNA to DNA.

Over time, increased functionality and stability of genetic material were achieved via Darwinian evolution. The changes were based on two different chemical modifications of genetic material, which require common adenosine-based coribozymes or cofactors from the "RNA-based World". Both modifications utilize SAM; however, their mechanisms are different and are based on methyl group transfer (SAM serves as a donor during the 2′-*O*-methylation of the D-ribose moiety within an RNA molecule) or potent radical chemistry (part of ribonucleotide reductases utilize SAM to generate deoxyribonucleotides; SAM is reductively cleaved to generate a radical). The first modification is based on the 2′-*O*-methylation of the ribose moieties in RNA molecules [104]. It is of note that all four canonical, as well as noncanonical, nucleotides can be modified in that manner [104]. This simple, usually local modification significantly changes the chemical properties of the methylated RNA. In particular, the nucleophilic character of the hydroxyl oxygen atom 2′-*O* is eradicated via methylation. Consequently, 2′-*O*-methylated ribonucleotide polymers are stable and resistant to alkaline or enzymatic hydrolysis (Figure 2b) [105,106]. Additionally, interactions between methylated RNA and other macromolecules are impaired, as the lack of a 2′-hydroxyl proton in the methylated nucleoside limits the formation of an intermolecular hydrogen-bond network [107–109]. Notably, 2′-*O* methylated RNA forms an A-type RNA double helix with base-pairing observed in non-modified RNA helices [110,111]. It remains unclear when this mechanism was adapted to modify RNA and how this reaction was initially catalyzed (by ribozymes or enzymes). However, numerous methyltransferase ribozymes that utilize SAM as a methyl group donor are well known and characterized [100,112–114]. Thus, it is probable that 2′-*O*-methylation modification of RNA arose before the existence of methyltransferase enzymes. Moreover, nowadays, observations that such a chemical modification stabilizes various RNA molecules such as tRNA and rRNA, especially at higher temperatures [115–120], may indicate that 2′-*O*-methylation was acquired at the initial steps of biogenesis, before the appearance and evolution of protein-mediated catalysis, to ensure the stabilization of RNA molecules under extreme conditions [104]. It is of note that SAM-dependent methylations are very common chemical modifications of RNA molecules and are not only restricted to 2′-*O*-methylation of the sugar moiety. Indeed, methylations of nucleobases are also abundant, and nowadays, the number of known such modifications is constantly increasing [121,122].

The significant Darwinian evolution toward increasing the stability of the coding material was based on the transition from the RNA to DNA. The change was based on the removal of the 2′-hydroxyl group of the ribose moiety in each ribonucleotide (Figure 2c) [123–127]. These reactions were and still are catalyzed by ribonucleotide reductases. Interestingly, many of these enzymes are AdoCbl- or SAM-dependent reductases; that is their activity depends on cofactors that are already present in the "RNA-based World". Enzymatic reduction of ribonucleotides to 2′-deoxyribonucleotides contributed to the change in the genetic information carrier from less stable RNA to DNA, and the very same reactions are still used by all living organisms. Furthermore, the replacement of uracil (a product of spontaneous cytosine deamination) with thymine in DNA enabled the development of repair mechanisms applicable to DNA-stored genetic material. Therefore, the change in the carrier of genetic information increased and enabled more accurate replication, which, in turn, significantly improved the accuracy of transmitted and stored genetic data.

A glimpse of the above two chemical modifications of nucleotides indicates the biological success of the latter in the formation of stable genetic material. Although the methylation ensures a higher stability of the modified RNA molecules, its character is local, as it is limited to certain nucleotide(s) of previously synthesized non-modified RNA oligonucleotides. Thus, the modification of all 2′-hydroxyl groups within an RNA molecule would require a high number of transmethylation catalysts (methyltransferase ribozyme or/and enzymes), as they are specific to particular elements of the modified RNA molecule. Moreover, the RNA double helix containing methylated 2′-*O*-methylated nucleotides is more stable than its non-methylated counterpart, which may affect its replication. On the other hand, the removal of the 2′-hydroxyl group of the sugar moiety in each ribonucleotide occurs prior to the synthesis of the DNA chain. Consequently, only a few ribonucleotide reductases are needed to synthesize deoxynucleotide polymers. Notably, the formation of a chimeric DNA-methylated RNA duplex is not favored [128], which may suggest that these two mechanisms of stabilization of the genetic material evolved separately and did not interfere with each other during their evolution. This conclusion may be supported by the fact that the reduction of ribonucleotides requires potent radical chemistry powered by redox-active transition metal cations, which could disturb an RNA structure and function [127,129]. Thus, such catalytic function probably did not evolve in the "RNA-based World" but arose with enzyme catalysis. Indeed, no ribonucleotide reductase ribozymes have been identified so far.

2.4. The Evolution of Metabolism toward the Last Universal Common Ancestor (LUCA)

The evolution of metabolism from catalytic, self-replicating RNA molecules to simple, single-cell organisms that are capable of information transfer from DNA through RNA to proteins took 200–500 million years. This period was determined based on the analysis of sedimentary rocks containing traces of cellular life fossils, the age of which was estimated to be about 3.5–3.8 billion years [11,130–132]. These organisms (whether it was a single organism or a group of organisms [133–135]) constitute a hypothetical link between the "RNA-based World" and all contemporary cells [136]. They are referred to as "the last universal common ancestor" (LUCA) of all currently living organisms [135,137]. It should be noted that most authors consider LUCA to be a living organism(s). However, defining LUCA is not so obvious [135], and according to some authors, LUCA existed before life began [138], e.g., as a chemical reaction-based system forming a part of inanimate matter or was a unique entity, which was different from any life form we observe today [139]). Bioinformatics reconstruction of the LUCA genome enabled identification of several hundred proteins involved in potential physiological processes occurring in the LUCA cell [136,140]. Considering the hot environment containing CO_2, H_2S H_2, N_2 and a variety of transition metal cations [136,141] that were prevalent at the time, the LUCA was likely a thermophilic, anaerobic autotroph [142–145]. Its metabolism was based on numerous reactions, which required the involvement of specific cofactors. Interestingly, most of them had already been present in the "RNA-based World", acting as coribozymes. However, a significant difference is that despite the involvement of the same cofactors in identical chemical reactions in the LUCA, the reactions were mainly enzyme-catalyzed, as is the case today. The analysis of the LUCA metabolism demonstrates the key role of cofactors, especially adenosine cofactors [136,146–149], in numerous physiological processes, including but not limited to CO_2 assimilation and fatty acid synthesis (CoA, ATP and NAD^+/NADH); N_2 assimilation (ATP, NAD^+/NADH, $NADP^+$/NADPH, FAD/$FADH_2$); numerous redox reactions that are unrelated to gas assimilation (NAD^+/NADH, $NADP^+$/NADPH, FAD/$FADH_2$ and free radical reactions based on FeS-SAM system) or chemical RNA base modification through SAM-dependent methylation. Interestingly, about 30% of several hundred potential LUCA proteins were *S*-adenosyl-L-methionine-dependent (SAM-dependent methyltransferases and reductases). The cofactors present in the "RNA-based World" and LUCA are present in all living organisms, regardless of their habitat [97].

3. Biological Success of *N*-Glycosyl Derivatives of β-D-Ribofuranose

3.1. *N-Glycosyl-β-*D*-Ribosides Have a Unique Structure That Renders Them Biologically Effective*

Studies on biogenesis indicate that *N*-glycosidic derivatives of β-D-ribofuranose significantly contributed to the formation and development of life on Earth. Furthermore, they are present in the cells of all known organisms as a key element of their metabolism, that is, both catalytic processes and the encoding of genetic information. What contributed to their biological success? Their chemical structure very likely did, from the early stages of metabolic development [56]. The nucleic acid make-up consists of three covalently bonded structures: (i) the nitrogenous base involved in molecular recognition processes through the formation of canonical or noncanonical Watson–Crick base pairs, (ii) the D-aldopentose residue serving as a bi-(D-deoxyribose) or trifunctional (D-ribose) conformationally flexible connector on which each nucleotide is formed, and (iii) ionized phosphate residue linking subsequent nucleotides. Some conformational flexibility of the sugar residue enables nucleic acid molecules to take different forms, depending on the nucleotide sequence or environmental conditions, without interfering with the nucleotide base pair geometry. This is particularly noticeable in the case of DNA molecules, for which many different structural forms have been identified [150], with the right-handed double helix (A-DNA or B-DNA) and left-handed double helix (Z-DNA) reported most commonly in the literature. The presence of numerous polar groups and phosphate anions enables non-covalent bonding with additional structures, such as water molecules or cations. A similar structural pattern can be observed in nucleotide (ATP, CoA) and dinucleotide (NAD^+/NADH, $NADP^+$/NADPH or FAD/$FADH_2$) adenosine-based cofactors. In the "RNA-based World", nucleic acid molecules have created motifs that enable selective binding of free nucleosides or nucleotides, e.g., adenosine-based cofactors or coribozymes. The shift to enzyme-mediated catalysis did not significantly affect the range of cofactors used, especially adenosine cofactors, which may prove their perfect adaptation to metabolic processes. What changed, though, was only the cofactor and macromolecular interaction type. Unlike coribozymes present on the RNA surface or grooves that are easily accessible from the solvent access region (Figure 3a) [112–114,151,152], coenzymes are commonly bound in deep catalytic cavities (Figure 3b) or within cavities formed within the protein surface region (Figure 3c) [153,154].

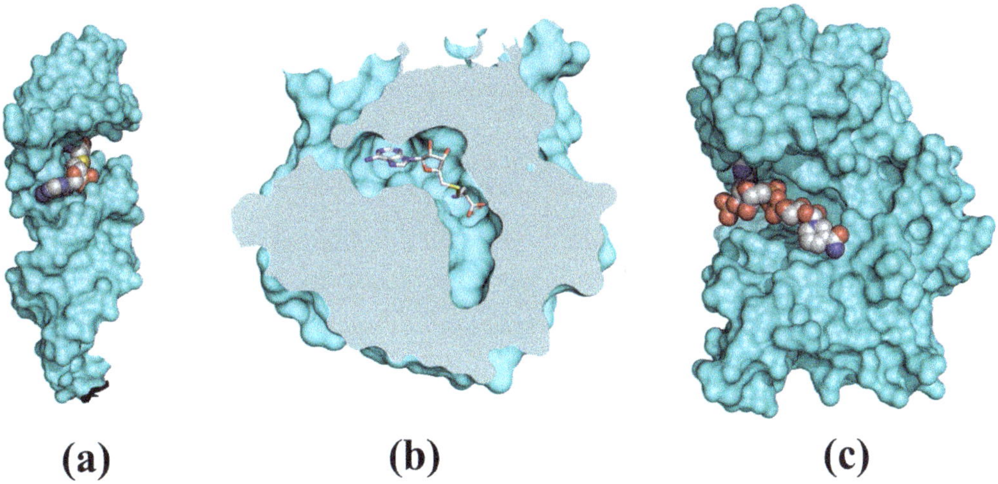

Figure 3. Illustration of small molecule binding area in selected macromolecules (cyan): (**a**) methyltransferase ribozyme complexed with SAM (PDB code 7DWH, [112]), (**b**) methyltransferase from *Bradyrhizobium japonicum* complexed with SAH (PDB code 3OFK, chain A, [155]), (**c**) glutamate dehydrogenase from *Thermococcus profundus* in complex with NADP (PDB code 8HHO, unpublished).

Not surprisingly, in proteins, there are numerous interactions (hydrogen and ionic bonds, as well as hydrophobic, van der Waals or dipole interactions) between a macromolecular environment and coenzyme molecule within the ligand-binding area. The involvement of multiple interactions in the binding cofactors somehow follows their structure, enabling their strong bonding to an active site of an enzyme. It should be noted that during evolution, numerous enzymes that utilize adenosine-based cofactors or substrates have evolutionarily developed the ability to coordinate metal ions, most often in the active site region or in its close proximity. The bond may affect alkaline metal ions [156], mainly Na^+ and K^+ or a number of transition metal cations, such as Zn^{2+}, Mn^{2+}, Fe^{2+} and others [157–159]. Moreover, activities of these enzymes can be regulated by using the coordination of both alkaline and transition metal cations in the active site area, as observed for pyruvate kinases [160], *S*-adenosyl-L-methionine synthetases [161] or *S*-adenosyl-L-homocysteine hydrolases [157,162,163]. As a result, not only was the range of catalyzed reactions broadened, but additional mechanisms were also created to positively or negatively regulate enzymatic activity, e.g., through the effect of cation on substrate bonding [156].

3.2. *N-Glycosyl Derivatives of β-D-Ribofuranose Are Ubiquitously Utilized in a Cellular Metabolism*

The biological success of ribose-based nucleotides and nucleosides is not limited to encoding genetic information and cellular catalysis nowadays. In addition, these compounds play numerous other roles in metabolic processes that occur in prokaryotic and eukaryotic cells. Some of them are very common, are present in all three domains of life, and are described abundantly in the literature. Briefly, ATP, NADH and $FADH_2$, serve as chemical energy storage molecules. Cyclic adenosine monophosphate (cAMP) is a second messenger molecule that is involved in intracellular signal transduction [164]. Furthermore, cAMP controls sugar catabolism in prokaryotic cells, regulating gene expression on appropriate operons [165]. Biological methylation within the cell is controlled by the concentrations of SAM and S-adenosyl-L-homocysteine (SAH, a by-product of SAM-dependent methylation reactions) in a cell. The SAM:SAH ratio is perceived as an indicator of the transmethylation activities of the cell [166] and plays a crucial role in numerous metabolic processes [167], including the regulation of biological rhythms [168,169].

An interesting function of nucleotides and nucleosides is regulating expression that is facilitated through non-covalent or covalent interactions with mRNA molecules. In bacterial cells, a riboswitch, a 5′-UTR region of mRNA molecule, could control the translation process. The riboswitch can bind non-covalently and specifically to appropriate ligands, which induces conformational changes within this 5′-UTR region and, consequently, can modulate transcription termination signals or the availability of ribosome binding sites in these mRNAs [170–173]. Among numerous small-molecule ligands that are sensed by riboswitches, *N*-glycosyl derivatives of β-D-ribofuranose, such as SAM, SAH, cAMP and NAD^+, are the most abundant [174–178]. Moreover, it was revealed that some adenosine-based nucleotides, such as CoA, NAD^+/NADH and FAD, might be covalently attached to prokaryotic and eukaryotic mRNA to form noncanonical 5′-cap structures [179]. Their biological significance is unclear nowadays. However, it has been revealed that these noncanonical modifications may stabilize the capped mRNA and influence the translation process in prokaryotic and eukaryotic cells [179–183].

3.3. Nucleotides and Nucleosides as Cofactors and Substrates in the Protein World

3.3.1. The Rossmann Fold Is Responsible for Binding Various Adenosine-Based Nucleosides and Nucleotides

The capacity for selective protein–ligand interactions is closely related to the presence of characteristic structural motifs that are responsible for binding individual low-molecular-weight compounds or macromolecules. As it was with the coribozyme-binding aptamers, characteristic motifs/domains selectively binding individual adenine-based nucleotides or nucleosides have evolved as a result of cofactor involvement in enzymatic catalysis [184]. A recent study on an evolutionary chronology of networks of domain organization suggests that these motifs/domains are the oldest evolutionary units of proteins [185]. Currently, the Rossmann fold is the most common protein motif responsible for binding these cofactors [186–192]. Depending on the enzyme group, it can be either canonical [186] (a sheet formed by six parallel β chains surrounded by four α-helices) or noncanonical (containing modifications within the β sheet, altered helical environment, etc.). The Rossmann fold can be found in numerous groups of enzymes that are involved in catalyzing more than three hundred chemical reactions (Figure 4a–h) [193].

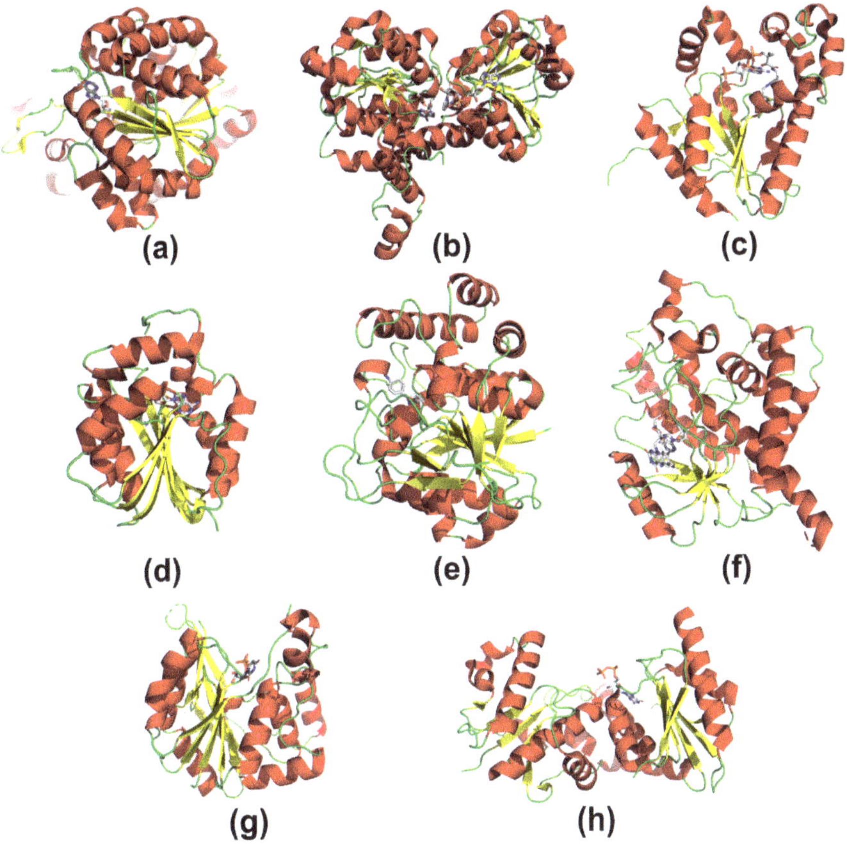

Figure 4. Selected members of enzymes with Rossmann fold, which utilize adenosine-based cofactors and substrates: (**a**) ADP-dependent glucokinase from *Pyrococcus horikoshii* complexed with AMP derivative (PDB code 5O0J, [194]), (**b**) *S*-adenosyl-L-homocysteine hydrolase from *Pseudomonas aeruginosa* complexed with NAD$^+$ and Ado (PDB code 6F3M, chain A, [157]), (**c**) coenzyme A synthase from *Mus musculus* complexed with acetyl CoA (PDB code 2F6R, unpublished), (**d**) methyltransferase from *Bradyrhizobium japonicum* complexed with SAH (PDB code 3OFK, chain A, [155]), (**e**) purine nucleotide hydrolase from *Crithidia fasciculata* complexed with a nucleoside inhibitor (PDB code 2MAS, chain A [195]), (**f**) FAD synthetase form *Saccharomyces cerevisiae* complexed with FAD (PDB code 2WSI, [196]), (**g**) glycosyltransferase SpsA from *Bacillus subtilis* complexed with UDP (PDB code 1QGS, [197]), and (**h**) Fucosyltransferase NodZ from *Bradyrhizobium* sp. WM9 complexed with GDP (PDB code 3SIX, [198]). The color code indicates the secondary structure elements as follows: red—helical regions; yellow—β-strands and the central β-sheet of the Rossmann fold; green—loop regions.

For instance, in SAM-dependent methyltransferases, numerous oxidoreductases utilizing nicotinamide cofactor, some ligases, *N*-ribohydrolyses and transferases utilizing adenosine phosphates are present [199,200]. The Rossmann fold is also present in enzymes, which bind adenine-based substrate reaction products, for example, in *S*-adenosyl-L-homocysteine (SAH) hydrolysis enzymes that regulate SAM-dependent cellular methylation reactions [153]. In all organisms, SAH hydrolysis is mediated by one or two methylation-regulating enzymes. The first is bifunctional methylthioadenosine/*S*-adenosyl-L-homocysteine (MTA/SAH) nucleosidase, which is a single-domain protein hydrolyzing the MTA (Figure 5a) to S-methyl-5-thio-D-ribose and adenine or decompose the SAH (Figure 5b) to adenine and S-ribosyl-L-homocysteine. The second enzyme is S-adenosyl-L-homocysteinase (SAHase), which converts SAH to adenosine and homocysteine. It contains two Rossmann folds, one located in the substrate/product (SAH or adenosine)-binding domain and the other in the key cofactor (NAD^+)-binding domain (Figure 4b).

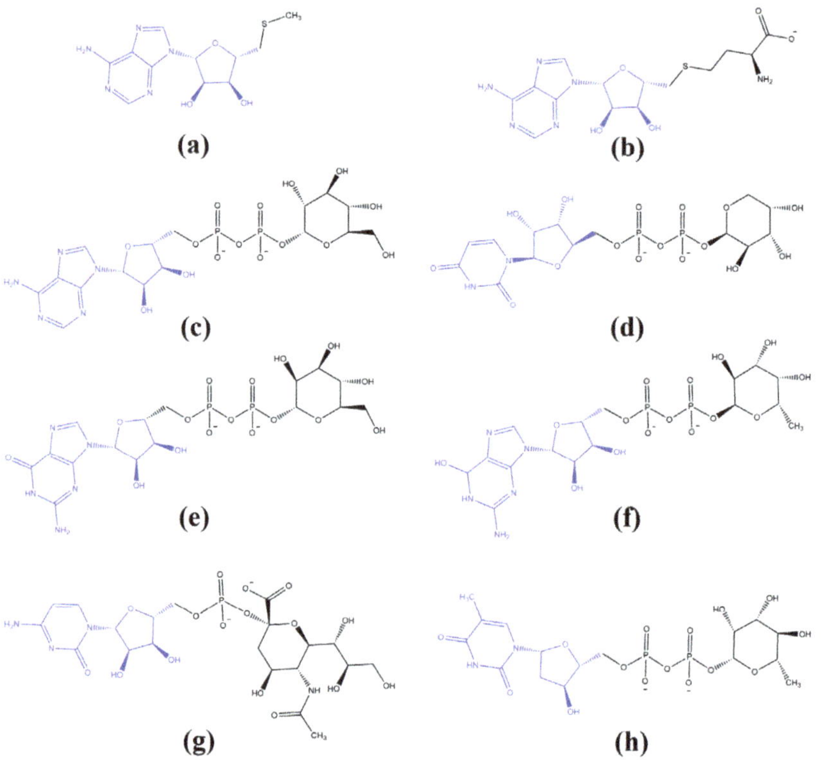

Figure 5. Other than the primordial C5′-substituted adenosine derivatives from the "RNA-based World", substrates commonly utilized by Rossmann-fold-based enzymes include the following: (**a**) methylthioadenosine (MTA), (**b**) S-adenosyl-L-homocysteine (SAH), (**c**) ADP-α-D-Glu, (**d**) UDP-β-L-arabinose, (**e**) GDP-α-D-Man, (**f**) GDP-β-L-Fuc, (**g**) CMP-sialic acid and (**h**) dTDP-β-L-rhamnose; C5′-substituted nucleoside moiety is shown in blue.

3.3.2. Proteins Can Utilize the Rossmann Fold to Bind a Variety of Purine- and Pyrimidine-Based Nucleotides and Nucleosides

During evolution, groups of enzymes have emerged that use nucleotide-based, yet not adenine-based, cofactors or co-substrates while preserving the functional (canonical or noncanonical) Rossmann fold. Examples of those are glycosyltransferases (GT, Figure 4g,h), a numerous group of enzymes catalyzing the transfer of the sugar residue from the donor molecule to a series of macromolecular or low-molecular-weight acceptors. Among glycosyltransferases, more than a hundred families have been identified so far [201], which typically use donor molecules consisting of sugar that is usually combined with mono- (rarely) or diphosphate (commonly) of the appropriate β-D-ribonucleoside (commonly) or β-D-deoxyribonucleotide (rarely), which are shown in Figure 5c–h [202–205]. Interestingly, only a few glycosyltransferases use adenosine-5′-diphosphate (ADP)-based donor molecules, and these catalyze only the transfer of glucose (Glu) moiety from ADP-α-D-Glu. The enzymes using guanosine-5′-diphosphate (GDP)-based donor molecules involved in the transfer of the mannose (Man) residue from GDP-α-D-Man, the fucose (Fuc) residue from GDP-β-L-Fuc and occasionally glucose (Glu) from GDP-α-D-Glu are slightly more common. However, the vast majority of glycosyltransferases utilize pyrimidine nucleoside phosphate-based donors, predominantly uridine-5′-diphosphate (UDP) derivatives, for transferring numbering sugar residues (e.g., glucose, glucuronic acid, galactose, rhamnose, arabinose and others). The formation of cofactors containing uracil to replace adenine probably occurred at the LUCA stage [136]. Glycosyltransferases sporadically use donors based on other pyrimidine nucleotides, namely, cytidine-5′-monophosphate (CMP, e.g., CMP-sialic acid) or deoxythymidine-5′-diphosphate (dTDP, e.g., dTDP-α-D-glucose, dTDP-β-L-rhamnose and dTDP-6-deoxy-4-keto-α-D-glucose). Glycosyltransferases are an extremely diverse group of enzymes, both in terms of sugar residue donors and amino acid sequences that are used. Furthermore, the transfer of the sugar residue can follow two different mechanisms, encompassing either inversion or maintained configuration at the anomeric carbon in the sugar ring. Despite their wide variety, glycosyltransferases utilizing sugar derivatives of nucleoside mono- and diphosphates as donors usually adopttwo structural forms: (i) GT-A containing a single domain (Figure 4g) or (ii) GT-B containing two domains (Figure 4h) [201,206]. Interestingly, there is no correlation between donor type, reaction mechanism and glycosyltransferase structure. The Rossmann fold is the only structural element present in each glycosyltransferase domain, which indicates that both structural forms must have evolved from a common ancestor [205]. It should also be noted that the Rossmann motif, initially responsible for binding adenine nucleotides, is used in glycosyltransferases for binding nucleotide derivatives containing other, both purine and pyrimidine, bases.

Other examples of Rossmann-fold-based enzymes with diverse nucleoside specificity are prokaryotic and eukaryotic single-domain *N*-ribohydrolases (Figure 4e) [200]. They are involved in the *N*-glycosidic bond hydrolysis of numerous nucleosides such as adenosine, SAM and MTA with the release of adenine. However, they can also metabolize a variety of other non-adenosine-based purine/pyrimidine nucleosides/deoxynucleosides, including cytidine, inosine, uridine and xanthosine [200,207,208], as well as NAD^+ dinucleotide [209]. Similarly to glycosyltransferases, the fold of *N*-ribohydrolases is highly conserved; however, a wide substrate selection results from some substitutions and insertions.

4. Conclusions

Nucleotides and nucleosides derived from β-D-ribofuranose have unique chemical and structural features that probably facilitated their selection as building blocks in the initial stages of the origins of life. These compounds are involved in numerous essential metabolic processes, including encoding genetic information, cellular catalysis, energy production and storage and many others. Looking into the past, the presence of *N*-ribosides in an environment was probably crucial to initiate primary catalytic processes that were initially based on RNA, peptides, ions and small molecule cofactors, which are based mostly on adenosine moiety. During evolution, the enzyme-driven metabolic pathways replaced RNA-based catalysis. However, numerous enzymes still require the same primordial cofactors from the "RNA-based World(s)." With time, the complexity of the metabolism reaction forced the enzymes to utilize nucleotide cofactors other than adenosine-based derivatives of β-D-ribose.

Author Contributions: Conceptualization, K.B.; writing—original draft preparation, K.W. and K.B.; writing—review and editing, K.W. and K.B.; visualization, K.W. and K.B.; supervision, K.B.; project administration, K.B. All authors have read and agreed to the published version of the manuscript.

Funding: This research received no external funding.

Institutional Review Board Statement: Not applicable.

Informed Consent Statement: Not applicable.

Data Availability Statement: Not applicable.

Conflicts of Interest: The authors declare no conflict of interest.

References

1. Koshland, D.E. Special Essay. The Seven Pillars of Life. *Science* **2002**, *295*, 2215–2216. [CrossRef] [PubMed]
2. Benner, S.A. Defining Life. *Astrobiology* **2010**, *10*, 1021–1030. [CrossRef] [PubMed]
3. Deamer, D.W.; Fleischaker, G.R. *Origins of Life: The Central Concepts*; Jones and Bartlett Publishers: Boston, MA, USA, 1994; ISBN 978-0-86720-181-9.
4. Blackburn, E.H.; Greider, C.W.; Szostak, J.W. Telomeres and Telomerase: The Path from Maize, Tetrahymena and Yeast to Human Cancer and Aging. *Nat. Med.* **2006**, *12*, 1133–1138. [CrossRef]
5. Baltimore, D. Viral RNA-Dependent DNA Polymerase: RNA-Dependent DNA Polymerase in Virions of RNA Tumour Viruses. *Nature* **1970**, *226*, 1209–1211. [CrossRef]
6. Yarus, M. Darwinian Behavior in a Cold, Sporadically Fed Pool of Ribonucleotides. *Astrobiology* **2012**, *12*, 870–883. [CrossRef] [PubMed]
7. Yarus, M. Efficient Heritable Gene Expression Readily Evolves in RNA Pools. *J. Mol. Evol.* **2017**, *84*, 236–252. [CrossRef]
8. Gradstein, P.F.M. *Concise Geologic Time Scale-2016*; Elsevier: Amsterdam, The Netherlands, 2016; ISBN 978-0-444-63771-0.
9. Rosing, M.T. 13C-Depleted Carbon Microparticles in >3700-Ma Sea-Floor Sedimentary Rocks from West Greenland. *Science* **1999**, *283*, 674–676. [CrossRef]
10. Grassineau, N.V.; Abell, P.; Appel, P.W.U.; Lowry, D.; Nisbet, E.G. Early Life Signatures in Sulfur and Carbon Isotopes from Isua, Barberton, Wabigoon (Steep Rock), and Belingwe Greenstone Belts (3.8 to 2.7 Ga). In *Memoir 198: Evolution of Early Earth's Atmosphere, Hydrosphere, and Biosphere—Constraints from Ore Deposits*; Geological Society of America: Boulder, CO, USA, 2006; Volume 198, pp. 33–52. ISBN 978-0-8137-1198-0.
11. Arndt, N.T.; Nisbet, E.G. Processes on the Young Earth and the Habitats of Early Life. *Annu. Rev. Earth Planet. Sci.* **2012**, *40*, 521–549. [CrossRef]
12. Martin, W.; Russell, M.J. On the Origins of Cells: A Hypothesis for the Evolutionary Transitions from Abiotic Geochemistry to Chemoautotrophic Prokaryotes, and from Prokaryotes to Nucleated Cells. *Philos. Trans. R. Soc. Lond. B Biol. Sci.* **2003**, *358*, 59–83; discussion 83–85. [CrossRef]
13. Martin, W.F.; Sousa, F.L.; Lane, N. Evolution. Energy at Life's Origin. *Science* **2014**, *344*, 1092–1093. [CrossRef]
14. Corliss, J.B. An Hypothesis Concerning the Relationships Between Submarine Hot Springs and the Origin of Life on Earth. *Oceanologica Acta* **1981**, *4*, 59–69.
15. Martin, W.; Baross, J.; Kelley, D.; Russell, M.J. Hydrothermal Vents and the Origin of Life. *Nat. Rev. Microbiol.* **2008**, *6*, 805–814. [CrossRef] [PubMed]
16. Damer, B.; Deamer, D. Coupled Phases and Combinatorial Selection in Fluctuating Hydrothermal Pools: A Scenario to Guide Experimental Approaches to the Origin of Cellular Life. *Life* **2015**, *5*, 872–887. [CrossRef] [PubMed]
17. Broecker, F. Genome Evolution from Random Ligation of RNAs of Autocatalytic Sets. *Int. J. Mol. Sci.* **2021**, *22*, 13526. [CrossRef]
18. Mulkidjanian, A.Y.; Bychkov, A.Y.; Dibrova, D.V.; Galperin, M.Y.; Koonin, E.V. Origin of First Cells at Terrestrial, Anoxic Geothermal Fields. *Proc. Natl. Acad. Sci. USA* **2012**, *109*, E821–E830. [CrossRef] [PubMed]
19. Nutman, A.P.; Bennett, V.C.; Friend, C.R.L.; Van Kranendonk, M.J.; Chivas, A.R. Rapid Emergence of Life Shown by Discovery of 3,700-Million-Year-Old Microbial Structures. *Nature* **2016**, *537*, 535–538. [CrossRef]
20. Djokic, T.; Van Kranendonk, M.J.; Campbell, K.A.; Walter, M.R.; Ward, C.R. Earliest Signs of Life on Land Preserved in ca. 3.5 Ga Hot Spring Deposits. *Nat. Commun.* **2017**, *8*, 15263. [CrossRef]
21. Woese, C.R.; Dugre, D.H.; Dugre, S.A.; Kondo, M.; Saxinger, W.C. On the Fundamental Nature and Evolution of the Genetic Code. *Cold Spring Harb. Symp. Quant. Biol.* **1966**, *31*, 723–736. [CrossRef]
22. Orgel, L.E. Evolution of the Genetic Apparatus. *J. Mol. Biol.* **1968**, *38*, 381–393. [CrossRef]
23. Crick, F.H. The Origin of the Genetic Code. *J. Mol. Biol.* **1968**, *38*, 367–379. [CrossRef]
24. Guerrier-Takada, C.; Gardiner, K.; Marsh, T.; Pace, N.; Altman, S. The RNA Moiety of Ribonuclease P Is the Catalytic Subunit of the Enzyme. *Cell* **1983**, *35*, 849–857. [CrossRef] [PubMed]
25. Cech, T.R. A Model for the RNA-Catalyzed Replication of RNA. *Proc. Natl. Acad. Sci. USA* **1986**, *83*, 4360–4363. [CrossRef]
26. Gilbert, W. Origin of Life: The RNA World. *Nature* **1986**, *319*, 618. [CrossRef]
27. Cech, T.R. Structural Biology. The Ribosome Is a Ribozyme. *Science* **2000**, *289*, 878–879. [CrossRef] [PubMed]
28. Stribling, R.; Miller, S.L. Energy Yields for Hydrogen Cyanide and Formaldehyde Syntheses: The HCN and Amino Acid Concentrations in the Primitive Ocean. *Orig. Life Evol. Biosph.* **1987**, *17*, 261–273. [CrossRef] [PubMed]
29. Yang, Y.-W.; Zhang, S.; McCullum, E.O.; Chaput, J.C. Experimental Evidence That GNA and TNA Were Not Sequential Polymers in the Prebiotic Evolution of RNA. *J. Mol. Evol.* **2007**, *65*, 289–295. [CrossRef]
30. Chen, J.J.; Tsai, C.-H.; Cai, X.; Horhota, A.T.; McLaughlin, L.W.; Szostak, J.W. Enzymatic Primer-Extension with Glycerol-Nucleoside Triphosphates on DNA Templates. *PLoS ONE* **2009**, *4*, e4949. [CrossRef]
31. Mutschler, H.; Holliger, P. Non-Canonical 3′-5′ Extension of RNA with Prebiotically Plausible Ribonucleoside 2′,3′-Cyclic Phosphates. *J. Am. Chem. Soc.* **2014**, *136*, 5193–5196. [CrossRef]

32. Zhang, W.; Kim, S.C.; Tam, C.P.; Lelyveld, V.S.; Bala, S.; Chaput, J.C.; Szostak, J.W. Structural Interpretation of the Effects of Threo-Nucleotides on Nonenzymatic Template-Directed Polymerization. *Nucleic Acids Res.* **2021**, *49*, 646–656. [CrossRef] [PubMed]
33. Kim, S.C.; Zhou, L.; Zhang, W.; O'Flaherty, D.K.; Rondo-Brovetto, V.; Szostak, J.W. A Model for the Emergence of RNA from a Prebiotically Plausible Mixture of Ribonucleotides, Arabinonucleotides, and 2′-Deoxynucleotides. *J. Am. Chem. Soc.* **2020**, *142*, 2317–2326. [CrossRef]
34. Banfalvi, G. Prebiotic Pathway from Ribose to RNA Formation. *Int. J. Mol. Sci.* **2021**, *22*, 3857. [CrossRef] [PubMed]
35. Drozdzal, P.; Manszewski, T.; Gilski, M.; Brzezinski, K.; Jaskolski, M. Right-Handed Z-DNA at Ultrahigh Resolution: A Tale of Two Hands and the Power of the Crystallographic Method. *Acta Crystallogr. D Struct. Biol.* **2023**, *79*, 133–139. [CrossRef] [PubMed]
36. Xu, Y.; Zhu, T.F. Mirror-Image T7 Transcription of Chirally Inverted Ribosomal and Functional RNAs. *Science* **2022**, *378*, 405–412. [CrossRef]
37. Rypniewski, W.; Vallazza, M.; Perbandt, M.; Klussmann, S.; Delucas, L.J.; Betzel, C.; Erdmann, V.A. The First Crystal Structure of an RNA Racemate. *Acta Crystallogr. D Biol. Crystallogr.* **2006**, *62*, 659–664. [CrossRef] [PubMed]
38. Kiliszek, A.; Błaszczyk, L.; Bejger, M.; Rypniewski, W. Broken Symmetry between RNA Enantiomers in a Crystal Lattice. *Nucleic Acids Res.* **2021**, *49*, 12535–12539. [CrossRef]
39. Kondepudi, D.K.; Nelson, G.W. Weak Neutral Currents and the Origin of Biomolecular Chirality. *Nature* **1985**, *314*, 438–441. [CrossRef]
40. Drew, K.N.; Zajicek, J.; Bondo, G.; Bose, B.; Serianni, A.S. 13C-Labeled Aldopentoses: Detection and Quantitation of Cyclic and Acyclic Forms by Heteronuclear 1D and 2D NMR Spectroscopy. *Carbohydr. Res.* **1998**, *307*, 199–209. [CrossRef]
41. Dass, A.V.; Georgelin, T.; Westall, F.; Foucher, F.; De Los Rios, P.; Busiello, D.M.; Liang, S.; Piazza, F. Equilibrium and Non-Equilibrium Furanose Selection in the Ribose Isomerisation Network. *Nat. Commun.* **2021**, *12*, 2749. [CrossRef]
42. Bloomfield, V.A.; Crothers, D.M.; Tinoco, I. *Nucleic Acids: Structures, Properties, and Functions*; University Science Books: Sausalito, CA, USA, 2000; ISBN 978-0-935702-49-1.
43. Banfalvi, G. Why Ribose Was Selected as the Sugar Component of Nucleic Acids. *DNA Cell Biol.* **2006**, *25*, 189–196. [CrossRef]
44. Banfalvi, G. Ribose Selected as Precursor to Life. *DNA Cell Biol.* **2020**, *39*, 177–186. [CrossRef]
45. Carter, C.W. What RNA World? Why a Peptide/RNA Partnership Merits Renewed Experimental Attention. *Life* **2015**, *5*, 294–320. [CrossRef]
46. Bowman, J.C.; Hud, N.V.; Williams, L.D. The Ribosome Challenge to the RNA World. *J. Mol. Evol.* **2015**, *80*, 143–161. [CrossRef] [PubMed]
47. Carter, C.W. An Alternative to the RNA World. *Nat. Hist.* **2016**, *125*, 28–33. [PubMed]
48. Kunnev, D.; Gospodinov, A. Possible Emergence of Sequence Specific RNA Aminoacylation via Peptide Intermediary to Initiate Darwinian Evolution and Code Through Origin of Life. *Life* **2018**, *8*, 44. [CrossRef]
49. Wills, P.R.; Carter, C.W. Insuperable Problems of the Genetic Code Initially Emerging in an RNA World. *Biosystems* **2018**, *164*, 155–166. [CrossRef] [PubMed]
50. Kunnev, D. Origin of Life: The Point of No Return. *Life* **2020**, *10*, 269. [CrossRef]
51. Radakovic, A.; Wright, T.H.; Lelyveld, V.S.; Szostak, J.W. A Potential Role for Aminoacylation in Primordial RNA Copying Chemistry. *Biochemistry* **2021**, *60*, 477–488. [CrossRef]
52. Tirumalai, M.R.; Rivas, M.; Tran, Q.; Fox, G.E. The Peptidyl Transferase Center: A Window to the Past. *Microbiol. Mol. Biol. Rev.* **2021**, *85*, e0010421. [CrossRef]
53. Müller, F.; Escobar, L.; Xu, F.; Węgrzyn, E.; Nainytė, M.; Amatov, T.; Chan, C.-Y.; Pichler, A.; Carell, T. A Prebiotically Plausible Scenario of an RNA-Peptide World. *Nature* **2022**, *605*, 279–284. [CrossRef]
54. Martínez Giménez, J.A.; Tabares Seisdedos, R. A Cofactor-Based Mechanism for the Origin of the Genetic Code. *Orig. Life Evol. Biosph.* **2022**, *52*, 149–163. [CrossRef]
55. Tagami, S.; Li, P. The Origin of Life: RNA and Protein Co-Evolution on the Ancient Earth. *Dev. Growth Differ.* **2023**, *65*, 167–174. [CrossRef] [PubMed]
56. Hud, N.V.; Cafferty, B.J.; Krishnamurthy, R.; Williams, L.D. The Origin of RNA and "My Grandfather's Axe". *Chem. Biol.* **2013**, *20*, 466–474. [CrossRef]
57. Szostak, J.W.; Bartel, D.P.; Luisi, P.L. Synthesizing Life. *Nature* **2001**, *409*, 387–390. [CrossRef] [PubMed]
58. Roy, D.; Najafian, K.; von Ragué Schleyer, P. Chemical Evolution: The Mechanism of the Formation of Adenine under Prebiotic Conditions. *Proc. Natl. Acad. Sci. USA* **2007**, *104*, 17272–17277. [CrossRef] [PubMed]
59. Powner, M.W.; Gerland, B.; Sutherland, J.D. Synthesis of Activated Pyrimidine Ribonucleotides in Prebiotically Plausible Conditions. *Nature* **2009**, *459*, 239–242. [CrossRef] [PubMed]
60. Szostak, J.W. Origins of Life: Systems Chemistry on Early Earth. *Nature* **2009**, *459*, 171–172. [CrossRef] [PubMed]
61. Ritson, D.; Sutherland, J.D. Prebiotic Synthesis of Simple Sugars by Photoredox Systems Chemistry. *Nat. Chem.* **2012**, *4*, 895–899. [CrossRef]
62. Stairs, S.; Nikmal, A.; Bučar, D.-K.; Zheng, S.-L.; Szostak, J.W.; Powner, M.W. Divergent Prebiotic Synthesis of Pyrimidine and 8-Oxo-Purine Ribonucleotides. *Nat. Commun.* **2017**, *8*, 15270. [CrossRef]

63. Ferris, J.P.; Hill, A.R.; Liu, R.; Orgel, L.E. Synthesis of Long Prebiotic Oligomers on Mineral Surfaces. *Nature* **1996**, *381*, 59–61. [CrossRef] [PubMed]
64. Joyce, G.F.; Orgel, L.E. The RNA World, 2nd ed. In *the Nature of Modern RNA Suggests a Prebiotic RNA World;* Cold Spring Harbor Laboratory Press: New York, NY, USA, 1999.
65. Jauker, M.; Griesser, H.; Richert, C. Spontaneous Formation of RNA Strands, Peptidyl RNA, and Cofactors. *Angew. Chem. Int. Ed.* **2015**, *54*, 14564–14569. [CrossRef]
66. Powner, M.; Sutherland, J.; Szostak, J. The Origins of Nucleotides. *Synlett* **2011**, *2011*, 1956–1964. [CrossRef]
67. Fuller, W.D.; Sanchez, R.A.; Orgel, L.E. Studies in Prebiotic Synthesis. VI. Synthesis of Purine Nucleosides. *J. Mol. Biol.* **1972**, *67*, 25–33. [CrossRef] [PubMed]
68. Orgel, L.E. Prebiotic Chemistry and the Origin of the RNA World. *Crit. Rev. Biochem. Mol. Biol.* **2004**, *39*, 99–123. [CrossRef] [PubMed]
69. Kim, H.-J.; Benner, S.A. Prebiotic Stereoselective Synthesis of Purine and Noncanonical Pyrimidine Nucleotide from Nucleobases and Phosphorylated Carbohydrates. *Proc. Natl. Acad. Sci USA* **2017**, *114*, 11315–11320. [CrossRef]
70. Becker, S.; Feldmann, J.; Wiedemann, S.; Okamura, H.; Schneider, C.; Iwan, K.; Crisp, A.; Rossa, M.; Amatov, T.; Carell, T. Unified Prebiotically Plausible Synthesis of Pyrimidine and Purine RNA Ribonucleotides. *Science* **2019**, *366*, 76–82. [CrossRef]
71. Rajamani, S.; Vlassov, A.; Benner, S.; Coombs, A.; Olasagasti, F.; Deamer, D. Lipid-Assisted Synthesis of RNA-like Polymers from Mononucleotides. *Orig. Life Evol. Biosph.* **2008**, *38*, 57–74. [CrossRef]
72. Pressman, A.; Blanco, C.; Chen, I.A. The RNA World as a Model System to Study the Origin of Life. *Curr. Biol.* **2015**, *25*, R953–R963. [CrossRef]
73. Joyce, G.F.; Szostak, J.W. Protocells and RNA Self-Replication. *Cold Spring Harb. Perspect. Biol.* **2018**, *10*, a034801. [CrossRef]
74. Hordijk, W.; Naylor, J.; Krasnogor, N.; Fellermann, H. Population Dynamics of Autocatalytic Sets in a Compartmentalized Spatial World. *Life* **2018**, *8*, 33. [CrossRef]
75. Piette, B.M.A.G.; Heddle, J.G. A Peptide-Nucleic Acid Replicator Origin for Life. *Trends Ecol. Evol.* **2020**, *35*, 397–406. [CrossRef]
76. Higgs, P.G.; Lehman, N. The RNA World: Molecular Cooperation at the Origins of Life. *Nat. Rev. Genet.* **2015**, *16*, 7–17. [CrossRef] [PubMed]
77. Tjhung, K.F.; Shokhirev, M.N.; Horning, D.P.; Joyce, G.F. An RNA Polymerase Ribozyme That Synthesizes Its Own Ancestor. *Proc. Natl. Acad. Sci. USA* **2020**, *117*, 2906–2913. [CrossRef]
78. Lincoln, T.A.; Joyce, G.F. Self-Sustained Replication of an RNA Enzyme. *Science* **2009**, *323*, 1229–1232. [CrossRef] [PubMed]
79. Attwater, J.; Raguram, A.; Morgunov, A.S.; Gianni, E.; Holliger, P. Ribozyme-Catalysed RNA Synthesis Using Triplet Building Blocks. *Elife* **2018**, *7*, e35255. [CrossRef] [PubMed]
80. Zaug, A.J.; Grabowski, P.J.; Cech, T.R. Autocatalytic Cyclization of an Excised Intervening Sequence RNA Is a Cleavage-Ligation Reaction. *Nature* **1983**, *301*, 578–583. [CrossRef]
81. Maurel, M.-C.; Leclerc, F.; Hervé, G. Ribozyme Chemistry: To Be or Not To Be under High Pressure. *Chem. Rev.* **2020**, *120*, 4898–4918. [CrossRef]
82. Cojocaru, R.; Unrau, P.J. Processive RNA Polymerization and Promoter Recognition in an RNA World. *Science* **2021**, *371*, 1225–1232. [CrossRef]
83. Adamala, K.; Engelhart, A.E.; Szostak, J.W. Generation of Functional RNAs from Inactive Oligonucleotide Complexes by Non-Enzymatic Primer Extension. *J. Am. Chem. Soc.* **2015**, *137*, 483–489. [CrossRef]
84. Zhou, L.; Kim, S.C.; Ho, K.H.; O'Flaherty, D.K.; Giurgiu, C.; Wright, T.H.; Szostak, J.W. Non-Enzymatic Primer Extension with Strand Displacement. *Elife* **2019**, *8*, e51888. [CrossRef]
85. Zhou, L.; O'Flaherty, D.K.; Szostak, J.W. Template-Directed Copying of RNA by Non-Enzymatic Ligation. *Angew. Chem. Int. Ed. Engl.* **2020**, *59*, 15682–15687. [CrossRef] [PubMed]
86. Longo, L.M.; Blaber, M. Protein Design at the Interface of the Pre-Biotic and Biotic Worlds. *Arch. Biochem. Biophys.* **2012**, *526*, 16–21. [CrossRef]
87. Fox, G.E. Origin and Evolution of the Ribosome. *Cold Spring Harb. Perspect. Biol.* **2010**, *2*, a003483. [CrossRef] [PubMed]
88. Zhang, B.; Cech, T.R. Peptide Bond Formation by in Vitro Selected Ribozymes. *Nature* **1997**, *390*, 96–100. [CrossRef]
89. Simonović, M.; Steitz, T.A. A Structural View on the Mechanism of the Ribosome-Catalyzed Peptide Bond Formation. *Biochim. Et Biophys. Acta (BBA)-Gene Regul. Mech.* **2009**, *1789*, 612–623. [CrossRef]
90. Kumar, P.K.; Ellington, A.D. Artificial Evolution and Natural Ribozymes. *FASEB J.* **1995**, *9*, 1183–1195. [CrossRef] [PubMed]
91. Chen, X.; Li, N.; Ellington, A.D. Ribozyme Catalysis of Metabolism in the RNA World. *Chem. Biodivers.* **2007**, *4*, 633–655. [CrossRef]
92. Janzen, E.; Blanco, C.; Peng, H.; Kenchel, J.; Chen, I.A. Promiscuous Ribozymes and Their Proposed Role in Prebiotic Evolution. *Chem. Rev.* **2020**, *120*, 4879–4897. [CrossRef]
93. Maurel, M.C.; Ninio, J. Catalysis by a Prebiotic Nucleotide Analog of Histidine. *Biochimie* **1987**, *69*, 551–553. [CrossRef]
94. Mir, A.; Golden, B.L. Two Active Site Divalent Ions in the Crystal Structure of the Hammerhead Ribozyme Bound to a Transition State Analogue. *Biochemistry* **2016**, *55*, 633–636. [CrossRef]
95. Alam, S.; Grum-Tokars, V.; Krucinska, J.; Kundracik, M.L.; Wedekind, J.E. Conformational Heterogeneity at Position U37 of an All-RNA Hairpin Ribozyme with Implications for Metal Binding and the Catalytic Structure of the S-Turn. *Biochemistry* **2005**, *44*, 14396–14408. [CrossRef]

96. Athavale, S.S.; Petrov, A.S.; Hsiao, C.; Watkins, D.; Prickett, C.D.; Gossett, J.J.; Lie, L.; Bowman, J.C.; O'Neill, E.; Bernier, C.R.; et al. RNA Folding and Catalysis Mediated by Iron (II). *PLoS ONE* **2012**, *7*, e38024. [CrossRef] [PubMed]
97. White, H.B. Coenzymes as Fossils of an Earlier Metabolic State. *J. Mol. Evol.* **1976**, *7*, 101–104. [CrossRef] [PubMed]
98. Jadhav, V.R.; Yarus, M. Coenzymes as Coribozymes. *Biochimie* **2002**, *84*, 877–888. [CrossRef] [PubMed]
99. Saran, D.; Frank, J.; Burke, D.H. The Tyranny of Adenosine Recognition among RNA Aptamers to Coenzyme A. *BMC Evol. Biol.* **2003**, *3*, 26. [CrossRef]
100. Breaker, R.R. Imaginary Ribozymes. *ACS Chem. Biol.* **2020**, *15*, 2020–2030. [CrossRef]
101. Samuelian, J.S.; Gremminger, T.J.; Song, Z.; Poudyal, R.R.; Li, J.; Zhou, Y.; Staller, S.A.; Carballo, J.A.; Roychowdhury-Saha, M.; Chen, S.-J.; et al. An RNA Aptamer That Shifts the Reduction Potential of Metabolic Cofactors. *Nat. Chem. Biol.* **2022**, *18*, 1263–1269. [CrossRef] [PubMed]
102. Eigen, M. Selforganization of Matter and the Evolution of Biological Macromolecules. *Naturwissenschaften* **1971**, *58*, 465–523. [CrossRef] [PubMed]
103. Leu, K.; Kervio, E.; Obermayer, B.; Turk-MacLeod, R.M.; Yuan, C.; Luevano, J.-M.; Chen, E.; Gerland, U.; Richert, C.; Chen, I.A. Cascade of Reduced Speed and Accuracy after Errors in Enzyme-Free Copying of Nucleic Acid Sequences. *J. Am. Chem. Soc.* **2013**, *135*, 354–366. [CrossRef]
104. Ayadi, L.; Galvanin, A.; Pichot, F.; Marchand, V.; Motorin, Y. RNA Ribose Methylation (2′-O-Methylation): Occurrence, Biosynthesis and Biological Functions. *Biochim. Biophys. Acta Gene Regul. Mech.* **2019**, *1862*, 253–269. [CrossRef] [PubMed]
105. Lane, B.G.; Tamaoki, T. Studies of the Chain Termini and Alkali-Stable Dinucleotide Sequences in 16 s and 28 s Ribosomal RNA from L Cells. *J. Mol. Biol.* **1967**, *27*, 335–348. [CrossRef]
106. Trim, A.R.; Parker, J.E. Preparation, Purification and Analyses of Thirteen Alkali-Stable Dinucleotides from Yeast Ribonucleic Acid. *Biochem. J.* **1970**, *116*, 589–598. [CrossRef]
107. Bardwell, V.J.; Wickens, M.; Bienroth, S.; Keller, W.; Sproat, B.S.; Lamond, A.I. Site-Directed Ribose Methylation Identifies 2′-OH Groups in Polyadenylation Substrates Critical for AAUAAA Recognition and Poly(A) Addition. *Cell* **1991**, *65*, 125–133. [CrossRef] [PubMed]
108. Andersen, C.B.F.; Ballut, L.; Johansen, J.S.; Chamieh, H.; Nielsen, K.H.; Oliveira, C.L.P.; Pedersen, J.S.; Séraphin, B.; Le Hir, H.; Andersen, G.R. Structure of the Exon Junction Core Complex with a Trapped DEAD-Box ATPase Bound to RNA. *Science* **2006**, *313*, 1968–1972. [CrossRef] [PubMed]
109. Mishler, D.M.; Christ, A.B.; Steitz, J.A. Flexibility in the Site of Exon Junction Complex Deposition Revealed by Functional Group and RNA Secondary Structure Alterations in the Splicing Substrate. *RNA* **2008**, *14*, 2657–2670. [CrossRef] [PubMed]
110. Prusiner, P.; Yathindra, N.; Sundaralingam, M. Effect of Ribose O(2′)-Methylation on the Conformation of Nucleosides and Nucleotides. *Biochim. Biophys. Acta* **1974**, *366*, 115–123. [CrossRef] [PubMed]
111. Kawai, G.; Ue, H.; Yasuda, M.; Sakamoto, K.; Hashizume, T.; McCloskey, J.A.; Miyazawa, T.; Yokoyama, S. Relation between Functions and Conformational Characteristics of Modified Nucleosides Found in TRNAs. *Nucleic Acids Symp. Ser.* **1991**, *25*, 49–50.
112. Jiang, H.; Gao, Y.; Zhang, L.; Chen, D.; Gan, J.; Murchie, A.I.H. The Identification and Characterization of a Selected SAM-Dependent Methyltransferase Ribozyme That Is Present in Natural Sequences. *Nat. Catal.* **2021**, *4*, 872–881. [CrossRef]
113. Deng, J.; Wilson, T.J.; Wang, J.; Peng, X.; Li, M.; Lin, X.; Liao, W.; Lilley, D.M.J.; Huang, L. Structure and Mechanism of a Methyltransferase Ribozyme. *Nat. Chem. Biol.* **2022**, *18*, 556–564. [CrossRef]
114. Scheitl, C.P.M.; Mieczkowski, M.; Schindelin, H.; Höbartner, C. Structure and Mechanism of the Methyltransferase Ribozyme MTR1. *Nat. Chem. Biol.* **2022**, *18*, 547–555. [CrossRef]
115. Agris, P.F.; Koh, H.; Söll, D. The Effect of Growth Temperatures on the in Vivo Ribose Methylation of Bacillus Stearothermophilus Transfer RNA. *Arch. Biochem. Biophys.* **1973**, *154*, 277–282. [CrossRef]
116. Kowalak, J.A.; Dalluge, J.J.; McCloskey, J.A.; Stetter, K.O. The Role of Posttranscriptional Modification in Stabilization of Transfer RNA from Hyperthermophiles. *Biochemistry* **1994**, *33*, 7869–7876. [CrossRef] [PubMed]
117. Noon, K.R.; Bruenger, E.; McCloskey, J.A. Posttranscriptional Modifications in 16S and 23S RRNAs of the Archaeal Hyperthermophile Sulfolobus Solfataricus. *J. Bacteriol.* **1998**, *180*, 2883–2888. [CrossRef] [PubMed]
118. Kirpekar, F.; Hansen, L.H.; Rasmussen, A.; Poehlsgaard, J.; Vester, B. The Archaeon Haloarcula Marismortui Has Few Modifications in the Central Parts of Its 23S Ribosomal RNA. *J. Mol. Biol.* **2005**, *348*, 563–573. [CrossRef] [PubMed]
119. Guymon, R.; Pomerantz, S.C.; Crain, P.F.; McCloskey, J.A. Influence of Phylogeny on Posttranscriptional Modification of RRNA in Thermophilic Prokaryotes: The Complete Modification Map of 16S RRNA of Thermus Thermophilus. *Biochemistry* **2006**, *45*, 4888–4899. [CrossRef] [PubMed]
120. Dennis, P.P.; Tripp, V.; Lui, L.; Lowe, T.; Randau, L. C/D Box SRNA-Guided 2′-O-Methylation Patterns of Archaeal RRNA Molecules. *BMC Genom.* **2015**, *16*, 632. [CrossRef]
121. Motorin, Y.; Helm, M. RNA Nucleotide Methylation. *WIREs RNA* **2011**, *2*, 611–631. [CrossRef]
122. Motorin, Y.; Helm, M. RNA Nucleotide Methylation: 2021 Update. *WIREs RNA* **2022**, *13*, e1691. [CrossRef]
123. Reichard, P. From RNA to DNA, Why so Many Ribonucleotide Reductases? *Science* **1993**, *260*, 1773–1777. [CrossRef]
124. Reichard, P. The Evolution of Ribonucleotide Reduction. *Trends Biochem. Sci.* **1997**, *22*, 81–85. [CrossRef]
125. Freeland, S.J.; Knight, R.D.; Landweber, L.F. Do Proteins Predate DNA? *Science* **1999**, *286*, 690–692. [CrossRef]
126. Stubbe, J.; Ge, J.; Yee, C.S. The Evolution of Ribonucleotide Reduction Revisited. *Trends Biochem. Sci.* **2001**, *26*, 93–99. [CrossRef] [PubMed]

127. Lundin, D.; Berggren, G.; Logan, D.T.; Sjöberg, B.-M. The Origin and Evolution of Ribonucleotide Reduction. *Life* **2015**, *5*, 604–636. [CrossRef] [PubMed]
128. Iribe, H.; Miyamoto, K.; Takahashi, T.; Kobayashi, Y.; Leo, J.; Aida, M.; Ui-Tei, K. Chemical Modification of the SiRNA Seed Region Suppresses Off-Target Effects by Steric Hindrance to Base-Pairing with Targets. *ACS Omega* **2017**, *2*, 2055–2064. [CrossRef] [PubMed]
129. Sofia, H.J.; Chen, G.; Hetzler, B.G.; Reyes-Spindola, J.F.; Miller, N.E. Radical SAM, a Novel Protein Superfamily Linking Unresolved Steps in Familiar Biosynthetic Pathways with Radical Mechanisms: Functional Characterization Using New Analysis and Information Visualization Methods. *Nucleic Acids Res.* **2001**, *29*, 1097–1106. [CrossRef]
130. Schopf, J.W. Microfossils of the Early Archean Apex Chert: New Evidence of the Antiquity of Life. *Science* **1993**, *260*, 640–646. [CrossRef]
131. Mojzsis, S.J.; Arrhenius, G.; McKeegan, K.D.; Harrison, T.M.; Nutman, A.P.; Friend, C.R. Evidence for Life on Earth before 3,800 Million Years Ago. *Nature* **1996**, *384*, 55–59. [CrossRef]
132. Ueno, Y.; Yurimoto, H.; Yoshioka, H.; Komiya, T.; Maruyama, S. Ion Microprobe Analysis of Graphite from ca. 3.8 Ga Metasediments, Isua Supracrustal Belt, West Greenland: Relationship between Metamorphism and Carbon Isotopic Composition. *Geochim. Et Cosmochim. Acta* **2002**, *66*, 1257–1268. [CrossRef]
133. Glansdorff, N.; Xu, Y.; Labedan, B. The Last Universal Common Ancestor: Emergence, Constitution and Genetic Legacy of an Elusive Forerunner. *Biol. Direct* **2008**, *3*, 29. [CrossRef]
134. Acevedo-Rocha, C.G.; Fang, G.; Schmidt, M.; Ussery, D.W.; Danchin, A. From Essential to Persistent Genes: A Functional Approach to Constructing Synthetic Life. *Trends Genet.* **2013**, *29*, 273–279. [CrossRef] [PubMed]
135. Cornish-Bowden, A.; Cárdenas, M.L. Life before LUCA. *J. Theor. Biol.* **2017**, *434*, 68–74. [CrossRef]
136. Weiss, M.C.; Sousa, F.L.; Mrnjavac, N.; Neukirchen, S.; Roettger, M.; Nelson-Sathi, S.; Martin, W.F. The Physiology and Habitat of the Last Universal Common Ancestor. *Nat. Microbiol.* **2016**, *1*, 16116. [CrossRef] [PubMed]
137. Woese, C.R.; Fox, G.E. The Concept of Cellular Evolution. *J. Mol. Evol.* **1977**, *10*, 1–6. [CrossRef] [PubMed]
138. Di Giulio, M. The Last Universal Common Ancestor (LUCA) and the Ancestors of Archaea and Bacteria Were Progenotes. *J. Mol. Evol.* **2011**, *72*, 119–126. [CrossRef] [PubMed]
139. Woese, C.R. Bacterial Evolution. *Microbiol. Rev.* **1987**, *51*, 221–271. [CrossRef] [PubMed]
140. Martin, W.F.; Weiss, M.C.; Neukirchen, S.; Nelson-Sathi, S.; Sousa, F.L. Physiology, Phylogeny, and LUCA. *Microb. Cell* **2016**, *3*, 582–587. [CrossRef]
141. Martin, W.F.; Nagies, F.S.P.; do Nascimento Vieira, A. To What Inanimate Matter Are We Most Closely Related and Does the Origin of Life Harbor Meaning? *Philosophies* **2021**, *6*, 33. [CrossRef]
142. Kikuchi, A.; Asai, K. Reverse Gyrase—A Topoisomerase Which Introduces Positive Superhelical Turns into DNA. *Nature* **1984**, *309*, 677–681. [CrossRef] [PubMed]
143. Cohen, B.A.; Swindle, T.D.; Kring, D.A. Support for the Lunar Cataclysm Hypothesis from Lunar Meteorite Impact Melt Ages. *Science* **2000**, *290*, 1754–1756. [CrossRef]
144. Nisbet, E.G.; Sleep, N.H. The Habitat and Nature of Early Life. *Nature* **2001**, *409*, 1083–1091. [CrossRef]
145. Say, R.F.; Fuchs, G. Fructose 1,6-Bisphosphate Aldolase/Phosphatase May Be an Ancestral Gluconeogenic Enzyme. *Nature* **2010**, *464*, 1077–1081. [CrossRef]
146. Griesser, H.; Bechthold, M.; Tremmel, P.; Kervio, E.; Richert, C. Amino Acid-Specific, Ribonucleotide-Promoted Peptide Formation in the Absence of Enzymes. *Angew. Chem. Int. Ed.* **2017**, *56*, 1224–1228. [CrossRef]
147. Kirschning, A. Coenzymes and Their Role in the Evolution of Life. *Angew. Chem. Int. Ed. Engl.* **2021**, *60*, 6242–6269. [CrossRef] [PubMed]
148. Kirschning, A. The Coenzyme/Protein Pair and the Molecular Evolution of Life. *Nat. Prod. Rep.* **2021**, *38*, 993–1010. [CrossRef]
149. Fiore, M.; Chieffo, C.; Lopez, A.; Fayolle, D.; Ruiz, J.; Soulère, L.; Oger, P.; Altamura, E.; Popowycz, F.; Buchet, R. Synthesis of Phospholipids Under Plausible Prebiotic Conditions and Analogies with Phospholipid Biochemistry for Origin of Life Studies. *Astrobiology* **2022**, *22*, 598–627. [CrossRef] [PubMed]
150. Ghosh, A.; Bansal, M. A Glossary of DNA Structures from A to Z. *Acta Crystallogr. Sect. D Biol. Crystallogr.* **2003**, *59*, 620–626. [CrossRef] [PubMed]
151. Guo, F.; Gooding, A.R.; Cech, T.R. Structure of the Tetrahymena Ribozyme: Base Triple Sandwich and Metal Ion at the Active Site. *Mol. Cell* **2004**, *16*, 351–362. [CrossRef]
152. Klein, D.J.; Wilkinson, S.R.; Been, M.D.; Ferré-D'Amaré, A.R. Requirement of Helix P2.2 and Nucleotide G1 for Positioning the Cleavage Site and Cofactor of the GlmS Ribozyme. *J. Mol. Biol.* **2007**, *373*, 178–189. [CrossRef]
153. Brzezinski, K. S-Adenosyl-l-Homocysteine Hydrolase: A Structural Perspective on the Enzyme with Two Rossmann-Fold Domains. *Biomolecules* **2020**, *10*, 1682. [CrossRef]
154. Grzechowiak, M.; Sliwiak, J.; Jaskolski, M.; Ruszkowski, M. Structural Studies of Glutamate Dehydrogenase (Isoform 1) From Arabidopsis Thaliana, an Important Enzyme at the Branch-Point Between Carbon and Nitrogen Metabolism. *Front. Plant Sci.* **2020**, *11*, 754. [CrossRef]
155. Cakici, O.; Sikorski, M.; Stepkowski, T.; Bujacz, G.; Jaskolski, M. Crystal Structures of NodS N-Methyltransferase from Bradyrhizobium Japonicum in Ligand-Free Form and as SAH Complex. *J. Mol. Biol.* **2010**, *404*, 874–889. [CrossRef]

156. Gohara, D.W.; Di Cera, E. Molecular Mechanisms of Enzyme Activation by Monovalent Cations. *J. Biol. Chem.* **2016**, *291*, 20840–20848. [CrossRef] [PubMed]
157. Czyrko, J.; Sliwiak, J.; Imiolczyk, B.; Gdaniec, Z.; Jaskolski, M.; Brzezinski, K. Metal-Cation Regulation of Enzyme Dynamics Is a Key Factor Influencing the Activity of S-Adenosyl-L-Homocysteine Hydrolase from Pseudomonas Aeruginosa. *Sci. Rep.* **2018**, *8*, 11334. [CrossRef]
158. Guengerich, F.P. Thematic Series: Metals in Biology. *J. Biol. Chem.* **2009**, *284*, 709. [CrossRef]
159. Guengerich, F.P. Metals in Biology 2016: Molecular Basis of Selection of Metals by Enzymes. *J. Biol. Chem.* **2016**, *291*, 20838–20839. [CrossRef] [PubMed]
160. Larsen, T.M.; Laughlin, L.T.; Holden, H.M.; Rayment, I.; Reed, G.H. Structure of Rabbit Muscle Pyruvate Kinase Complexed with Mn2+, K+, and Pyruvate. *Biochemistry* **1994**, *33*, 6301–6309. [CrossRef]
161. Komoto, J.; Yamada, T.; Takata, Y.; Markham, G.D.; Takusagawa, F. Crystal Structure of the S-Adenosylmethionine Synthetase Ternary Complex: A Novel Catalytic Mechanism of S-Adenosylmethionine Synthesis from ATP and Met. *Biochemistry* **2004**, *43*, 1821–1831. [CrossRef]
162. Czyrko, J.; Jaskolski, M.; Brzezinski, K. Crystal Structure of S-Adenosyl-L-Homocysteine Hydrolase from Cytophaga Hutchinsonii, a Case of Combination of Crystallographic and Non-Crystallographic Symmetry. *Croat. Chem. Acta* **2018**, *91*, 153–163. [CrossRef]
163. Li, M.; Li, Y.; Chen, J.; Wei, W.; Pan, X.; Liu, J.; Liu, Q.; Leu, W.; Zhang, L.; Yang, X.; et al. Copper Ions Inhibit S-Adenosylhomocysteine Hydrolase by Causing Dissociation of NAD+ Cofactor. *Biochemistry* **2007**, *46*, 11451–11458. [CrossRef] [PubMed]
164. Beavo, J.A.; Brunton, L.L. Cyclic Nucleotide Research—Still Expanding after Half a Century. *Nat. Rev. Mol. Cell Biol.* **2002**, *3*, 710–717. [CrossRef]
165. Malan, T.P.; Kolb, A.; Buc, H.; McClure, W.R. Mechanism of CRP-CAMP Activation of Lac Operon Transcription Initiation Activation of the P1 Promoter. *J. Mol. Biol.* **1984**, *180*, 881–909. [CrossRef]
166. Chiang, P.K. Biological Effects of Inhibitors of S-Adenosylhomocysteine Hydrolase. *Pharmacol. Ther.* **1998**, *77*, 115–134. [CrossRef] [PubMed]
167. Tehlivets, O.; Malanovic, N.; Visram, M.; Pavkov-Keller, T.; Keller, W. S-Adenosyl-L-Homocysteine Hydrolase and Methylation Disorders: Yeast as a Model System. *Biochim. Biophys. Acta* **2013**, *1832*, 204–215. [CrossRef] [PubMed]
168. Fustin, J.-M.; Ye, S.; Rakers, C.; Kaneko, K.; Fukumoto, K.; Yamano, M.; Versteven, M.; Grünewald, E.; Cargill, S.J.; Tamai, T.K.; et al. Methylation Deficiency Disrupts Biological Rhythms from Bacteria to Humans. *Commun. Biol.* **2020**, *3*, 211. [CrossRef] [PubMed]
169. Greco, C.M.; Cervantes, M.; Fustin, J.-M.; Ito, K.; Ceglia, N.; Samad, M.; Shi, J.; Koronowski, K.B.; Forne, I.; Ranjit, S.; et al. S-Adenosyl-l-Homocysteine Hydrolase Links Methionine Metabolism to the Circadian Clock and Chromatin Remodeling. *Sci. Adv.* **2020**, *6*, eabc5629. [CrossRef]
170. Winkler, W.; Nahvi, A.; Breaker, R.R. Thiamine Derivatives Bind Messenger RNAs Directly to Regulate Bacterial Gene Expression. *Nature* **2002**, *419*, 952–956. [CrossRef] [PubMed]
171. Mironov, A.S.; Gusarov, I.; Rafikov, R.; Lopez, L.E.; Shatalin, K.; Kreneva, R.A.; Perumov, D.A.; Nudler, E. Sensing Small Molecules by Nascent RNA. *Cell* **2002**, *111*, 747–756. [CrossRef] [PubMed]
172. Brantl, S. Bacterial Gene Regulation: From Transcription Attenuation to Riboswitches and Ribozymes. *Trends Microbiol.* **2004**, *12*, 473–475. [CrossRef]
173. Tucker, B.J.; Breaker, R.R. Riboswitches as Versatile Gene Control Elements. *Curr. Opin. Struct. Biol.* **2005**, *15*, 342–348. [CrossRef]
174. Wang, J.X.; Breaker, R.R. Riboswitches That Sense S-Adenosylmethionine and S-Adenosylhomocysteine. *Biochem. Cell Biol.* **2008**, *86*, 157–168. [CrossRef]
175. Breaker, R.R. Riboswitches and the RNA World. *Cold Spring Harb. Perspect. Biol.* **2012**, *4*, a003566. [CrossRef] [PubMed]
176. McCown, P.J.; Corbino, K.A.; Stav, S.; Sherlock, M.E.; Breaker, R.R. Riboswitch Diversity and Distribution. *RNA* **2017**, *23*, 995–1011. [CrossRef]
177. Panchapakesan, S.S.S.; Corey, L.; Malkowski, S.N.; Higgs, G.; Breaker, R.R. A Second Riboswitch Class for the Enzyme Cofactor NAD. *RNA* **2021**, *27*, 99–105. [CrossRef] [PubMed]
178. Kavita, K.; Breaker, R.R. Discovering Riboswitches: The Past and the Future. *Trends Biochem. Sci.* **2023**, *48*, 119–141. [CrossRef]
179. Mattay, J. Noncanonical Metabolite RNA Caps: Classification, Quantification, (de)Capping, and Function. *WIREs RNA* **2022**, *13*, e1730. [CrossRef] [PubMed]
180. Bird, J.G.; Zhang, Y.; Tian, Y.; Panova, N.; Barvík, I.; Greene, L.; Liu, M.; Buckley, B.; Krásný, L.; Lee, J.K.; et al. The Mechanism of RNA 5′ Capping with NAD+, NADH and Desphospho-CoA. *Nature* **2016**, *535*, 444–447. [CrossRef] [PubMed]
181. Jiao, X.; Doamekpor, S.K.; Bird, J.G.; Nickels, B.E.; Tong, L.; Hart, R.P.; Kiledjian, M. 5′ End Nicotinamide Adenine Dinucleotide Cap in Human Cells Promotes RNA Decay through DXO-Mediated DeNADding. *Cell* **2017**, *168*, 1015–1027.e10. [CrossRef] [PubMed]
182. Wang, Y.; Li, S.; Zhao, Y.; You, C.; Le, B.; Gong, Z.; Mo, B.; Xia, Y.; Chen, X. NAD+−Capped RNAs Are Widespread in the Arabidopsis Transcriptome and Can Probably Be Translated. *Proc. Natl. Acad. Sci. USA* **2019**, *116*, 12094–12102. [CrossRef]
183. Zhang, Y.; Kuster, D.; Schmidt, T.; Kirrmaier, D.; Nübel, G.; Ibberson, D.; Benes, V.; Hombauer, H.; Knop, M.; Jäschke, A. Extensive 5′-Surveillance Guards against Non-Canonical NAD-Caps of Nuclear MRNAs in Yeast. *Nat. Commun.* **2020**, *11*, 5508. [CrossRef]

184. Lupas, A.N.; Ponting, C.P.; Russell, R.B. On the Evolution of Protein Folds: Are Similar Motifs in Different Protein Folds the Result of Convergence, Insertion, or Relics of an Ancient Peptide World? *J. Struct. Biol.* **2001**, *134*, 191–203. [CrossRef]
185. Aziz, M.F.; Caetano-Anollés, G. Evolution of Networks of Protein Domain Organization. *Sci. Rep.* **2021**, *11*, 12075. [CrossRef]
186. Rossmann, M.G.; Moras, D.; Olsen, K.W. Chemical and Biological Evolution of Nucleotide-Binding Protein. *Nature* **1974**, *250*, 194–199. [CrossRef]
187. Aravind, L.; Mazumder, R.; Vasudevan, S.; Koonin, E.V. Trends in Protein Evolution Inferred from Sequence and Structure Analysis. *Curr. Opin. Struct. Biol.* **2002**, *12*, 392–399. [CrossRef]
188. Xie, L.; Bourne, P.E. Detecting Evolutionary Relationships across Existing Fold Space, Using Sequence Order-Independent Profile-Profile Alignments. *Proc. Natl. Acad. Sci. USA* **2008**, *105*, 5441–5446. [CrossRef]
189. Edwards, H.; Abeln, S.; Deane, C.M. Exploring Fold Space Preferences of New-Born and Ancient Protein Superfamilies. *PLoS Comput. Biol.* **2013**, *9*, e1003325. [CrossRef]
190. Bukhari, S.A.; Caetano-Anollés, G. Origin and Evolution of Protein Fold Designs Inferred from Phylogenomic Analysis of CATH Domain Structures in Proteomes. *PLoS Comput. Biol.* **2013**, *9*, e1003009. [CrossRef]
191. Shin, W.-H.; Kihara, D. 55 Years of the Rossmann Fold. *Methods Mol. Biol.* **2019**, *1958*, 1–13. [CrossRef] [PubMed]
192. Medvedev, K.E.; Kinch, L.N.; Dustin Schaeffer, R.; Pei, J.; Grishin, N.V. A Fifth of the Protein World: Rossmann-like Proteins as an Evolutionarily Successful Structural Unit. *J. Mol. Biol.* **2021**, *433*, 166788. [CrossRef]
193. Tóth-Petróczy, Á.; Tawfik, D.S. The Robustness and Innovability of Protein Folds. *Curr. Opin. Struct. Biol.* **2014**, *26*, 131–138. [CrossRef] [PubMed]
194. Grudnik, P.; Kamiński, M.M.; Rembacz, K.P.; Kuśka, K.; Madej, M.; Potempa, J.; Dawidowski, M.; Dubin, G. Structural Basis for ADP-Dependent Glucokinase Inhibition by 8-Bromo-Substituted Adenosine Nucleotide. *J. Biol. Chem.* **2018**, *293*, 11088–11099. [CrossRef] [PubMed]
195. Degano, M.; Almo, S.C.; Sacchettini, J.C.; Schramm, V.L. Trypanosomal Nucleoside Hydrolase. A Novel Mechanism from the Structure with a Transition-State Inhibitor. *Biochemistry* **1998**, *37*, 6277–6285. [CrossRef]
196. Leulliot, N.; Blondeau, K.; Keller, J.; Ulryck, N.; Quevillon-Cheruel, S.; van Tilbeurgh, H. Crystal Structure of Yeast FAD Synthetase (Fad1) in Complex with FAD. *J. Mol. Biol.* **2010**, *398*, 641–646. [CrossRef] [PubMed]
197. Charnock, S.J.; Davies, G.J. Structure of the Nucleotide-Diphospho-Sugar Transferase, SpsA from Bacillus Subtilis, in Native and Nucleotide-Complexed Forms. *Biochemistry* **1999**, *38*, 6380–6385. [CrossRef] [PubMed]
198. Brzezinski, K.; Dauter, Z.; Jaskolski, M. Structures of NodZ A1,6-Fucosyltransferase in Complex with GDP and GDP-Fucose. *Acta Crystallogr. D Biol. Crystallogr.* **2012**, *68*, 160–168. [CrossRef] [PubMed]
199. Laurino, P.; Tóth-Petróczy, Á.; Meana-Pañeda, R.; Lin, W.; Truhlar, D.G.; Tawfik, D.S. An Ancient Fingerprint Indicates the Common Ancestry of Rossmann-Fold Enzymes Utilizing Different Ribose-Based Cofactors. *PLoS Biol.* **2016**, *14*, e1002396. [CrossRef]
200. Degano, M. Structure, Oligomerization and Activity Modulation in N-Ribohydrolases. *Int. J. Mol. Sci.* **2022**, *23*, 2576. [CrossRef]
201. Coutinho, P.M.; Deleury, E.; Davies, G.J.; Henrissat, B. An Evolving Hierarchical Family Classification for Glycosyltransferases. *J. Mol. Biol.* **2003**, *328*, 307–317. [CrossRef]
202. Breton, C.; Šnajdrová, L.; Jeanneau, C.; Koča, J.; Imberty, A. Structures and Mechanisms of Glycosyltransferases. *Glycobiology* **2006**, *16*, 29R–37R. [CrossRef]
203. Henrissat, B.; Sulzenbacher, G.; Bourne, Y. Glycosyltransferases, Glycoside Hydrolases: Surprise, Surprise! *Curr. Opin. Struct. Biol.* **2008**, *18*, 527–533. [CrossRef]
204. Ünligil, U.M.; Rini, J.M. Glycosyltransferase Structure and Mechanism. *Curr. Opin. Struct. Biol.* **2000**, *10*, 510–517. [CrossRef]
205. Breton, C.; Fournel-Gigleux, S.; Palcic, M.M. Recent Structures, Evolution and Mechanisms of Glycosyltransferases. *Curr. Opin. Struct. Biol.* **2012**, *22*, 540–549. [CrossRef]
206. Campbell, J.A.; Davies, G.J.; Bulone, V.; Henrissat, B. A Classification of Nucleotide-Diphospho-Sugar Glycosyltransferases Based on Amino Acid Sequence Similarities. *Biochem. J.* **1997**, *326 Pt 3*, 929–939. [CrossRef] [PubMed]
207. Parkin, D.W. Purine-Specific Nucleoside N-Ribohydrolase from Trypanosoma Brucei Brucei. *J. Biol. Chem.* **1996**, *271*, 21713–21719. [CrossRef] [PubMed]
208. Jung, B.; Hoffmann, C.; Möhlmann, T. Arabidopsis Nucleoside Hydrolases Involved in Intracellular and Extracellular Degradation of Purines: Arabidopsis Nucleoside Hydrolase Family. *Plant J.* **2011**, *65*, 703–711. [CrossRef] [PubMed]
209. Martin, R.; Qi, T.; Zhang, H.; Liu, F.; King, M.; Toth, C.; Nogales, E.; Staskawicz, B.J. Structure of the Activated ROQ1 Resistosome Directly Recognizing the Pathogen Effector XopQ. *Science* **2020**, *370*, eabd9993. [CrossRef] [PubMed]

biomolecules MDPI

Article

nWASP Inhibition Increases Wound Healing via TrKb/PLCγ Signalling

Bethan A. Frugtniet [1], Fiona Ruge [1], Andrew J. Sanders [2], Sioned Owen [3], Keith G. Harding [4], Wen G. Jiang [1] and Tracey A. Martin [1,*]

1 Division of Cancer and Genetics, Cardiff University School of Medicine, Cardiff University, Cardiff CF14 4XN, UK
2 Institute of Biomedical Science, University of Gloucestershire, Cheltenham GL50 2RH, UK
3 School of Applied Sciences, University of South Wales, Pontypridd CF37 4AT, UK
4 Wound Healing Research Unit, Cardiff University School of Medicine, Cardiff University, Cardiff CF14 4XN, UK
* Correspondence: martintal@cardiff.ac.uk; Tel.: +44-(0)202-068-7209

Abstract: (1) Background: Chronic wounds represent a major burden to patients and healthcare systems and identifying new therapeutic targets to encourage wound healing is a significant challenge. This study evaluated nWASP as a new therapeutic target in human wound healing and determined how this can be regulated. (2) Methods: Clinical cohorts from patients with chronic wounds were tested for the expression of nWASP and cell models were employed to evaluate the influence of nWASP on cellular functions that are key to the healing process following knockdown and/or the use of nWASP-specific inhibitors. (3) Results: nWASP was significantly elevated at transcript levels in human non-healing chronic wounds versus healing tissues. nWASP inhibitors, wiskostatin and 187-1, along with the knockdown of nWASP, modified both HaCaT and HECV cell behaviour. We then identified two signalling pathways affected by nWASP inhibition: TrkB signalling and downstream PLCγ1 phosphorylation were impaired by nWASP inhibition in HaCaT cells. The healing of wounds in a diabetic murine model was significantly improved with an nWASP inhibitor treatment. (4) Conclusions: This study showed that nWASP activity was related to the non-healing behaviour of chronic wounds and together with the findings in the in vivo models, it strongly suggested nWASP as a therapeutic target in non-healing wounds that are regulated via TrkB and PLCγ1 signalling.

Keywords: wound healing; nWASP; therapy; chronic wounds; motility

1. Introduction

Wound healing can be described as the process of the repair of cutaneous tissue damage or breakdown. It consists of a complex cascade of events that can be summarised into several overlapping stages, each requiring coordination of a variety of signals and cell types [1–6]. Wound remodelling involves many cells and signalling pathways and consists of the critical re-epithelialisation of the wound by keratinocytes as they proliferate and migrate in order to close the wound. This is achieved through cell polarisation in the direction of movement according to numerous signals, such as cytokines and growth factors, and the extension of actin-rich membrane structures [7]. Furthermore, adhesion dynamics, where a coordinated balance in the formation of adhesions at the leading edge of the cell and disassembly at the rear, allows cells to gain traction by connecting the actin cytoskeleton to the surrounding matrix, subsequently enabling the cell to move. An optimal level of adhesion is key in allowing the motility of migrating cells [6,8–10].

In most cases, acute cutaneous wounds, which typically occur suddenly or following surgery, close within a reasonable period. However, underlying pathological complications or other factors that impair the normal healing process can cause chronic or non-healing wounds to form. Many factors, such as vascular compromise, repetitive insult to the tissue or chronic inflammation, can lead to chronic wound formation. This makes aged individuals or patients with diabetes, vascular diseases and auto-immune diseases more vulnerable to developing chronic wounds, the majority of which are classified as leg, pressure or diabetic foot ulcers [11–16]. Chronic non-healing wounds present a substantial economic burden to the healthcare system; significant reductions in quality of life for those affected; and often precede serious events, such as limb amputations or even premature deaths [17,18]. In addition to the significant costs associated with these wounds, they also have a large impact on the quality of life of patients. Physical symptoms from chronic wounds include leakage, odour and pain, as well as impairments to daily living [19–21]. The lack of appropriate investigation and classification of chronic wounds was identified by several studies and healthcare professionals currently often rely on limited, traditional wound care approaches, such as dressings, debridement and compression treatment [22–24]. There is a clear need for effective diagnosis and identification of new therapeutic approaches to treat chronic wounds. The complex and diverse underlying genetic and molecular processes involved in the wound healing process and chronic wound development make identifying the cause of a particular chronic wound a difficult task. However, these genetic and protein expression deficiencies in chronic wounds also represent potential diagnostic and therapeutic opportunities. Several studies explored these molecular changes and, as a result, highlighted many novel proteins that could potentially be used to explain the development of chronic wounds, offer prognoses and even influence wound repair [25–27]. In one study, nWASP, amongst other molecular markers, was shown to have differential expression in acute, chronic healing and chronic non-healing wound tissues [27].

nWASP (neural Wiskott–Aldrich syndrome protein) belongs to the WASP/WAVE (WASP family verprolin-homologous protein) family. Mutations in the first WASP family member to be identified, namely, WASP itself, were found to be the cause of WAS, which is a recessive disorder that was initially described as a triad of symptoms, namely, thrombocytopenia, eczema and immunodeficiency [28–30]; furthermore, they disrupt the activity of important functional domains, leading to more severe phenotypes [31,32]. nWASP was the second protein to be classified as a WASP family member due to the detection of several functional motifs shared with WASP. This 65 kDa protein was named neural WASP due to its shared homology with WASP and abundance in the brain, although it is widely expressed in different tissues throughout the body [33]. Under resting conditions, nWASP exists in an inactive, auto-inhibited confirmation whereby the main catalytic domain is shielded by the N-terminus regions. The WASP/WAVE family mediates the signals between the Rho GTPase family members, such as Rho, Rac and Cdc42, and the factors that modulate the actin cytoskeleton, in particular the actin-related protein 2/3 (Arp2/3) complex. Cdc42 can bind to the GBD domain, destabilising the folded confirmation of nWASP and exposing its catalytic domain [34,35]. Arp2/3 becomes activated when bound to the C-terminus CA region on WASP family proteins, allowing actin polymerisation to be initiated if an actin monomer is bound in conjunction with the V region [36–39].

Through its role as an organiser of the actin cytoskeleton, nWASP was also shown to be involved in the formation of membrane protrusions that are important for cell movement and its expression was shown to correlate with certain cancer phenotypes [34,40–43]. Consequently, nWASP has been recognised as a potential therapeutic target in a range of contexts, including wound healing.

The aims of this study were to validate nWASP as a novel therapeutic target for the treatment of chronic wounds by examining expression levels in human wound tissues. The therapeutic potential of targeting nWASP was then explored by examining the effect of inhibiting nWASP using the agents 187-1 and wiskostatin, which act to maintain nWASP in its inactive, auto-inhibited state (small-molecule inhibitors of nWASP, namely, wiskostatin and 187-1, bind to nWASP and allosterically block its activity by stabilising the closed, auto-inhibited conformation of nWASP. Both inhibitors act in the same way to stabilise the autoinhibited state of nWASP, blocking activation and hence actin polymerisation, but wiskostatin was used more extensively in research to examine the effects of nWASP activity. Wiskostatin is a cell-permeable N-alkylated carbazole that interacts with a cleft in the regulatory GBD of nWASP in the solution structure of the complex [44], as well as on human skin cell behaviour in vitro and in vivo on the closure of wounds in mouse models with impaired healing abilities. We also investigated the potential effects of these treatments on downstream signalling changes in skin cells in order to elucidate mechanistic effects.

2. Materials and Methods

2.1. Cell Lines and Culture Conditions

Human keratinocyte cells (HaCaT cells) from the German Cancer Institute/Cell Service, Germany, human vascular endothelial cells (HECV cells) from Interlab, Italy, and TE 354.T cells (LCG Standards, Bury, UK) were cultured in Dulbecco's Modified Eagle Medium (DMEM) (Sigma-Aldrich, Dorset, UK) supplemented with 10% foetal bovine serum (FBS; Sigma-Aldrich, Dorset, UK), 100 units/mL penicillin and 100 µg/mL streptomycin (Sigma-Aldrich, Dorset, UK). The cells were incubated at 37 °C, 5% CO_2 and 95% humidity.

2.2. Tissue Collection

Fresh tissues from chronic wounds (n = 14), acute wounds (n = 10) and normal skin from healthy volunteers (n = 10) were collected under the approval of the local ethical committee (ethical approval ID: 04/WSE03/92). Chronic wound tissues were obtained from patients with chronic leg ulcers that had been present for a minimum of 6 months prior to their biopsy with no evidence of healing 6 weeks before the biopsy. Acute wound tissues were obtained from patients with acute surgical wounds following the excision of pilonidal disease and normal tissues were from healthy, unwounded skin. A second cohort of chronic wound tissues was collected from chronic venous leg ulcer patients following ethical approval from a local committee (South East Wales Ethics Committee reference number 09/WSE02/51). Wound size was recorded at the time of the initial biopsy and again after 3 months, during which wounds were treated as per best medical practise (n-77). Wounds that reduced in size over this period were described as 'healing/healed' and those that grew or did not reduce in size were described as 'non-healing'. All tissues were snap-frozen following collection and stored at −80 °C until processing. Written informed consent was obtained from each patient who agreed for a biopsy to be taken. Wound cohorts were previously described [45–47].

2.3. Tissue Processing for the RNA Extraction and Reverse Transcription

Tissues were sectioned on a cryostat (Leica, Microsystems Ltd., Milton Keynes, UK) with either a 7 µm thickness for immune-histochemical analysis or at 20 µm thickness for the extraction of RNA. Approximately 20 sections from the same patient sample biopsy were pooled and homogenised in an ice-cold TRI reagent (Sigma-Aldrich, Poole, UK) using a hand-held homogeniser (Cole Palmer, London, UK). RNA was extracted from the tissues following the manufacturer's instructions, and the same protocol was also used to extract RNA from cultured cell lines. Samples were standardised and cDNA was subsequently generated (BioRad, Hemel Hempstead, UK).

2.4. Tissue Processing for the Ex Vivo Model

The *ex vivo* effects of the nWASP inhibitor 187-1 in chronic human wound tissues were examined using a method previously established [48]. Briefly, fresh biopsies from human chronic wounds were immediately placed in a buffer that mimics physiological fluid and contains a mixture of antibiotics. The tissues were finely minced using a sterile scalpel to sizes less than 1 mm in diameter. After extensive washing in the buffer, the living tissues were immediately embedded in an extracellular matrix gel. The gels and the topping solution included the treatments (nWASP inhibitor). The tissues were photographed daily. The degree of expansion from the implanted tissues was calculated using the imager, as previously reported [48].

2.5. Quantitative, Real-Time Polymerase Chain Reaction (qPCR) and Conventional PCR

Analysis of gene transcript expression was carried out using qPCR with cDNA produced from a human wound and skin tissues and conventional PCR using cDNA produced from RNA extracted from cell lines. This study adopted Ampliflor quantitation technology in which one set of gene-specific primers (designed using Beacon Design software, PREMIER Biosoft, Palo Alto, CA, USA) and Uniprimer probes (Intergen Inc., New York, NY, USA) were used. The reaction was carried out using the ICycleIQ (BioRad, UK). The real-time qPCR conditions were 95 °C for 15 min, followed by 60 cycles of 95 °C for 20 s, 55 °C for 30 s and 72 °C for 20 s. In all the assays, GAPDH and actin were amplified and used as housekeeping controls, and an internal standard was also employed for quantitation purposes. The nWASP primers were as follows: BNDF (F8: TTCATACTTTGGTTGCATGA, R8: TTCAGTTGGCCTTTTGATAC), GAPDH (F8: GGCTGCTTTTAACTCTGGTA, R8: GACTGTGGTCATGAGTCCTT), nWASP (F8: AGTCCCTCTTCACTTTCCTC, R8: GCTTTTCCCTTCTTCTTTTC), TrkB (F2a: CCCACTCACATGAACAATGG, R2a: TCAGTGACGTCTGTGGAAGG), Actin (F8: GGACCTGACTGACTACCTCA, zR8: ACTGAACCTGACCGTACAAGCTTCTCCTTAATGTCACG), BDNF (F8: TTCATACTTTGGTTGCATGA, Rz8: ACTGAACCTGACCGTACACTCTTGAACCTGCCTTGG), GAPDH (F: CTAGTACGTCGTGGAGTC, zR: ACTGAACCTGACCGTACACAGAGATGATGATGACCCTTTTG), PDPL (F: GAATCATCGTTGTGGTTATG, zR: ACTGAACCTGACCGTACACTTTCATTTGCCTATCACAT) and nWASPKD (F8: AGTCCCTCTTCACTTTCCTC, zR8a: ACTGAACCTGACCGTACAAGATCTCTGTGGATTGTCCT).

2.6. Reagents and Treatments

nWASP inhibitors, wiskostatin (Merck Pharmaceuticals, Watford, UK) and 187-1 (TOCRIS, Bristol, UK) were used. For the in vitro studies, wiskostatin was dissolved in 30% dimethyl sulfoxide (DMSO, Sigma-Aldrich, Dorset, UK) in normal cell culture medium to a stock concentration of 300 μM, whereas 187-1 was diluted in BSS. For the in vivo studies, nWASP inhibitor compounds, 187-1 and wiskostatin were formulated for systemic and topical application. For systemic application, 187-1 was dissolved and diluted in BSS to the required concentration. Wiskostatin was first dissolved in DMSO (Sigma-Aldrich, Gillingham, UK) at a concentration of 5 mg/mL. The DMSO solution was then gradually diluted in BSS in order to avoid precipitation. The solutions were prepared so that each 100 μL contained the correct amount of compounds and was stored at −20 °C until used. The primary antibodies used were as follows: actin (sc-16515, Insight Biotechnology, Middlesex, UK), Erk1/2 (v114A, Promega, Southampton, UK), GAPDH (sc32233, Insight Biotechnology, Middlesex, UK), PLCγ1 (sc81, Insight Biotechnology, Middlesex, UK), p-PLCγ1 (sc22141, Insight Biotechnology, Middlesex, UK), nWASP (NBP1-82512, Novus Biologicals, Abingdon, UK), TrkB Y816 (ABN1381, EMD-Millipore, Watford, UK), TrkB pan (07-225, EMD-Millipore, Watford, UK) and TrkB pan (sc-136990, Insight Biotechnology, Middlesex, UK). The secondary antibodies were as follows: goat anti-mouse IgG (A4416, Sigma-Aldrich, Dorset, UK), goat anti-rabbit IgG (A6154, Sigma-Aldrich, Dorset, UK), rabbit anti-goat IgG (A5420, Sigma-Aldrich, Dorset, UK), DAPI (D1306, Life Technologies, Warrington, UK), anti-rabbit AlexaFluor 488 (A21206, Life Technologies, Warrington, UK), anti-mouse AlexaFluor 488 (A21202, Life Technologies, Warrington, UK), anti-rabbit AlexaFluor 594 (A21207, Life Technologies, Warrington, UK), anti-mouse AlexaFluor 594 (A21203, Life Technologies, Oxford, UK) and anti-goat AlexaFluor 594 (A11058, Life Technologies, Warrington, UK).

2.7. Functional Assays

2.7.1. Electric Cell–Substrate Impedance Sensing (ECIS)

ECIS instruments (Applied Biophysics Inc., New York, NJ, USA) were used to electrically monitor the coverage of gold electrodes by cells by measuring the resistance and impedance at a frequency of 4 kHz unless otherwise specified. Cells were seeded in a culture medium containing treatments. Cells were then constantly monitored following seeding and electrical wounds were applied where described. Wound settings were a time of 30 s, current of 2400 μA and frequency of 60,000 Hz. Increased resistance and impedance illustrate a greater number of cells present on the electrode. This can therefore be used to assess attachment and motility.

2.7.2. In Vitro Cell Viability Test

Briefly, cells were seeded in a 96-well plate in varying treatment concentrations/controls and incubated for 92 h. After a further 4 h incubation in 1:10 MTT (Thiazolyl Blue Tetrazolium Bromide, Sigma-Aldrich, Gillingham, UK) solution (5 mg/mL in phosphate buffered solution (PBS)), acidified isopropanol (Sigma-Aldrich, Dorset, UK) was applied and the optical density (OD) was determined using an absorbance reader (Biotek ELx800, Swindon, UK) at 540 nm and used to calculate relative cell viability at each treatment concentration [49].

2.7.3. In Vitro Growth Assay

Cells were seeded into a 96-well plate with appropriate treatments and fixed after 1/2/4 day incubation periods using 4% formalin (Sigma-Aldrich, Dorset, UK). Plates were stained with 1% crystal violet (Sigma-Aldrich, Gillingham, UK), which was then extracted from cells using 10% acetic acid (Sigma-Aldrich, Dorset, UK) in distilled water (v/v). Absorbance was determined at a 540 nm wavelength on an absorbance plate reader (Biotek ELx800, Swindon, UK).

2.7.4. In Vitro Carrier Bead Assay

A total of 100 μL of Cytodex-2 carrier beads (Sigma-Aldrich Ltd., Gillingham, UK) were added to 10 mL of 1×10^4 cells/mL and gently mixed. The cells were then left at 37 °C for 2 h. Following a wash with medium, cells were allowed to settle until aliquoted into a 96-well plate and treated with inhibitors. After incubation overnight, the beads were washed off in PBS and the cells that had migrated onto the culture vessel floor were counted after visualisation with crystal violet.

2.7.5. In Vitro Scratch Assay

Cells were seeded in appropriate treatments into each well on a 24-well plate. Upon reaching confluence, the monolayer was scratched to create a linear wound. The plate was placed in an EVOS® FL Auto Imaging System (Life Technologies, Oxford, UK), which maintained the plate in normal culture conditions throughout the experiment. Images were captured of the wound every 30 min for up to 24 h.

2.7.6. In Vitro Adhesion Assay

Wells were pre-coated with Matrigel basement membrane matrix (BD Biosciences, Oxford, UK) at 50 μg/mL in a normal culture medium. Cells were then seeded into each well onto the Matrigel membrane in treatments and incubated for 25 min. Adherent cells were fixed, stained and quantified as above.

2.7.7. Tubule Formation Assay

A Matrigel matrix (10 mg/mL) was gently pipetted into the bottom of each well of a 96-well plate. The plate was placed at 37 °C for 40 min whilst the Matrigel polymerised. Then, 4×10^4 cells (and treatments/control reagents where appropriate) were plated on top of the Matrigel in each well in triplicate and incubated at 37 °C, 5% CO_2 and 95% humidity. Images of each well were captured at X5 magnification and intervals ranging from 30 to 1170 min using a Leica DMi1 microscope equipped with an MC120 HD camera and Leica Application Suite version 3.0.0 software (Leica Microsystems, Milton Keynes, UK). Analysis of the images was carried out using ImageJ, where the lengths of cell structures that form part of a tubule structure were measured to track the progress of tubule formation.

2.8. In Vivo Tolerance Test

The main tolerance tests were conducted using CD-1 athymic mice (Charles River Laboratories, Oxford, UK), owing to their slow and steady rate of growth and the ease of observing changes in the skin (hairless) and other possible side effects. Briefly, CD-1 mice that were 4–6 weeks old and 20 g in weight were housed in filter-topped cages. 187-1 (MW 1784, dissolved in BSS buffer) and wiskostatin (MW 426, dissolved in DMSO and diluted in BSS) were injected via the intraperitoneal route on a daily basis. Dosages administered were 1 and 10 µM in 100 µL volumes for each compound, equivalent to 1.8 g/kg/day and 17.8 g/kg/day for 187-1 and 0.43 g/kg/day and 4.3 g/kg/day for wiskostatin. CD-1 mice were observed daily and weighed twice weekly based on the previous in vitro experiments. An additional tolerance and efficacy test was carried out using the diabetic db/db strain obtained from Harlan Laboratories (Cambridge, UK), which exhibited impaired wound healing abilities. These mice were 4–6 weeks old and when their body weight reached 20 g they were used for tolerance and efficacy tests.

2.9. In Vivo Efficacy Test and Wound Healing

An ear punch method previously described [50] was used to create wounds in the mice that were 1 mm in diameter. Treatment was given systemically or topically. For topical application, two carrier gels that are currently used in wound care were used. The inhibitors were diluted from the master stock in the gel at a concentration of 1 mg/g followed by low-speed homogenisation using a hand-held homogeniser. For use, small amounts (150 µL) of the gel were rubbed into the wound area. Both treatments were given every other day. For systemic application, 187-1 was applied at 0.5 and 5 µM (equivalent to 0.89 g/kg/day and 8.9 g/kg/day) and wiskostatin was applied at 1 and 10 µM (equivalent to 0.43 g/kg/day and 4.3 g/kg/day). Images were obtained weekly. The sizes of the wounds were determined using image analysis software. Data are presented in two ways: as the area of the wounds in pixels, where two sample Student t-tests were used for statistical analysis, or as the change in the size of the wound from the starting point calculated using the following formula: (area at a given point—area at the starting point)/(area of the starting point) × 100, in which case, the Bonferroni model was used for data analysis.

2.10. Microarray Analysis

This study used the Kinexus^{TM} KAM880 protein array service provided by Kinexus Bioinformatics Ltd. (Vancouver, BC, Canada). Signal quantification was performed with ImaGene 8.0 by Kinexus Bioinformatics Ltd., which has predetermined settings for spot segmentation and background correction. The background-corrected raw intensity data were then globally normalised by summing the intensities of all the net signal median values for a sample to obtain the globally normalised signal intensities for each protein. The percentage change of the treated samples from the control was calculated based on the globally normalised intensity for each protein using the following calculation: %CFC = (globally normalised treated − globally normalised control)/globally normalised control) × 100. The percentage error range was also calculated to examine how tightly the globally normalised net signal intensity varied for duplicate spots of the same protein in the sample. The z-scores were also calculated by subtracting the overall average intensity of all replicate spots from the raw intensity for each spot and then dividing it by the standard deviation of all the measured intensities within each sample. The z-ratio was calculated by dividing the differences between the observed z-scores by the standard deviation of all the differences for that comparison. Several factors were used to determine the most important changes in protein expression and phosphorylation including %CFC, error ranges, value for the globally normalised intensity of one of the samples of >1500 and significance based on z-ratios of <−1.64 or >1.64.

2.11. Immunofluorescence

Cells were cultured in Millicell EZ 8-well chamber slides (Merck Millipore, Watford, UK). To fix the cells, the culture medium was removed and the cells were washed with PBS and then fixed in 100% ice-cold ethanol. To proceed with immunofluorescence staining, the cells were washed 3 times in PBS and then permeabilised using 0.1% Triton X-100 (Sigma-Aldrich, Dorset, UK). Blocking buffer (5–10% donkey serum (D9663, Sigma-Aldrich, Dorset, UK)) in PBS was added to each well. The slide was left for 3 h at room temperature in a blocking buffer. The chamber slide was then incubated with primary antibodies for an hour on the bench or at 4 °C overnight. Secondary antibodies were (1:500) with the addition of DAPI (1:100), and each primary antibody was incubated with the corresponding secondary for a further hour. Following three washes in PBS, the slides were mounted in FluorSave™ (Calbiochem, Nottingham, UK) and allowed to dry before being visualised using an Olympus BX51 microscope with a Hamamatsu Orca ER digital camera at ×40. Images were analysed using ImageJ.

2.12. Immunohistochemistry (IHC)

Immunohistological analysis using an avidin–biotin peroxidase technique was performed on human tissue samples collected from cohort 2. The frozen sections were fixed in dried acetone (10162180, Fisher Scientific, Loughborough, UK), air-dried and washed in PBS. Sections were then incubated in 0.1% BSA/10% horse serum in PBS (referred to as blocking solution) in a humidified box at room temperature, followed by primary antibody solutions (diluted in blocking buffer to a final concentration of 2 µg/mL). Following washing with PBS, biotinylated horse anti-mouse/rabbit IgG secondary antibody (Vector Laboratories, Oxford, UK) was applied for 30 min, followed by ABC reagent for 30 min, both of which were provided in the VECTASTAIN® Elite ABC Kit (Vector Laboratories, Peterborough, UK). 3,3′-Diaminobenzidine (DAB)(Sigma-Aldrich, Gillingham, UK) substrate (5 mg/mL) was used to develop the final reaction and the sections were counterstained with Gill's haematoxylin (Vector Laboratories, Oxford, UK). Following dehydration and clearing in xylene, sections were mounted in Distyrene Plasticizer Xylene (DPX, Merck Pharmaceuticals, Gillingham, UK). Staining was visualised using a Leica DM1000LED microscope with an MC120 HD camera and Leica Application Suite (version 3.0.0) software (Leica Microsystems, Milton Keynes, UK). The localisation and intensity of staining were judged blindly by two people independently. Positive staining was seen as a brown/black deposit, whilst negatively stained cells could be clearly distinguished using a blue nucleated stain.

2.13. Protein Extraction, SDS-PAGE and Western Blotting

A lysis buffer was used to extract protein from cells, which was then used for SDS-PAGE. Proteins were transferred onto Immobilon® PVDF membranes (Merck Millipore, Watford, UK), which were blocked and probed with primary antibodies and then incubated with the corresponding peroxidase-conjugated secondary antibodies. Proteins were visualised using an EZ-ECL Kit (Biological Industries, Beit Haemk, Israel).

2.14. siRNA Silencing of nWASP

nWASP siRNA (sc36006) was obtained from Insight Biotechnology (UK) and non-targeting siRNA (NT) was obtained from Dharmacon (D001810; Layfayette, CO, USA). Cells were transfected with nWASP siRNA (sc360006; Insight Biotechnology, UK) at varying concentrations or NT control siRNA at the same concentration, as described. siRNA, which was diluted according to the desired end concentration in SFM, was combined with equal volumes of Lipofectamine 3000 reagent (ThermoFisher Scientific, Boston, MA, USA). This siRNA/Lipofectamine 3000 mix in SFM (with no antibiotics) was allowed to stand at room temperature for 30–40 min. Antibiotic-free DMEM supplemented with 5% FBS was then added to the siRNA/Lipofectamine 3000 solution to achieve the desired end concentration of reagents and make up the required volume for application to cells; then, the final solution was gently applied to each well. Cells were incubated under normal culture conditions and a normal culture medium was used 24 h after transfection for any further culturing or assays unless stated otherwise.

2.15. Statistical Analysis

Statistical analysis was conducted using Minitab, SPSS, GraphPad 6.0 Prism (PRISM, Boston, MA, USA) and an online chi-square service tool. Transcript levels from qPCR experiments are reported as median ± SEM and the Mann–Whitney test was used to analyse qPCR data. Representative data are presented. A p-value < 0.05 was considered statistically significant.

3. Results

3.1. nWASP Expression in the Human Chronic Wound Tissues and Cell Lines

qPCR analysis of human chronic wound tissues demonstrated that nWASP transcript expression was significantly increased in the non-healing chronic wounds compared with those defined as healing in the 12 weeks following biopsy collection. In cohort 1, nWASP expression normalised with actin in healing/healed wounds (n = 20) was significantly lower than in non-healing wounds (n = 49) (p = 0.0028) (Figure 1A). Likewise, cohort 2 showed a similar trend where the nWASP transcript levels were significantly lower (p = 0.0001) in the healing chronic wounds (n = 77) than in the non-healing chronic wounds (n = 32) (Figure 1B). nWASP protein expression was examined in two human cell lines, representative of typical skin cells, HaCaT keratinocyte cells and HECV vascular endothelial cells (Figure 1C).

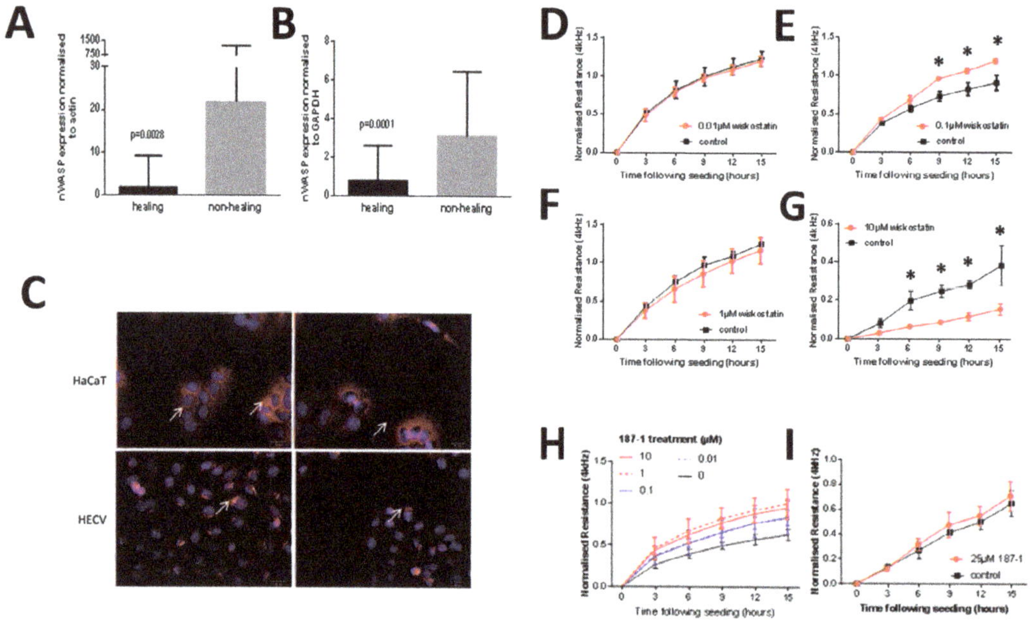

Figure 1. Analysis of nWASP in tissues and cells with the effect of the inhibitor treatment on the cell attachment. qPCR analysis of wound tissues from cohort 1 (**A**) and cohort 2 (**B**). Immunofluorescence staining for nWASP in keratinocyte and endothelial cell lines; the scale bar is 100 μm (**C**). The effect of wiskostatin on HaCaT attachment and spreading, as measured using ECIS (**D–G**). The effect of 187-1 treatment on HaCaT attachment and spreading, as measured using ECIS (**H**,**I**). * indicates statistical difference (p < 0.05).

3.2. The Effect of nWASP Inhibition on the HaCaT Cell Attachment and Spreading

Initial ECIS experiments were carried out over a range of inhibitor concentrations. At 0.01 µM, no change was observed (Figure 1D). However, at 0.1 µM levels, significantly higher resistances were detected in the wiskostatin-treated cells compared with the control (Figure 1E). This increase in resistance did not continue with increased inhibitor treatments (Figure 1F,G). At high levels of wiskostatin, the opposite effect was found (Figure 1G). The ECIS analysis of HaCaT cells treated with a range of concentrations of 187-1 demonstrated that the resistance at 4 kHz was significantly increased with the inhibitor between concentrations of 0.01–10 µM 187-1 (Figure 1H). The increase in resistance compared with the control in the 15 h following seeding was not found with a higher level of 187-1 at 25 µM (Figure 1I).

3.3. ECIS Analysis of nWASP Knockdown in HaCaT Cells

nWASP was knocked down using siRNA, with a successful knockdown shown in Figure 2A. There was no difference in the resistance or capacitance in cells treated with the lower concentration of siRNA (0.5 µg/mL) (Figure 2B,D). A significant difference in the resistance and capacitance was found between the cells treated with 1.0 µg/mL siRNA and the NT cells from the point where the measurements reached a plateau, at approximately 30 h from seeding, and the cells were assumed to have reached confluency (Figure 2C,E).

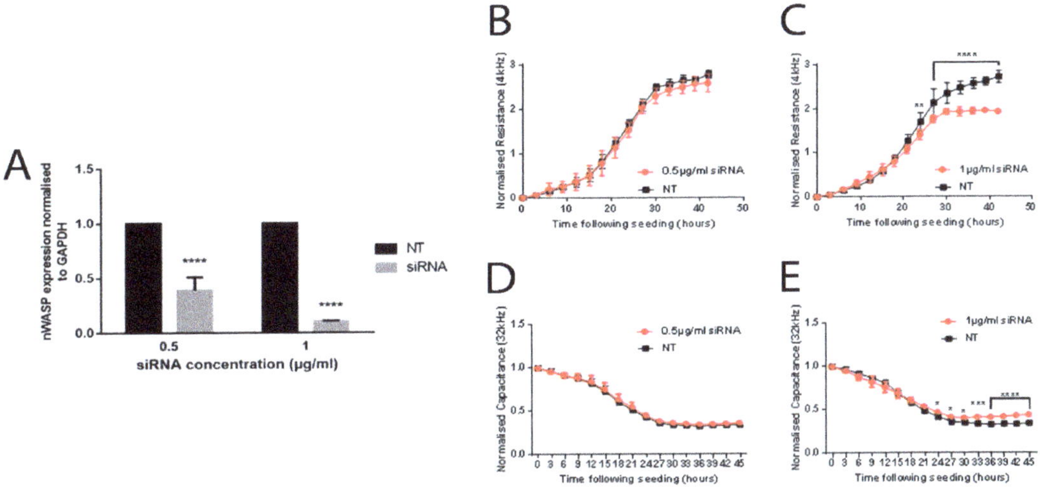

Figure 2. nWASP knockdown in the HaCaT cells. qPCR and PCR analysis of nWASP transcript expression (**A**). ECIS analysis of nWASP knockdown HaCaT cells. HaCaT cells treated with nWASP siRNA (**B–E**). * indicates statistical significance (* $p < 0.05$, ** $p < 0.01$, *** $p < 0.001$ **** $p < 0.0001$).

3.4. Inhibition of Growth following the nWASP Inhibitor Treatment

No significant inhibition in growth was found following treatment with the full range of 187-1 tested and hence no IC_{50} value was calculated (Figure 3A,B). There was also little difference in the control and inhibitor-treated cells treated with 0.1 µM wiskostatin and 10 µM 187-1 (Figure 3C–E, respectively).

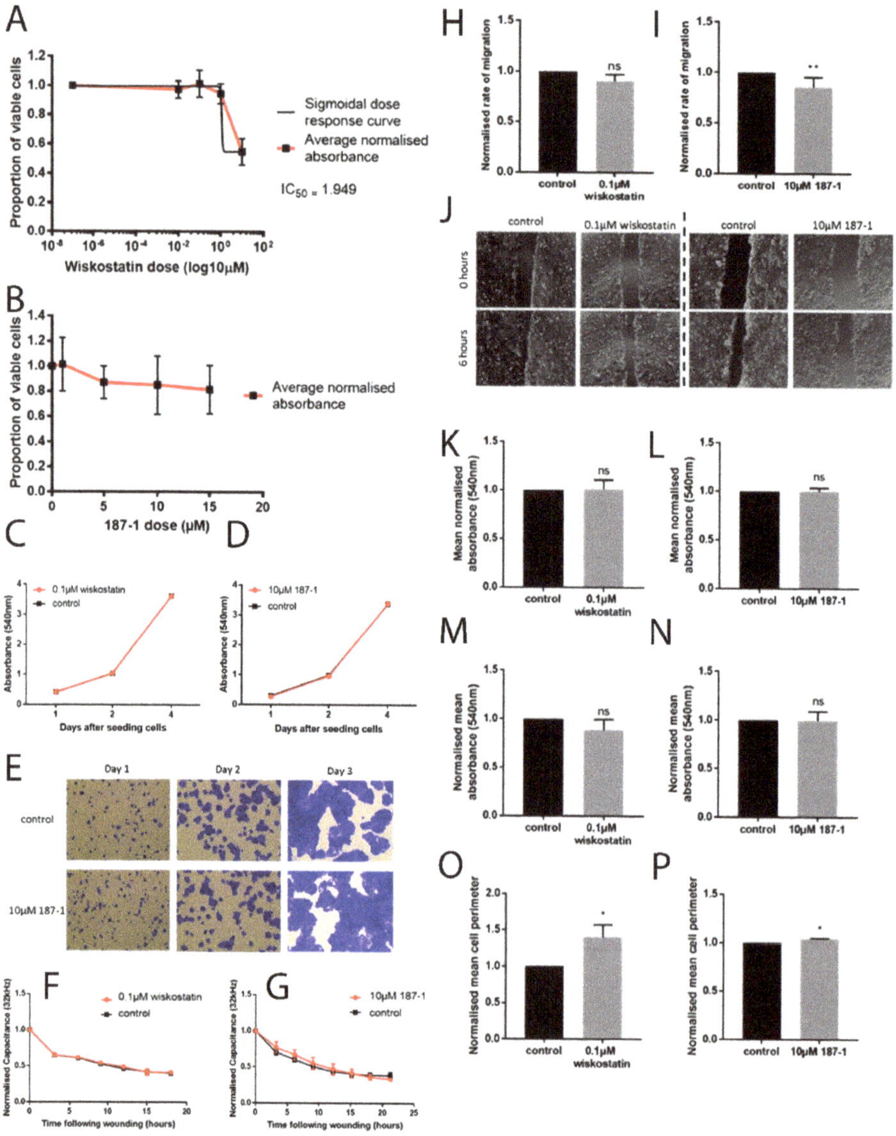

Figure 3. The effect of nWASP inhibition on the HaCaT cell behaviour. Dose–response curves (**A**,**B**). HaCaT cells treated with 0.1 μM wiskostatin (**C**) or 10 μM 187-1 (**D**). Average absorbance ± SD is presented, representative data are shown and $n = 5$ replicates were in each independent repeat. Representative images of crystal-violet-stained cells (**E**). HaCaT cells following an electrical wounding treated with 0.1 μM wiskostatin (**F**) or 10 μM 187-1 (**G**). The effect of wiskostatin and 187-1 on the rate of migration of HaCaT cells: 0.1 μM wiskostatin (**H**) and 10 μM 187-1 (**I**). Normalised rate of migration + SD is presented. Representative images of the scratch wound at 0 and 6 h are also shown (**J**). The 0.1 μM wiskostatin (**K**) and 10 μM 187-1 (**L**) treatment results are shown. The change from the control is presented with the mean + SD shown. The effects of 0.1 μM wiskostatin (**M**) and 10 μM 187-1 (**N**) on the normalised absorbance, and the effects of 0.1 μM wiskostatin (**O**) and 10 μM 187-1 (**P**) on the spread of HaCaT cells. * indicates statistical significance from 0 h to each subsequent time point (* $p < 0.05$, ** $p < 0.01$, ns: not significant).

3.5. The Effect of nWASP Inhibition on the Keratinocyte Cell Migration and Adhesion

Several methods were utilised to examine the effect of nWASP inhibitors on cell motility/migration. Where HaCaT cells were treated with 0.1 μM wiskostatin, the rate of migration assessed using ECIS was found to be, on average, 90% of the control when treated with wiskostatin (p = 0.0572; Figure 3F–H). In 187-1 treated cells, the rate of migration was found to be significantly decreased (p = 0.0234) at 85% of the control migration rate (Figure 3I,J). When using a micro-carrier bead assay, there were no observable differences after the treatment (Figure 3K,L). The effect on the ability of HaCaT cells to initially adhere to a Matrigel matrix in response to the nWASP inhibitor treatment showed no significant difference in the number of cells adhered to the matrix following either the 0.1 μM wiskostatin treatment or the 10 μM 187-1 treatment (Figure 3M,N). An alternative measure of cell adhesion through measuring the spread of HaCaT cells several hours after seeding in nWASP inhibitor treatments was used. The spread according to the perimeter of cells 6 h after seeding was found to be significantly increased in the cells treated with 0.1 μM wiskostatin compared with the controls (p = 0.0137, Figure 3O). The cell perimeter was also found to be increased in cells treated with 10 μM 187-1 compared with the controls, but this increase was very small (p = 0.01, Figure 3P).

3.6. The Effect of nWASP Inhibition on the HECV Cell Behaviour

The effect of the wiskostatin treatment was initially examined using ECIS (Figure 4A–C). The 10 μM wiskostatin treatments significantly delayed the formation of a HECV monolayer, as shown by the decrease in the resistance in the 10 μM wiskostatin treated cells (Figure 4D). Wiskostatin caused a significant reduction in the resistance of the HECV cells compared with the control during the initial attachment and spreading phase following seeding as the resistance increased prior to reaching a plateau. This significant reduction in the resistance was also found in the HECV cells treated with higher levels of the inhibitor (1 and 10 μM). Low levels of the inhibitor did not have a significant effect on the electric wounding (Figure 4E–G). However, at 10 μM wiskostatin, the capacitance was consistently significantly increased, suggesting a delay in the healing of the electrically wounded inhibitor-treated cells (Figure 4H). The 187-1 treatment at 0.1, 1 and 10 μM concentrations caused the capacitance at 32 kHz to consistently decrease (Figure 4I,J). The inhibitors did not appear to affect the growth of the HECV cells, as no change in the growth of more than 12% was observed (Figure 4K,L). The lengths of the cell structures that formed part of microtubules in the tubule formation assay were measured following seeding and the mean total length for each image was calculated (Figure 4M). The wiskostatin treatment significantly inhibited the formation of tubule structures immediately following seeding but did not prevent the formation of tubule structures, which typically developed after about 8 h. Following the maturity of these tubule formations, the HECV cells treated with the control medium began to cluster and tubule formations were lost. However, in the wiskostatin-treated cells, although the tubules became thicker and some clustering was also evident, tubule formation was still evident after 20 h (Figure 4N). There was a significant decrease in the tubule structures in the wiskostatin-treated cells following seeding with a subsequent significant increase in tubule structures after 8 h, where the tubule structures were maintained in wiskostatin treatment and lost in the control cells as the HECV cells formed clusters (Figure 4O).

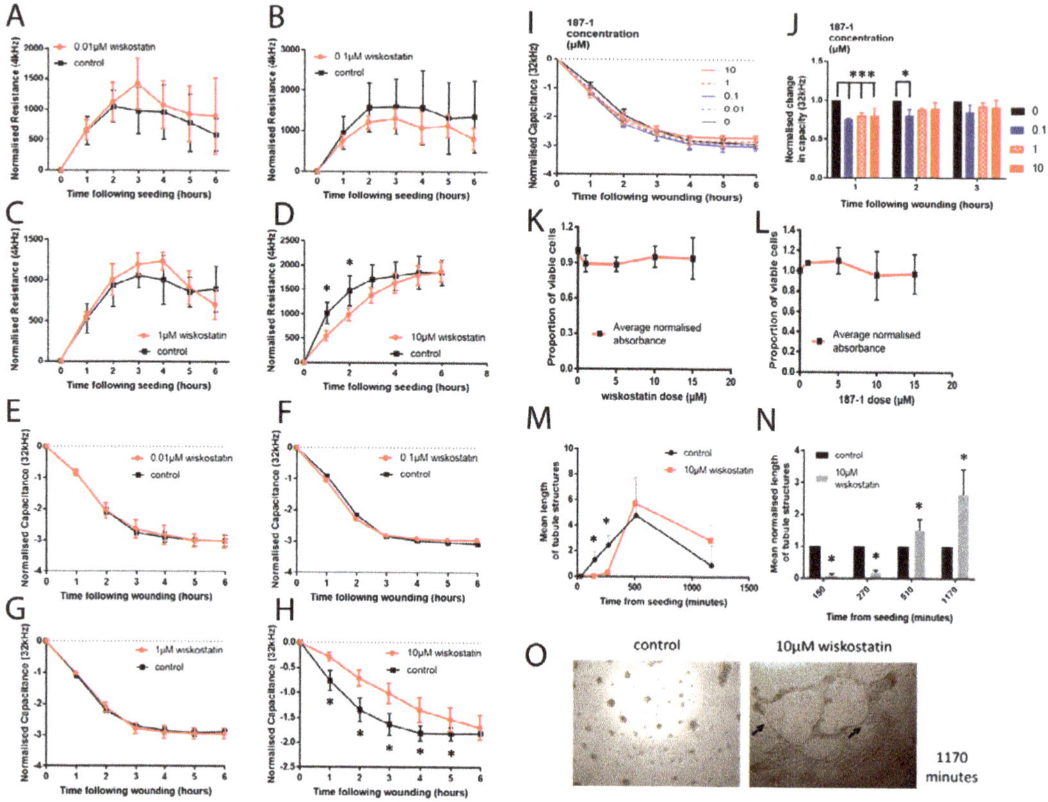

Figure 4. The effect of wiskostatin on the HECV cell behaviour. HECV cells were treated with wiskostatin at 0.01, 0.1, 1 and 10 μM concentrations (**A–D**). HECV cells were treated with wiskostatin at 0.01, 0.1, 1 and 10 μM concentrations (**E–H**). HECV cells were treated with 187-1 at 0, 0.1, 1 and 10 μM concentrations and monitored for 6 h following an electrical wounding event (**I**). The mean changes in capacity in the 187-1-treated wells are also shown (**J**). Following 48 h with wiskostatin and 187-1 (**K**,**L**). Mean lengths of the tubule structures of 10 μM wiskostatin-treated and control HECV cells (**M**). Tubule structure lengths at time points of 150–1170 min (**N**,**O**) in terms of the number of pixels. * indicates statistical significance (* $p < 0.05$, *** $p < 0.001$).

3.7. The 187-1 and Wiskostatin In Vivo Effects

The 187-1 and wiskostatin were delivered systemically on a daily basis as part of a tolerance assay for two weeks in athymic CD-1 and db/db mouse strains. Two concentrations, namely, 0.5 and 5 μM of 187-1, were given systemically (interperitoneal injections daily in a volume of 0.5 mL; mice ears were photographed twice weekly and weighed to assess health, together with general health and appearance). Following this, 5 mm hole punches were made in each ear and treated twice weekly, either with a topical injection (ears massaged twice weekly or interperitoneal injections as per tolerance testing). Healing of the wounds was significantly increased after the systemic and topical administration of 187-1 (Figure 5A,B). After 24 days, the wounds treated with 0.5 and 5 μM 187-1 were significantly smaller than the control group ($p < 0.0001$; Figure 5C). The topical application of 187-1 in both carrier gels A and B also showed a significant effect after 24 days ($p < 0.0001$). The 187-1 in topical carrier gel B appeared to encourage wound healing at a faster rate compared with the other treatments, as the wound size was significantly smaller than the control group after only 7 days, whereas the other treatments saw a significant difference after 15 days. Two concentrations of wiskostatin (1 and 10 μM) were also given systemically. Healing of the wounds was increased following both methods of administration of wiskostatin after 24 days (Figure 5D,E). After 24 days, the wounds treated with 1 and 10 μM wiskostatin were significantly smaller than the control group ($p < 0.0001$; Figure 5F). The topical application of wiskostatin in both carrier gels A and B also showed a significant effect after 24 days ($p < 0.0001$). The wiskostatin treatment appeared to encourage healing at a faster rate than the 187-1, as a significant reduction in the wound size compared with the control wounds was observed in all the treatment methods after only 7 days from wounding.

3.8. Protein Signalling Changed following the Wiskostatin Treatment

A protein array analysis was carried out to examine the protein signalling changes that took place in HaCaT cells in response to the nWASP inhibitor treatment. The most significantly decreased protein signals based on the z-ratio are presented in Figure 6A. A full description of the changes in protein expression or phosphorylation of the most significantly increased or decreased signals (i.e., $-1.64 < \text{z-ratio} < 1.64$) can be found in Table 1. Amongst the most significantly increased or decreased proteins, several membrane-bound proteins and receptors were identified and are highlighted in grey. Through further experiments to examine the protein signalling changes following the 0.1 μM wiskostatin treatment, TrkB signalling was found to be significantly affected by inhibitor treatment, as described in the following sections.

Table 1. Significant changes in protein expression and phosphorylation following the wiskostatin treatment.

Protein Name	Phospho Site	%CFC	Protein Name	Phospho Site	%CFC
	Decreased		**Increased**		
PKCI	Pan-specific	−95	RSK1	Pan-specific	248
RSK1/2/3	T573	−92	HSF4	Pan-specific	228
Rb	S795	−91	Fos	T232	227
Trail	Pan-specific	−89	alF2a	S52	225
Tyrosine hydrolase	S40	−89	Mnk2	Pan-specific	239
Progesterone receptor	S294	−85	Integrin α4	S988	325
PTEN	Pan-specific	−81	P70 S6K	T421/S424	337
Smac/DIABLO	Pan-specific	−78	Yes	Pan-specific	327
TYK2	Pan-specific	−81	Akt2 (PkBb)	Pan-specific	386
VEGFR2	Y1214	−74	PTP-PEST	Pan-specific	669
GSK3a	T19+pS21	−71			
SPHK2	Pan-specific	−76			
TTK	Pan-specific	−75			
VGFR3	Pan-specific	−76			
Src	Pan-specific	−74			
Rad17	S645-72	−72			
p38a MAPK	T180/Y182	−65			
Smad2/3	Pan-specific	−67			

Table 1. *Cont.*

Protein Name	Phospho Site	%CFC	Protein Name	Phospho Site	%CFC
	Decreased		Increased		
MEK6	Pan-specific	−66			
Sik2	Pan-specific	−68			
WNK3	Pan-specific	−67			
Smad1	S465	−59			
ERK2	Pan-specific	−68			
RONa	Pan-specific	−62			
Tau	S422	−62			
ZIPK	Pan-specific	−65			
SHIP2	Pan-specific	−64			
PLC R(PLCγ2)	Pan-specific	−61			
STAT3	Pan-specific	−61			
EGFR	Pan-specific	−63			
SG2NA	Pan-specific	−61			
DUSP3	Pan-specific	−60			
Tau	S199/202	−57			
Jun	S243	−53			
Rb	T821	−57			

Matched Western blot analysis using protein extracts from HaCaT cells highlighted that TrkB phosphorylation at tyrosine 816 (Y816) was significantly downregulated in the wiskostatin-treated samples (p = 0.0011, n = 3; Figure 6B). This signal was lost in the samples treated with CIP, a reagent which removes phosphorylation from protein samples, whereas no effect on total protein expression was seen in the actin-positive control bands. This demonstrated that this signal, which was detected using the ABN1381 TrkB Y816 antibody, was a phosphorylation signal, as intended. Furthermore, no significant difference between the TrkB Y816 expression in samples from wiskostatin and control-treated cells that were exposed to hydrogen peroxide and sodium orthovanadate was detected (Figure 6C). This suggested that the differential expression of TrkB Y816 that was detected between wiskostatin and control-treated samples was due to a change in the phosphorylation of TrkB at tyrosine 816 rather than a change in total protein. The TrkB total protein was found to have a 23% CFC with a z-ratio of 0.27, which was not deemed significant (Figure 6E). The change in TrkB Y816 expression in the wiskostatin-treated cells from the controls shown in Figure 6D demonstrates that the tyrosine 816 phosphorylation of TrkB was affected by wiskostatin treatment.

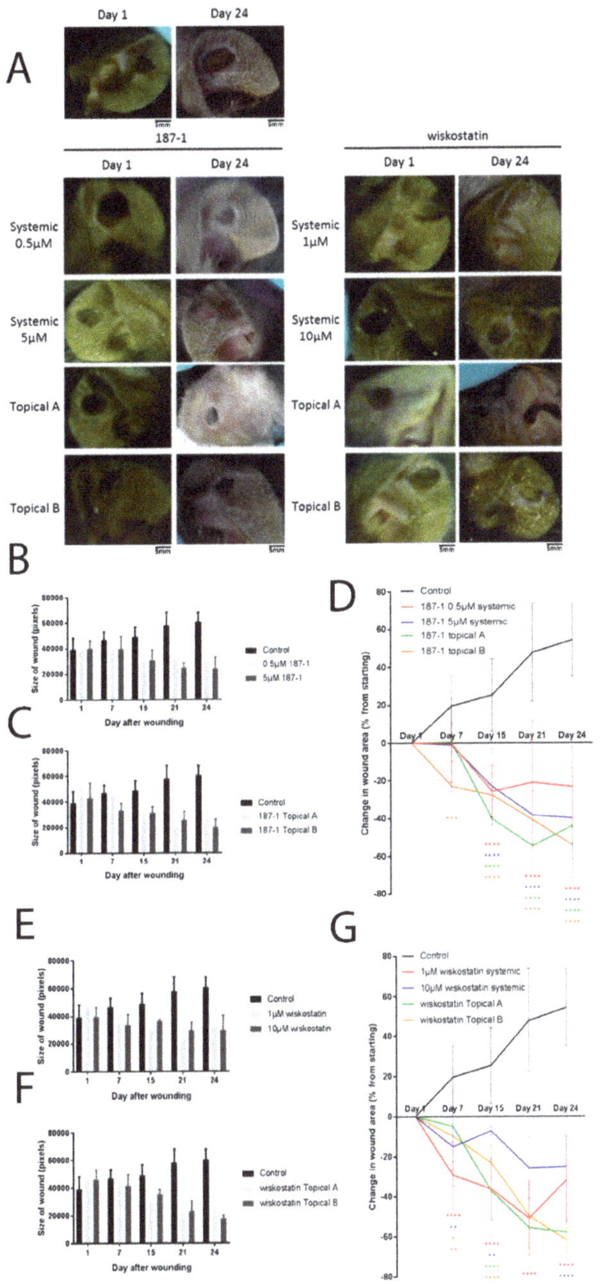

Figure 5. nWASP inhibition and the change in healing rates in a murine mouse model. The effect on the size of the hole punch wound in pixels (mean +/− SD, $n = 6$) following the systemic (**A**) and topical (**B**) application of 187-1. Percentage changes in wound area from the first measurement after wounding (day 1) (**C**). The effect of the administration of wiskostatin is shown in (**D–G**). * indicates statistical significance (* $p < 0.05$, ** $p < 0.01$, *** $p < 0.001$ **** $p < 0.0001$).

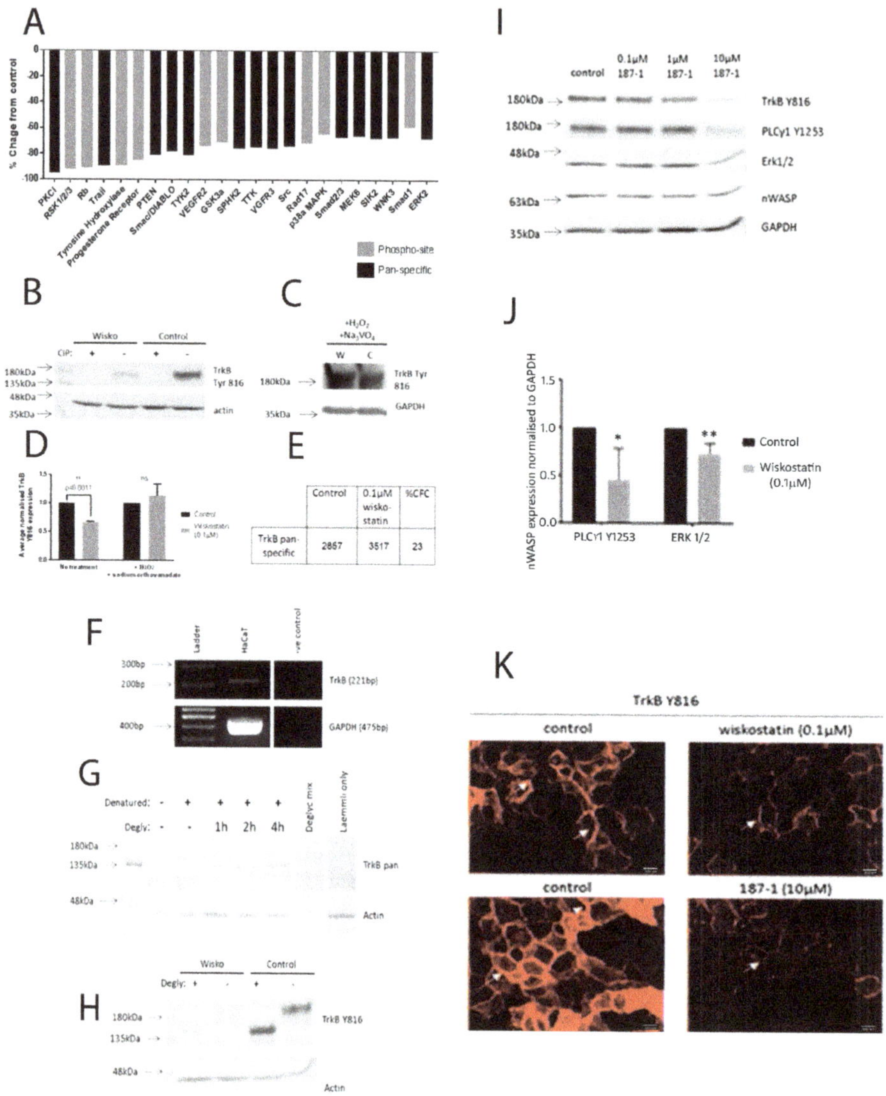

Figure 6. Protein signalling changes in response to the wiskostatin treatment. (**A**) Changes in the total protein expression or phosphorylation sites in response to the 0.1 μM wiskostatin treatment for 1 h according to the %CFC. Protein samples that were treated with CIP at 0.6 units/μg for 1 h (**B**). Positive controls (**C**). Integrated densities as a change from the control (**D**). Globally normalised expression of total TrkB protein and %CFC according to protein array analysis (**E**). TrkB transcript expression in the HaCaT cell line (**F**). Protein extracts from HaCaT cells were treated with denaturation and deglycosylation reagents (**G**). Samples from cells treated with 0.1 μM wiskostatin/control were denatured and then deglycosylated to examine the TrkB Y816 activity (**H**). Western blot analysis of TrkB Y816, PLCγ1 Y1253 and Erk1/2 of HACAT cells treated with 0.1 μM wiskostatin or 10 μM 187-1 (**I**,**J**). Immunofluorescence analysis of TrkB Y816 expression in HaCaT cells serum starved for 4 h and treated with nWASP inhibitors; scale bar indicates 50 μm (**K**). * indicates statistical significance (* $p < 0.05$, ** $p < 0.01$,.ns: not significant).

3.9. TrkB Was Expressed in the HaCaT Cells and Was Glycosylated

In order to explain the larger-than-expected band observed in the Western blot results, experiments were carried out to validate whether TrkB was expressed in the HaCaT cell line and then to explore whether the TrkB protein may have existed in a glycosylated state. TrkB was found to be expressed in HaCaT cells at the transcript level, as shown in Figure 6F. Further experiments were carried out to determine why the observed band size for TrkB Y816 was at 180 kDa, higher than expected based on the literature, and why no corresponding band size could be seen using TrkB-pan antibodies. HaCaT protein samples were treated with deglycosylation reagents for 0–4 h and then Western blotting using TrkB-pan antibodies was carried out (Figure 6G). After 1 h of deglycosylation treatment, TrkB expression could be detected at 140 kDa using a TrkB-pan antibody, whereas no signal could be detected without deglycosylation treatment. This signal was not detectable using the usual, previously described protein extraction and denaturation techniques, which did not include a deglycosylation step. This experiment showed that TrkB was in fact expressed in HaCaT cells but was not previously detected using the ABN6180 TrkB antibody due to glycosylation, which may have blocked access of the antibody to the target site. Having shown that TrkB was present in HaCaT cells with a size of 140 kDa following deglycosylation, it was hypothesised that the TrkB Y816 signal should shift size from 180 kDa to 140 kDa after deglycosylation. Protein samples collected from HaCaT cells treated with 0.1 μM wiskostatin/vehicle control were denatured and treated with deglycosylation reagents for 1 h and then probed for TrkB-Y816 and actin as a positive control. Following the deglycosylation treatments, there was an obvious shift in the size of the protein detected using the TrkB-Y816 antibody from 180 kDa to 140 kDa (Figure 6H). No significant change in the level of phosphorylation of Y816 was observed following deglycosylation but the reduced phosphorylation at Y816 in the wiskostatin-treated cells compared with the control was still apparent, demonstrating that TrkB was glycosylated in the HaCaT cells' larger protein band.

3.10. The TrkB Signalling Pathway Was Affected by Confluency, Serum Availability and nWASP Inhibitor Treatment in the HaCaT Cells

Having identified that wiskostatin treatment can alter the phosphorylation of TrkB at tyrosine 816 in HaCaT cells, an investigation into other protein signals that are related to the TrkB pathway was carried out. The effect of the additional nWASP inhibitor 187-1 over a range of concentrations (0, 0.1, 1 and 10 μM) on TrkB phosphorylation and signalling was also examined (Figure 6I). The effect on PLCγ1 phosphorylation at tyrosine 1253 and Erk1/2 following 1 h of 0.1 μM wiskostatin or a range of 187-1 treatments after 4 h serum starvation mirrored those observed at the TrkBY816 site in that the nWASP inhibitor treatment significantly decreased the expressions of TrkB Y816, PLCγ1 Y1253 and Erk1/2 (Figure 6H). This effect on TrkB Y816 signalling was only observed at the 10 μM 187-1 concentration. No effect on nWASP expression was found. PLCγ1 Y1253 and Erk1/2 expressions were significantly decreased in the wiskostatin-treated cells (Figure 6J). No significant difference in the PLCγ1 Y1253 expression between the wiskostatin and control cells was seen (Figure 6J). This suggested that the differential expression of PLCγ1 Y1253 that was detected between the wiskostatin-treated and control samples was due to a change in the phosphorylation status rather than a change in the total protein. Immunofluorescence studies confirmed the effect of nWASP inhibitors (Figure 6K).

3.11. BDNF Was Linked to TrkB Signalling and Functional Changes in the HaCaT Cells

BDNF, as a putative ligand for TrkB and TrkB transcript expression, was examined using PCR in HaCaT cells under 0, 2 and 4 h serum starvation with/without 0.1 μM wiskostatin treatments for 1 h (Figure 7A). Analysis using conventional PCR revealed that the BDNF expression decreased according to the serum starvation in the HaCaT cells ($n = 2$, Figure 7A). The qPCR analysis results were normalised to the 0 h serum starvation (Figure 7B) and to the controls to compare with the 0.1 μM wiskostatin-treated samples

(Figure 7C). BDNF was expressed in the HaCaT cells under normal, serum-supplemented culture conditions but following serum withdrawal, the expression of the BDNF transcript was significantly reduced. The treatment with 0.1 μM wiskostatin appeared to have a slight effect on reducing the BDNF transcript expression during normal culture conditions and on increasing the BDNF expression following serum starvation in the qPCR analysis (Figure 7C). The TrkB transcript expression was found to increase following the serum withdrawal—significantly so after 4 h (Figure 7D). The 0.1 μM wiskostatin treatment had no effect on the TrkB transcript under normal culture conditions, but after 2 h, TrkB expression was found to increase when cells were treated with wiskostatin (Figure 7E). Under normal serum conditions, TrkB Y816 phosphorylation was found to be relatively low. The addition of BDNF to HaCaT cells growing at low confluency under these conditions induced TrkB Y816 signalling (Figure 7F). After serum starvation, where TrkB Y816 phosphorylation increased, the addition of BDNF reduced the TrkB Y816 signalling in a similar fashion to the nWASP inhibitor application (Figure 7F). The addition of BDNF between 50 and 150 ng/mL concentrations increased the resistance measured at 4 kHz following cell seeding (Figure 7G,H). The effect of dual treatments with 0.1 μM wiskostatin and 50 ng/mL or 100 ng/mL BDNF was also observed (Figure 7J,K, respectively).

3.12. nWASP Inhibitor Treatment May Have Affected the TrkB Signalling through Grb2

In an attempt to explore how nWASP inhibitors may be able to alter the phosphorylation status of TrkB, the effect of SOS1, which is a small-molecule inhibitor that acts to inhibit the activity of Grb2, on TrkB Y816 was examined. The SOS1 treatment at 0.1 and 1 μM concentrations inhibited TrkB Y816 phosphorylation in the HaCaT cells at low confluency and under serum starvation (Figure 7L); the MSOS1 treatments appeared to increase the attachment and spreading of HaCaT cells (Figure 7N).

3.13. nWASP Knockdown and TrkB Signalling

The effect of nWASP knockdown on TrkB signalling in HaCaT cells was examined. The cells were serum starved for 4 h and the effect on TrkB Y816 and PLCγ1 Y1253 phosphorylation was then examined using Western blotting (Figure 7O). nWASP knockdown was found to increase TrkB signalling through tyrosine 816 under serum-starved and low confluency conditions. TrkB Y816 activity was examined following prolonged nWASP inhibitor and serum starvation treatment for 4 h. These results demonstrated that the nWASP inhibitor treatment had the same effect of inhibiting TrkB Y816 phosphorylation even when applied from the beginning of the serum withdrawal period (Figure 7).

3.14. TrkB Expression in Human Chronic Wound Tissues

Following the discovery of a link between nWASP and TrkB and that nWASP is overexpressed in non-healing chronic wounds, the expression of TrkB and BDNF in chronic wounds was examined using IHC on human chronic wound samples. Overall, BDNF showed increased expression in healing (distal) compared with non-healing wounds (Figure 8A,B). TrkB was found to be expressed at higher levels in the basal epidermal cells at sites distal from the wound edge in samples taken from non-healing wound tissues compared with healing tissues (Figure 8C,D).

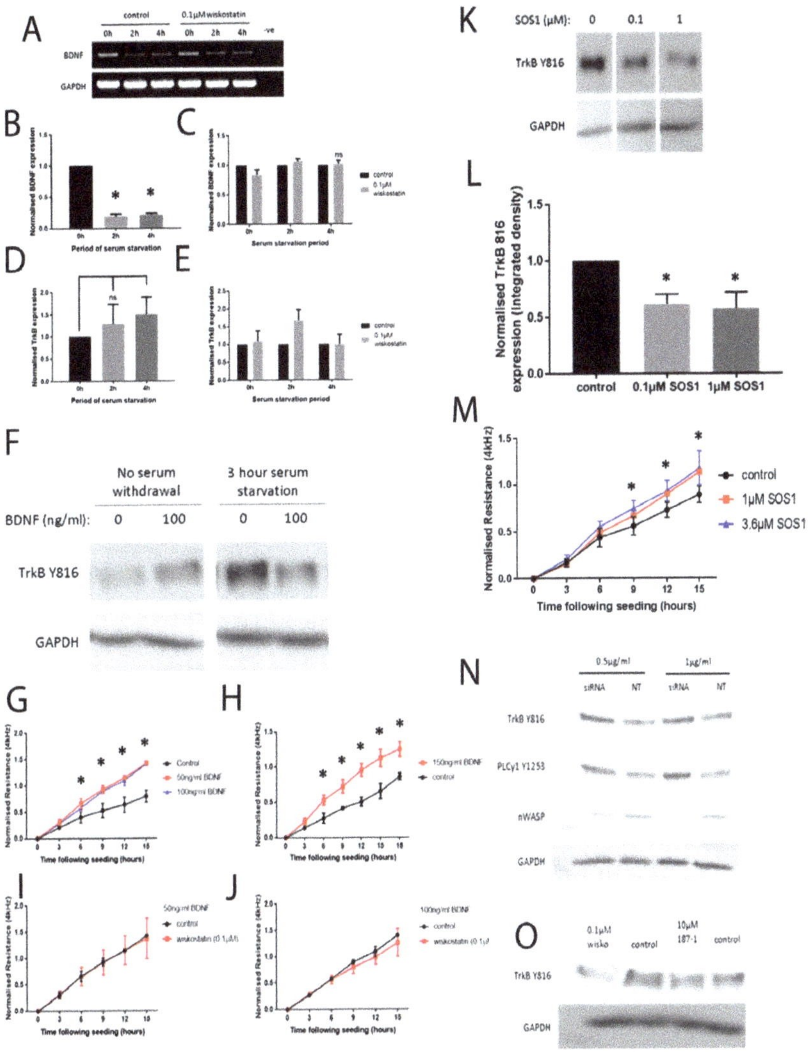

Figure 7. BDNF and TrkB transcript expression in the HaCaT cells and the effects of the BDNF, SOS1 and knockdown models. PCR/qPCR results for BDNF and TrkB expression in HaCaT cells treated with 0.1 µM wiskostatin (**A**,**B**); wiskostatin-treated samples at each serum starvation time point (**C**). qPCR analysis results of the TrkB transcript expression (**D**) and controls (**E**). Western blot analysis of TrkB Y816 (**F**). HaCaT cells were treated with rhBDNF at 50 and 100 ng/mL (**G**) and 150 ng/mL (**H**). Cells treated with 50 or 100 ng/mL BDNF and 0.1 µM wiskostatin (**I**,**J**). TrkB phosphorylation and SOS1 (**K**) and the integrated density (**L**). HaCaT cells treated with SOS1 (**M**). The effect of nWASP inhibition on TrkB Y816 phosphorylation (**N**). The effect of 0.1 µM wiskostatin and 10 µM 187-1 on TrkB Y816 phosphorylation (**O**). * indicates statistical significance ($p < 0.05$). ns: not significant.

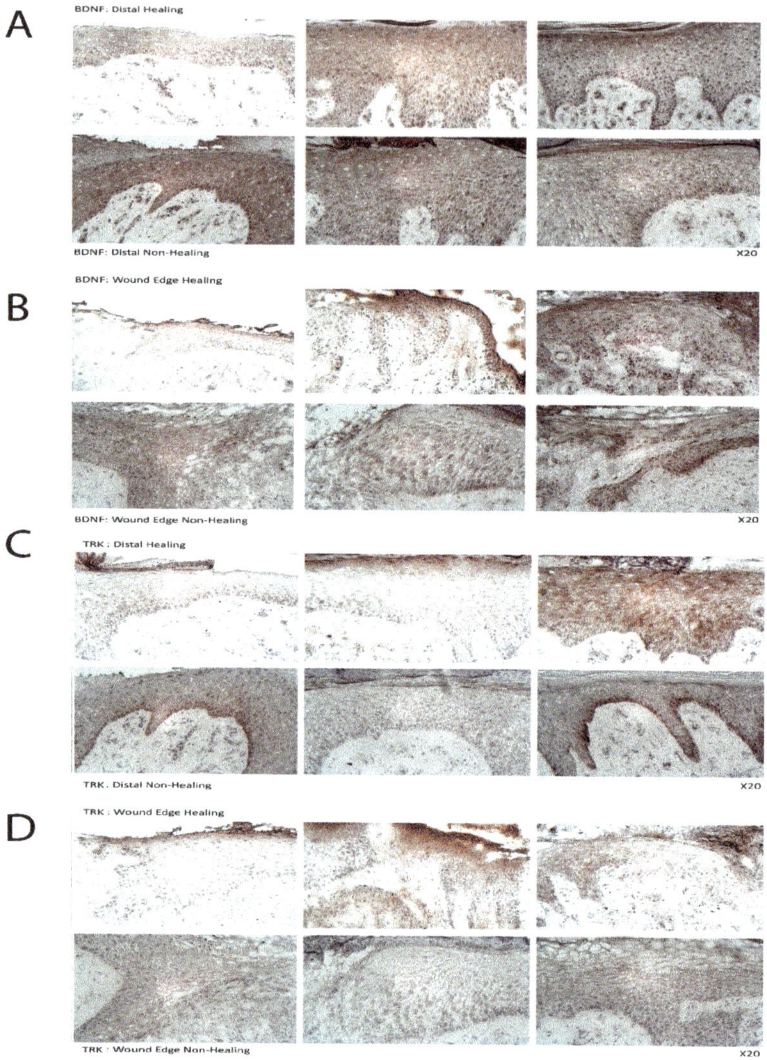

Figure 8. BDNF and TrkB expression in human chronic wound tissues. IHC of BDNF protein expression in healing and non-healing chronic wound tissues at distal (**A**) and wound-edge (**B**) locations. IHC of TrkB at distal (**C**) and wound-edge (**D**) locations. Arrows indicate areas of high-intensity staining. Two representative images of each group are shown; ×20 magnification was used.

4. Discussion

This study identified nWASP as a molecule that is important in human wound healing and recognises nWASP as a new molecular target to encourage healing in human chronic wounds. Non-healing chronic wounds showed higher expression of nWASP than healing chronic wounds. This quantitative transcript analysis suggests that a balance of nWASP activity in human wounds may be key for healthy wound healing behaviour with over-expression above a certain level indicative of impaired wound healing. Hence, this study proposes that targeting nWASP in a clinical setting may provide a means to encourage healing behaviour in chronic wounds.

In vitro work carried out in this study supported the idea that altering nWASP activity through inhibitor treatment and at a transcript expression level can affect human keratinocyte and endothelial behaviour. The reduction of nWASP activity through nWASP inhibition and transcript expression knockdown affected the cell spreading and attachment properties of cells. Decreased nWASP activity from wild-type keratinocyte levels through transcript knockdown or inhibition appeared to increase the adherent properties of cells; furthermore, the knockdown of nWASP in endothelial cells caused the opposite effect in that attachment and spreading appeared to be impaired. This suggests that nWASP may have a role in the attachment properties of cells to a surface. This could explain how a balance of nWASP expression is important for correct wound healing behaviour, as the optimal level of adhesion and attachment of cells were shown to be a key factor in their migratory properties [9,10]. The implication that nWASP may be involved in affecting these properties of cells may explain how this molecule is involved in wound healing in humans. The molecular mechanisms involved in nWASP activity affecting the attachment properties of keratinocytes in this context are currently unknown but the diverse roles of nWASP and the effects of the actin polymerisation that can result from its activation offer many potential explanations for how nWASP may influence cell attachment and motility.

Using *ex vivo* and in vivo models, this study also demonstrated how nWASP inhibition can encourage wound healing behaviour in tissues with impaired healing abilities and thus provides a new management model for chronic wounds. Using mouse models to study the effect of nWASP inhibitor treatment on wound healing has not only supported previous findings that nWASP has potential as a target in wound healing but highlights a very simple but possibly life-changing therapy option that has potential in human cases. The fact that commercially available nWASP inhibitors can be applied so simply using carrier gels that are already used to treat chronic wounds in the clinic but with such a dramatic effect on encouraging healing in wounds in mice that naturally exhibit impaired wound healing abilities is very promising.

To begin to understand how nWASP inhibitors affect cell behaviour and may act in the context of a chronic wound, an investigation into the effect of altering nWASP activity on the signalling mechanisms in cell models that represent the skin and wound environment was carried out. Initially, a protein array highlighted changes in the expression and phosphorylation of hundreds of kinases and other common signalling proteins following treatment with wiskostatin in HaCaT cells. Of the most significantly altered proteins, several proteins belonging to very common signalling pathways were highlighted, including members of the Ras–Raf–Mek–Erk, Akt-mTOR and Jak–STAT pathways. Numerous receptors and membrane-bound proteins were also identified as being significantly altered by wiskostatin treatment, including VEGFR2 and -3, EGFR and the progesterone receptor, which have been shown to be upstream of these common signalling cascades. TrkB signalling was found to be significantly altered by nWASP inhibitor treatments in HaCaT cells and, as such, the effect of nWASP activity on TrkB signalling became the focus. The findings that were reported here appeared to demonstrate that the TrkB activity in HaCaT cells was extremely sensitive to confluency, serum starvation, BDNF application and nWASP inhibition. Furthermore, the TrkB activity is found to be increased in basal keratinocytes in chronic wound tissues. How TrkB, BDNF and nWASP activity may be linked in the context of chronic wounds requires significant further investigation, especially considering the sensitivity to factors such as confluency and serum availability of the TrkB pathway. The novel relationship between nWASP and TrkB and the downstream PLCγ1 signalling pathway that is affected by changes in signalling may have implications in terms of cell differentiation. As such, further investigation into the factors that affect this signalling and how this may translate into the wound environment needs to be explored.

It is clear that nWASP and TrkB signalling are linked in HaCaT cells and possibly in the chronic wound environment. How nWASP activity alters TrkB signalling and why this pathway is sensitive to factors such as confluency and serum starvation are yet to be answered. The established role of nWASP in endocytosis and receptor trafficking could represent one such potential link. nWASP and TrkB were shown to share numerous interaction partners that may facilitate this activity, for instance, Grb2, pacsin and Nck. A link between nWASP and TrkB through the common binding partner Grb2 has begun to be explored through the use of SOS1. This is an inhibitor of Grb2 activity that acts via blocking the SH3 binding domain of the guanine nucleotide exchange factor SOS, which is the ligand for the adaptor protein Grb2. This blocks the SOS/Grb2 interaction and prevents Ras activation via receptor tyrosine kinases, such as TrkB. SOS1 was found to act in the same way as nWASP inhibition in that TrkB Y816 phosphorylation was significantly reduced at low confluency and after 4 h serum starvation in HaCaT cells. Similarly, Grb2 inhibition was found to increase the resistance of HaCaT cells using ECIS, similar to nWASP inhibitors. This is only the first step in exploring the link between TrkB and nWASP, but this work identified an avenue for further study.

In summary, this study demonstrated that nWASP activity in human wounds can be indicative of its ability to heal effectively and, as a result, using nWASP inhibitors can influence the healing behaviour of wounds. Functional assays indicated that the mechanism through which nWASP may influence wound healing behaviour is potentially via affecting the attachment properties of cells and that this can be mediated through the BDNf/TrkB pathway. This study recognised nWASP as an important therapeutic target in human chronic wounds and proposes the use of nWASP inhibitors as a simple and effective means to encourage wound healing in difficult wounds.

Author Contributions: Conceptualisation, T.A.M., K.G.H. and W.G.J.; methodology, B.A.F., T.A.M., S.O., F.R. and A.J.S.; formal analysis, B.A.F., W.G.J., T.A.M. and A.J.S.; investigation, B.A.F., T.A.M. and W.G.J.; data curation, B.A.F., W.G.J. and T.A.M.; writing—original draft preparation, B.A.F. and T.A.M.; writing—review and editing, T.A.M., W.G.J. and F.R.; supervision, T.A.M. and W.G.J.; project administration, T.A.M.; funding acquisition, T.A.M. and W.G.J. and K.G.H. All authors have read and agreed to the published version of the manuscript.

Funding: This research was funded by the Life Sciences Network Wales, grant number Jiang2014.

Institutional Review Board Statement: The study was conducted in accordance with the Declaration of Helsinki and approved by the Ethics Committee of the University Hospital of Wales with ethical approval from the SouthEast Wales Research Ethics Committee (Panel C) with ethical approval IDs: 04/WSE03/92 and 09/WSE02/5.

Informed Consent Statement: Informed consent was obtained from all subjects whose tissues were involved in the study.

Data Availability Statement: Not applicable.

Conflicts of Interest: The authors declare no conflict of interest. The funders had no role in the design of the study; in the collection, analyses, or interpretation of data; in the writing of the manuscript; or in the decision to publish the results.

References

1. Shaw, T.; Martin, P. Wound repair at a glance. *J. Cell Sci.* **2009**, *122*, 3209–3213. [CrossRef] [PubMed]
2. Bryant, R.; Nix, D. *Acute & Chronic Wounds: Current Management Concepts*, 4th ed.; Elsevier Mosby: St. Louis, MO, USA, 2012.
3. Barrientos, S.; Stojadinovic, O.; Golinko, M.S.; Brem, H.; Tomic-Canic, M. Growth factors and cytokines in wound healing. *Wound Repair Regen.* **2008**, *16*, 585–601. [CrossRef] [PubMed]
4. Jacinto, A.; Martinez-Arias, A.; Martin, P. Mechanisms of epithelial fusion and repair. *Nat. Cell Biol.* **2001**, *3*, E117–E123. [CrossRef] [PubMed]
5. Behm, B.; Babilas, P.; Landthaler, M.; Schreml, S. Cytokines, chemokines and growth factors in wound healing. *J. Eur. Acad. Dermatol. Venereol.* **2012**, *26*, 812–820. [CrossRef]
6. Velnar, T.; Bailey, T.; Smrkolj, V. The Wound Healing Process: An Overview of the Cellular and Molecular Mechanisms. *J. Int. Med. Res.* **2009**, *37*, 1528–1542. [CrossRef]
7. Ridley, A.J.; Schwartz, M.A.; Burridge, K.; Firtel, R.A.; Ginsberg, M.H.; Borisy, G.; Parsons, J.T.; Horwitz, A.R. Cell Migration: Integrating Signals from Front to Back. *Science* **2003**, *302*, 1704–1709. [CrossRef]
8. Mierke, C.T. The role of focal adhesion kinase in the regulation of cellular mechanical properties. *Phys. Biol.* **2013**, *10*, 065005. [CrossRef]

9. Palecek, S.P.; Loftus, J.C.; Ginsberg, M.H.; Lauffenburger, D.A.; Horwitz, A.F. Integrin-ligand binding properties govern cell migration speed through cell-substratum adhesiveness. *Nature* **1997**, *385*, 537–540. [CrossRef]
10. Holly, S.P.; Larson, M.K.; Parise, L.V. Multiple Roles of Integrins in Cell Motility. *Exp. Cell Res.* **2000**, *261*, 69–74. [CrossRef]
11. Brissett, A.; Hom, D. The effects of tissue sealants, platelet gels, and growth factors on wound healing. *Curr. Opin. Otolaryngol. Head Neck Surg.* **2003**, *11*, 245–250. [CrossRef]
12. Pradhan, L.; Nabzdyk, C.; Andersen, N.D.; LoGerfo, F.W.; Veves, A. Inflammation and neuropeptides: The connection in diabetic wound healing. *Expert Rev. Mol. Med.* **2009**, *11*, e2. [CrossRef] [PubMed]
13. Guo, S.; DiPietro, L.A. Factors Affecting Wound Healing. *J. Dent. Res.* **2010**, *89*, 219–229. [CrossRef] [PubMed]
14. Reed, B.R.; Clark, R.A. Cutaneous tissue repair: Practical implications of current knowledge. II. *J. Am. Acad. Dermatol.* **1985**, *13*, 919–941. [CrossRef] [PubMed]
15. Moreo, K. Understanding and overcoming the challenges of effective case management for patients with chronic wounds. *Case Manag.* **2005**, *16*, 62–67. [CrossRef] [PubMed]
16. Clark, J.J. Wound Repair and Factors Influencing Healing. *Crit. Care Nurs. Q.* **2002**, *25*, 1–12. [CrossRef]
17. Järbrink, K.; Ni, G.; Sönnergren, H.; Schmidtchen, A.; Pang, C.; Bajpai, R.; Car, J. The humanistic and economic burden of chronic wounds: A protocol for a systematic review. *Syst. Rev.* **2017**, *6*, 15. [CrossRef]
18. Posnett, J.; Franks, P.J. The burden of chronic wounds in the UK. *Nurs. Times* **2008**, *104*, 44–45.
19. Franks, P.; Morgan, P. Health-related quality of life with chronic leg ulceration. *Expert Rev. Pharm. Outcomes Res.* **2003**, *3*, 611–622. [CrossRef]
20. E Price, P.; Fagervik-Morton, H.; Mudge, E.J.; Beele, H.; Ruiz, J.C.; Nystrøm, T.H.; Lindholm, C.; Maume, S.; Melby-Østergaard, B.; Peter, Y.; et al. Dressing-related pain in patients with chronic wounds: An international patient perspective. *Int. Wound J.* **2008**, *5*, 159–171. [CrossRef]
21. Green, J.; Jester, R. Health-related quality of life and chronic venous leg ulceration: Part 2. *Br. J. Community Nurs.* **2012**, *15*, S4–S14. [CrossRef]
22. Drew, P.; Posnett, J.; Rusling, L. Wound Care Audit T: The cost of wound care for a local population in England. *Int. Wound J.* **2007**, *4*, 149–155. [CrossRef] [PubMed]
23. Gottrup, F.; Holstein, P.; Jørgensen, B.; Lohmann, M.; Karlsmar, T. A New Concept of a Multidisciplinary Wound Healing Center and a National Expert Function of Wound Healing. *Arch. Surg.* **2001**, *136*, 765–772. [CrossRef] [PubMed]
24. Harding, K.G.; Morris, H.L.; Patel, G.K. Science, medicine, and the future: Healing chronic wounds. *BMJ* **2002**, *324*, 160–163. [CrossRef] [PubMed]
25. Fernandez, M.; Broadbent, J.; Shooter, G.; Malda, J.; Upton, Z. Development of an enhanced proteomic method to detect prognostic and diagnostic markers of healing in chronic wound fluid. *Br. J. Dermatol.* **2008**, *158*, 281–290. [CrossRef]
26. Huang, C.-M.; Wang, C.-C.; Barnes, S.; Elmets, C.A. In vivo detection of secreted proteins from wounded skin using capillary ultrafiltration probes and mass spectrometric proteomics. *Proteomics* **2006**, *6*, 5805–5814. [CrossRef]
27. Jiang, W.; Harding, K. Method and kit for the classification and prognosis of wounds. Patent number EP10754777A, 6 November 2023.
28. Derry, J.M.; Ochs, H.D.; Francke, U. Isolation of a novel gene mutated in Wiskott-Aldrich syndrome. *Cell* **1994**, *78*, 635–644. [CrossRef]
29. Aldrich, R.A.; Steineberg, A.G.; Campbell, D.C. Pedigree demonstrating a sex-linked recessive condition characterized by draining ears, eczematoid dermatitis and bloody diarrhea. *Pediatrics* **1954**, *13*, 133–139. [CrossRef]
30. Sullivan, K.E.; Mullen, C.A.; Blaese, R.; Winkelstein, J.A. A multiinstitutional survey of the Wiskott-Aldrich syndrome. *J. Pediatr.* **1994**, *125*, 876–885. [CrossRef]
31. Orange, J.; Stone, K.; Turvey, S.; Krzewski, K. The Wiskott-Aldrich syndrome. *Cell Mol. Life Sci.* **2004**, *61*, 2361–2385. [CrossRef]
32. Zhu, Q.; Watanabe, C.; Liu, T.; Hollenbaugh, D.; Blaese, R.; Kanner, S.; Aruffo, A.; Ochs, H. Wiskott-Aldrich syndrome/X-linked thrombocytopenia: WASP gene mutations, protein expression, and phenotype. *Blood* **1997**, *90*, 2680–2689. [CrossRef]
33. Miki, H.; Miura, K.; Takenawa, T. N-WASP, a novel actin-depolymerizing protein, regulates the cortical cytoskeletal rearrangement in a PIP2-dependent manner downstream of tyrosine kinases. *EMBO J.* **1996**, *15*, 5326–5335. [CrossRef]
34. Takenawa, T.; Suetsugu, S. The WASP–WAVE protein network: Connecting the membrane to the cytoskeleton. *Nat. Rev. Mol. Cell Biol.* **2007**, *8*, 37–48. [CrossRef] [PubMed]
35. Prehoda, K.; Scott, J.; Mullins, R.; Lim, W. Integration of multiple signals through cooperative regulation of the N-WASP-Arp2/3 complex. *Science* **2010**, *290*, 801–806. [CrossRef] [PubMed]
36. Machesky, L.; Insall, R. Scar1 and the related Wiskott-Aldrich syndrome protein, WASP, regulate the actin cytoskeleton through the Arp2/3 complex. *Curr. Biol.* **1998**, *8*, 1347–1356. [CrossRef] [PubMed]
37. Takenawa, T.; Miki, H. WASP and WAVE family proteins: Key molecules for rapid rearrangement of cortical actin filaments and cell movement. *J. Cell Sci.* **2001**, *114*, 1801–1809. [CrossRef] [PubMed]
38. Rohatgi, R.; Ma, L.; Miki, H.; Lopez, M.; Kirchhausen, T.; Takenawa, T.; Kirschner, M. The interaction between N-WASP and the Arp2/3 complex links Cdc42-dependent signals to actin assembly. *Cell* **1999**, *97*, 221–231. [CrossRef] [PubMed]
39. Yarar, D.; To, W.; Abo, A.; Welch, M. The Wiskott-Aldrich syndrome protein directs actin-based motility by stimulating actin nu-cleation with the Arp2/3 complex. *Curr. Biol.* **1999**, *9*, 555–558. [CrossRef] [PubMed]
40. Raftopoulou, M.; Hall, A. Cell migration: Rho GTPases lead the way. *Dev. Biol.* **2004**, *265*, 23–32. [CrossRef]

41. Martin, T.A.; Frugtniet, B.; Jiang, W.G. Role of the WASP and WAVE family proteins in breast cancer invasion and metastasis. *Breast Cancer* **2015**, *7*, 99–109. [CrossRef]
42. Kurisu, S.; Takenawa, T. WASP and WAVE family proteins: Friends or foes in cancer invasion? *Cancer Sci.* **2012**, *101*, 2093–2104. [CrossRef]
43. Albiges-Rizo, C.; Destaing, O.; Fourcade, B.; Planus, E.; Block, M.R. Actin machinery and mechanosensitivity in invadopodia, podosomes and focal adhesions. *J. Cell Sci.* **2009**, *122*, 3037–3049. [CrossRef] [PubMed]
44. Peterson, J.R.; Bickford, L.C.; Morgan, D.; Kim, A.S.; Ouerfelli, O.; Kirschner, M.W.; Rosen, M.K. Chemical inhibition of N-WASP by stabilization of a native autoinhibited conformation. *Nat. Struct. Mol. Biol.* **2004**, *11*, 747–755. [CrossRef] [PubMed]
45. Conway, K.; Ruge, F.; Price, P.; Harding, K.G.; Jiang, W.G. Hepatocyte growth factor regulation: An integral part of why wounds become chronic. *Wound Repair Regen.* **2007**, *15*, 683–692. [CrossRef] [PubMed]
46. Conway, K.P.; Price, P.; Harding, K.G.; Jiang, W.G. The role of vascular endothelial growth inhibitor in wound healing. *Int. Wound J.* **2007**, *4*, 55–64. [CrossRef]
47. Sanders, A.J.; Jiang, D.G.; Jiang, W.G.; Harding, K.G.; Patel, G.K. Activated leukocyte cell adhesion molecule impacts on clinical wound healing and inhibits HaCaT migration. *Int. Wound J.* **2001**, *8*, 500–507. [CrossRef]
48. Harding, K.; Jiang, W. Enhancement of wound tissue expansion and angiogenesis by matrix-embedded fibroblast (dermagraft), a role of hepatocyte growth factor/scatter factor. *Int. J. Mol. Med.* **1998**, *2*, 203–413. [CrossRef]
49. Van Meerloo, J.; Kaspers, G.; Cloos, J. Cell sensitivity assays: The MTT assay. *Methods Mol. Biol.* **2001**, *731*, 237–245.
50. Cho, C.-H.; Sung, H.-K.; Kim, K.-T.; Cheon, H.G.; Oh, G.T.; Hong, H.J.; Yoo, O.-J.; Koh, G.Y. COMP-angiopoietin-1 promotes wound healing through enhanced angiogenesis, lymphangiogenesis, and blood flow in a diabetic mouse model. *Proc. Natl. Acad. Sci. USA* **2006**, *103*, 4946–4951. [CrossRef]

biomolecules

MDPI

Article

The Role of a Cytokinin Antagonist in the Progression of Clubroot Disease

Jana Bíbová [1], Veronika Kábrtová [1], Veronika Večeřová [1], Zuzana Kučerová [2], Martin Hudeček [1], Lenka Plačková [1], Ondřej Novák [1], Miroslav Strnad [1,*] and Ondřej Plíhal [1,*]

[1] Laboratory of Growth Regulators, Faculty of Science, Institute of Experimental Botany of the Czech Academy of Sciences, Palacký University, Šlechtitelů 27, CZ-78371 Olomouc, Czech Republic

[2] Department of Biophysics, Faculty of Science, Palacký University, Šlechtitelů 27, CZ-78371 Olomouc, Czech Republic

* Correspondence: miroslav.strnad@upol.cz (M.S.); ondrej.plihal@upol.cz (O.P.)

Abstract: *Plasmodiophora brassicae* is an obligate biotrophic pathogen causing clubroot disease in cruciferous plants. Infected plant organs are subject to profound morphological changes, the roots form characteristic galls, and the leaves are chlorotic and abscise. The process of gall formation is governed by timely changes in the levels of endogenous plant hormones that occur throughout the entire life cycle of the clubroot pathogen. The homeostasis of two plant hormones, cytokinin and auxin, appears to be crucial for club development. To investigate the role of cytokinin and auxin in gall formation, we used metabolomic and transcriptomic profiling of *Arabidopsis thaliana* infected with clubroot, focusing on the late stages of the disease, where symptoms were more pronounced. Loss-of-function mutants of three cytokinin receptors, AHK2, AHK3, and CRE1/AHK4, were employed to further study the homeostasis of cytokinin in response to disease progression; *ahk* double mutants developed characteristic symptoms of the disease, albeit with varying intensity. The most susceptible to clubroot disease was the *ahk3 ahk4* double mutant, as revealed by measuring its photosynthetic performance. Quantification of phytohormone levels and pharmacological treatment with the cytokinin antagonist PI-55 showed significant changes in the levels of endogenous cytokinin and auxin, which was manifested by both enhanced and reduced development of disease symptoms in different genotypes.

Keywords: clubroot; *Plasmodiophora brassicae*; cytokinin; auxin; PI-55; photosynthesis

1. Introduction

Plasmodiophora brassicae is an obligate biotrophic protist that is the causal agent of clubroot disease, one of the most devastating diseases in the Brassicaceae family. The clubroot disease is spread worldwide, probably as a result of the transport of diseased plant material by early colonists [1]. To date, considerable progress has been made in understanding the biology of this pathogen [2–5], but the complex mechanisms used by the pathogen during the infection process are still not fully understood [6]. It is becoming clear that the clubroot pathogen modulates host immunity responses, as suggested by the identification of a *P. brassicae* methyltransferase responsible for the methylation of salicylic acid (SA) and subsequent production of methyl salicylate (MeSA), leading to the control of free SA levels in infected galls [7].

As a biotrophic parasite, the clubroot pathogen requires a plant host to reproduce, and the infection process can be divided into two main phases: (I) the primary infection and (II) the secondary infection of the root cortex [8]. While the primary infection is restricted to the root hairs of the plant host, the secondary phase is primarily targeted to the cortex of hypocotyl and roots [9]. Infection by clubroot leads to profound changes in meristematic activity that imply changes in plant anatomy, and disease symptoms appear mostly at later stages, including increasing hypertrophy and hyperplasia of infected roots; the roots become swollen and form characteristic galls [10].

The clubroot pathogen and Brassicaceae crops establish a specific relationship where plant hormones are involved in triggering immune responses and orchestrating disease-phase-dependent transcriptomic changes [11]. In addition to SA, other groups of plant hormones are involved in different stages of disease progression, especially cytokinin (CK) and auxin, two key players in plant organ development [12,13]. By manipulating CK and auxin levels, the pathogen promotes hypertrophy and cell division by reprogramming meristematic activity in infected roots, leading to the formation of root galls [13–15]. Increased CK levels promoting cell division were observed in the early stage of clubroot disease in Arabidopsis. Conversely, reduced CK levels and concomitant downregulation in the expression of host plant CK sensing and metabolism-related genes were observed in the late stage of infection [12,13,16]. Shifts in CK and auxin homeostasis in developing galls can potentially influence the progression of the clubroot disease; it was reported by Siemens et al. that plants constitutively overexpressing cytokinin oxidase/dehydrogenase show reduced disease symptoms [13].

The plant hormone cytokinin plays a crucial role in various processes of plant growth and development [17]. In addition, in close cooperation with other plant hormones, CKs also play a role in plant defense, affecting many biotic and abiotic stress responses [18,19]. Their homeostasis is controlled by an interplay of multiple processes—the biosynthesis, irreversible degradation, formation, and breakdown of CK conjugates, and CK transport [20,21]. CK catabolism is primarily controlled by cytokinin oxidase/dehydrogenase (CKX; EC 1.5.99.12; [22]), a key enzyme in CK irreversible degradation. On the other hand, isoprenoid CK biosynthesis is catalyzed by adenosine phosphate isopentenyl transferase (IPT; EC 2.5.1.27; [23]), and this reaction represents the rate-limiting step in CK biosynthesis.

The perception of CKs in *Arabidopsis thaliana* is associated with three membrane-bound sensor histidine kinases (HKs), AHK2, AHK3, CRE1/AHK4, based on a two-component phosphorelay signal transduction system [24–28]. The signal perception in Arabidopsis is redundant; individual receptors contribute to different developmental and other processes, and double mutant combinations of *ahk* mutant alleles are viable, but show changes to root morphology or chlorophyll content [29]. AHK2/AHK3 combination in particular appears to play a dominant role in controlling root vs. shoot growth. The rosette of the *ahk2 ahk3* double mutant was reduced approximately to half the size of the wild type; this mutant line also formed fewer cells in the shoot and showed reduced chlorophyll content and faster primary root growth. While *CRE1/AHK4* is predominantly expressed in the root and can be found in the vascular cylinder, *AHK2* and *AHK3* show much broader expression patterns among different plant organs, especially in aerial parts.

Another approach to studying CK perception in plants is the use of CK agonists or antagonists, i.e., compounds that activate HK receptors or prevent agonist binding to the receptor, respectively, leading to the occupancy and blocking of the receptor's active site. PI-55, 6-(2-hydroxy-3-methylbenzylamino)purine, is closely related to naturally occurring 6-benzylaminopurine (BAP), an aromatic CK commonly used in laboratory practice and plant micropropagation techniques. PI-55 is the first example of an anticytokinin compound that specifically interacts with the CRE1/AHK4 receptor and antagonizes the function of natural CKs [30]. In this study, PI-55 was used together with *ahk* double mutant lines to assess whether changes in CK homeostasis may affect the process of clubroot development.

2. Materials and Methods

2.1. Biological Material and Growth Conditions

The Columbia (Col-0) ecotype of *Arabidopsis thaliana* (L.) Heynh was used. *ahk2-2 ahk3-3*, *ahk2-5 cre1-2*, and *ahk3-7 cre1-2* mutant lines were kindly provided by Thomas Schmülling [29]. The studied plants were grown in a cultivation chamber on soil (mixture with perlite 1:1) at 22 °C under long-day conditions (16 h light/8 h dark). Clubroot-infected roots of the Chinese cabbage variety "Granaat" were kindly provided by Pavel Kopecký (Crop Research Institute, Olomouc, Czechia).

2.2. Preparation of the Cytokinin Antagonist PI-55

Synthesis of the PI-55 compound and its biological testing was performed as previously described [30]. Details of the chemical synthesis are provided in the Supplementary Information.

2.3. Inoculation of A. thaliana Plants and Pharmacological Treatment

The clubroot galls were stored at −20 °C. Resting spores were extracted by homogenizing mature clubroot galls in distilled water. The suspension was filtrated through gauze (25 μm pore width) and subsequently centrifuged three times (1000× *g*, 7 min). Then, 14-day-old Arabidopsis plants were inoculated with 0.5 mL of the prepared inoculum of *Plasmodiophora brassicae* (10^7 spores mL^{-1}). Subsequently, non-infected or infected Arabidopsis plants were treated with PI-55 three days after inoculation (dai) for the first time, or the control group was mock-treated with DMSO (0.1%). Treatments with the cytokinin antagonist (or the mock treatments) were performed 9 times over the period of 3 weeks by injecting the soil around each plant with 5 mL of either 1 or 10 μM solution of PI-55.

2.4. Quantification of Plant Hormones

Fresh plant material (roots + hypocotyls) was homogenized under liquid nitrogen and divided into CK and auxin parts with weights of approximately 10 mg per biological sample. CK samples were extracted in 1 mL of modified Bieleski buffer (methanol/water/formic acid, 15/4/ 1, *v*/*v*/*v*; [31]) using an internal standard of stable isotopically labeled CKs (0.4 pmol per sample of CK bases, ribosides, 9-glucoside, and 7-glucoside, and 1 pmol per sample of *O*-glucosides and nucleotides) to check the recovery of the purification step and to validate the determination. Samples were analyzed by ultraperformance liquid chromatography (Acquity UPLC® I-class system; Waters, Milford, MA, USA) coupled to a triple quadrupole mass spectrometer equipped with an electrospray interface (Xevo TQ-S, Waters, Manchester, UK), via the method described previously [32]. Quantification was performed by multiple reaction monitoring of $[M + H]^+$ and the appropriate product ion [32,33]; Masslynx software was used using a standard isotope dilution method. Auxin samples were extracted using 0.8 mL of 50 mM Na-phosphate buffer (pH 7.0; 4 °C) containing 0.1% diethyldithiocarbamic acid salt, using an internal standard of stable isotopically labeled auxins (5 pmol per sample; as previously described [34]). Samples were dissolved in 30 μL of 10% MeOH, injected onto a reversed-phase column (Kinetex 1.7 μm C18 100 Å, 50 × 2.1 mm; Phenomenex), and analyzed using an Acquity UPLC® I-class system (Waters, Milford, MA, USA) coupled to XevoTM TQ-S (Waters, Manchester, UK) equipped with an electrospray interface (ESI), via the method described previously [34].

2.5. Quantitative Real-Time PCR

The Spectrum Plant Total RNA Kit (Merck, Darmstadt, Germany) was used for the isolation of total RNA from 100 mg (fresh weight) of the biological material (hypocotyls + leaves of 28 dai plants), according to the manufacturer's instructions. To minimize contamination with genomic DNA, total RNA was treated with 1.5 μL TURBO DNase (Thermo Fisher Scientific, Waltham, MA, USA) with 1 μL of the enzyme for 30 min. Subsequently, 1.5 μg of extracted RNA was used for first-strand cDNA synthesis by RevertAid H Minus M-MuLV Reverse Transcriptase (Thermo Fisher Scientific, Waltham, MA, USA), according to the manufacturer's instructions, and diluted to a final volume of 200 μL. qPCR analyses were performed in the StepOnePlus Real-Time PCR System (Applied Biosystem, Waltham, MA, USA). The reaction mixture consisted of 5 μL of gb SG PCR Master Mix (Generi Biotech, Hradec Králové, Czechia), 1 μL of gb Passive Reference Dye (Generi Biotech, Hradec Králové, Czechia), 0.6 μL forward primer (5 μM), 0.6 μL reverse primer (5 μM), and diluted cDNA (2 μL) in a final volume of 10 μL. PCR was performed according to standard procedures using the following amplification program: 95 °C for 10 min, followed by 40 cycles of 95 °C for 10 s, 65 °C for 15 s, and 72 °C for 25 s. Ct values were normalized to actin 7 and final values were calculated using $2^{-\Delta\Delta Ct}$ [35].

2.6. Measurement of Photosynthetic Activity

Control and clubroot-infected plants at 28 dai, 33 dai, and 41 dai were used for the maximum quantum yield of photosystem II photochemistry measurements ($F_V/F_M = (F_M - F_0)/F_M$) using FluorCamFC 800 (PSI, Drásov, Czechia). Prior to the measurement, plants were dark-adapted for 20 min. The minimal fluorescence of the dark-adapted sample (F_0) was determined by applying several μ-seconds-long measuring flashes (red light, 0.1 μmol photons m^{-2} s^{-1}). The maximal fluorescence of the dark-adapted sample (F_M) was measured using the 800 ms saturating pulse (white light, 2700 μmol photons m^{-2} s^{-1}).

3. Results and Discussion

3.1. Phenotypic Characterization of Clubroot-Infected Cytokinin Receptor Mutant Lines

To better understand how modulations of cytokinin sensing and metabolism may contribute to the control of the proliferation of clubroot disease in the model organism *A. thaliana*, we took advantage of available *ahk* double mutant lines that exhibit homeostatic control of endogenous cytokinin levels that differ between genotypes [29]. In addition, we further modulated endogenous hormone levels by repeated application of a cytokinin antagonist, PI-55, during the infection phase of Arabidopsis growth. PI-55 represents a powerful tool for modulating cytokinin sensing, and its effects in controlling the root architecture through the manipulation of cytokinin status in affected plants have been demonstrated [30].

In this study, 14-day-old Arabidopsis wild-type plants and double mutant lines *ahk2 ahk3*, *ahk2 ahk4*, and *ahk3 ahk4* were inoculated with *P. brassicae* and disease progression was monitored at selected time points: 28, 33, or 41 days after inoculation (dai). In primary infection, the so-called root hair infection, there are visible subtle changes in the lower part of the root system. At 41 dai, we could observe prominent symptoms associated with later stages—the ground leaf rosette was deformed and the leaves were chlorotic, with already forming necrotic parts (Figure 1). The application of the cytokinin antagonist in the higher concentration used (10 μM) alleviated some of the symptoms of the disease. Both the wild type and the *ahk2 ahk4* mutant line appeared to be more vital, whereas the *ahk3 akh4* double mutant was less responsive to PI-55 treatment (Figure 1). To obtain more convincing evidence of the observed effects, we further measured the photosynthetic performance of control and clubroot-infected shoots, as described in the following section.

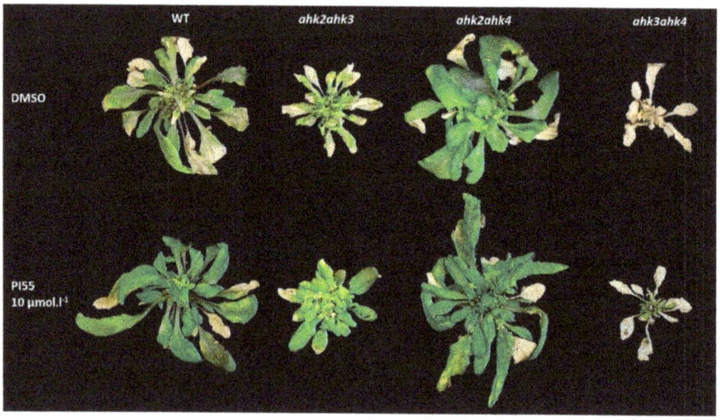

Figure 1. Comparison of clubroot disease symptoms in wild-type (Col-0) and *ahk2 ahk3*, *ahk2 ahk4*, and *ahk3 ahk4* double mutant lines. The 14-day-old plants were infected with *P. brassicae* and mock-treated

with 0.1% DMSO or the cytokinin antagonist PI-55 at a concentration of 10 μmol L^{-1}. Images of representative plants 41 days after inoculation (dai) are shown.

3.2. Photosynthetic Performance in Infected Plants and Effects of PI-55 Treatment

Infection with the clubroot disease in Col-0 and mutant genotypes progressively led to a decrease in photosystem II (PSII) function represented by a reduction in the maximum quantum yield of PSII photochemistry (F_V/F_M) in all Arabidopsis plants (Figure 2). Accordingly, the downregulation of photosynthetic genes, including genes of the Calvin cycle, chlorophyll biosynthesis, and genes of photosynthesis light reactions, was observed in the shoots of clubroot-infected plants in a transcriptomic study that also showed specific subsets of DEGs in the infected roots and shoots from 17 to 24 dai [3].

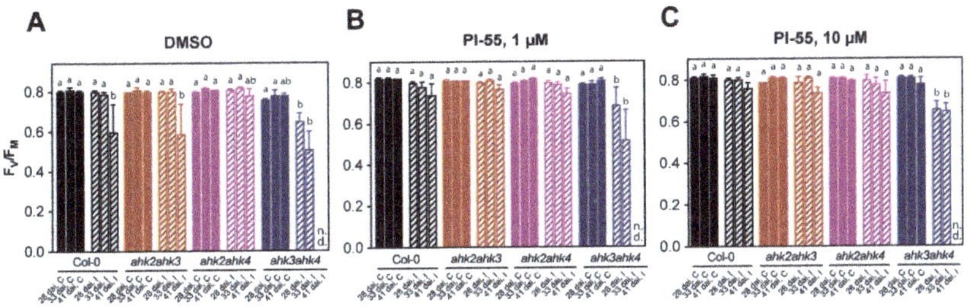

Figure 2. The maximum quantum yield of PSII photochemistry (F_V/F_M) in non-infected control wild-type (Col-0) or mutant Arabidopsis plants (C) and in respective plants infected with *Plasmodiophora brassicae* (I) 28, 33, and 41 days after infection (dai). Infected Col-0 and double mutant lines are indicated by diagonal hatching. Plants were mock-treated with (**A**) 0.1% DMSO or compound PI-55 at a concentration (**B**) 1 μmol L^{-1} or (**C**) 10 μmol L^{-1}. Medians and quartiles are presented (n = 3–5). Different letters indicate a statistically significant difference (Tukey's test; $p < 0.05$) between individual genotypes treated with DMSO or PI-55 (1 or 10 μmol L^{-1}) within dai. n.d., not detected.

Representative images of F_V/F_M imaging in individual genotypes revealed that among cytokinin receptor double mutants, the *ahk2 ahk4* mutant plants showed the highest resistance to fungal pathogen proliferation. Similar to WT plants, in the *ahk2 ahk4* double mutant, we observed only mild disease symptoms in plants at 28 and 33 dai, indicating relatively slow disease progression under our experimental conditions (Figure 3). In stark contrast, in the case of *ahk3 ahk4* mutants, the impairment of PSII function was severe (Figures 2 and 3). At 41 dai, photosynthetic activity in the *ahk3 ahk4* mutant plants expressed as F_V/F_M was below the limit of detection, and already at 28 dai, plants showed severe stunting, leaf curling, and chlorosis together with PSII impairment (Figure 3). In the *ahk2 ahk3* mutant plants, disease symptoms caused by *P. brassicae* infection were similar, but the decline in F_V/F_M was not as pronounced (Figures 2 and 3).

Application of the cytokinin antagonist PI-55 at both concentrations (1 and 10 μmol L^{-1}) alleviated the decrease in F_V/F_M values in both WT and *ahk2 ahk3* double mutants (Figures 2B,C and 3C), while there was no positive effect of PI-55 treatment in *ahk3 ahk4* double mutants, as F_V/F_M ratios were comparable to those of mock-treated (DMSO) plants (Figures 2B,C and 3C).

Collectively, these results indicate that *P. brassicae* infection generates disease symptoms and affects PSII function and photosynthetic parameters primarily in plants where CRE1/AHK4 and AHK2 cytokinin receptors are present. This means that these receptors may represent primary targets that *P. brassicae* uses to modulate host responses via the cytokinin signaling pathway. It was previously reported by Siemens et al. that *CRE1/AHK4* is significantly upregulated in the early stage of infection, which was consistent with the increased level of the cytokinin response gene *ARR5* [13]. Interestingly, the CRE1/AHK4 and AHK2 receptors share a similar ligand binding spectrum and appear to be at least partially functionally redundant and significantly distinct from the AHK3 receptor. Compared with infected Col-0 plants, the *ahk3 ah4* double mutant line showed more pronounced disease symptoms and, importantly, did not respond to PI-55 treatment. Since PI-55 primarily targets the CRE1/AHK4 receptor [30], it can be assumed that the effect of the compound on the possible modulation of disease progression can be observed preferentially in WT and *ahk2 ahk3* double mutants treated with PI-55.

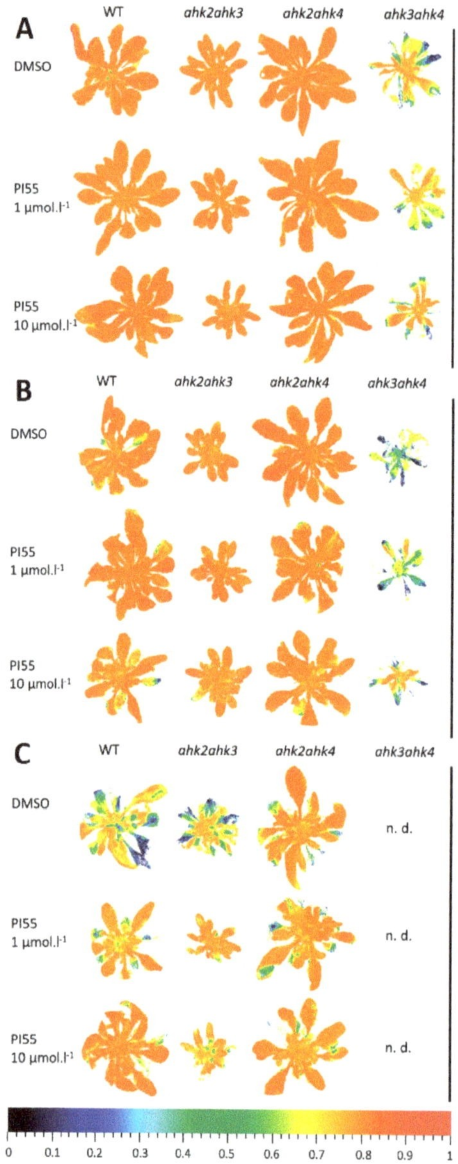

Figure 3. Imaging of maximum quantum yield of PSII photochemistry (F_V/F_M) in representative Arabidopsis wild-type (WT) or mutant plants at (**A**) 28, (**B**) 33, and (**C**) 41 days after infection (dai) with *Plasmodiophora brassicae*. Plants were mock-treated with 0.1% DMSO or the cytokinin antagonist PI-55 at a concentration of 1 or 10 µmol L^{-1}. n.d., not detected.

3.3. The Hormonal Content in Infected Galls and Effects of PI-55 Treatment

We have previously successfully developed a procedure for the highly reproducible detection of a complex pool of CK metabolites in milligram amounts of tissue (root + hypocotyl) in clubroot-infected plant material. As expected, *trans*-zeatin (*t*Z) and isopentenyladenine (iP) were confirmed as the most abundant active CKs in 28 dai plants, with *t*Z being slightly more abundant in both uninfected and infected tissues in all genotypes used in our study (Figure 4). This was in line with our previous observations, where these two CK free bases comprised the majority of the active CK forms, and the abundance of other free bases, *cis*-zeatin (*c*Z) and dihydrozeatin (DHZ), was ~100 fold lower [16]. The overall pattern in CK free base abundance at 28 dai was evidently decreasing compared to uninfected plants of the same age, corresponding to the strong downregulation of cytokinin homeostasis genes, as previously reported [13,16]. Similar trends were seen with other CK forms, although not with the same impact among the studied genotypes (Figure 4). While the concentrations of both CK ribosides and CK nucleotides only showed small changes after infection in Col-0 plants, all other genotypes, and *ahk2 ahk4* in particular, showed significant reductions. In contrast, the most abundant inactive CKs, CK *N*-glucosides, were reduced to a similar extent in all genotypes. Consistent with our previous observations, we could observe a small increase in CK *O*-glucoside conjugates, which was evident in the *ahk2 ahk3* double mutant (Figure 4).

All receptor double mutants showed significantly higher CK content (both before and after infection) compared to Col-0 plants, consistent with previous observations [29]. However, unlike the aforementioned study of Riefler et al., where younger Arabidopsis seedlings were used for CK quantification, we could observe highly elevated levels of both CK free bases, *t*Z and iP, primarily in the *ahk3 ahk4* double mutant (Figure 4). Similar trends were observed for CK precursors (ribosides and nucleotides), where CK levels in the *ahk3 ahk4* mutant plants were followed by the *ahk2 ahk3* genotype. Interestingly, the application of the cytokinin antagonist PI-55 led to different effects among the studied genotypes. Importantly, we observed significant changes in the content of CK free bases, *t*Z and iP—whereas, in clubroot-infected Col-0 and the *ahk2 ahk3* double mutant, the application of PI-55 led to a significant decrease in active CK levels, in the *ahk3 ahk4* mutant, the trend was opposite (Figure 4). In the case of the *ahk2 ahk4* mutant line, the concentrations of active bases were generally low in infected tissues compared to other mutant genotypes and similar to infected Col-0 plants. Notably, the *ahk3 ahk4* mutant showed a similar pattern to most other forms of CK—both 1 μmol L^{-1} and 10 μmol L^{-1} PI-55 treatment induced an increase in hormonal content in infected tissues compared to DMSO-treated plants.

Next, we evaluated the profiles of auxin, another plant hormone that may play an important role in gall formation (Figure 5). Auxin and cytokinin concentrations have previously been shown to change rapidly in developing root galls following *P. brassicae* infection [36]. Many genes of auxin homeostasis were upregulated in club development; it appears that the balance of auxins and cytokinins may play a regulatory role during disease proliferation [13]. The previous results suggested that free IAA levels do not change dramatically in infected tissues at later stages of infection [16]. Consistent with this, the IAA content of wild-type tissues was not significantly affected by clubroot infection (Figure 5). Similarly, *ahk* double mutants showed relatively little change, except for *ahk3 ahk4*, where approximately a 25% increase in IAA levels was observed. Treatment with PI-55 (10 μmol L^{-1}) in infected tissues resulted in a relatively high, approximately 2.5-fold increase in auxin content in the *ahk2 ahk3* and *ahk2 ahk4* double mutant lines and approximately a two-fold increase in wild-type plants.

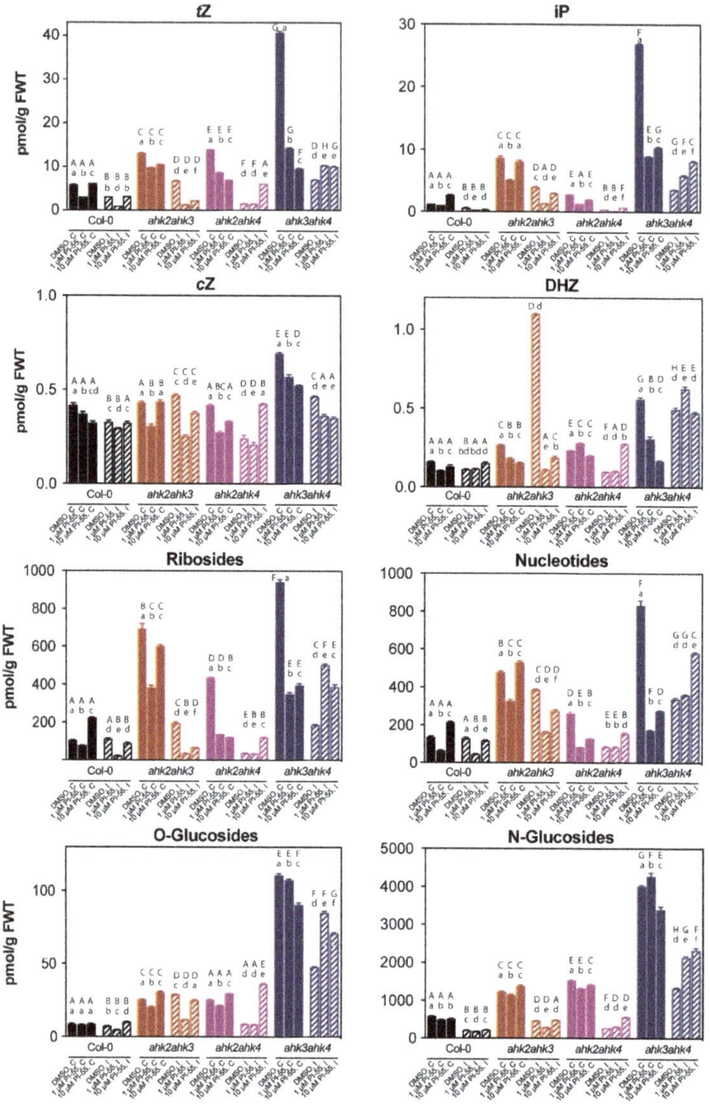

Figure 4. The cytokinin content of control and clubroot-infected wild type (Col-0) and the *ahk2 ahk3*, *ahk2 ahk4*, and *ahk3 ahk4* double mutant lines. Active (*t*Z, iP, *c*Z, and DHZ), precursor (ribosides and nucleotides), and conjugated (*O*-glucosides and *N*-glucosides) cytokinin content of control (C) and infected (I) root + hypocotyl tissue in 28 days after infection (dai) plants are shown. Plants were mock-treated with 0.1% DMSO or the cytokinin antagonist PI-55 at a concentration of 1 or 10 µmol L^{-1}. Infected Col-0 and double mutant lines are indicated by diagonal hatching. Results are means (±SD) of 3 independent replicates expressed as pmol g^{-1} fresh weight. Different letters indicate a statistically significant difference (Tukey's test; $p < 0.05$) within individual genotypes (lowercase letters) or between treatments with DMSO or 1 or 10 µmol L^{-1} PI-55 (uppercase letters).

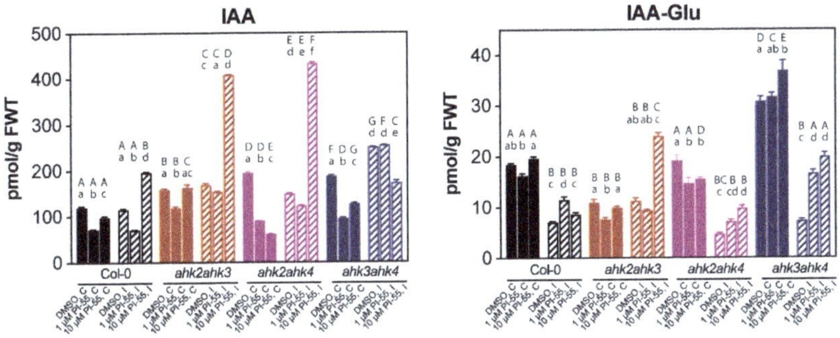

Figure 5. IAA and IAA-Glu profiles in wild type (Col-0) and the *ahk2 ahk3*, *ahk2 ahk4*, and *ahk3 ahk4* double mutant lines. Content of active and conjugated forms of IAA of control (C) and infected (I) root + hypocotyl tissue in 28 days after infection (dai) plants are shown. Plants were mock-treated with 0.1% DMSO or the cytokinin antagonist PI-55 at a concentration of 1 or 10 µmol L^{-1}. Infected Col-0 and double mutant lines are indicated by diagonal hatching. Results are means (±SD) of 3 independent replicates expressed as pmol g^{-1} fresh weight. Different letters indicate a statistically significant difference (Tukey's test; $p < 0.05$) within individual genotypes (lowercase letters) or between treatments with DMSO or 1 or 10 µmol L^{-1} PI-55 (uppercase letters).

Taken together, our results indicate that the balance of cytokinin and, somewhat unexpectedly, also auxin, is affected in response to the cytokinin antagonist treatment and that these changes may be related to the cytokinin status of infected plants. In general, compared to Col-0 plants, our results show higher overall CK content, especially *t*Z and iP free bases, in all CK receptor double mutants. This was particularly evident in the *ahk3 ahk4* double mutant, where the free base content was relatively high regardless of whether PI-55 treatment was applied. Unexpectedly, photosynthetic parameters and plant fitness were relatively worse in this receptor mutant (Figures 2 and 3), and PI-55 treatment could not rescue the photosynthetic performance. Conversely, the relatively low content of CK free bases (especially after PI-55 treatment) in the *ahk2 ahk4* double mutant correlated with better plant fitness and preservation of photosynthetic performance in treated shoots. This leads us to believe that the content of cytokinins, especially the most abundant active forms, *t*Z and iP, but also reflected in inactive CK forms (ribosides, nucleotides, and glucosides), is directly correlated with the progression of this disease, which further confirms the positive role of CK in the process. With the exception of the *ahk3 ahk4* double mutant, treatment with PI-55 mostly resulted in further reductions in free base content, although some variability was observed between the concentrations used. However, the effect of concentration did not seem to affect photosynthetic parameters, as both 1 and 10 µmol L^{-1} PI-55 treatments produced similar results (Figure 2). Unlike cytokinin, the exact function of auxin in later stages of infection is still not clear, but it has been hypothesized that elevated auxin may stimulate cell growth and promote cell wall extensibility, associated with cell expansion [37]. In our experiments, the levels of free IAA were only mildly affected in response to the infection process, but treatments with 10 µmol L^{-1} PI-55 seemed to produce more significant effects. This suggests that auxin homeostasis may also be affected by CK antagonist treatment, but further experiments will be needed to understand the hormone's role in the process, as IAA levels change quite dynamically from early to later stages of infection [12].

3.4. Expression Analysis of Cytokinin-Related Genes in Clubroot-Infected Plants

In our experiments, the wild type and the *ahk2 ahk3* double mutant appeared to respond to PI-55 treatment, as reflected in the evaluation of their photosynthetic parameters (Figures 2 and 3). While the treatment with PI-55 had no positive effect on disease progression in the *ahk3 ahk4* mutant line, in the case of *ahk2 ahk3* and Col-0 lines, 10 µM PI-55 treatment partially prevented chlorosis and increased the chlorophyll content in infected shoots (Figure 1). Furthermore, the wild type and the *ahk2 ahk4* double mutant appear to share similar phenotypes and photosynthetic parameters, as well as similar profiles of active CK forms. Therefore, to further evaluate the possible role of the cytokinin antagonist in the later stages of clubroot development, we used the wild type and the *ahk2 ahk3* double mutant line to determine the transcript levels of selected CK pathway-related genes previously associated with disease progression (Figure 6). At 28 dai, quantitative real-time PCR of infected and uninfected plant samples revealed that the general expression pattern clearly showed signs of overall downregulation of the cytokinin pathway. Although some *ARR* genes in wild-type infected plants were slightly upregulated, the changes in the expression levels of these genes were rather low and close to the detection limit (Figure 6 and Supplemental Table S2). Notably, the expression of *ARR5* was slightly upregulated in Col-0 infected plants, in agreement with previous transcriptomic analysis where higher expression was observed for *CRE1/AHK4*, *ARR5*, and *ARR10* [13]. By contrast, the *ahk2 ahk3* double mutant showed an approximately twofold reduction in *ARR5* expression levels in the infected tissues. Additionally, in the *ahk2 ahk3* double mutant, we could observe the downregulation of most cytokinin pathway genes, in agreement with previously published data [13,16]. PI-55 treatment in wild-type infected plants showed further downregulation of the expression levels of cytokinin oxidase/dehydrogenase gene *CKX1* and two cytokinin de novo biosynthesis genes, *IPT3* and *IPT5* (Figure 6). Interestingly, *CKX4* appeared to be relatively strongly upregulated by around fivefold in the wild type after PI-55 treatment; slight upregulation of *CKX4* at a later stage of infection was also reported in a transcriptomic study by Siemens et al. [13].

These results are in general agreement with previous studies that conclusively demonstrated that cytokinin homeostasis begins to be downregulated already at early stages of infection [13], and the expression of genes associated with cytokinin metabolism and signaling was strongly suppressed [16]. Differential analysis of genes in clubroot-infected resistant vs. susceptible lines of *Brassica rapa* L. and subsequent metabolomic analyses further explored the function of different classes of plant hormones in infection and highlighted their importance in triggering immune responses in the plant host [11]. Interestingly, regarding active CKs, the authors report high variability, especially in *tZ* levels during gall formation, which is consistent with *tZ* acropetal transport through the xylem [11]. It is therefore possible that an imbalance in *tZ* levels can significantly affect cambium activity and development.

Taken together, the relatively strong repression of genes related to cytokinin metabolism and sensing both in the wild type and the *ahk2 ahk3* double mutant line appears to be consistent with our metabolomic data, where we could observe a strong reduction in CK content after infection that was further promoted by PI-55 treatment. In particular, *IPT3* and *IPT5*, together with *IPT7*, are the major players in root growth and lateral root organogenesis [38,39]. Temporal regulation of isopentenyl transferase genes in clubs of *B. rapa* has been described, which is associated with disease development [40]. Importantly, although the quadruple *ipt1;3;5;7* mutant was able to develop some symptoms of clubroot disease, club formation was severely compromised [16]. Thus, it has been speculated that CKs may directly regulate plasmodial development, and reduced CK levels may interfere with the development of the vascular cambium, which is essential for gall formation [16].

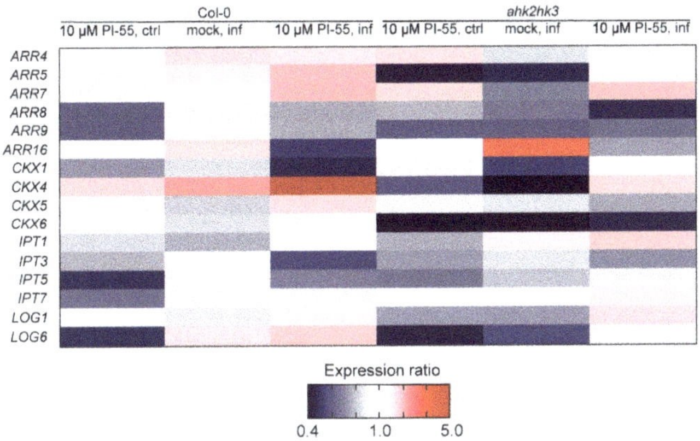

Figure 6. Expression profile of selected genes related to CK metabolism and sensing in the wild type (Col-0) and the *ahk2 ahk3* double mutant. The 28 days after infection (dai) control (ctrl) or clubroot-infected (inf) plants were mock-treated with 0.1% DMSO (mock) or the cytokinin antagonist PI-55 at a concentration of 10 μmol L^{-1}. A heat map showing the relative expression of up- (red) and downregulated (blue) genes is color-coded according to the scheme shown in the figure. Expression changes are presented relative to either uninfected Col-0 or *ahk2 ahk3* plants that were mock-treated with 0.1% DMSO.

4. Conclusions

Our data from *ahk* double mutants infected with *P. brassicae* and PI-55 treatments strongly support a role of the plant hormone cytokinin in the development of clubroot disease. Higher cytokinin content in infected roots, as observed in the *ahk3 ahk4* double mutant, coincided with better disease progression. PI-55 treatment led to changes in the internal pools of cytokinin and auxin, which further affected disease symptoms. Thus, it seems that cytokinin antagonist treatment may lead to a further decrease in endogenous CK content and a temporarily increased auxin status in infected plants, and the disrupted balance of these two phytohormones in developing clubs may modulate the growth of the biotrophic pathogen at later stages of infection. Our results demonstrate that changes in cytokinin perception through genetic or pharmacological approaches are a promising strategy to alter endogenous cytokinin (and possibly also auxin) levels, potentially allowing increased resistance to the infection and delayed progression of clubroot disease symptoms.

Supplementary Materials: The following supporting information can be downloaded at: https://www.mdpi.com/article/10.3390/biom13020299/s1, Supplementary Methods S1: Chemical synthesis and testing of PI-55; Table S1: Primers for qRT-PCR; Table S2: Expression profiles of genes related to cytokinin signaling and metabolism.

Author Contributions: J.B. and Z.K., conceptualization; J.B., O.N., and Z.K., data curation; J.B., L.P., V.V., and Z.K., formal analysis; J.B., V.K., V.V., Z.K., M.H., and L.P., investigation; O.P., O.N., and M.S., project administration; O.P. and M.S., validation; O.P. and Z.K., visualization; J.B. and O.P., writing—original draft; O.P., writing—review and editing; J.B. and O.P., supervision; M.S., funding acquisition. All authors have read and agreed to the published version of the manuscript.

Funding: This work was supported by the student project IGA_PrF_2023_031 of the Palacký University.

Data Availability Statement: The data that support the findings of this study are available from the corresponding authors upon reasonable request.

Acknowledgments: We thank Marek Zatloukal for providing compound PI-55 and Hana Svobodová for the technical assistance.

Conflicts of Interest: The authors declare no conflict of interest.

References

1. Dixon, G.R. The Occurrence and Economic Impact of *Plasmodiophora brassicae* and Clubroot Disease. *J. Plant Growth Regul.* **2009**, *28*, 194–202. [CrossRef]
2. Ciaghi, S.; Schwelm, A.; Neuhauser, S. Transcriptomic response in symptomless roots of clubroot infected kohlrabi (*Brassica oleracea var. gongylodes*) mirrors resistant plants. *BMC Plant Biol.* **2019**, *19*, 288. [CrossRef] [PubMed]
3. Irani, S.; Trost, B.; Waldner, M.; Nayidu, N.; Tu, J.; Kusalik, A.J.; Todd, C.D.; Wei, Y.; Bonham-Smith, P.C. Transcriptome analysis of response to *Plasmodiophora brassicae* infection in the Arabidopsis shoot and root. *BMC Genom.* **2018**, *19*, 23. [CrossRef] [PubMed]
4. Rolfe, S.A.; Strelkov, S.E.; Links, M.G.; Clarke, W.E.; Robinson, S.J.; Djavaheri, M.; Malinowski, R.; Haddadi, P.; Kagale, S.; Parkin, I.A.P.; et al. The compact genome of the plant pathogen *Plasmodiophora brassicae* is adapted to intracellular interactions with host Brassica spp. *BMC Genom.* **2016**, *17*, 272. [CrossRef]
5. Schwelm, A.; Fogelqvist, J.; Knaust, A.; Jülke, S.; Lilja, T.; Rosso, G.B.; Karlsson, M.; Shevchenko, A.; Dhandapani, V.; Choi, S.R.; et al. The *Plasmodiophora brassicae* genome reveals insights in its life cycle and ancestry of chitin synthases. *Sci. Rep.* **2015**, *5*, 11153. [CrossRef] [PubMed]
6. Javed, M.A.; Schwelm, A.; Zamani-Noor, N.; Salih, R.; Vañó, M.S.; Wu, J.; García, M.G.; Heick, T.M.; Luo, C.; Prakash, P.; et al. The clubroot pathogen *Plasmodiophora brassicae*: A profile update. *Mol. Plant Pathol.* **2022**, *24*, 89–106. [CrossRef]
7. Ludwig-Müller, J.; Jülke, S.; Geiß, K.; Richter, F.; Mithöfer, A.; Šola, I.; Rusak, G.; Keenan, S.; Bulman, S. A novel methyltransferase from the intracellular pathogen *Plasmodiophora brassicae* methylates salicylic acid. *Mol. Plant Pathol.* **2015**, *16*, 349–364. [CrossRef] [PubMed]
8. Kageyama, K.; Asano, T. Life Cycle of Plasmodiophora brassicae. *J. Plant Growth Regul.* **2009**, *28*, 203–211. [CrossRef]
9. Aist, J.R.; Williams, P.H. The cytology and kinetics of cabbage root hair penetration by Plasmodiophora brassicae. *Can. J. Bot.* **1971**, *49*, 2023–2034. [CrossRef]
10. Malinowski, R.; Truman, W.; Blicharz, S. Genius Architect or Clever Thief—How *Plasmodiophora brassicae* Reprograms Host Development to Establish a Pathogen-Oriented Physiological Sink. *Mol. Plant Microbe Interact.* **2019**, *32*, 1259–1266. [CrossRef]
11. Wei, X.; Zhang, Y.; Zhao, Y.; Xie, Z.; Hossain, M.R.; Yang, S.; Shi, G.; Lv, Y.; Wang, Z.; Tian, B.; et al. Root Transcriptome and Metabolome Profiling Reveal Key Phytohormone-Related Genes and Pathways Involved Clubroot Resistance in *Brassica rapa* L. *Front. Plant Sci.* **2021**, *12*, 759623. [CrossRef] [PubMed]
12. Devos, S.; Laukens, K.; Deckers, P.; Van Der Straeten, D.; Beeckman, T.; Inzé, D.; Van Onckelen, H.; Witters, E.; Prinsen, E. A Hormone and Proteome Approach to Picturing the Initial Metabolic Events During *Plasmodiophora brassicae* Infection on Arabidopsis. *Mol. Plant Microbe Interact.* **2006**, *19*, 1431–1443. [CrossRef]
13. Siemens, J.; Keller, I.; Sarx, J.; Kunz, S.; Schuller, A.; Nagel, W.; Schmülling, T.; Parniske, M.; Ludwig-Müller, J. Transcriptome Analysis of Arabidopsis Clubroots Indicate a Key Role for Cytokinins in Disease Development. *Mol. Plant Microbe Interact.* **2006**, *19*, 480–494. [CrossRef]
14. Boivin, S.; Fonouni-Farde, C.; Frugier, F. How Auxin and Cytokinin Phytohormones Modulate Root Microbe Interactions. *Front. Plant Sci.* **2016**, *7*, 1240. [CrossRef]
15. Malinowski, R.; Smith, J.A.; Fleming, A.J.; Scholes, J.D.; Rolfe, S.A. Gall formation in clubroot-infected Arabidopsis results from an increase in existing meristematic activities of the host but is not essential for the completion of the pathogen life cycle. *Plant J.* **2012**, *71*, 226–238. [CrossRef]
16. Malinowski, R.; Novak, O.; Borhan, M.H.; Spíchal, L.; Strnad, M.; Rolfe, S.A. The role of cytokinins in clubroot disease. *Eur. J. Plant Pathol.* **2016**, *145*, 543–557. [CrossRef]
17. Kieber, J.J.; Schaller, G.E. Cytokinins. *Arab. Book* **2014**, *12*, e0168. [CrossRef] [PubMed]
18. Choi, J.; Choi, D.; Lee, S.; Ryu, C.-M.; Hwang, I. Cytokinins and plant immunity: Old foes or new friends? *Trends Plant Sci.* **2011**, *16*, 388–394. [CrossRef]
19. Bryksová, M.; Dabravolski, S.; Kučerová, Z.; Kokáš, F.Z.; Špundová, M.; Plíhalová, L.; Takáč, T.; Grúz, J.; Hudeček, M.; Hloušková, V.; et al. Aromatic Cytokinin Arabinosides Promote PAMP-like Responses and Positively Regulate Leaf Longevity. *ACS Chem. Biol.* **2020**, *15*, 1949–1963. [CrossRef] [PubMed]
20. Mok, D.W.; Mok, M.C. Cytokinin metabolism and action. *Annu. Rev. Plant Biol.* **2001**, *52*, 89–118. [CrossRef] [PubMed]
21. Kudo, T.; Kiba, T.; Sakakibara, H. Metabolism and Long-distance Translocation of Cytokinins. *J. Integr. Plant Biol.* **2010**, *52*, 53–60. [CrossRef] [PubMed]
22. Chatfield, J.M.; Armstrong, D.J. Regulation of Cytokinin Oxidase Activity in Callus Tissues of *Phaseolus vulgaris* L. cv Great Northern. *Plant Physiol.* **1986**, *80*, 493–499. [CrossRef]
23. Kakimoto, T. Identification of Plant Cytokinin Biosynthetic Enzymes as Dimethylallyl Diphosphate:ATP/ADP Isopentenyltransferases. *Plant Cell Physiol.* **2001**, *42*, 677–685. [CrossRef] [PubMed]
24. Inoue, T.; Higuchi, M.; Hashimoto, Y.; Seki, M.; Kobayashi, M.; Kato, T.; Tabata, S.; Shinozaki, K.; Kakimoto, T. Identification of CRE1 as a cytokinin receptor from Arabidopsis. *Nature* **2001**, *409*, 1060–1063. [CrossRef]
25. Ueguchi, C.; Sato, S.; Kato, T.; Tabata, S. The AHK4 Gene Involved in the Cytokinin-Signaling Pathway as a Direct Receptor Molecule in Arabidopsis thaliana. *Plant Cell Physiol.* **2001**, *42*, 751–755. [CrossRef] [PubMed]
26. Ueguchi, C.; Koizumi, H.; Suzuki, T.; Mizuno, T. Novel family of sensor histidine kinase genes in Arabidopsis thaliana. *Plant Cell Physiol.* **2001**, *42*, 231–235. [CrossRef]

27. Suzuki, T.; Miwa, K.; Ishikawa, K.; Yamada, H.; Aiba, H.; Mizuno, T. The Arabidopsis Sensor His-kinase, AHK4, Can Respond to Cytokinins. *Plant Cell Physiol.* **2001**, *42*, 107–113. [CrossRef]
28. Yamada, H.; Suzuki, T.; Terada, K.; Takei, K.; Ishikawa, K.; Miwa, K.; Yamashino, T.; Mizuno, T. The Arabidopsis AHK4 Histidine Kinase is a Cytokinin-Binding Receptor that Transduces Cytokinin Signals Across the Membrane. *Plant Cell Physiol.* **2001**, *42*, 1017–1023. [CrossRef] [PubMed]
29. Riefler, M.; Novak, O.; Strnad, M.; Schmülling, T. ArabidopsisCytokinin Receptor Mutants Reveal Functions in Shoot Growth, Leaf Senescence, Seed Size, Germination, Root Development, and Cytokinin Metabolism. *Plant Cell* **2006**, *18*, 40–54. [CrossRef]
30. Spíchal, L.; Werner, T.; Popa, I.; Riefler, M.; Schmülling, T.; Strnad, M. The purine derivative PI-55 blocks cytokinin action via receptor inhibition. *FEBS J.* **2009**, *276*, 244–253. [CrossRef]
31. Hoyerova, K.; Gaudinova, A.; Malbeck, J.; Dobrev, P.; Kocabek, T.; Solcova, B.; Trávníčková, A.; Kaminek, M. Efficiency of different methods of extraction and purification of cytokinins. *Phytochemistry* **2006**, *67*, 1151–1159. [CrossRef] [PubMed]
32. Svačinová, J.; Novák, O.; Plačková, L.; Lenobel, R.; Holík, J.; Strnad, M.; Doležal, K. A new approach for cytokinin isolation from Arabidopsis tissues using miniaturized purification: Pipette tip solid-phase extraction. *Plant Methods* **2012**, *8*, 17. [CrossRef]
33. Novák, O.; Hauserová, E.; Amakorová, P.; Doležal, K.; Strnad, M. Cytokinin profiling in plant tissues using ultra-performance liquid chromatography–electrospray tandem mass spectrometry. *Phytochemistry* **2008**, *69*, 2214–2224. [CrossRef]
34. Pěnčík, A.; Rolčík, J.; Novák, O.; Magnus, V.; Barták, P.; Buchtík, R.; Salopek-Sondi, B.; Strnad, M. Isolation of novel indole-3-acetic acid conjugates by immunoaffinity extraction. *Talanta* **2009**, *80*, 651–655. [CrossRef]
35. Livak, K.J.; Schmittgen, T.D. Analysis of relative gene expression data using real-time quantitative PCR and the $2^{-\Delta\Delta CT}$ Method. *Methods* **2001**, *25*, 402–408. [CrossRef]
36. Grsic-Rausch, S.; Kobelt, P.; Siemens, J.M.; Bischoff, M.; Ludwig-Müller, J. Expression and Localization of Nitrilase during Symptom Development of the Clubroot Disease in Arabidopsis. *Plant Physiol.* **2000**, *122*, 369–378. [CrossRef] [PubMed]
37. Devos, S.; Vissenberg, K.; Verbelen, J.; Prinsen, E. Infection of Chinese cabbage by *Plasmodiophora brassicae* leads to a stimulation of plant growth: Impacts on cell wall metabolism and hormone balance. *New Phytol.* **2005**, *166*, 241–250. [CrossRef] [PubMed]
38. Miyawaki, K.; Tarkowski, P.; Matsumoto-Kitano, M.; Kato, T.; Sato, S.; Tarkowska, D.; Tabata, S.; Sandberg, G.; Kakimoto, T. Roles of Arabidopsis ATP/ADP isopentenyltransferases and tRNA isopentenyltransferases in cytokinin biosynthesis. *Proc. Natl. Acad. Sci. USA* **2006**, *103*, 16598–16603. [CrossRef]
39. Bielach, A.; Podlešáková, K.; Marhavý, P.; Duclercq, J.; Cuesta, C.; Müller, B.; Grunewald, W.; Tarkowski, P.; Benková, E. Spatiotemporal Regulation of Lateral Root Organogenesis in Arabidopsis by Cytokinin. *Plant Cell* **2012**, *24*, 3967–3981. [CrossRef]
40. Ando, S.; Asano, T.; Tsushima, S.; Kamachi, S.; Hagio, T.; Tabei, Y. Changes in gene expression of putative isopentenyltransferase during clubroot development in Chinese cabbage (*Brassica rapa* L.). *Physiol. Mol. Plant Pathol.* **2005**, *67*, 59–67. [CrossRef]

biomolecules

MDPI

Article

Heparin-Induced Changes of Vascular Endothelial Growth Factor ($VEGF_{165}$) Structure

Ekaterina L. Nemashkalova [1], Marina P. Shevelyova [1], Andrey V. Machulin [2], Dmitry D. Lykoshin [3], Roman S. Esipov [3] and Evgenia I. Deryusheva [1,*]

1 Institute for Biological Instrumentation, Pushchino Scientific Center for Biological Research of the Russian Academy of Sciences, Pushchino 142290, Russia
2 Skryabin Institute of Biochemistry and Physiology of Microorganisms, Pushchino Scientific Center for Biological Research of the Russian Academy of Sciences, Pushchino 142290, Russia
3 Shemyakin-Ovchinnikov Institute of Bioorganic Chemistry, Russian Academy of Sciences, Moscow 117997, Russia
* Correspondence: evgenia.deryusheva@gmail.com

Abstract: Vascular endothelial growth factor-A (VEGF-A), a secreted homodimeric glycoprotein, is a critical regulator of angiogenesis in normal and pathological states. The binding of heparin (HE) to $VEGF_{165}$ (the major form of VEGF-A) modulates the angiogenesis-related cascade, but the mechanism of the observed changes at the structural level is still insufficiently explored. In the present study, we examined the effect of HE on the structural and physicochemical properties of recombinant human $VEGF_{165}$ ($rhVEGF_{165}$). The HE binding results in an increase of hydrophobic surface exposure in $rhVEGF_{165}$ without changes in its secondary structure. Differential scanning calorimetry measurements for intact and HE-bound $rhVEGF_{165}$ reveals the absence of any pronounced thermally induced transitions in the protein in the temperature range from 20 to 100 °C. The apolar area increase during the heparin binding explains the pronounced HE-induced oligomerization/aggregation of $rhVEGF_{165}$, as studied by chemical glutaraldehyde cross-linking and dynamic light scattering. Molecular modeling and docking techniques were used to model the full structure of dimeric $VEGF_{165}$ and to reveal putative molecular mechanisms underlying the function of the $VEGF_{165}$/HE system. In general, the results obtained can be a basis for explaining the modulating effect of HE on the biological activity of VEGF-A.

Keywords: vascular endothelial growth factor; heparin; protein aggregation; differential scanning calorimetry

1. Introduction

Vascular endothelial growth factors (VEGFs) are a family of secreted signaling proteins playing a critical role in vasculo- and angiogenesis. Both processes are essential for physiologic homeostasis and the pathogenesis of tumor growth, metastasis and ophthalmic diseases [1–3]. The VEGF levels in the plasma and serum of healthy people are 19–47 $pg \cdot mL^{-1}$ and 119–238 $pg \cdot mL^{-1}$, respectively, while a marked increased concentration have been observed in various types of cancer (rev in [4]). The VEGF gene family includes VEGF-A, VEGF-B, VEGF-C, VEGF-D, virally encoded VEGF-E, and placenta growth factor, differing in their expression pattern, biological functions and receptor specificity [5].

VEGF proteins, belonging to the large PDGF (platelet-derived growth factor) family (SCOP ID 4003427), are members of the cystine knot growth factor superfamily (SCOP ID: 3000846). VEGF-A plays a dominant role in the regulation of angiogenesis. It binds to two tyrosine kinase receptors: vascular endothelial growth factor receptor 1 (VEGFR1) and 2 (VEGFR2) [6–8]. VEGF binding to VEGFRs promotes a dimerization of the intracellular domain of VEGFRs, which subsequently leads to the activation of a cascade of downstream signaling molecules [9]. For efficient signal transduction, VEGF requires the binding of co-receptors, neuropilin-1 [10,11] and heparin (HE) or heparan sulfate (HS) [12]. HE and

HS are glycosaminoglycans (GAGs), which are anionic linear periodic polysaccharides composed of repetitive disaccharide units containing an uronic acid and an amino sugar ring. GAGs are located on cell surfaces and in the extracellular matrix and participate in diverse signaling processes via interactions with their protein targets [13,14]. The range of plasma HE concentration in clinically normal individuals is 1–2.4 mg·L^{-1} [15].

The alternative splicing of VEGF-A gives rise to peptides of 121 ($VEGF_{121}$), 165 ($VEGF_{165}$), 189 ($VEGF_{189}$), and 206 ($VEGF_{206}$) amino acids. These variants share a common N-terminal receptor-binding domain (RBD) of 115 residues and have a similar affinity for VEGFR2 [16]. X-ray crystallography of the RBD (residues 8-109) has shown that VEGF forms an antiparallel homodimer covalently linked by two disulfide bridges between Cys-51 and Cys-60 with a central four-stranded β-sheet and cystine knot motif within the monomer [17,18]. The key difference between spliced variants of VEGF is their ability to bind HE and HS proteoglycans [19]. The affinity of the VEGF forms to HE is increased in the following order: $VEGF_{165}$, $VEGF_{189}$, and $VEGF_{206}$, whereas $VEGF_{121}$ does not bind heparin [19].

VEGF-A forms capable of binding HE contain the same C-terminal heparin-binding domain (HBD). Nuclear magnetic resonance (NMR) spectroscopy of $VEGF_{165}$ HBD (PDB:1VGH, NMR, 55 residues) revealed two subdomains, each containing two disulfide bridges and a short two-stranded antiparallel β-sheet; the carboxy-terminal subdomain also contains a short α helix [20]. The loss of the carboxyl-terminal domain, whether due to proteolysis or alternative splicing, correlates with a substantial decrease in endothelial cell mitogenic activity of VEGF [21]. The isolated HBD was shown to have high affinity towards HE, comparable to that observed for the full $VEGF_{165}$ [20–22]. Surface electrostatic potential calculations revealed that positively charged residues on one side of the HBD surface could make up a potential binding site for the negatively charged HE. The GAGs binding site in the HBD is mapped out by Lys and Arg residues located towards the C-terminus of the domain [11,20], but an experimental structure for HBD in complex with HE is not available. Additionally, three-dimensional structure that includes both RBD and HBD has not yet been obtained either due to the challenges of the crystallization procedure or to the highly flexible linker that connects them. Several models were proposed for the full VEGF: some of them are able to bind VEGFR2, for others, such binding is sterically hindered. In some models, HE binds to VEGF through its HBD and is able to redistribute the conformational ensemble of the protein, whereas the binding of HE to RBD is not favorable. It is also assumed that HE binding could induce the formation of a sandwich structure, where one HE molecule is symmetrically bound by two HBDs [11,23,24].

The apparent dissociation constant (K_D) value for the HE–VEGF-A complex is 11–80 nM [25,26]. HE binding is relevant to VEGF activity within endothelial cells. The cell surface-associated heparin-like molecules were shown to be required for the interaction of VEGF with its cell surface receptors [12]. The amount of $VEGF_{165}$ bound to VEGFR2 in the presence of HE increases more than three-fold, while the apparent affinity of protein for the receptor does not change [21]. At the same time, the affinity of $VEGF_{165}$ to VEGFR1 decreases in the presence of heparin, but the remaining binding results in more efficient kinase activation [27]. The difference between HE-modulating actions on VEGFRs may be explained by the presence of heparin-binding peptides in the extracellular domain of VEGFR2 but not VEGFR1 [28]. Notably, $VEGF_{110}$ (the form of $VEGF_{165}$, which lacks HBD) maintains the ability of the native protein to bind to the VEGF receptors, but the endothelial cell mitogenic potency of homodimeric $VEGF_{110}$ is decreased more than 100-fold compared to that of $VEGF_{165}$ [21].

VEGF binding to Avastin (commercial name of bevacizumab), the first humanized anti-VEGF monoclonal antibody approved by the US Food and Drug Administration for the first-line treatment of metastatic colorectal cancer [29], is significantly enhanced in the presence of HE, with a maximal effect (>2-fold) being observed in the presence of 0.1 μg·mL^{-1} HE [30]. Moreover, the reduced Avastin–VEGF binding at an acidic pH is rescued by HE [30]. Ko et al. have showed that $VEGF_{189}$ exhibits a high affinity to HE, and is preferentially localized in cancer cell-derived small extracellular vesicles (sEVs) via heparin-binding. sEV-associated VEGF is highly stable and is not neutralized by the therapeutic VEGF antibody bevacizumab, raising the possibility that elevated levels of sEV-VEGF contribute in part to the resistance of tumors to bevacizumab [31].

Despite the plethora of works devoted to the effect of HE/HS on the biological activity of VEGF, the mechanism of the observed changes at the structural level is still insufficiently explored. In the present study, we examined the effect of HE on the structural and physicochemical properties of human $VEGF_{165}$, the most abundant and biologically active form of VEGF. In this work, a modeling approach in combination with circular dichroism (CD) spectroscopy, dynamic light scattering (DLS) and differential scanning calorimetry (DSC) was applied to characterize the putative full-length $VEGF_{165}$ structure and to reveal the impact of HE binding on it.

2. Materials and Methods

2.1. Materials

Recombinant human $VEGF_{165}$ produced in *E. coli* was obtained from the Institute of Bioorganic Chemistry of the Russian Academy of Sciences. Recombinant human $VEGF_{165}$ produced in Chinese Hamster Ovary (CHO) cells (#PSG010-10) was purchased from SCI-store (Moscow, Russia). Unfractioned sodium heparin from the intestinal mucosae of pig (HE) 5000 IU/mL was from Velpharm (Kurgan, Russia). Sodium chloride, sodium di-hydrogen phosphate 2-hydrate and sodium dodecyl sulfate (SDS) were from Panreac Applichem (Darmstadt, Germany). Ultra-grade Tris was purchased from VWR Life Science AMRESCO (Vienna, Austria). Bis-ANS (4,4′-dianilino-1,1′-binaphthyl-5,5′-disulfonic acid), grade II glutaraldehyde, and Coomassie R-250 were from Merck (Darmstadt, Germany). All buffers and other solutions were prepared using ultrapure water.

The freeze-dried recombinant human $VEGF_{165}$ ($rhVEGF_{165}$) and HE were dialyzed exhaustively against a suitable buffer (cutoff 12–14 kDa and 1 kDa were used for rhVEGF-165 and HE, respectively) before measurements. Protein concentrations were measured spectrophotometrically using molar extinction coefficients at 280 nm of 13,920 $M^{-1}cm^{-1}$ calculated according to previous work [32]. The concentration of water stock solution of bis-ANS was evaluated using the molar extinction coefficient at 385 nm of 16,790 $M^{-1}cm^{-1}$ [33]. The mass concentration of the HE stock solution after dialysis was calculated from the freeze-dried mass of HE. The average molar concentration was estimated using the average molecular mass of HE 13.8 kDa; the molar concentration of disaccharide residues was calculated using their molecular mass. The average molecular mass of HE was estimated using the average hydrodynamic diameter measured by DLS method, using a set of linear polymers of different molar masses (from 8 to 400 kDa) as a reference. The evaluated diameter value agreed with the literature data for similar pharmaceutical HEs of the same origin [34].

2.2. Circular Dichroism Studies

Far-UV CD measurements were performed using a J-810 spectropolarimeter (JASCO, Inc., Tokyo, Japan) equipped with a Peltier-controlled cell holder. A quartz cell with pathlength of 1 mm was used. Protein concentration was 1.8–1.9 μM. HE was added up to an equimolar ratio to the protein. The contribution of buffer (pH 7.4, 10 mM Na-phosphate, 50 mM NaCl) in the absence/presence of HE was subtracted from the experimental spectra. Estimations of the content of the secondary structure elements were made using the CDPro software package with SDP48 and SMP56 reference protein sets [35].

2.3. Scanning Calorimetry Measurements

The scanning microcalorimetry studies were carried out on a Nano DSC microcalorimeter (TA Intruments Inc., New Castle, DE, USA) at a 1 K·min^{-1} heating rate and excess pressure of 4 bars (pH 7.4, 10 mM Na-phosphate, 50 mM NaCl) in the presence/absence of 3.8 μM of HE (88.2 μM of disaccharide units). Protein concentrations were 1.0–1.4 μM. The protein specific heat capacity (c_p) was calculated as described by Privalov and Potekhin [36]. The partial molar volume and specific heat capacity of a fully unfolded protein were estimated according to Häckel et al. [37,38] and Makhatadze and Privalov [39,40].

2.4. Fluorescence Studies

Fluorescence spectra were measured using a Cary Eclipse spectrofluorometer (Varian Inc., Palo Alto, CA, USA), equipped with a Peltier-controlled cell holder, using quartz cells at 25 °C. Buffer conditions were the following: pH 7.4, 10 mM Na–phosphate, 50 mM NaCl. The rhVEGF$_{165}$ and HE concentration was 25 μM (or 580 μM of disaccharide units). Fluorescence of bis-ANS (2.5 μM) was excited at 385 nm. The emission band width was 5 nm. All spectra were fitted to log-normal curves [41] using LogNormal software (IBI RAS, Pushchino, Russia). Fluorescence emission maximum positions and quantum yields were obtained from these fits.

2.5. Dynamic Light Scattering Studies

DLS measurements were carried out using a Zetasizer Nano ZS (Malvern Instruments Ltd., Malvern, UK) system. The backscattered light from a 4 mW He-Ne laser 632.8 m was collected at an angle of 173°. The VEGF concentration was 11 μM (256 μM of disaccharide units). Buffer conditions were the following: pH 7.4, 10 mM Na–phosphate, 50 mM NaCl. Sample temperature was 25 °C. The acquisition time for the single autocorrelation function was 150 s. The resulting autocorrelation functions are averaged values from three measurements. The volume-weighted size distributions were calculated using the following parameters for the sodium phosphate buffer: refractive index of 1.333 measured with RL3 refractometer (PZO, Warszawa, Poland); the viscosity value for sodium phosphate buffer η = 0.882 mPa·s measured using micro-rheology method with a water suspension of standard latex nanoparticles. The hydrodynamic diameter for natively unfolded protein was calculated from the correlation equation for hydrodynamic volume V_h $\log(V_h) = f(\log(N))$, where N = 330 is the number of amino acid residues of dimeric VEGF$_{165}$ [42].

2.6. Chemical Crosslinking of Proteins

Crosslinking of rhVEGF$_{165}$ (25 μM) with 0.02% glutaraldehyde was performed at a pH of 7.4, with 10 mM Na–phosphate and 50 mM NaCl, in the absence/presence of 25 μM HE (or 580 μM of disaccharide units) as mainly described in [43]. The samples were subjected to sodium dodecyl sulfate–polyacrylamide gel electrophoresis (SDS-PAGE) under non-reducing conditions (5% concentrating and 12% resolving gels).

2.7. Modeling the Full Structure of Human VEGF$_{165}$ and Its Complex with HE

The tertiary structures of the RBD and HBD domains of VEGF$_{165}$ were taken as the basis for modeling (PDB entries 2VPF and 1VGH, respectively). The homodimeric form of the VEGF$_{165}$ protein was predicted by the AlphaFold server (https://alphafold.ebi.ac.uk/, accessed on 1 November 2022) in the multimer model option based on the VEGF$_{165}$ sequence (UniProt ID: P15692-4). The signal peptide was excluded from the protein amino acid sequence for the prediction. I-TASSER server (https://zhanggroup.org/I-TASSER/, accessed on 1 November 2022) predicted the monomeric subunits of VEGF$_{165}$ (HBD, linker, HBD). The alignment, refinement and visualization of 3D structures were performed using the Pairwise Structure Alignment RCSB PDB online service (https://www.rcsb.org/alignment, accessed on 1 November 2022), Stride server (http://webclu.bio.wzw.tum.de/cgi-bin/stride/stridecgi.py, accessed on 1 November 2022) and PyMOL v. 2.5 (https://pymol.org, accessed on 1 November 2022). The calculation of the content of secondary structure elements in the $VEGF_{165}$ model, calculation of the surface areas (solvent-accessible surface area, polar and apolar surface area) and the measurement of linear parameters were carried out using scripts in PyMOL v. 1.6 (https://pymol.org, accessed on 1 November 2022).

A total of 25 models of the $VEGF_{165}$–HE complex were generated using ClusPro docking server [44] in the heparin-binding mode [45]. The contact residues in the docking models of $VEGF_{165}$–HE complex were calculated using Python 3.5 programming language (implemented in PyCharm Community Edition 2020 development environment), Matplotlib Python plotting library and NumPy numerical mathematics extension. The numbering of the contact residues is according to the protein data bank (PDB) entries. The tertiary structure models were drawn with the molecular visualization system PyMOL v. 1.6.

3. Results and Discussion

3.1. The Effect of Heparin on the Secondary and Tertiary Structure of $VEGF_{165}$

It has been previously shown that the secondary structure of $VEGF_{165}$ consists predominantly of unordered (43%) and β-sheet (32%) parts [26]. Since the functional properties of $VEGF_{165}$ expressed in pro- and eukaryotes are identical, we have compared the secondary structure of the proteins derived from various sources. An analysis of far-UV CD spectra for intact $rhVEGF_{165}$, expressed in eukaryotic (CHO) and prokaryotic (*E. coli*) cells, reveals a similar content of these elements in the secondary structure (Table 1). An addition of equimolar amount of HE does not change the secondary structure of $rhVEGF_{165}$ (Table 1). Notably, the two-fold increase of HE concentrations also does not change the distribution of the secondary structure fractions. Early, Wijelath et al. showed for $VEGF_{165}$ that HE induces a partial transition of β-sheets to α-helices, but the source and preparation method of the protein, as well as buffer conditions of the measurements were not specified [26], which makes it difficult to compare the results.

Table 1. The secondary structure fractions estimated from the CD data using CDPro software package [35] for $rhVEGF_{165}$ expressed in CHO or *E. coli* at 25 °C (pH 7.4, 10 mM Na–phosphate, 50 mM NaCl). Protein concentration was 1.8–1.9 μM. HE was added up to an equimolar ratio to the protein.

Protein	α-Helices, %	β-Sheets, %	Turns, %	Unordered Structure, %
$rhVEGF_{165}$-CHO	8.8 ± 3.5	30.2 ± 4.5	22.7 ± 2.8	38.4 ± 4.8
$rhVEGF_{165}$-CHO + HE	8.1 ± 2.4	31.9 ± 2.9	22.3 ± 2.1	37.7 ± 3.2
$rhVEGF_{165}$-*E. coli*	8.7 ± 2.7	29.4 ± 2.8	22.3 + 2.7	39.6 ± 5.3
$rhVEGF_{165}$-*E. coli + HE*	7.0 ± 1.7	30.0 ± 2.2	22.7 ± 2.2	40.2 ± 4.6

To assess the heparin-induced alterations in $rhVEGF_{165}$ tertiary structure, we used DSC and hydrophobic probe bis-ANS fluorescence methods. The thermogram for $rhVEGF_{165}$ in the absence of heparin has no cooperative peaks characteristic of the protein thermal denaturation (Figure 1A) in the temperature range from 20 to 100 °C. The decrease in the thermogram at temperatures above 100 °C seems to be caused by a protein aggregation. The high level of specific heat capacity for $VEGF\text{-}A_{165}$ and the absence of cooperative transitions in the wide temperature region suggest the absence of a rigid tertiary structure in this protein. The addition of a two-fold amount of HE has practically no effect on the thermal behavior of the protein, but the increase of absolute heat capacity up to the heat capacity of fully unfolded VEGF evidences an increase of relative proportion of solvent accessible apolar areas.

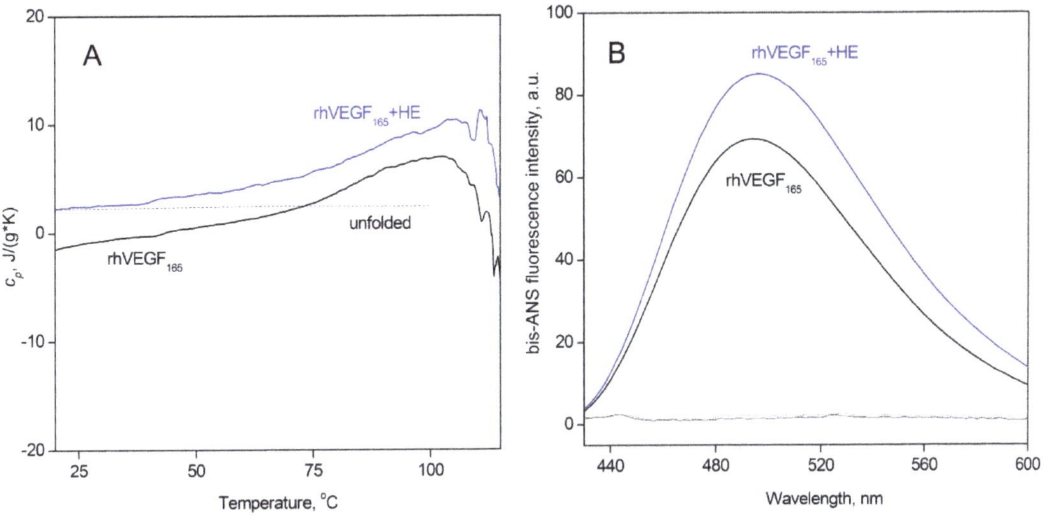

Figure 1. (**A**) Temperature dependencies of specific heat capacity for rhVEGF$_{165}$ (1 µM) in the absence (blue line) and the presence of 3.8 µM HE (black line). Grey dashed line corresponds to unfolded VEGF$_{165}$, calculated according to [38]; (**B**) Fluorescence spectra of bis-ANS (2.5 µM) in the absence (black line) and in the presence of heparin (blue line) in the solution containing 2.5 µM rhVEGF$_{165}$, monitored by fluorescence emission spectrum of the dye at 25 °C. Thin solid and dashed lines correspond to spectra of bis-ANS in buffer in the presence (grey line) or absence (dashed line) of HE. Buffer conditions: 10 mM Na-phosphate, 50 mM NaCl, pH 7.4.

Since the methods used hereinafter in our work (spectrofluorimetry, DLS, chemical crosslinking) require a large amount of protein, in some experiments we used rhVEGF$_{165}$ expressed in *E. coli*. The bacterial expression systems are well-suited to produce significant amounts of rhVEGF$_{165}$ with secondary structures identical to that of the protein expressed in eukaryotic cells (Table 1). Moreover, N-linked glycosylation of rhVEGF$_{165}$ was shown not to affect its functional properties [21,46,47].

The results of spectrofluorimetric monitoring of hydrophobic probe bis-ANS interactions with rhVEGF$_{165}$ in the absence/presence of HE are shown in Figure 1B. Interactions with nonpolar solvent-exposed cavities of proteins result in an increase of fluorescence quantum yield and a blue shift of the bis-ANS fluorescence spectrum maximum [48]. Heparin binding to rhVEGF$_{165}$ is accompanied by a 28% increase of bis-ANS quantum yield with a slight (2 nm) red shift of the spectrum, apparently reflecting heparin-induced exposure of the hydrophobic surface of rhVEGF$_{165}$. Similar effect was found early for ANS binding to bovine serum albumin in the presence of heparin and was explained by the formation of protein aggregates [49]. It can be assumed that the process of HE binding by two HBDs domains may promote a VEGF rearrangement with the opening of the cavity leading to an increase in the area of hydrophobic surfaces. The slight red shift in probe fluorescence may suggest that HE-binding to rhVEGF$_{165}$ induces the filling of additional sites for bis-ANS, in which the polarity of the probe environment is increased.

3.2. *HE-Dependent Changes in the Quaternary Structure of rhVEGF$_{165}$*

Since the observed HE-induced structural alteration of rhVEGF$_{165}$ may partly be a consequence of changes in oligomeric state of the proteins, we have examined this possibility using chemical crosslinking of the protein with 0.02% glutaraldehyde (37 °C for 1 h). We did not use reducing agents in the following SDS-PAGE analysis since all SH groups in the VEGF$_{165}$ homodimer are included into the intermolecular disulfide bridges. About a third of intact rhVEGF$_{165}$ exists as dimers and third of the protein could not penetrate the 12% resolving gels (high-molecular weight forms) (Figure 2A). Heparin binding to rhVEGF$_{165}$ leads to approximately a two-fold increase of high-molecular-weight forms and an almost complete disappearance of dimers.

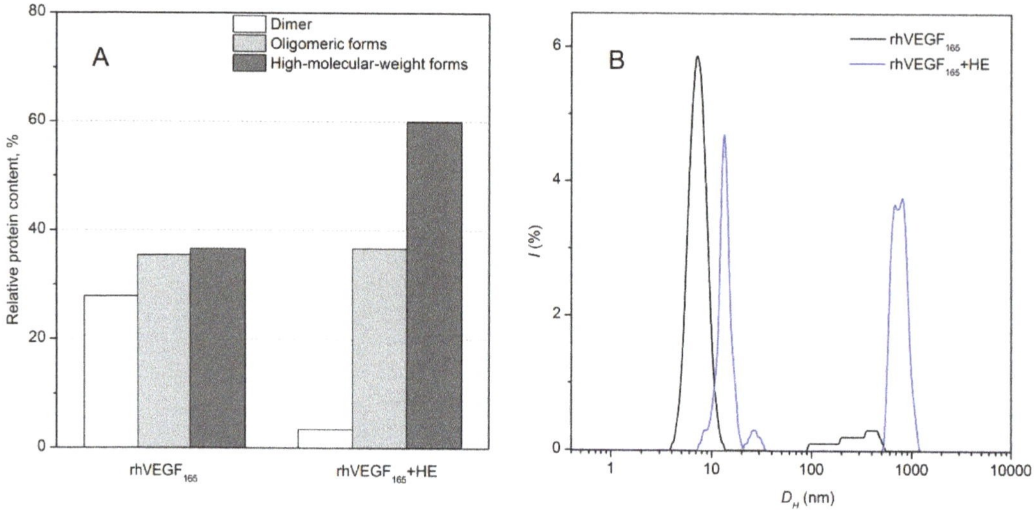

Figure 2. HE-dependent changes in quaternary structure of rhVEGF$_{165}$ (10 mM Na-phosphate, 50 mM NaCl, pH 7.4). (**A**) Effects of HE (25 μM) on distribution of rhVEGF$_{165}$ (25 μM) multimeric forms at 37 °C, as judged from glutaraldehyde crosslinking experiments at 37 °C; (**B**) DLS data on size distribution (hydrodynamic diameter D_h) of rhVEGF$_{165}$ (10 μM) in solutions in the absence (black line) and presence of 11 μM HE (blue line) at 25 °C.

DLS spectroscopy confirmed HE-induced oligomerization of rhVEGF$_{165}$ at 25 °C, which is clearly seen in the increase in its hydrodynamic diameter, D_h (Figure 2B). Reference rhVEGF$_{165}$ exhibits a size distribution of D_h = (6.5 ± 1.3) nm, which is satisfactory with that calculated for such a protein in the natively unfolded state [42], and the addition of equimolar quantity of HE causes a shift of the main scattered peak towards higher D_h values: (12.8 ± 1.4) nm together with aggregates formation detected at nearly 1 μm.

Heparin-induced aggregation of various proteins has been shown earlier (reviewed in [49]). Study of this process using bovine serum albumin as a model protein has revealed the downhill polymerization mechanism of the HE-induced protein aggregation [49]. Moreover, HS presents in most, if not all, intra- and extracellular protein aggregates that accumulate in neurodegenerative diseases, including those made of Aβ, tau, and α-synuclein (reviewed in [50]).

3.3. *Model of VEGF$_{165}$–HE Complex*

Since the three-dimensional structure of full VEGF$_{165}$ is still not clear, we used AlphaFold server for modeling the structure of homodimeric VEGF$_{165}$, Figure 3A. Two structural domains (RBD and HBD) are clearly seen in the figure. Moreover, the spatial arrangement of monomeric subunits during the dimerization contributes to the formation of a structural cavity in the center of the molecule. The linear parameters of the VEGF$_{165}$ homodimer model 65 × 73 × 81 Å agree with the D_h value. The secondary structure content calculated from the model using PyMOL (8.5%, 36.5% and 55 % for α-helices, β-sheets, and unordered structure, respectively) is consistent with the experimental data (Table 1). Taken together, these data indicate that our model adequately reflects the structure of VEGF$_{165}$.

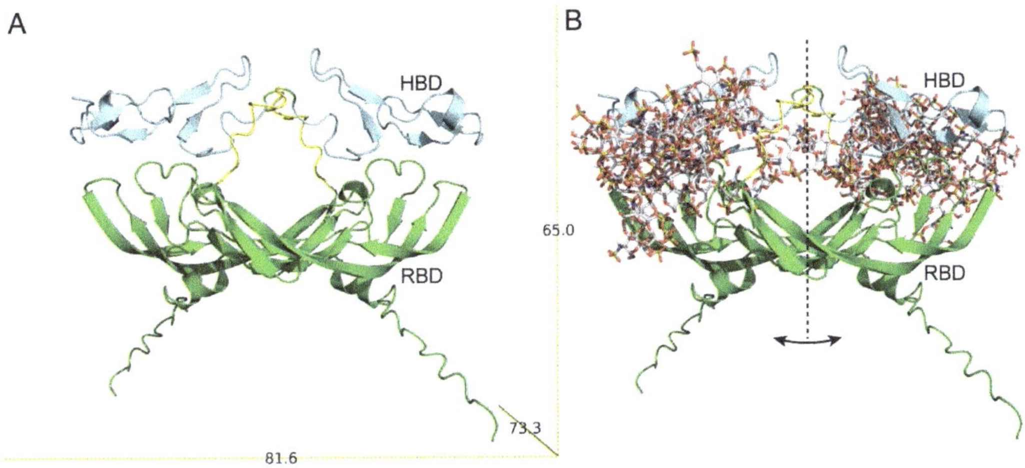

Figure 3. (**A**) The model of homodimeric $VEGF_{165}$. RBD and HBD domains are highlighted by green and cyan, respectively; (**B**) Overlay of the 25 representative models of the $VEGF_{165}$–HE complexes calculated using ClusPro docking server [44]. HE is shown in the stick representation.

In accordance with the literature data, the HBD of homodimeric VEGF would be preferred for HE binding, whereas a negative electrostatic potential of RBD is not prone to HE interactions [20]. This finding is partially confirmed by the docking results obtained for VEGF–HE models generated using the ClusPro docking server [44]: 25 top ranked solutions indicate that HE binds predominantly to HBD (Figure 3B). It should be noted that the model proposed in our work differs from the one proposed in [23] by a more compact arrangement of the HBDs. However, for some models, the binding sites include residues from RBD. Notably, our proposed model allows binding both by one HBD and by two HBDs (Figure 3B). Out of 25 models, only in 6 models HE interacts with only one chain of the homodimer $VEGF_{165}$. The following $VEGF_{165}$ residues contribute most to the HE binding: Lys48, Asn62, Glu64, Glu67, Gly65, and His86 from RBD, and Arg156, Arg159, Lys162, and Arg164 from HBD. Some of these residues (Lys48, His86) are part of the receptor-binding sites (PDB IDs 1FLT and 3V2A for VEGFR1 and VEGFR2, respectively). Notably, the bevacizumab-binding site in $VEGF_{165}$(PDB ID 6BFT) overlaps with some model $VEGF_{165}$-HE sites (His86, Gln87, Gly88, Gln89). In several models, contacts of HE with both linkers (between RBD and HBD) belonging to two chains were found.

The PyMOL program allows us to evaluate the solvent-accessible and buried areas in $VEGF_{165}$ and its complex with HE. SAS (solvent-accessible surface) polar and apolar areas are equivalent to the contact area between polar or non-polar amino acid residues and a solvent, while the buried surface area corresponds to the amino acid area excluded from the solvent interaction [51]. Since the exterior amino acids participate in protein–protein and protein–solvent interactions, a change in SASap (solvent-accessible apolar area) and SASpol (solvent accessible polar area) together with BSA (area buried from the solvent) correlates with driving forces in protein folding and conformational rearrangement upon binding [52]. It was shown that for intrinsically disordered proteins the higher the flexibility of the protein when compared to a folded protein of the same molecular mass, the deeper the conformational changes upon binding [52–54]. SAS of homodimeric $VEGF_{165}$ is composed predominantly of polar amino acids, forming SASpol. Notably, the SASpol is almost three times more than SASap while normal area ratio is about 55% of SASap and 45% of SASpol [55]. Upon HE binding, a part of polar positively charged amino acid residues becomes shielded, so SASpol's relative contribution to SAS decreases (Figure 3B and Table 2), which is qualitatively agreed with calorimetric and bis-ANS fluorescence data (Figure 1).

Table 2. Solvent-accessible area parameters calculated from the full structure models of $VEGF_{165}$ and $VEGF_{165}$ + HE using PyMOL v. 1.6.

	Total Area, Å^2	BSA, Å^2	SAS, Å^2	SASap, Å^2	SASpol, Å^2
$VEGF_{165}$	39,586	17,422	22,166	5713	16,453
$VEGF_{165}$ + HE	41,094	20,154	20,940	4856	11,254

SAS—solvent-accessible surface area; BSA—area buried from the solvent; SASap—solvent-accessible apolar area; SASpol—solvent-accessible polar area.

4. Conclusions

The effect of HE and its derivatives on the biological activity of VEGF may be partly explained by the HE-induced structural changes of VEGF revealed in our work. The experimental and modeling results obtained in our study allow us to consider $VEGF_{165}$ as an intrinsically disordered protein with abnormally large solvent-accessible surface areas. The interaction with HE does not affect the secondary structure of $VEGF_{165}$, but increases its hydrophobic areas exposed to the solvent and changes its quaternary structure. The aggregation of $rhVEGF_{165}$ can partly explain the increase of VEGFR2-bound $VEGF_{165}$ in the presence of HE [21]. Moreover, the predicted HE-binding site of VEGF shares several amino acids with bevacizumab and receptor-binding sites, which may explain the HE modulation of $VEGF_{165}$'s interaction with antibodies/VEGFRs [30]. In particular, the HE-shielding of positively charged residues of $VEGF_{165}$ in the VEGFR1 binding site could result in a decrease in protein–receptor binding efficiency, found earlier [27]. The presented modeling data allow us to propose a molecular mechanism for the participation of HE in the VEGF binding processes. In the absence of HE, $VEGF_{165}$ apparently can take on several conformations, which differ in the mutual orientation of HBDs due to flexible linkers. This flexibility allows the binding of $VEGF_{165}$ to the receptor and thus the activation of the signaling pathway [6]. Further investigations of these HE-mediated mechanisms could be of practical use for the production of novel biomaterials with well-understood and controlled properties for their implementation in tissue regeneration therapies.

Author Contributions: Conceptualization, M.P.S., E.L.N. and E.I.D.; methodology, M.P.S., E.L.N. and E.I.D.; software, A.V.M. and E.I.D.; validation, M.P.S., E.L.N. and E.I.D.; formal analysis, M.P.S., E.L.N., E.I.D., D.D.L., R.S.E.; investigation, M.P.S., E.L.N. and E.I.D.; resources, D.D.L. and R.S.E.; data curation, M.P.S., E.L.N., A.V.M. and E.I.D.; writing-original draft preparation, M.P.S., E.L.N., D.D.L. and E.I.D.; writing-review and editing, A.V.M.; visualization, M.P.S., E.L.N., A.V.M. and E.I.D.; supervision, E.I.D.; project administration, E.I.D.; funding acquisition, E.I.D. All authors have read and agreed to the published version of the manuscript.

Funding: This research was funded by the Russian Science Foundation, Grant Number 22-24-00083.

Institutional Review Board Statement: Not applicable.

Informed Consent Statement: Not applicable.

Data Availability Statement: Not applicable.

Acknowledgments: We are indebted to Eugene Permyakov (IBI RAS) for the fruitful discussions of the manuscript.

Conflicts of Interest: The authors declare no conflict of interest.

References

1. Apte, R.S.; Chen, D.S.; Ferrara, N. VEGF in Signaling and Disease: Beyond Discovery and Development. *Cell* **2019**, *176*, 1248–1264. [CrossRef] [PubMed]
2. Adamis, A.P.; Shima, D.T. The role of vascular endothelial growth factor in ocular health and disease. *Retina* **2005**, *25*, 111–118. [CrossRef] [PubMed]
3. Carmeliet, P.; Jain, R.K. Molecular mechanisms and clinical applications of angiogenesis. *Nature* **2011**, *473*, 298–307. [CrossRef] [PubMed]
4. Kut, C.; Mac Gabhann, F.; Popel, A.S. Where is VEGF in the body? A meta-analysis of VEGF distribution in cancer. *Br. J. Cancer* **2007**, *97*, 978–985. [CrossRef]

5. Holmes, D.I.R.; Zachary, I. The vascular endothelial growth factor (VEGF) family: Angiogenic factors in health and disease. *Genome Biol.* **2005**, *6*, 209. [CrossRef]
6. Ferrara, N.; Gerber, H.-P.; LeCouter, J. The biology of VEGF and its receptors. *Nat. Med.* **2003**, *9*, 669–676. [CrossRef]
7. Ferrara, N. VEGF-A: A critical regulator of blood vessel growth. *Eur. Cytokine Netw.* **2009**, *20*, 158–163. [CrossRef]
8. Nagy, J.A.; Dvorak, A.M.; Dvorak, H.F. VEGF-A and the Induction of Pathological Angiogenesis. *Annu. Rev. Pathol. Mech. Dis.* **2007**, *2*, 251–275. [CrossRef]
9. Olsson, A.-K.; Dimberg, A.; Kreuger, J.; Claesson-Welsh, L. VEGF receptor signalling - in control of vascular function. *Nat. Rev. Mol. Cell Biol.* **2006**, *7*, 359–371. [CrossRef]
10. Soker, S.; Takashima, S.; Miao, H.Q.; Neufeld, G.; Klagsbrun, M. Neuropilin-1 is expressed by endothelial and tumor cells as an isoform-specific receptor for vascular endothelial growth factor. *Cell* **1998**, *92*, 735–745. [CrossRef]
11. Krilleke, D.; Ng, Y.-S.E.; Shima, D.T. The heparin-binding domain confers diverse functions of VEGF-A in development and disease: A structure-function study. *Biochem. Soc. Trans.* **2009**, *37*, 1201–1206. [CrossRef] [PubMed]
12. Gitay-Goren, H.; Soker, S.; Vlodavsky, I.; Neufeld, G. The binding of vascular endothelial growth factor to its receptors is dependent on cell surface-associated heparin-like molecules. *J. Biol. Chem.* **1992**, *267*, 6093–6098. [CrossRef] [PubMed]
13. Gandhi, N.S.; Mancera, R.L. The structure of glycosaminoglycans and their interactions with proteins. *Chem. Biol. Drug Des.* **2008**, *72*, 455–482. [CrossRef] [PubMed]
14. Morla, S. Glycosaminoglycans and Glycosaminoglycan Mimetics in Cancer and Inflammation. *Int. J. Mol. Sci.* **2019**, *20*, 1963. [CrossRef] [PubMed]
15. Engelberg, H. Plasma heparin levels in normal man. *Circulation* **1961**, *23*, 578–581. [CrossRef]
16. Delcombel, R.; Janssen, L.; Vassy, R.; Gammons, M.; Haddad, O.; Richard, B.; Letourneur, D.; Bates, D.; Hendricks, C.; Waltenberger, J.; et al. New prospects in the roles of the C-terminal domains of VEGF-A and their cooperation for ligand binding, cellular signaling and vessels formation. *Angiogenesis* **2013**, *16*, 353–371. [CrossRef]
17. Muller, Y.A.; Christinger, H.W.; Keyt, B.A.; de Vos, A.M. The crystal structure of vascular endothelial growth factor (VEGF) refined to 1.93 A resolution: Multiple copy flexibility and receptor binding. *Structure* **1997**, *5*, 1325–1338. [CrossRef]
18. Muller, Y.A.; Li, B.; Christinger, H.W.; Wells, J.A.; Cunningham, B.C.; de Vos, A.M. Vascular endothelial growth factor: Crystal structure and functional mapping of the kinase domain receptor binding site. *Proc. Natl. Acad. Sci. USA* **1997**, *94*, 7192–7197. [CrossRef]
19. Houck, K.A.; Leung, D.W.; Rowland, A.M.; Winer, J.; Ferrara, N. Dual regulation of vascular endothelial growth factor bioavailability by genetic and proteolytic mechanisms. *J. Biol. Chem.* **1992**, *267*, 26031–26037. [CrossRef]
20. Fairbrother, W.J.; Champe, M.A.; Christinger, H.W.; Keyt, B.A.; Starovasnik, M.A. Solution structure of the heparin-binding domain of vascular endothelial growth factor. *Structure* **1998**, *6*, 637–648. [CrossRef]
21. Keyt, B.A.; Berleau, L.T.; Nguyen, H.V.; Chen, H.; Heinsohn, H.; Vandlen, R.; Ferrara, N. The carboxyl-terminal domain (111-165) of vascular endothelial growth factor is critical for its mitogenic potency. *J. Biol. Chem.* **1996**, *271*, 7788–7795. [CrossRef] [PubMed]
22. Stauffer, M.E.; Skelton, N.J.; Fairbrothe, W.J. Refinement of the solution structure of the heparin-binding domain of vascular endothelial growth factor using residual dipolar couplings. *J. Biomol. NMR* **2002**, *23*, 57–61. [CrossRef] [PubMed]
23. Uciechowska-Kaczmarzyk, U.; Babik, S.; Zsila, F.; Bojarski, K.K.; Beke-Somfai, T.; Samsonov, S.A. Molecular dynamics-based model of VEGF-A and its heparin interactions. *J. Mol. Graph. Model.* **2018**, *82*, 157–166. [CrossRef] [PubMed]
24. Guryanov, I.; Tennikova, T.; Urtti, A. Peptide Inhibitors of Vascular Endothelial Growth Factor A: Current Situation and Perspectives. *Pharmaceutics* **2021**, *13*, 1337. [CrossRef] [PubMed]
25. Zhao, W.; McCallum, S.A.; Xiao, Z.; Zhang, F.; Linhardt, R.J. Binding affinities of vascular endothelial growth factor (VEGF) for heparin-derived oligosaccharides. *Biosci. Rep.* **2012**, *32*, 71–81. [CrossRef] [PubMed]
26. Wijelath, E.; Namekata, M.; Murray, J.; Furuyashiki, M.; Zhang, S.; Coan, D.; Wakao, M.; Harris, R.B.; Suda, Y.; Wang, L.; et al. Multiple mechanisms for exogenous heparin modulation of vascular endothelial growth factor activity. *J. Cell. Biochem.* **2010**, *111*, 461–468. [CrossRef]
27. Ito, N.; Claesson-Welsh, L. Dual effects of heparin on VEGF binding to VEGF receptor-1 and transduction of biological responses. *Angiogenesis* **1999**, *3*, 159–166. [CrossRef]
28. Dougher, A.M.; Wasserstrom, H.; Torley, L.; Shridaran, L.; Westdock, P.; Hileman, R.E.; Fromm, J.R.; Anderberg, R.; Lyman, S.; Linhardt, R.J.; et al. Identification of a heparin binding peptide on the extracellular domain of the KDR VEGF receptor. *Growth Factors* **1997**, *14*, 257–268. [CrossRef]
29. Ferrara, N.; Hillan, K.J.; Novotny, W. Bevacizumab (Avastin), a humanized anti-VEGF monoclonal antibody for cancer therapy. *Biochem. Biophys. Res. Commun.* **2005**, *333*, 328–335. [CrossRef]
30. Tsiros, D.; Sheehy, C.E.; Pecchia, S.; Nugent, M.A. Heparin potentiates Avastin-mediated inhibition of VEGF binding to fibronectin and rescues Avastin activity at acidic pH. *J. Biol. Chem.* **2019**, *294*, 17603–17611. [CrossRef]
31. Ko, S.Y.; Lee, W.; Kenny, H.A.; Dang, L.H.; Ellis, L.M.; Jonasch, E.; Lengyel, E.; Naora, H. Cancer-derived small extracellular vesicles promote angiogenesis by heparin-bound, bevacizumab-insensitive VEGF, independent of vesicle uptake. *Commun. Biol.* **2019**, *2*, 386. [CrossRef]
32. Pace, C.N.; Vajdos, F.; Fee, L.; Grimsley, G.; Gray, T. How to measure and predict the molar absorption coefficient of a protein. *Protein Sci.* **1995**, *4*, 2411–2423. [CrossRef]

33. Farris, F.J.; Weber, G.; Chiang, C.C.; Paul, I.C. Preparation, crystalline structure, and spectral properties of the fluorescent probe 4,4′-bis-1-phenylamino-8-naphthalenesulfonate. *J. Am. Chem. Soc.* **1978**, *100*, 4469–4474. [CrossRef]
34. Fu, L.; Li, G.; Yang, B.; Onishi, A.; Li, L.; Sun, P.; Zhang, F.; Linhardt, R.J. Structural characterization of pharmaceutical heparins prepared from different animal tissues. *J. Pharm. Sci.* **2013**, *102*, 1447–1457. [CrossRef] [PubMed]
35. Sreerama, N.; Venyaminov, S.Y.; Woody, R.W. Estimation of protein secondary structure from circular dichroism spectra: Inclusion of denatured proteins with native proteins in the analysis. *Anal. Biochem.* **2000**, *287*, 243–251. [CrossRef]
36. Privalov, P.L.; Potekhin, S.A. Scanning microcalorimetry in studying temperature-induced changes in proteins. *Methods Enzymol.* **1986**, *131*, 4–51. [CrossRef] [PubMed]
37. Häckel, M.; Hinz, H.J.; Hedwig, G.R. Partial molar volumes of proteins: Amino acid side-chain contributions derived from the partial molar volumes of some tripeptides over the temperature range 10-90 degrees C. *Biophys. Chem.* **1999**, *82*, 35–50. [CrossRef]
38. Häckel, M.; Hinz, H.J.; Hedwig, G.R. A new set of peptide-based group heat capacities for use in protein stability calculations. *J. Mol. Biol.* **1999**, *291*, 197–213. [CrossRef]
39. Privalov, P.L.; Makhatadze, G.I. Heat capacity of proteins. II. Partial molar heat capacity of the unfolded polypeptide chain of proteins: Protein unfolding effects. *J. Mol. Biol.* **1990**, *213*, 385–391. [CrossRef]
40. Makhatadze, G.I.; Privalov, P.L. Heat capacity of proteins. I. Partial molar heat capacity of individual amino acid residues in aqueous solution: Hydration effect. *J. Mol. Biol.* **1990**, *213*, 375–384. [CrossRef]
41. Burstein, E.A.; Permyakov, E.A.; Emelyanenko, V.I.; Bushueva, T.L.; Pechere, J.F. Investigation of some physico-chemical properties of muscular parvalbumins by means of the luminescence of their phenylalanyl residues. *Biochim. Biophys. Acta* **1975**, *400*, 1–16. [CrossRef] [PubMed]
42. Uversky, V.N. Natively unfolded proteins: A point where biology waits for physics. *Protein Sci.* **2002**, *11*, 739–756. [CrossRef] [PubMed]
43. Nemashkalova, E.L.; Permyakov, E.A.; Permyakov, S.E.; Litus, E.A. Modulation of linoleic acid-binding properties of human serum albumin by divalent metal cations. *Biometals* **2017**, *30*, 341–353. [CrossRef]
44. Desta, I.T.; Porter, K.A.; Xia, B.; Kozakov, D.; Vajda, S. Performance and Its Limits in Rigid Body Protein-Protein Docking. *Structure* **2020**, *28*, 1071–1081.e3. [CrossRef]
45. Mottarella, S.E.; Beglov, D.; Beglova, N.; Nugent, M.A.; Kozakov, D.; Vajda, S. Docking server for the identification of heparin binding sites on proteins. *J. Chem. Inf. Model.* **2014**, *54*, 2068–2078. [CrossRef] [PubMed]
46. Claffey, K.P.; Senger, D.R.; Spiegelman, B.M. Structural requirements for dimerization, glycosylation, secretion, and biological function of VPF/VEGF. *Biochim. Biophys. Acta* **1995**, *1246*, 1–9. [CrossRef]
47. Walter, D.H.; Hink, U.; Asahara, T.; Van Belle, E.; Horowitz, J.; Tsurumi, Y.; Vandlen, R.; Heinsohn, H.; Keyt, B.; Ferrara, N.; et al. The in vivo bioactivity of vascular endothelial growth factor/vascular permeability factor is independent of N-linked glycosylation. *Lab. Investig.* **1996**, *74*, 546–556. Available online: http://www.ncbi.nlm.nih.gov/pubmed/8780172 (accessed on 1 November 2022). [PubMed]
48. Hawe, A.; Sutter, M.; Jiskoot, W. Extrinsic fluorescent dyes as tools for protein characterization. *Pharm. Res.* **2008**, *25*, 1487–1499. [CrossRef]
49. Ahanger, I.A.; Parray, Z.A.; Nasreen, K.; Ahmad, F.; Hassan, M.I.; Islam, A.; Sharma, A. Heparin Accelerates the Protein Aggregation via the Downhill Polymerization Mechanism: Multi-Spectroscopic Studies to Delineate the Implications on Proteinopathies. *ACS Omega* **2021**, *6*, 2328–2339. [CrossRef]
50. Maïza, A.; Chantepie, S.; Vera, C.; Fifre, A.; Huynh, M.B.; Stettler, O.; Ouidja, M.O.; Papy-Garcia, D. The role of heparan sulfates in protein aggregation and their potential impact on neurodegeneration. *FEBS Lett.* **2018**, *592*, 3806–3818. [CrossRef]
51. Mitternacht, S. FreeSASA: An open source C library for solvent accessible surface area calculations. *F1000Research* **2016**, *5*, 189. [CrossRef] [PubMed]
52. Marsh, J.A.; Teichmann, S.A. Relative Solvent Accessible Surface Area Predicts Protein Conformational Changes upon Binding. *Structure* **2011**, *19*, 859–867. [CrossRef]
53. Rambo, R.P.; Tainer, J.A. Characterizing flexible and intrinsically unstructured biological macromolecules by SAS using the Porod-Debye law. *Biopolymers* **2011**, *95*, 559–571. [CrossRef] [PubMed]
54. Marsh, J.A. Buried and Accessible Surface Area Control Intrinsic Protein Flexibility. *J. Mol. Biol.* **2013**, *425*, 3250–3263. [CrossRef] [PubMed]
55. Freire, E. Thermal denaturation methods in the study of protein folding. *Methods Enzymol.* **1995**, *259*, 144–168. [CrossRef] [PubMed]

biomolecules

MDPI

Article

Proteome Profile Changes Induced by Heterologous Overexpression of *Mycobacterium tuberculosis*-Derived Antigens PstS-1 (Rv0934) and Ag85B (Rv1886c) in *Mycobacterium microti*

Viridiana García-Ruiz [1], Patricia Orduña [2], Antonia I. Castillo-Rodal [1], Teresa J. Flores-Rodríguez [1] and Yolanda López-Vidal [1,*]

1 Programa de Inmunología Molecular Microbiana, Departamento de Microbiología y Parasitología, Facultad de Medicina, Universidad Nacional Autónoma de México, Av. Ciudad Universitaria 3000, Coyoacán, Ciudad de México 04510, CP, Mexico

2 Laboratorio de Microbioma, División de Investigación, Facultad de Medicina, Universidad Nacional Autónoma de México, Av. Ciudad Universitaria 3000, Coyoacán, Ciudad de México 04510, CP, Mexico

* Correspondence: lvidal@unam.mx; Tel.: +52-(55)-5623-2148

Abstract: The development of new tuberculosis vaccines remains a global priority, and recombinant vaccines are a frequently investigated option. These vaccines follow a molecular strategy that may enhance protective efficacy. However, their functional differences, particularly with respect to glycosylation, remain unknown. Recent studies have shown that glycosylation plays a key role in the host-pathogen interactions during immune recognition. The aim of this study was to determine the differences in the glycosylation profiles of two recombinant strains of *Mycobacterium microti*, overexpressing Ag85B (Rv1886c) and PstS-1 (Rv0934) antigens of *M. tuberculosis*. For each strain, the glycosylation profile was determined by Western blotting with lectins. The results showed the presence of mannosylated proteins and evidence of linked sialic acid proteins. Interestingly, different proteome and glycoproteome profiles were observed between the two recombinant strains and the wild-type strain. We have shown here that the construction of the recombinant strains of *M. microti* has altered the proteome and glycosylation profiles of these strains, leading us to ask what impact these changes might have on the immune response.

Keywords: vaccine; antigen; glycosylation; *Mycobacterium microti*; *Mycobacterium tuberculosis*; BCG; proteome profile

1. Introduction

Tuberculosis (TB) has been officially recognized as a global emergency since 1993 [1,2] and is caused by airborne intracellular bacilli of the *Mycobacterium tuberculosis* (Mtb) complex [3]. According to the Global Tuberculosis Report 2019, TB remains one of the top ten causes of death worldwide [4] and is the leading cause of death associated with an infectious disease in adults [1], particularly in low-income countries [2]. Every day, nearly 4000 people die due to TB [4,5], and part of this high number is due the lack of effective vaccines and diagnostic tools [6,7].

The bacillus Calmette-Guerin (BCG) of *Mycobacterium bovis* is currently the only vaccine available worldwide for the prevention of tuberculosis, with a protective efficacy of up to 80% against severe forms of TB (i.e., miliary and meningeal) [8,9]. However, against pulmonary TB, the most common form of the disease, this vaccine is only 50–80% effective [10]. These variations in the protective efficacy have inspired efforts to develop new vaccines, including subunit vaccines [11–13], live vaccines against bacteria within the same complex [14] and recombinant vaccines consisting of immunogenic antigens [15,16].

So far, none of these vaccines has successfully elicited a better immune response than the BCG vaccine [17,18].

Other attempts to achieve better results than the BCG vaccine have been investigated, in particular live bacterial vaccines, with *Mycobacterium vaccae* or *Mycobacterium microti*, which belong to the *Mycobacterium tuberculosis* complex (MTBC) [19,20]. *M. microti* causes pulmonary TB in voles but is naturally attenuated in humans [19]. Between 1950 and 1969, this strain was used as a TB vaccine, and showed approximately a 77% protective efficacy with the same immunogenicity against TB [21]. However, about 3–17% of patients that were vaccinated with *M. microti* developed skin reactions with a more lupoid aspect and a greater severity than those vaccinated with BCG, which led to discontinuation of the *M. microti* vaccine [22]. This effect was attributed to the administration of the *M. microti* vaccine with multiple pricks. *M. microti* has been used as a vaccine in animal models and for alternative routes of administration, such as intraperitoneal, orogastric and intravenous administration. These studies demonstrated that *M. microti* conferred similar protection levels as the post-infection BCG vaccine when tested against the *M. tuberculosis* strain CDC1551 in mice [19,23].

The increase in multi-drug resistance of Mtb compared to classical treatments such as rifampicin and isoniazid [24,25], estimated to cost USD 13 billion annually worldwide [4], has made the development of more effective vaccines and diagnostics more urgent. In this sense, the research into Mtb virulence factors such as immunogenic antigens are attractive, as they have shown promising results when used for these purposes [18].

Mtb produces several antigenic proteins, some of which are post-translationally modified, notably by mannose type glycan attachments [26–29] which contribute to disease persistence by allowing bacteria to colonize and invade the host cell [28,30–32]. These secreted Mtb proteins, including PstS-1 and Ag85B, trigger antigen-specific host immune responses [33]. Korycka-M et al., postulated that the PPE51 protein participates in the uptake of disaccharides from *M. tuberculosis* [34].

PstS-1 (Rv0934) is a 38 kDa cell wall glycol lipoprotein with attached mannoses. This antigen has been described as an adhesin that induces the proliferation of CD4+ T cells and induces the expression of high levels of interferon (IFN)-γ and immunoglobulin (Ig)G2a [35]. In addition, PstS-1 induces the upregulated production of interleukin (IL)-12 by macrophages and stimulates the CD8+ cytotoxic T-cells [36–39].

Ag85B (Rv1886c) is a 30 kDa mycolil transferase that plays a role in the synthesis of the cell wall components. It is one of the main components of the culture filtrates in both the *M. tuberculosis* and *M. bovis* BCG vaccine [40]. Ag85B also belongs to a group of three proteins known as complex 85, which is a component of the cell wall in all species of mycobacteria [41]. This antigen has been extensively studied and tested during the production of DNA, protein and recombinant subunit vaccines [42–45].

Glycosylation is a post-translational modification (PTM) and refers to the covalent linkage between carbohydrates and a polypeptide, lipid, polynucleotide or organic compound [46–49]. In Mtb, this PTM is involved in various cellular mechanisms, including pathogenesis, virulence and antigenic properties such as the stimulation of the immune response against various pathogens [31,39,50].

The use of mycobacterial glycoproteins is an option for the development of a TB vaccine. Studies have shown that glycosylation plays a key role in the T-cell antigenicity, suggesting that the use of glycosylated proteins in vaccines could enhance the induction of a protective immune response [51]. This effect has been associated, for example, with the native glycosylation pattern of APA (Rv1860), also known as 45/47 kDa. When the pattern is altered, a lower stimulation of T-lymphocyte responses is observed [52,53]. However, there is little research about the role of glycosylation on recombinant antigens and its importance in the immunological response against TB.

The first glycosylated proteins in *M. tuberculosis* were described in the 1970s and 1980s, and recent quantitative glycoproteomic analyses have revealed 101 differentially glycosylated sites in 67 proteins in four *M. tuberculosis* strains [54]. Most proteins that have been described for Mtb are O-mannosylated proteins. However, the presence of other types of glycosylation, glycan types, glycan occupancy and glycan length are poorly understood.

Most of the identified glycosylated proteins/lipids in mycobacteria are located in the cell wall or are secreted. These glycoproteins are involved in several biological mechanisms, such as cell adhesion, invasion, cellular signaling, intracellular bacterial survival, antigenicity, pathogenicity and virulence, among others [50,54]. Nandakumar et al. (2013) have shown that mannosylation of the adhesin APA in *M. tuberculosis* is essential for the human T-cell activation in a murine model of subcutaneous infection. In addition, the differences in cytokine production, including IFN-γ, the tumor necrosis factor (TNF)-α and IL-2, have been demonstrated following the immunization with the APA protein, independent of the mannosylated state [27]. The importance of glycosylation for various *M. tuberculosis* proteins and the effects of glycosylation on the interactions with the immune response need to be further explored to understand their implications for vaccine development [51].

Recently, our research group studied the *M. microti* strain ATCC 11,152 in a murine model of pulmonary TB, and we observed that the bacterial burden of *M. tuberculosis* was significantly reduced, resulting in a lower percentage of pneumonia in lung tissue after it was challenged with *M. tuberculosis* H37Rv (data not shown). Based on these results and the positive outcomes of recombinant vaccines, we decided to use *M. microti* as a vehicle for the overexpression of the *M. tuberculosis* antigens PstS-1 and Ag85B. These antigens have been widely studied and are known to stimulate T- and B-lymphocyte proliferation and elicit a T-helper (Th)1-type response in animal models [37,38,55–57]. These antigens are encoded in the genome of *M. microti*, but little is known about the in vitro and in vivo expression of the two antigens in this strain.

In the present study, we developed two recombinant strains of *M. microti*, overexpressing these PstS-1 and Ag85B antigens. The protein expression of these antigens was performed in a constitutive vector and analyzed with SDS-PAGE and 2D-PAGE. In addition, the proteomic profiles of the developed strains and the wild-type strain were examined and the differences in the proteomic profiles between the three strains were found, suggesting that the recombination event not only affected the protein expression but also changed the global proteomic profile of these strains.

2. Materials and Methods

2.1. Reagents

The culture media were purchased from Thermo Fisher Scientific™ (BD Difco™ Middlebrook 7H10 Agar and Middlebrook 7H9 Broth Base, Yeast Extract, Bacto™ Triptone; Becton Dickinson, Detroit, MI, USA), and JT Baker, Phillipsburg, NJ, USA (NaCl, KCl, $MgSO_4$, $CaCl_2$, Glycerol, KOAc, Glucose, L-Asparagine, KH_2PO_4, Ferric Ammonium Citrate, Citric Acid, Sodium Pyruvate).

The molecular biology reagents were purchased from Promega™, Madison, WI, USA (Wizard® SV Gel and PCR Clean-Up System), SigmaTM, St Louis, MO, USA (Ethidium Bromide, Agarose, Kanamycin, Tween® 80), InvitrogenTM, Carlsbad, CA, USA (TOPO™ Cloning Kit with pCR™2.1, Anza™ T4 DNA Ligase Master Mix) and New England BioLabs®, Ipswich, MA, USA (BamHI, EcoRI, HindIII). The pMV261 plasmid was provided by Clara Espitia and Luis Servín (Instituto de Investigaciones Biomédicas, UNAM, Mexico).

The proteomic reagents were purchased from Bio-Rad, Hercules, CA, USA (Bradford Assay, Bio-Lyte® 3/10 Ampholyte, ReadyStrip™ IPG Strips) and Amersham Biosciences, Buckinghamshire, UK (CyDye™ DIGE Fluor Labelling Kit).

The following reagents were obtained through TBVTRM Contract, Colorado State University, USA: purified protein PstS1 from *Mycobacterium tuberculosis* (NR-14859), monoclonal anti-*Mycobacterium tuberculosis* PhoS1/PstS1 (NR-13605) and purified protein Ag85B from *Mycobacterium tuberculosis* H37Rv (NR-53526).

2.2. Strains

For the transformation and expression of the antigens PstS-1 (Rv0934) and Ag85B (Rv1886c), we used *Mycobacterium microti* ATCC 11152, provided by Dr. Gutiérrez-Pabello (Faculty of Medicine UNAM). For the DNA extraction and cloning, we used the strain *Mycobacterium tuberculosis* H37Rv provided by Dra. Torres (INER). For the propagation, cloning and subcloning of the constructed vectors, we used *E. coli* DH5α competent cells provided by Dr. L. Servín (Instituto de Investigaciones Biomédicas, UNAM, Mexico).

2.3. DNA Extraction and Sequencing

The PstS-1 (Rv0934) and Ag85B (Rv1886c) coding genes in *M. microti* and *M. tuberculosis* were amplified by a polymerase chain reaction (PCR) and submitted to sequencing. Initially, *M. microti* and *M. tuberculosis* were grown in BD Difco™ Middlebrook 7H10 agar and incubated at 37 °C for 15 days. After the appearance of colonies, a single colony of each strain was taken and inoculated into 100 mL of Middlebrook 7H9 broth base supplemented with 0.05% Tween 80 and 10% ADC (Albumin-dextrose-catalase) in flasks, allowed to grow at 37 °C and shaken at 150 rpm until an A.U. of 0.6 (O.D. 600 nm) was achieved. The cells were harvested using centrifugation at 10,000 rpm for 15 min at 4 °C. The supernatant was discarded, and the pellet resuspended in 10 mL of glycerol 10%. The genomic DNA was extracted using the lysozyme and proteinase K method [58]. The DNA was resuspended in 50 µL of distilled H_2O and stored at −20 °C for further use. Specific primers were designed for each gene based on the genome of *M. tuberculosis* H37Rv (GeneBank: AL123456.3). For PstS-1 (Rv0934), the forward primer 5′-ATAGTACGAGGATCCTGAAAATTCGTTTGC-3′ and the reverse primer 5′-ATAGTACGAGAATTCCTAGCTGGAAATCGT-3′ were used with the restriction sites BamHI and EcoRI (underlined). For Ag85B (Rv1886c), the forward primer 5′-GGCTGAATTCATGACAGACGTGAGCCGAAA-3′ and the reverse primer 5′-CGCGAAGCTTCAGCCGGCGCCTAACGAAC-3′ were used with the restriction sites EcoRI and HindIII (underlined). The PCR conditions for PstS-1 (Rv0934) were used as follows: one initial denaturation cycle at 94 °C followed by 30 cycles of 1 min at 94 °C for denaturation, 1 min at 56 °C for annealing and 1 min at 72 °C for elongation. For Ag85B (Rv1886c), there was an initial denaturation cycle at 94 °C and then 35 cycles of 1 min at 94 °C for denaturation, 1 min at 55 °C for annealing and 1 min at 72 °C for elongation. The PCR amplicons were visualized with a 2% gel electrophoresis and stained with ethidium bromide. They were then cleaned using the Wizard® SV Gel and the PCR Clean-Up System and subcloned using a TOPO™ cloning kit with pCR™2.1 according to the manufacturer's instructions. Subsequently, the resulting vector constructs were sequenced using commercially available primers: M13 in an ABI PRISM 3100 Genetic Analyzer (Applied Biosystems™, Foster City, CA, USA) at the Biology Institute, UNAM. The sequences were assembled using the SnapGene® viewer 6.1 bioinformatics software (Dotmatics, Boston, MA, USA) and analyzed using the Basic Local Alignment Search Tool (BLAST; NCBI, Bethesda, MD, USA) and the Clustal Omega [59] bioinformatics tools. The constructed vectors of *M. microti*-TOPO-PstS-1, *M. microti*-TOPO-Ag85B, *M. tuberculosis*-TOPO-PstS1 and *M. tuberculosis*-TOPO-Ag85B were transformed into the competent *E. coli* DH5α by heat shock using standard procedures [60] and stored at −20 °C for further use.

2.4. Vector Construction

The vectors *M. tuberculosis*-TOPO-PstS1 and *M. tuberculosis*-TOPO-Ag85B were purified from *E. coli* GM2199 and digested with the restriction enzymes BamHI/EcoRI and EcoRI/HindIII, respectively, to obtain the desired sequences. The PstS-1 and Ag85B gene inserts were then resolved with 1% agarose electrophoresis and purified using the Wizard® SV Gel and the PCR Clean-Up System. A reaction of the previously BamHI/EcoRI and EcoRI/HindIII digested vectors pMV261 was ligated with the purified gene inserts PstS-1 and Ag85B using the Anza™ T4 DNA Ligase Master Mix according to the manufacturer's instructions. The correct cloning of the inserts in the vectors designated pMV261-PstS-1 and pMV261-Ag85B, in which the promoter was Hsp60, was confirmed by the sequencing and then transformed into the competent *E. coli* DH5α as described above. The isolation of the plasmids from the *E. coli* strains was performed using the alkaline lysis method. The transformation of the plasmids pMV261 (control), pMV261-PstS-1 and pMV261-Ag85B into *M. microti* was achieved by electroporation. Briefly, 1 µg of purified DNA from each plasmid was added to the competent *M. microti* cells, and the mixture was incubated on ice for 10 min. The mixture was then placed in a Gene Pulser™ 0.2 cm electroporation cuvette (Bio-Rad, Richmond, CA, USA) and a 2.5 kV current pulse was applied. The cuvette was placed on ice for 4 min. Then 1 mL of Middlebrook 7H9 broth base was added for both strains. The culture was incubated overnight at 37 °C with stirring at 150 rpm. Finally, the cells were harvested using centrifugation and the different dilutions were spread on the BD Difco™ Middlebrook 7H10 Agar containing 30 µg/mL kanamycin. The plates were incubated at 37 °C for 3 to 4 weeks until the presence of colonies was observed. The colonies were selected and the corresponding inserts were confirmed by sequencing, as described above. The strains were designated *M. microti*-pMV261 (as control), *M. microti*-PstS-1 and *M. microti*-Ag85 and cryopreserved at −80 °C for further experiments.

2.5. Antigen Expression in M. microti Strains

The strains *M. microti*-pMV261, *M. microti*-PstS-1 and *M. microti*-Ag85 were grown in 250 mL flasks containing 50 mL of Sauton media (composition per liter: 4 g L-asparagine, 0.5 g monopotassium phosphate, 0.5 g magnesium sulphate, 50 mg ferric ammonium citrate, 2 g citric acid, 60 mL glycerol and 0.05% Tween 80), supplemented with 40 mM of glucose and 40 mM of sodium pyruvate at 37 °C with constant stirring at 150 rpm. The kinetic growth was determined every 48 h for 196 h by measuring the O.D. (600 nm) and expressed in absorbance units (A.U.). At the end of the culture, 1 mM of PMSF (phenylmethylsulphonyl fluoride) as a protease inhibitor, 100 µM of EDTA and 0.05% sodium azide were added to the supernatant. The supernatant was concentrated to a volume of 10 mL using an Amicon® Stirred Cell (Merck™) system and ultrafiltration discs with a nominal molecular weight of 3 kDa (Merck™). It was then filtered again using an Amicon® Ultra-15 Centrifugal Filter Unit (Merck™). The proteins from the cell extract were recovered by sonicating the centrifuged bacterial pellet with 15 pulses of 100 mV (1 min on/1 min off) at 4 °C in an Ultrasonic Processor (Cole-Parmer®). To prevent proteolysis, 1 mM of PMSF was added during the sonication. Finally, the concentrated supernatants and cellular extracts of the three strains were recovered and stored at −70 °C until further use.

2.6. SDS-PAGE, Immunoblot and Lectin Blot Analysis

The proteins from the cell extracts and supernatants of each strain were quantified using the Bradford Assay (Bio-Rad, Hercules, CA, USA) and 20 µg of each sample was resolved with SDS-PAGE 12% at 100 V for 2 h. For the immunoblot detection, the proteins were transferred to PVDF Amersham™ Hybond® P Western blotting membranes (Amersham Biosciences, Piscataway, NJ, USA) in a Trans-Blot® SD Semi-Dry Transfer Cell (Bio-Rad, Richmond, CA, USA) at 20 V for 30 min. Subsequently, the PstS-1 and Ag85B antigens were immunodetected with TB71 and α-Ag85 monoclonal antibodies (BEI Resources, TBVTRM Contract, Colorado State University, USA), respectively. Briefly, the transferred membrane was blocked with blocking solution (5.0% skimmed milk in TBS-Tween-20, 136 mM NaOH, 2 mM KCl, 25 mM Tris and 0.05% Tween-20) at room temperature (RT) and stirred for 2 h. The membrane was then incubated three times, for 10 min each in TBS-Tween and incubated overnight with the diluted primary antibodies: 1:100 for TB71 and 1:1000 for α-Ag85 in blocking solution at 4 °C with shaking. The membranes were then washed three times with TBS-Tween 20 and incubated with secondary goat anti-mouse IgG (Jackson ImmunoResearch Inc., West Grove, PA, USA) for the PstS-1 membranes and anti-rabbit IgG (ZYMED, Carlsbad CA, USA) for the Ag85B membranes in blocking solution at RT for 2 h with stirring. The membranes were then washed three times for 10 min with TBS-Tween 20. Finally, the membranes were revealed with a substrate diluted with luminol and detected using a chemiluminescent substrate (DuoLuX; Vector Laboratories, Burlingame, CA, USA).

For the lectin blot analysis, 1 μg of the sample proteins was resolved by 12% PAGE, silver-stained and transferred to a PVDF membrane as previously described. The membranes were then analyzed for glycoprotein identification using a DIG Glycan Differentiation Kit™ (Roche, Mannheim, Germany) according to the manufacturer's instructions. Additionally, glycosylation prediction profiles were performed to evaluate the in silico potential glycosylated residues using predefined algorithms from the bioinformatic software NetOGlyc-4.0 [61], Glycosylation Predictor GPP [62] and GLYCOPP V1.0 [63]. The SDS-PAGE immunoblots and lectin blots were visualized using a GS-800™ Calibrated Imaging Densitometer (Bio-Rad). The image processing and further analysis was performed using the Quantity One™ Image Analysis Software v.4.6 (Bio-Rad, Hercules, CA, USA) and Image J 1.53k.

2.7. Protein Labelling and Two-Dimensional Fluorescence Differential Gel Electrophoresis

The extracted proteins from the recombinant *M. microti* strains were fluorescently labeled and analyzed by two-dimensional fluorescence differential gel electrophoresis (2-D DIGE) in duplicate. A CyDye™ DIGE Fluor Labelling Kit (Amersham Biosciences, Buckinghamshire, UK) was used to differentially label the cell extracts of *M. microti*-PMV261, *M. microti*-PstS-1 and *M. microti*-Ag85B with the following dyes: Cy3 (orange), Cy5 (red) and Cy2 (green), respectively. Briefly, the samples were incubated in 400 pmol/μL of dimethylformamide at 4 °C in the dark for 30 min. The reaction was stopped by adding 10 mM of lysine, followed by incubation at 4 °C in the dark for 10 min. The proteins were dissolved in a buffer solution containing 9 M of urea, 4% CHAPS (3-[(3-cholamidopropyl) dimethylammonio]-1-propane sulfonate), 100 mM of dithiothreitol, 0.1% Bio-Lyte® 3/10 Ampholyte (Bio-Rad) and 0.001% bromophenol blue. They were then placed on immobilized pH gradient ReadyStrip™ IPG Strips of pH 4–7, 11 cm (Bio-Rad) for 16 h. The isoelectric focusing was performed under the following conditions: first, a current of 500 V, 1 mA, 5 W, 1 V/h; second, 2500 V, 1 mA, 5 W, 1 V/h; and lastly, 2500 V, 1 mA, 5 W, 49,500 V/h were applied with a MultiPhor™ II Electrophoresis Unit (Pharmacia Biotech, Uppsala, Sweden) within 8 h. Then, the strips were placed in 70 mM of dithiothreitol and a buffer solution composed of 6M of urea, 30% *v/v* glycerol, 50 mM of Tris, 2% SDS and 0.002% bromophenol blue. It was then immediately placed in 193 mM of iodoacetamide. The proteins were separated by electrophoresis in a 12% SDS-PAGE using a SE600 Standard Dual Cooled Vertical Electrophoresis Unit (Hoefer SE600™ unit; Pharmacia Biotech, Uppsala, Sweden) and a buffer solution consisting of 192 mM of glycine, 25 mM pf Tris, and 0.1% SDS at 50 V for 30 min, 100 V for 2 h, and 200 V for 3 h. Finally, the images of the gels were captured using a Typhoon™ 9400 scanner imager (Amersham Biosciences, Piscataway, NJ, USA) and analyzed using the Quantity One™ Image Analysis Software (Bio-Rad, Hercules, CA, USA).

2.8. Tandem Mass Spectrometry Analysis

The tandem mass spectrometry (MS/MS) analysis of the selected proteins was performed in an AB Sciex 3200 QTRAP LC/MS System (Applied Biosystems) hybrid tandem mass spectrometer equipped with a Sciex Nanospray II (Applied Biosystems) electrospray ion source and a Sciex Micro Ion Spray Source (Applied Biosystems, Foster City, CA, USA). The proteins were identified based on their MS/MS spectral data sets using the MASCOT v1.6b9 (Matrix Science, London, UK) and the NCBI database previously described [64].

2.9. Statistical Analysis

The experiments are presented as mean with standard deviation of at least three independent replicates. The statistical analyses were performed using a one-way analysis of variance (ANOVA) with Tukey's correction and using the GraphPad Prism software V. 9.0. The differences were considered statistically significant at $p \leq 0.05$.

3. Results

3.1. Rv0934 (PstS-1) and Rv1886c (Ag85B) Gene Identification

The primers were designed based on the Rv0934 and Rv1886c genes of *M. tuberculosis*, as described previously, and were used to amplify and compare the same genes of M. microti using PCR. In Figure 1C,D, the PCR amplicons were detected for both strains that corresponded to the appropriate molecular weight (MW) of 1125 bp for Rv0934 and 978 bp for Rv1886c in *M. tuberculosis* (Figures 1C and 1D, lane 1) and *M. microti* (Figures 1C and 1D, lane 2). The sequencing and analysis of the Rv0934 gene from *M. microti* revealed a 99.9% identity to the sequence reported for *M. tuberculosis*, with a base change from T-to-C at nucleotide 1 055 bp, resulting in an Ala-to-Val mutation in the protein sequence (Figure S1A). For the Rv1886c gene, we found a 100% identity between the two strains (Figure S1B). These results confirmed that the Rv0934 and Rv1886c genes of M. microti are homologous to the previously described genes of *M. tuberculosis*.

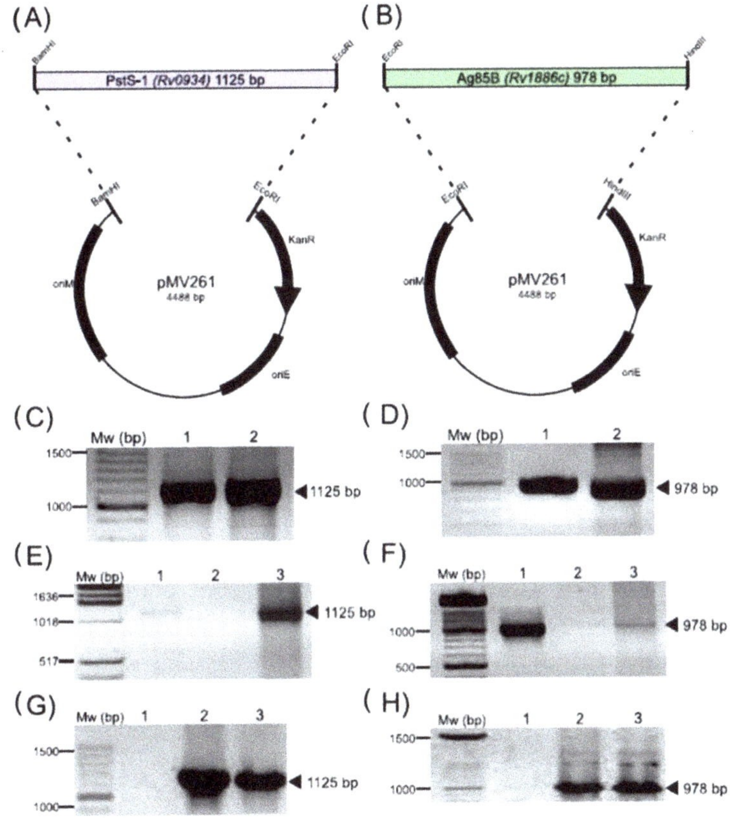

Figure 1. Plasmid design and construction of (**A**) pMV261-Rv0934 for PstS-1 and (**B**) pMV261-Rv1886c for Ag85B. (**C**,**D**) PCR amplification of Rv0934 and Rv1886c, respectively, from *M. tuberculosis* (lane 1) and *M. microti* (lane 2). (**E**,**F**) PCR amplification of Rv0934 and Rv1886c from plasmid isolated from *E. coli* colonies after transformation with pMV261-Rv0934 and pMV261-Rv1886c, respectively (lane 3); controls (lanes 1 and 2, respectively). (**G**,**H**) PCR amplification of Rv0934 and Rv1886c from plasmid isolated from *M. microti* colonies transformed with pMV261-Rv0934 and pMV261-Rv1886c, respectively (lane 3); controls (lanes 1 and 2). Black arrows indicate the expected amplicon size.

3.2. Recombinant Strains M. microti-PstS-1 and M. microti-Ag85B Construction

After the successful amplification and sequence analysis of Rv0934 and Rv1886 for both strains, we decided to construct vectors containing the corresponding genes of *M. tuberculosis* in order to overexpress these genes in *M. microti*. The vectors obtained were named pMV261-PstS-1 and pMV261-Ag85B (Figure 1A,B, respectively). After the cloning and propagation of the respective vectors in *E. coli*, the genes were again amplified to confirm their size (Figure 1E,F) and then sequenced (data not shown). Subsequently, the plasmids pMV261-PstS-1 and pMV261-Ag85B were independently cloned into *M. microti*. The resulting recombinant strains were designated *M. microti*-PstS-1 and *M. microti*-Ag85B. Both strains were analyzed for their corresponding gene plasmid transformation. To avoid the genomic amplification of the native genes, a PCR from the isolated plasmid was performed on both *M. microti* recombinant strains (Figure 1G,H). The resulting amplicons from the recombinant *M. microti* strains were verified using sequencing (data not shown). With these results, we confirmed the correct cloning and transformation of the *M. microti* strains.

3.3. Kinetic Growth of Recombinant Strains M. microti-PstS-1 and M. microti-Ag85B

After the selection of the positive clones for *M. microti*-PstS-1 and *M. microti*-Ag85B, the strains were cultured in triplicate for 196 h in shake flasks to determine if there were any growth changes due to the genetic loading and to recover any biomass and recombinant protein produced. An *M. microti*-PMV261 strain with the empty plasmid was used as a control in all cases. In Figure 2, the biomass accumulation, measured by optical density (O.D. 600 nm) and expressed in absorbance units (A.U.), is plotted against time (h). The samples were measured every 48 h and the biomass accumulation was plotted on a linear scale (Figure 2A). The biomass obtained after 196 h of culture was 0.71 ± 0.04, 0.76 ± 0.08 and 0.78 ± 0.05 for *M. microti*-PstS-1, *M. microti*-Ag85B and *M. microti* wild type, respectively, with no significant differences compared to the wild-type strain. To compare the specific growth rate (μ), a log-biomass diagram was constructed (Figure 2B) and μ was calculated for each strain. We obtained μ values of 0.031 ± 0.002, 0.030 ± 0.005 and 0.035 ± 0.003 for *M. microti*-PstS-1, *M. microti*-Ag85B and *M. microti* wild type, respectively, with no significant differences between the strains. Based on the results, we could conclude that no changes related to the growth of the recombinant *M. microti* strains occurred during the genetic transformation with the constructed vectors.

3.4. Production of PstS-1 and Ag85B and Analysis by Western Blotting

An analysis of the PstS-1 and Ag85B proteins in the cell extract and supernatant of the *M. microti* strains was determined through immunodetection with Western blotting using the monoclonal antibodies (mAbs) α-TB71 (PstS-1) and α-Ag85B (Ag85B) after 196 h of culture growth (Figure 3). The results showed that the expression of PstS-1 and Ag85B proteins in the wild-type *M. microti* strain (Figures 3A and 3C, lanes 1 and 4) was at a baseline level in both cellular extracts and supernatants. Surprisingly, the recombinant antigens PstS-1 were detected with MW differences between the cellular extracts (Figure 3A, lanes 1,2 and 3) with a MW of about 32 kDa compared to the secreted ones (Figure 3A, lanes 4, 5 and 7) with a MW of approximately 38 kDa. Moreover, the *M. microti*-PstS-1 strain overexpressed PstS-1 in the cell extract (Figure 3A, lane 3), with a 5.2-fold increase according to the densitometric analysis (Figure 3B), while the release of PstS-1 into the supernatant appeared to be inhibited (Figure 3A, lane 6). In contrast, the *M. microti*-Ag85B strain overexpressed Ag85B in both the cell extract and the supernatant (Figure 3C, lanes 3 and 6). Compared to the *M. microti* wild-type and a control consisting of an *M. microti* with an empty vector (ev), we found a 2.6- and 3.2-fold increase in the Ag85B antigen in the cell extract and supernatant, respectively (Figure 3D). These results show that the overexpression of the PstS-1 (Rv0934) and Ag85B (Rv1886c) antigens of *M. tuberculosis* could be successfully carried out in *M. microti* strains.

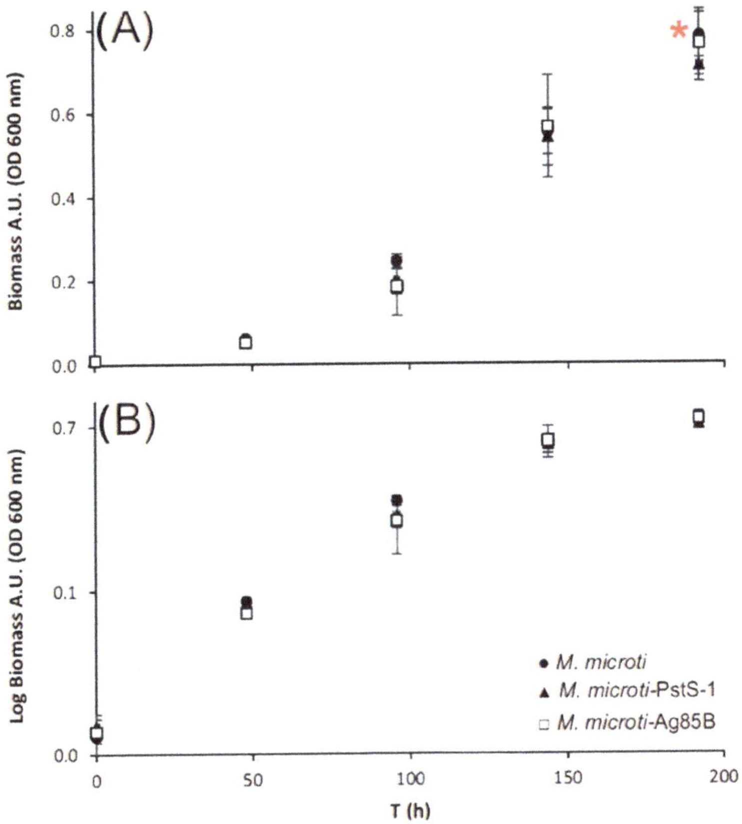

Figure 2. Kinetic growth of *M. microti* strains in Sauton media. The biomass accumulation in absorbance units (A.U.) measured by optical density (O.D.) during 196 h: (**A**) Lineal scale growth and (**B**) Log scale growth of the strains *M. microti* (closed circles), *M. microti*-PstS-1 (closed triangles) and *M. microti*-Ag85B (open squares). The red asterisk indicates the collection time for protein analysis.

3.5. Glycosylation of Proteins Produced by M. microti Recombinant Strains

To determine whether *M. microti*-PstS-1, *M. microti*-Ag85B and the *M. microti* wild type can still produce glycosylated proteins, we performed a lectin blot analysis using a DIG Glycan Differentiation Kit™ (Merck™). As suggested by our results, the recombinant *M. microti* strains produced glycosylated proteins (Figure S2A). These proteins are likely mannosylated, as we observed signal recognition with GNA lectin. On the other hand, to our surprise, we detected faint bands with an approximate MW of 50 kDa in the three strains when the lectin blot was performed with MAA lectin, which recognizes sialic acids (Figure S2B). To our knowledge, this is the first time that sialic acid has been detected in proteins produced by *M. microti* strains. However, a more in-depth analysis of the glycoproteome of *M. microti*, such as MS/MS and the complementation with other techniques, is required to confirm our results. To better understand the glycosylation profile of the PstS-1 and Ag85B antigens, we performed an in silico analysis to predict the potential glycosylation sites on both the recombinant antigens, using three different algorithms to compare the results. In Table 1, we summarize our results of the in silico analysis for the *O*-glycosylation of the PstS-1 and Ag85B proteins.

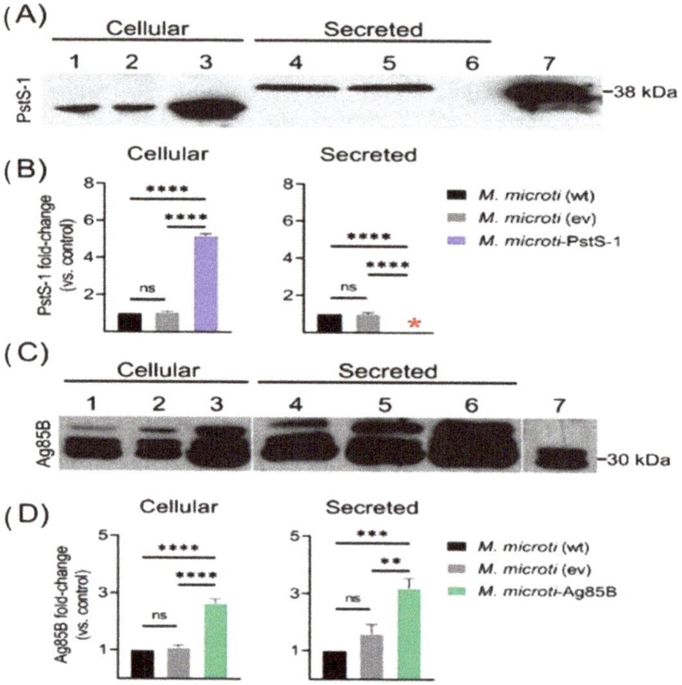

Figure 3. Protein analysis of PstS-1 and Ag85B, either cellular or secreted, produced by *M. microti* strains. (**A**,**C**) detection by Western blot of antigens PstS-1 and Ag85B, respectively. (**B**,**D**) fold expression analysis of PstS-1 and Ag85B, respectively. (**A**,**C**) lanes 1 and 4: *M. microti* wild type (wt), lanes 2 and 5: *M. microti* with empty vector (ev), lane 7 purified protein control. (**A**) lanes 3 and 5 *M. microti*-PstS-1; (**B**) lanes 3 and 5 *M. microti*-Ag85B. The fold change analysis performed by densitometric analysis by Image J. Red asterisk indicates the absence of protein. The differences were considered statistically significant with a $p \leq 0.05$. **, ***, ****, ns = not significant. Using a one-way analysis of variance (ANOVA) with Tukey's correction.

As our results show, we were able to predict the glycosylation sites of PstS-1 and Ag85B and compare them with the alanine and proline rich antigen APA. PstS-1 showed a positive glycosylation profile after all three analyses. However, differences were observed with respect to the glycosylated residues, as can be seen in Table 1. On the other hand, Ag85B showed positive glycosylation sites only with the Glycosylation Predictor GPP algorithm, with 17 hits. When analyzing the antigen APA and its mannosylation profile with the three algorithms, we also found differences in the number and location of the residue. The mannosylation profile of APA was determined by Dobos et. al. (1996) [65] and later by Smith et. al. (2013) [29], who together reported seven mannosylated residues (Thr 10, 18, 27, 274, 276, 277 and 279). In our results, we found that the most accurate analysis was performed by the GLYCOPP V 1.0 algorithm with six potentially glycosylated residues (Ser 17, Thr 18, 27, 274, 276 and 277). So far, no glycosylation has been demonstrated for Ag85B by other authors, and this was confirmed by our results with GLYCOPP V 1.0. Although PstS-1 has previously been described as a glycolipoprotein, its mannosylation profile has not previously been determined. Our results with GLYCOPP V 1.0 indicate that Thr 86, 160, 331 and Ser 88 are likely to be mannosylated in vivo. Based on our analysis using the antigen APA as a reference for the glycosylation profile, we suggest that prediction of glycosylation with the GLYCOPP V1.0 algorithm could be used to assess a glycosylation profile similar to that occurring in the glycoproteins produced by mycobacteria.

Table 1. Potential glycosylated residues in recombinant antigens PstS-1 and Ag85B.

Antigen	UniProt Acc. No.	NetOGlyc-4.0 [1]	Glycosylation Predictor GPP [2]	GLYCOPP V 1.0 [3]	Experimental Determination
PstS-1 (Rv0934)	B2MVV3	**S30**, **S32**, T35, T41, **T44**, **T45**, **S48**, **S49**, **S290**	T7, T13, S26, S30, S32, **T44**, **T45**, **S48**, **S49**, T52, T56, T59, T80, T99, S110, S128, S140, **T160**, S186, T196, T216, T236, **S290**, S299, **T331**, S365, T371, S373, S374	T86, S88, **T160**, **T331**	ND
Mod. AA		9	29	4	
Ag85B (Rv1886c)	P9WQP1	None predicted	T2, S5, S42, S57, N71, S116, S117, S120, S150, T158, S160, S166, S171, S202, T234, S320, S321	None predicted	ND
Mod. AA		0	17	0	
APA (Rv1860)	P9WIR7	**T10**, **T11**, **S14**, **S17**, **T18**, **T27**, T39, **T108**, S243, **T274**, **T276**, T277, T279, **T283**	**T10**, **T11**, **S14**, **S17**, **T18**, **T27**, S101, S105, T107, **T108**, T142, S144, S152, T164, S181, S183, S190, S193, T201, S243, **T274**, **T276**, **T283**	**S17**, **T18**, **T27**, **T274**, **T276**, **T277**	T10, T18, T27, T277 (Dobos et al., 1996 [65]) T274, T276, T277, T279 (Smith et al., 2014 [29])
Mod. AA		14	23	6	4

Acc. No.: Accession number, Mod. AA: Modified amino acids, ND: Not determined. APA antigen was used for comparison purposes and analysed in silico without the signal peptide composed of residues 1–39. Residues in bold are shared by two or more prediction tools. [1] Based on a neural network predictions of mucin type GalNAc O-glycosylation sites in mammalian proteins where a value higher than 0.5 was considered a positive hit [61]. [2] Prediction based on random forests algorithms [63]. [3] Prediction based on Composition Profile of Patterns (CPP) where a value higher than 0.9 was considered a positive hit [62].

3.6. Differential Proteome of M. microti Recombinant Strains

Based on the differences in the PstS-1 production and secretion (Figure 3A, lanes 3 and 6), we decided to perform a proteomic analysis to investigate what other changes might occur in native proteins in the *M. microti*-PstS-1 and *M. microti*-Ag85B recombinant strains and compare them with the *M. microti* wild type. The two-dimensional fluorescence differential gel electrophoresis was performed with labelled proteins from the cell extracts of the three strains, as mentioned in materials and methods. On the fluorescence images obtained for each strain, we observed different protein spots that were differentially expressed compared to each other: *M. microti* vs. *M. microti*-PstS-1, *M. microti* vs. *M. microti*-Ag85B and *M. microti*-PstS-1 vs. *M. microti*-Ag85B (Figure S3). The image analysis allowed us to perform a silver stained 2D SDS-PAGE (Figure 4) to identify the spots that were differentially expressed and to select the spots with the highest intensities for further sequencing using MS/MS. Our analysis results showed that 46 proteins were differentially expressed in the three examined *M. microti* strains and based on intensity, we selected 18 proteins for further analysis. These results suggest that changes in the genetic load of *M. microti*, such as the insertion of foreign plasmids and the subsequent overexpression of proteins, cause changes in the proteome profile of *M. microti*-PstS-1 and *M. microti*-Ag85B in the cell extract. However, further high throughput proteome and transcriptome analyses are still required to substantiate our results.

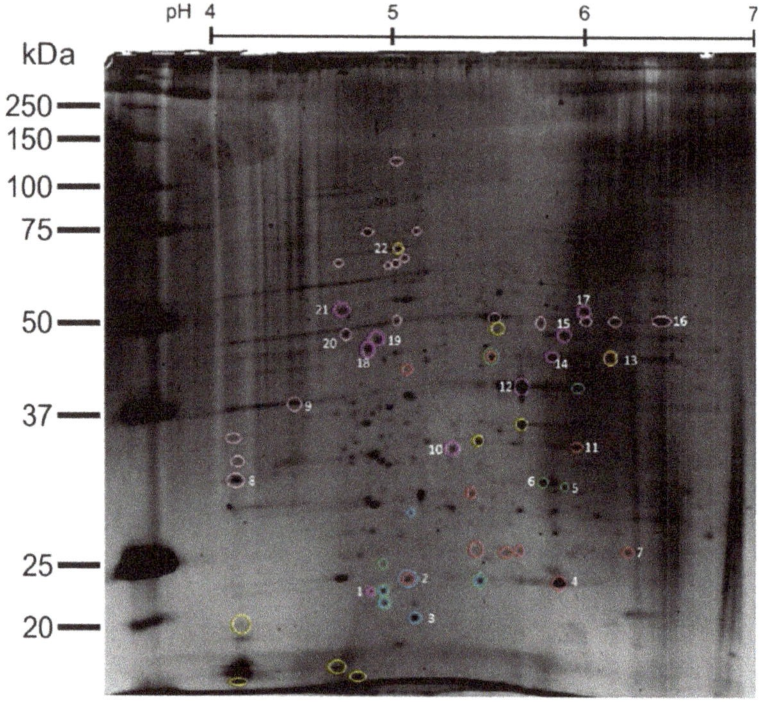

Figure 4. 2D SDS-PAGE cell extract proteome analysis of *M. microti* strains. Circled proteins were selected based on the fluorometric analysis according to the differential expression: *M. microti*-PMV261 (red circles), *M. microti*-PstS-1 (blue circles) and *M. microti*-Ag85 (green circles). Numbered proteins were sequenced using MS/MS (Table 2). Unnumbered pink and yellow circled spots represent glycosylated proteins.

3.7. MS/MS Identification of Differentially Expressed Proteins in M. microti Strains

The selected protein spots based on the intensity values from Figure 4 were analyzed and sequenced using MS/MS. The spectra data sets were processed using MASCOT v1.6b9 (Matrix Science) and BLAST from the NCBI database. The identified proteins are summarized in Table 2, together with their unique accession numbers from the UniProt database [66] and a brief description of each identified protein. Our results show that spots 2, 3, 4, 5, 6, 7, 10, 11, 13, 15, 16, 17, 18, 19 and 20 correspond to either enzymes or enzyme subunits, while spot 22 is involved in protein folding and spots 12 and 21 have an unknown function. Post-translational modifications (PTM) were present in spots 10, 12, 15, 16, 17, 20, 21 and 22 according to mass spectra data (not shown). These PTM's correspond to mannose glycans, as previously noted in our lectin blot analysis. As for the differential expression between the recombinant *M. microti* strains, we found that spots 2, 3, 4, 5, 6, 7, 11, 12, 13, 18 and 19 showed a differential expression between the three strains either by presence (+) or absence (-), as indicated in Table 2. Based on the differential expression of the three *M. microti* strains, a Venn diagram was constructed (Figure 5). Considering the different proteins present only in *M. microti*-PstS-1 and *M. microti*-Ag85B (MoxR, RpoA, Eno, SodA, DapA and MAP_3475c) and related to energy production and growth, we can expect to observe the overexpression of these types of proteins to cope with the metabolic burden of a recombinant protein.

Table 2. Proteins identified by MS/MS from *M. microti* strains.

Spot	Protein	UniProt	Function	MW (kDa)	Global Score *	*M. microti* strains			PTM
						wt	PstS-1	Ag85B	
2	ATP-dependent Clp protease proteolytic subunit 2 (ClpP2) [1]	Q73XM8	Cleaves peptides in various proteins in a process that requires ATP hydrolysis. Has a chymotrypsin-like activity. Plays a major role in the degradation of misfolded proteins.	21.665	297	+	-	-	ND
3	Alkyl hydroperoxide reductase (AhpD) [1]	Q73ZL4	Antioxidant protein with alkyl hydroperoxidase activity. Required for the reduction of the AhpC active site cysteine residues and for the regeneration of the AhpC enzyme activity.	18.842	151	+	-	+	ND
4	Superoxide dismutase (SodA) [2]	B1A036	Destroys superoxide anion radicals which are normally produced within the cells, and which are toxic to biological systems.	22.476	125	-	+	+	ND
5	AB hydrolase-1 domain-containing protein (MAP_1452) [1]	Q73ZZ9	Hydrolase activity.	32.404	303	+	+	-	ND
6	Peroxisomal multifunctional enzyme type 2 (MAV_5146) [3]	A0A0H2ZVK0	3-hydroxyacyl-CoA dehydrogenase activity. 3alpha, 7alpha, 12alpha-trihydroxy-5beta-cholest-24-enoyl-CoA hydratase activity.	29.906	945	+	+	-	ND
7	Nitroreductase domain-containing protein (MAP_3475c) [1]	Q73U93	Oxidoreductase activity: nitroreductase	24.012	186	-	+	+	ND
10	Fructose-bisphosphate aldolase class 1 (MAP_4308c) [3]	Q73RX0	Fructose-bisphosphate aldolase activity.	33.645	891	+	+	+	Yes
11	4-hydroxy-tetradihydrodipicolinate synthase (DapA) [1]	Q73VZ7	Catalyzes the condensation of (S)-aspartate-beta-semialdehyde [(S)-ASA] and pyruvate to 4-hydroxy-tetrahydrodipicolinate (HTPA).	30.973	350	-	+	+	ND
12	p40 protein [4]	Q9AIQ0	ND	36.232	1109	+	-	+	Yes

Table 2. *Cont.*

Spot	Protein	UniProt	Function	MW (kDa)	Global Score *	*M. microti* strains			PTM
						wt	PstS-1	Ag85B	
13	ATPase (MoxR) [1]	Q740Y7	ATP binding and hydrolysis activity	40.730	308	-	+	+	ND
15	AB hydrolase-1 domain-containing protein (MAP_3192) [1]	Q73V24	Hydrolase activity	43.256	674	+	+	+	Yes
16	3-ketoacyl-(acyl-carrier-protein) reductase (FabG) [5]	I2AKG1	3-oxoacyl-[acyl-carrier-protein] reductase (NADPH) activity.	47.139	1008	+	+	+	Yes
17	Serine hydroxymethyltransferase (GlyA) [1]	Q73WG1	Catalyzes the reversible interconversion of serine and glycine with tetrahydrofolate (THF) serving as the one-carbon carrier.	44.955	46	+	+	+	Yes
18	DNA-directed RNA polymerase subunit alpha (RpoA) [1]	Q73S43	DNA-dependent RNA polymerase catalyzes the transcription of DNA into RNA using the four ribonucleoside triphosphates as substrates.	37.704	1261	-	+	+	ND
19	Enolase (Eno) [1]	Q741U7	Catalyzes the reversible conversion of 2-phosphoglycerate into phosphoenolpyruvate. It is essential for the degradation of carbohydrates via glycolysis.	44.873	1242	-	+	+	ND
20	Enolase (Eno) [3]	A0QBX4	Catalyzes the reversible conversion of 2-phosphoglycerate into phosphoenolpyruvate. It is essential for the degradation of carbohydrates via glycolysis.	44.873	1096	+	+	+	Yes
21	Xyppx repeat family protein (MAV_0024) [3]	A0A0H3A0B1	ND	58.801	214	+	+	+	Yes
22	Chaperonin (GroEL) 2 [3]	A0QLP6	Together with its co-chaperonin GroES, plays an essential role in assisting protein folding.	56.615	17263	+	+	+	Yes

Detected peptides were matched against reference sequences: [1] *Mycobacterium avium* spp. *paratuberculosis* K10, [2] *M. avium* spp. *paratuberculosis*, [3] *M. avium* 104, [4] *M. avium*, [5] *Mycobacterium* spp. MOTT36Y. * Scores >25 are considered significant ($p < 0.05$) according to the Mascot Search Results algorithm. MW: molecular weight, kDa: kilodalton, WT: wild type PTM: post-translational modified, ND: not detected/determined.

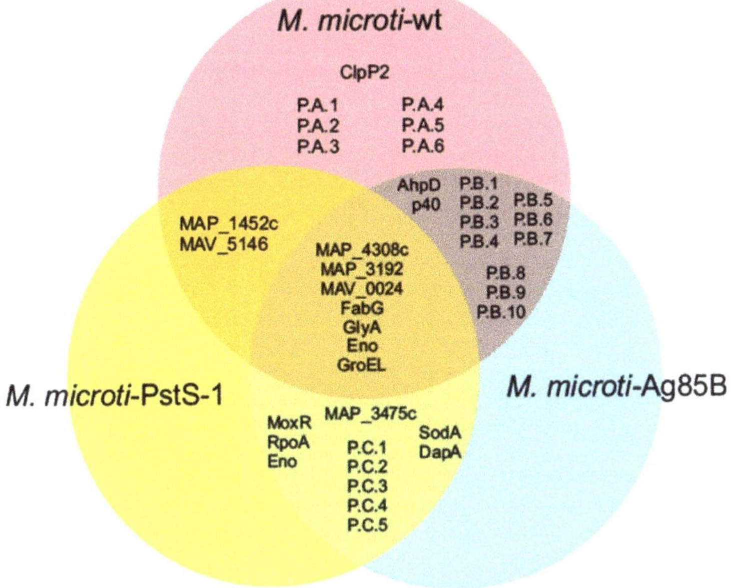

Figure 5. Venn diagram of the differential expression proteome of the cellular extract from *M. microti* strains. P.A, P.B and P.C: protein spots from Figure S3A–C, respectively.

4. Discussion

Technologies are currently being used to develop new vaccines and/or treatments to prevent TB, which is a global health priority. One of these is the use of proteomics and genetic engineering to discover and produce antigens with a known ability to elicit an immune response [67]. Our work focused on the genetic engineering of mycobacterial strains belonging to the Mtb complex to produce antigens of interest. Such antigens and their epitopes, either alone or in combination, may help to develop new ways to increase the efficacy of current TB vaccines, for example, the currently available *M. bovis* BCG live-attenuated vaccine [68–72]. Therefore, we generated and analyzed two recombinant strains of *M. microti* overexpressing the *M. tuberculosis* PstS-1 and Ag85B antigens. Both antigens are known to stimulate the immune system and act as potent immunomodulators in the production of cytokines and antibodies when used as multimeric vaccine components [37,68,70,71]. The complete genome of *M. microti* has not been released up to date. However, studies show that the RD1 region is absent in several *M. microti* strains [73]. In our work, a comparative genetic analysis of Rv0934 (PstS-1) and Rv1886c (Ag85B) revealed that these genes are present in our wild-type *M. microti* (ATCC 11152) strain with a homology of 99.9% and 100%, respectively. This confirms that these selected antigens from *M. microti* are very similar in sequence to those previously described in *M. tuberculosis* and may have a similar immunomodulatory function.

PstS-1 and Ag85B were overexpressed and detected in both *M. microti* strains (Figure 3). For the PstS-1 antigen, we observed differences in the MW size depending on whether the protein originated from a cell extract or from the secreted form. First, an analysis of the cellular extract of PstS-1 revealed a band showing a 5.2-fold increase in production of the antigen with a MW of ~32 kDa (Figure 3, lane 3). However, we did not observe a secretion of PstS-1 in this strain (Figure 3, lane 6). Instead, we detected a band size of 38 kDa in *M. microti* (wt) and *M. microti* (ev) (Figure 3, lanes 4 and 5). We hypothesize that this difference in the molecular mass of this protein is due to the detection of proteins prior to the post-translational modifications, as PstS-1 typically undergoes post-translational modifications, including glycosylation and lipoylation, which likely contribute to its molecular mass [36]. In addition, the glycosylation of *M. tuberculosis* proteins may also represent a mechanism for regulating protein export [74]. It is possible that the recombination influenced the transport of this protein outside the bacterial cell. However, this is not a problem, as the intracellular antigen could still be available through the processing performed using professional antigen-presenting cells, where the phagosome maturation leads to the lysis of the microorganism and the antigens are subsequently presented by the molecules of the major histocompatibility complex (MHC) [75–77]. On the other hand, the expression analysis of Ag85B showed that this antigen was successfully expressed in both the cellular and secreted forms. Western blotting showed the presence of the three bands (Figure 3C lanes 1–6) in both the cell extract and the supernatant. This is probably due to the fact that the antibody used in this study can detect the Ag85 complex (A, B and C) in addition to Ag85B (Figure 3C, lanes 1–6 compared to control lane 7).

These results led us to investigate whether the recombinant strains had different proteomes and/or glycosylation profiles. We found that each strain of *M. microti*-PstS-1 and *M. microti*-Ag85B had a different proteomic profile. We identified six proteins (SodA, hypothetical protein MAP3475c, DapA, MoxR, RpoA and Eno) involved in the energy production process, possibly due to an increased heterologous protein production, that are not expressed in the *M. microti* wild type. Similar observations were previously made by Cheng et al. who reported that the overexpression of proteins in recombinant *Escherichia coli* strains leads to a change in the protein profiles, resulting in a differential expression of different cellular proteins [78,79].

There are few studies describing the presence of glycosylation in the proteins of the genus *Mycobacterium*, and most existing studies have focused on mannosylation [54,80,81]. Examining the glycosylation profiles in the wild-type and the recombinant *M. microti* allowed us to identify two potential glycosylation types in these strains: mannosylation and sialylation, as shown in Figure S2. Mannose glycosylation is widely reported in the genus *Mycobacterium* [49,50,54,82]. In the present study, we identified and sequenced eight proteins that were potentially glycosylated with terminal mannose, including MAP_4308c, hypothetical protein p40, MAP_3192, FabG, GlyA, Eno, MAV_0024 and GroEL. The mannosylation has various functions in mycobacteria. For example, it is part of the cell wall and is primarily associated with virulent species and enables protein recognition by mannose and the Toll-like receptor (TLR) 2 expressed by macrophages and other cell types [83]. In addition, studies of proteins that were glycosylated with mannose, such as PstS-1 from *M. tuberculosis* H37Rv, have shown that these proteins promote phagocytosis [35]. Pan et al. developed ZXL1 (double-stranded DNA), a therapeutic target that prevents mannose from binding to host cells by inhibiting the immunosuppression of CD11c+ dendritic cells and enhancing the activity of antigen-presenting cells for the activation of naïve CD4+ Th1 cells [84]. For PstS-1 and Ag85B, we predicted the glycosylation profiles using three algorithms that allowed us to compare the different results in terms of the number and position of residues involved in mannosylation. Our results show that PstS-1 could be produced in its mannosylated form. In contrast, no glycosylation was predicted for Ag85B (Table 1) using the GLYCOPP V 1.0 algorithm.

We have identified proteins that may be glycosylated with sialic acid (Figure S2B). Currently, there is little evidence of proteins glycosylated with sialic acid in the genus Mycobacterium. Therefore, we searched the genome of *M. microti* for enzymes associated with sialic acid metabolism that have been identified in other bacteria, including *Neisseria meningitides*, *E. coli* K12 and *Haemophilus influenzae* [85–88]. We found that *M. microti* contains gene encoding proteins that are known to be involved in the de novo synthesis of sialic acid (UDP–*N*–acetylglucosamine–pyrophosphorylase/glucosamine–1–phosphate-*N*-acetyltransferase, L-serine dehydratase and the hypothetical protein B8A25_03135) in the case of *M. tuberculosis sp.* variant *microti* (data not shown). Overall, these results remain to be confirmed by mass spectrometry assays. The identification of the differences in the proteomic and glycosylation profiles between the wild type and recombinant strains of *M.*

microti may lead to alterations in the immune system recognition and evasion [89,90]. For example, the mannose glycosylation of the protein APA has been shown to play a key role in its antigenicity, while the deglycosylation of the same protein has been demonstrated to modify the immune response elicited in guinea pigs [53]. Taken together, these results demonstrate the need to thoroughly investigate the proteomic profiles of the different mycobacterial strains being considered for the development of new TB vaccines, as these events may alter the immune response and the effectiveness of these potential vaccines.

5. Conclusions

This study has shown that the overexpression of PstS-1 and Ag85B in the recombinant strains of *M. microti* alters their proteome, suggesting that the impact of these changes on the immune responses elicited by these and other newly generated strains, needs to be re-evaluated, especially when considering their potential use in the development of TB vaccines. In addition, we observed a differentially mannosylated profile in the analyzed proteome and we obtained evidence for the possible sialylation of the proteins expressed by *M. microti*, although this remains to be confirmed. These glycosylations play a key role in the interactions between bacteria and host cells. Therefore, studying the role of these PTMs in this genus is of the utmost importance to understand how these modifications affect their immunogenicity.

Supplementary Materials: The following supporting information can be downloaded at: https://www.mdpi.com/article/10.3390/biom12121836/s1, Figure S1: Protein sequence translation and alignments of genes (A) Rv0934 and (B) Rv1886c from strains *M. microti* and *M. tuberculosis* used in this study; Figure S2. Differential expression and glycosylation of proteins from wild-type and recombinant *M. microti* strains. (A) Mannose lectin blot, (B) Sialic acid lectin blot; Figure S3. Two-dimensional fluorescence difference gel electrophoresis with labelled proteins from the cell extract of (A) *M. microti* vs. *M. microti*-PstS-1, (B) *M. microti* vs. *M. microti*-AG85B and (C) *M. microti*-PstS-1 vs. *M. microti*-Ag85B.

Author Contributions: Conceptualization, Y.L.-V.; Formal analysis, V.G.-R., P.O. and Y.L.-V.; Methodology, V.G.-R. and T.J.F.-R.; Writing–original draft, V.G.-R. and P.O.; Writing–review and editing, A.I.C.-R. and Y.L.-V. All authors have read and agreed to the published version of the manuscript.

Funding: This work was supported by Fundación Gonzalo Rio Arronte S.590, DGAPA-PAPIIT IT202020, Universidad Nacional Autónoma de México.

Institutional Review Board Statement: Not applicable.

Informed Consent Statement: Not applicable.

Data Availability Statement: Not applicable.

Acknowledgments: Viridiana García-Ruiz received a fellowship from Conacyt, Programa de Doctorado en Ciencias Biomédicas de la Universidad Nacional Autónoma de México (261858/419838). Teresa de Jesús Flores-Rodríguez received a Conacyt-México master's degree scholarship at Programa de Maestría y Doctorado en Ciencias Bioquímicas de la Universidad Nacional Autónoma de México.

Conflicts of Interest: The authors declare no conflict of interest.

References

1. Furin, J.; Cox, H.; Pai, M. Tuberculosis. *Lancet* **2019**, *393*, 1642–1656. [CrossRef] [PubMed]
2. Harries, A.D.; Lin, Y.; Kumar, A.M.V.; Satyanarayana, S.; Takarinda, K.C.; Dlodlo, R.A.; Zachariah, R.; Olliaro, P. What can National TB Control Programmes in low- and middle-income countries do to end tuberculosis by 2030? *F1000Research* **2018**, *7*, 1011. [CrossRef]
3. Pai, M.; Behr, M.A.; Dowdy, D.; Dheda, K.; Divangahi, M.; Boehme, C.C.; Ginsberg, A.; Swaminathan, S.; Spigelman, M.; Getahun, H.; et al. Tuberculosis. *Nat. Rev. Dis. Prim.* **2016**, *2*, 16076. [CrossRef] [PubMed]
4. WHO. *World Health Organization. Geneva: 2021 Global Tuberculosis Report*; Licence: CC BY-NC-SA 3.0 IGO; WHO: Geneva, Switzerland, 2021. Available online: https://www.who.int/publications/i/item/9789240037021 (accessed on 14 September 2022).
5. Pai, M. Tuberculosis: The story after the Primer. *Nat. Rev. Dis. Prim.* **2020**, *6*, 29. [CrossRef] [PubMed]

6. Mustafa, A.S. Development of new vaccines and diagnostic reagents against tuberculosis. *Mol. Immunol.* **2002**, *39*, 113–119. [CrossRef]
7. Uma Devi, K.R.; Ramalingam, B.; Brennan, P.J.; Narayanan, P.R.; Raja, A. Specific and early detection of IgG, IgA and IgM antibodies to *Mycobacterium tuberculosis* 38kDa antigen in pulmonary tuberculosis. *Tuberculosis* **2001**, *81*, 249–253. [CrossRef]
8. Andersen, P.; Doherty, T.M. The success and failure of BCG—Implications for a novel tuberculosis vaccine. *Nat. Rev. Microbiol.* **2005**, *3*, 656–662. [CrossRef]
9. Aronson, J.D.; Schneider, P. The problem of standardization of BCG vaccine. *Am. J. Public Health Nations Health* **1950**, *40*, 533–544. [CrossRef]
10. Davenne, T.; McShane, H. Why don't we have an effective tuberculosis vaccine yet? *Expert Rev. Vaccines* **2016**, *15*, 1009–1013. [CrossRef]
11. Bellini, C.; Horvati, K. Recent Advances in the Development of Protein- and Peptide-Based Subunit Vaccines against Tuberculosis. *Cells* **2020**, *9*, 2673. [CrossRef]
12. Stylianou, E.; Harrington-Kandt, R.; Beglov, J.; Bull, N.; Pinpathomrat, N.; Swarbrick, G.M.; Lewinsohn, D.A.; Lewinsohn, D.M.; McShane, H. Identification and Evaluation of Novel Protective Antigens for the Development of a Candidate Tuberculosis Subunit Vaccine. *Infect. Immun.* **2018**, *86*, e00014-18. [CrossRef]
13. Ullah, I.; Bibi, S.; Ul Haq, I.; Safia, U.K.; Ge, L.; Shi, X.; Bin, M.; Niu, H.; Tian, J.; Zhu, B. The Systematic Review and Meta-Analysis on the Immunogenicity and Safety of the Tuberculosis Subunit Vaccines M72/AS01E and MVA85A. *Front. Immunol.* **2020**, *11*, 1806. [CrossRef]
14. Broset, E.; Calvet Seral, J.; Arnal, C.; Uranga, S.; Kanno, A.I.; Leite, L.C.C.; Martín, C.; Gonzalo-Asensio, J. Engineering a new vaccine platform for heterologous antigen delivery in live-attenuated *Mycobacterium tuberculosis*. *Comput. Struct. Biotechnol. J.* **2021**, *19*, 4273–4283. [CrossRef]
15. Counoupas, C.; Pinto, R.; Nagalingam, G.; Britton, W.J.; Triccas, J.A. Protective efficacy of recombinant BCG over-expressing protective, stage-specific antigens of *Mycobacterium tuberculosis*. *Vaccine* **2018**, *36*, 2619–2629. [CrossRef]
16. Khan, A.; Sayedahmed, E.E.; Singh, V.K.; Mishra, A.; Dorta-Estremera, S.; Nookala, S.; Canaday, D.H.; Chen, M.; Wang, J.; Sastry, K.J.; et al. A recombinant bovine adenoviral mucosal vaccine expressing mycobacterial antigen-85B generates robust protection against tuberculosis in mice. *Cell Rep. Med.* **2021**, *2*, 100372. [CrossRef]
17. Mangtani, P.; Abubakar, I.; Ariti, C.; Beynon, R.; Pimpin, L.; Fine, P.E.M.; Rodrigues, L.C.; Smith, P.G.; Lipman, M.; Whiting, P.F.; et al. Protection by BCG vaccine against tuberculosis: A systematic review of randomized controlled trials. *Clin. Infect. Dis.* **2014**, *58*, 470–480. [CrossRef]
18. Vekemans, J.; Gebreselassie, N.; Zignol, M.; Friede, M.; Kasaeva, T.; Swaminathan, S. A new tuberculosis vaccine: Breakthrough, challenges, and a call for collaboration. *Lancet Infect. Dis.* **2019**, *19*, 123–125. [CrossRef]
19. Manabe, Y.C.; Scott, C.P.; Bishai, W.R. Naturally attenuated, orally administered Mycobacterium microti as a tuberculosis vaccine is better than subcutaneous Mycobacterium bovis BCG. *Infect. Immun.* **2002**, *70*, 1566–1570. [CrossRef]
20. Riojas, M.A.; McGough, K.J.; Rider-Riojas, C.J.; Rastogi, N.; Hazbon, M.H. Phylogenomic analysis of the species of the *Mycobacterium tuberculosis* complex demonstrates that Mycobacterium africanum, Mycobacterium bovis, Mycobacterium caprae, Mycobacterium microti and Mycobacterium pinnipedii are later heterotypic synonyms of *Mycobacterium tuberculosis*. *Int. J. Syst. Evol. Microbiol.* **2018**, *68*, 324–332. [CrossRef]
21. Brodin, P.; Majlessi, L.; Brosch, R.; Smith, D.; Bancroft, G.; Clark, S.; Williams, A.; Leclerc, C.; Cole, S.T. Enhanced protection against tuberculosis by vaccination with recombinant Mycobacterium microti vaccine that induces T cell immunity against region of difference 1 antigens. *J. Infect. Dis.* **2004**, *190*, 115–122. [CrossRef]
22. Hart, P.D.; Sutherland, I. BCG and vole bacillus vaccines in the prevention of tuberculosis in adolescence and early adult life. *Br. Med. J.* **1977**, *2*, 293–295. [CrossRef]
23. Dannenberg, A.M.; Bishai, W.R.; Parrish, N.; Ruiz, R.; Johnson, W.; Zook, B.C.; Boles, J.W.; Pitt, L.M. Efficacies of BCG and vole bacillus (Mycobacterium microti) vaccines in preventing clinically apparent pulmonary tuberculosis in rabbits: A preliminary report. *Vaccine* **2000**, *19*, 796–800. [CrossRef] [PubMed]
24. Machado, D.; Couto, I.; Perdigã, J.O.; Rodrigues, L.; Portugal, I.; Baptista, P.; Veigas, B.; Amaral, L.; Viveiros, M. Contribution of efflux to the emergence of isoniazid and multidrug resistance in *Mycobacterium tuberculosis*. *PLoS ONE* **2012**, *7*, e34538. [CrossRef] [PubMed]
25. Palomino, J.C.; Martin, A. Drug Resistance Mechanisms in *Mycobacterium tuberculosis*. *Antibiotics* **2014**, *3*, 317–340. [CrossRef] [PubMed]
26. Gonzalez-Zamorano, M.; Mendoza-Hernandez, G.; Xolalpa, W.; Parada, C.; Vallecillo, A.J.; Bigi, F.; Espitia, C. *Mycobacterium tuberculosis* glycoproteomics based on ConA-lectin affinity capture of mannosylated proteins. *J. Proteome Res.* **2009**, *8*, 721–733. [CrossRef]
27. Nandakumar, S.; Kannanganat, S.; Dobos, K.M.; Lucas, M.; Spencer, J.S. O-mannosylation of the *Mycobacterium tuberculosis* adhesin Apa is crucial for T cell antigenicity during infection but is expendable for protection. *PLoS Pathog.* **2013**, *9*, e1003705. [CrossRef]
28. Ragas, A.; Roussel, L.; Puzo, G.; Riviere, M. The *Mycobacterium tuberculosis* cell-surface glycoprotein apa as a potential adhesin to colonize target cells via the innate immune system pulmonary C-type lectin surfactant protein A. *J. Biol. Chem.* **2007**, *282*, 5133–5142. [CrossRef]

29. Smith, G.T.; Sweredoski, M.J.; Hess, S. O-linked glycosylation sites profiling in *Mycobacterium tuberculosis* culture filtrate proteins. *J. Proteom.* **2014**, *97*, 296–306. [CrossRef]
30. Flores, L.L.; Steingart, K.R.; Dendukuri, N.; Schiller, I.; Minion, J.; Pai, M.; Ramsay, A.; Henry, M.; Laal, S. Systematic review and meta-analysis of antigen detection tests for the diagnosis of tuberculosis. *Clin. Vaccine Immunol.* **2011**, *18*, 1616–1627. [CrossRef]
31. Liu, C.F.L.; Malaga, W.; Beau, M.; Stella, A.; Bouyssié, D.; Jackson, M.C.; Nigou, J.; Puzo, G.; Guilhot, C.; Burlet-Schiltz, O.; et al. Bacterial protein-O-mannosylating enzyme is crucial for virulence of *Mycobacterium tuberculosis*. *Proc. Natl. Acad. Sci. USA* **2013**, *110*, 6560–6565. [CrossRef]
32. Villeneuve, C.; Etienne, G.; Rie Abadie, V.; Montrozier, H.; Bordier, C.; Oise Laval, F.; Daffe, M.; Maridonneau-Parini, I.; Astarie-Dequeker, C. Surface-exposed glycopeptidolipids of Mycobacterium smegmatis specifically inhibit the phagocytosis of mycobacteria by human macrophages. Identification of a novel family of glycopeptidolipids. *J. Biol. Chem.* **2003**, *278*, 51291–51300. [CrossRef]
33. Li, H.; Liu, L.; Zhang, W.J.; Zhang, X.; Zheng, J.; Li, L.; Zhu, X.; Yang, Q.; Zhang, M.; Liu, H.; et al. Analysis of the Antigenic Properties of Membrane Proteins of *Mycobacterium tuberculosis*. *Sci. Rep.* **2019**, *9*, 3042. [CrossRef] [PubMed]
34. Korycka-Machała, M.; Pawełczyk, J.; Borówka, P.; Dziadek, B.; Brzostek, A.; Kawka, M.; Bekier, A.; Rykowski, S.; Olejniczak, A.B.; Strapagiel, D.; et al. PPE51 Is Involved in the Uptake of Disaccharides by *Mycobacterium tuberculosis*. *Cells* **2020**, *9*, 603. [CrossRef]
35. Esparza, M.; Palomares, B.; Garcia, T.; Espinosa, P.; Zenteno, E.; Mancilla, R. PstS-1, the 38-kDa *Mycobacterium tuberculosis* glycoprotein, is an adhesin, which binds the macrophage mannose receptor and promotes phagocytosis. *Scand. J. Immunol.* **2015**, *81*, 46–55. [CrossRef]
36. Castanon-Arreola, M.; Lopez-Vidal, Y.; Espitia-Pinzon, C.; Hernandez-Pando, R. A new vaccine against tuberculosis shows greater protection in a mouse model with progressive pulmonary tuberculosis. *Tuberculosis* **2005**, *85*, 115–126. [CrossRef]
37. Fonseca, D.P.; Benaissa-Trouw, B.; van Engelen, M.; Kraaijeveld, C.A.; Snippe, H.; Verheul, A.F. Induction of cell-mediated immunity against *Mycobacterium tuberculosis* using DNA vaccines encoding cytotoxic and helper T-cell epitopes of the 38-kilodalton protein. *Infect. Immun.* **2001**, *69*, 4839–4845. [CrossRef]
38. Jung, S.B.; Yang, C.S.; Lee, J.S.; Shin, A.R.; Jung, S.S.; Son, J.W.; Harding, C.V.; Kim, H.J.; Park, J.K.; Paik, T.H.; et al. The mycobacterial 38-kilodalton glycolipoprotein antigen activates the mitogen-activated protein kinase pathway and release of proinflammatory cytokines through Toll-like receptors 2 and 4 in human monocytes. *Infect. Immun.* **2006**, *74*, 2686–2696. [CrossRef]
39. Lopez-Vidal, Y.; de Leon-Rosales, S.P.; Castanon-Arreola, M.; Rangel-Frausto, M.S.; Melendez-Herrada, E.; Sada-Diaz, E. Response of IFN-gamma and IgG to ESAT-6 and 38 kDa recombinant proteins and their peptides from *Mycobacterium tuberculosis* in tuberculosis patients and asymptomatic household contacts may indicate possible early-stage infection in the latter. *Arch. Med. Res.* **2004**, *35*, 308–317. [CrossRef]
40. Wiker, H.G.; Harboe, M. The antigen 85 complex: A major secretion product of *Mycobacterium tuberculosis*. *Microbiol. Rev.* **1992**, *56*, 648–661. [CrossRef]
41. Daffe, M. The mycobacterial antigens 85 complex—From structure to function and beyond. *Trends Microbiol.* **2000**, *8*, 438–440. [CrossRef]
42. Davila, J.; McNamara, L.A.; Yang, Z. Comparison of the predicted population coverage of tuberculosis vaccine candidates Ag85B-ESAT-6, Ag85B-TB10.4, and Mtb72f via a bioinformatics approach. *PLoS ONE* **2012**, *7*, e40882. [CrossRef] [PubMed]
43. Meerak, J.; Wanichwecharungruang, S.P.; Palaga, T. Enhancement of immune response to a DNA vaccine against *Mycobacterium tuberculosis* Ag85B by incorporation of an autophagy inducing system. *Vaccine* **2013**, *31*, 784–790. [CrossRef] [PubMed]
44. Xu, Y.; Yang, E.; Wang, J.; Li, R.; Li, G.; Liu, G.; Song, N.; Huang, Q.; Kong, C.; Wang, H. Prime-boost bacillus Calmette-Guerin vaccination with lentivirus-vectored and DNA-based vaccines expressing antigens Ag85B and Rv3425 improves protective efficacy against *Mycobacterium tuberculosis* in mice. *Immunology* **2014**, *143*, 277–286. [CrossRef] [PubMed]
45. Yuan, X.; Teng, X.; Jing, Y.; Ma, J.; Tian, M.; Yu, Q.; Zhou, L.; Wang, R.; Wang, W.; Li, L.; et al. A live attenuated BCG vaccine overexpressing multistage antigens Ag85B and HspX provides superior protection against *Mycobacterium tuberculosis* infection. *Appl. Microbiol. Biotechnol.* **2015**, *99*, 10587–10595. [CrossRef]
46. Jones, M.B.; Rosenberg, J.N.; Betenbaugh, M.J.; Krag, S.S. Structure and synthesis of polyisoprenoids used in N-glycosylation across the three domains of life. *Biochim. Biophys. Acta* **2009**, *1790*, 485–494. [CrossRef]
47. Nothaft, H.; Szymanski, C.M. Protein glycosylation in bacteria: Sweeter than ever. *Nat. Rev. Microbiol.* **2010**, *8*, 765–778. [CrossRef]
48. Sonawane, A.; Mohanty, S.; Jagannathan, L.; Bekolay, A.; Banerjee, S. Role of glycans and glycoproteins in disease development by *Mycobacterium tuberculosis*. *Crit. Rev. Microbiol.* **2012**, *38*, 250–266. [CrossRef]
49. Mehaffy, C.; Belisle, J.T.; Dobos, K.M. Mycobacteria and their sweet proteins: An overview of protein glycosylation and lipoglycosylation in *M. tuberculosis*. *Tuberculosis* **2019**, *115*, 1–13. [CrossRef]
50. van Els, C.A.C.M.; Corbière, V.; Smits, K.; van Gaans-van den Brink, J.A.M.; Poelen, M.C.M.; Mascart, F.; Meiring, H.D.; Locht, C. Toward Understanding the Essence of Post-Translational Modifications for the *Mycobacterium tuberculosis* Immunoproteome. *Front. Immunol.* **2014**, *5*, 361. [CrossRef]
51. Facciuolo, A.; Mutharia, L.M. Mycobacterial glycoproteins: A novel subset of vaccine candidates. *Front. Cell. Infect. Microbiol.* **2014**, *4*, 133. [CrossRef]

52. Horn, C.; Namane, A.; Pescher, P.; Rivière, M.; Romain, F.; Puzo, G.; Bârzu, O.; Marchal, G. Decreased capacity of recombinant 45/47-kDa molecules (Apa) of *Mycobacterium tuberculosis* to stimulate T lymphocyte responses related to changes in their mannosylation pattern. *J. Biol. Chem.* **1999**, *274*, 32023–32030. [CrossRef]
53. Romain, F.C.; Pescher, P.; Namane, A.; Riviere, M.; Puzo, G.; Barzu, O.; Marchal, G. Deglycosylation of the 45/47-kilodalton antigen complex of *Mycobacterium tuberculosis* decreases its capacity to elicit in vivo or in vitro cellular immune responses. *Infect. Immun.* **1999**, *67*, 5567–5572. [CrossRef] [PubMed]
54. Birhanu, A.G.; Yimer, S.A.; Kalayou, S.; Riaz, T.; Zegeye, E.D.; Holm-Hansen, C.; Norheim, G.; Aseffa, A.; Abebe, M.; Tønjum, T. Ample glycosylation in membrane and cell envelope proteins may explain the phenotypic diversity and virulence in the *Mycobacterium tuberculosis* complex. *Sci. Rep.* **2019**, *9*, 2927. [CrossRef]
55. D'Souza, S.; Rosseels, V.; Romano, M.; Tanghe, A.; Denis, O.; Jurion, F.; Castiglione, N.; Vanonckelen, A.; Palfliet, K.; Huygen, K. Mapping of murine Th1 helper T-Cell epitopes of mycolyl transferases Ag85A, Ag85B, and Ag85C from *Mycobacterium tuberculosis*. *Infect. Immun.* **2003**, *71*, 483–493. [CrossRef]
56. Gilleron, M.; Nigou, J.; Nicolle, D.; Quesniaux, V.; Puzo, G. The acylation state of mycobacterial lipomannans modulates innate immunity response through toll-like receptor 2. *Chem. Biol.* **2006**, *13*, 39–47. [CrossRef]
57. Kariyone, A.; Tamura, T.; Kano, H.; Iwakura, Y.; Takeda, K.; Akira, S.; Takatsu, K. Immunogenicity of Peptide-25 of Ag85B in Th1 development: Role of IFN-gamma. *Int. Immunol.* **2003**, *15*, 1183–1194. [CrossRef]
58. Flores-Trevino, S.; Morfín-Otero, R.; Rodríguez-Noriega, E.; González-Díaz, E.; Pérez-Gómez, H.R.; Bocanegra-García, V.; Vera-Cabrera, L.; Garza-González, E. Genetic diversity of *Mycobacterium tuberculosis* from Guadalajara, Mexico and identification of a rare multidrug resistant Beijing genotype. *PLoS ONE* **2015**, *10*, e0118095. [CrossRef]
59. Madeira, F.; Pearce, M.; Tivey, A.R.N.; Basutkar, P.; Lee, J.; Edbali, O.; Madhusoodanan, N.; Kolesnikov, A.; Lopez, R. Search and sequence analysis tools services from EMBL-EBI in 2022. *Nucleic Acids Res.* **2022**, *50*, W276–W279. [CrossRef]
60. Froger, A.; Hall, J.E. Transformation of plasmid DNA into E. coli using the heat shock method. *J. Vis. Exp.* **2007**, *6*, 253. [CrossRef]
61. Steentoft, C.; Vakhrushev, S.Y.; Joshi, H.J.; Kong, Y.; Vester-Christensen, M.B.; Schjoldager, K.T.B.G.; Lavrsen, K.; Dabelsteen, S.; Pedersen, N.B.; Marcos-Silva, L.; et al. Precision mapping of the human O-GalNAc glycoproteome through SimpleCell technology. *EMBO J.* **2013**, *32*, 1478–1488. [CrossRef]
62. Fletcher, C.M.; Coyne, M.J.; Comstock, L.E. Theoretical and experimental characterization of the scope of protein O-glycosylation in Bacteroides fragilis. *J. Biol. Chem.* **2011**, *286*, 3219–3226. [CrossRef] [PubMed]
63. Hamby, S.E.; Hirst, J.D. Prediction of glycosylation sites using random forests. *BMC Bioinform.* **2008**, *9*, 500. [CrossRef] [PubMed]
64. Orduna, P.; Cevallos, M.A.; de Leon, S.P.; Arvizu, A.; Hernandez-Gonzalez, I.L.; Mendoza-Hernandez, G.; Lopez-Vidal, Y. Genomic and proteomic analyses of Mycobacterium bovis BCG Mexico 1931 reveal a diverse immunogenic repertoire against tuberculosis infection. *BMC Genom.* **2011**, *12*, 493. [CrossRef]
65. Dobos, K.M.; Khoo, K.H.; Swiderek, K.M.; Brennan, P.J.; Belisle, J.T. Definition of the full extent of glycosylation of the 45-kilodalton glycoprotein of *Mycobacterium tuberculosis*. *J. Bacteriol.* **1996**, *178*, 2498–2506. [CrossRef] [PubMed]
66. UniProt, C. UniProt: The universal protein knowledgebase in 2021. *Nucleic Acids Res.* **2021**, *49*, D480–D489. [CrossRef]
67. Singh, Y.; Beamer, G.; Sun, X.; Shukla, P. Recent developments in systems biology and genetic engineering toward design of vaccines for TB. *Crit. Rev. Biotechnol.* **2022**, *42*, 532–547. [CrossRef]
68. Fan, X.; Li, X.; Wan, K.; Zhao, X.; Deng, Y.; Chen, Z.; Luan, X.; Lu, S.; Liu, H. Construction and immunogenicity of a T cell epitope-based subunit vaccine candidate against *Mycobacterium tuberculosis*. *Vaccine* **2021**, *39*, 6860–6865. [CrossRef] [PubMed]
69. Hu, Z.; Lu, S.H.; Lowrie, D.B.; Fan, X.Y. Research Advances for Virus-vectored Tuberculosis Vaccines and Latest Findings on Tuberculosis Vaccine Development. *Front. Immunol.* **2022**, *13*, 895020. [CrossRef]
70. Sarmiento, M.E.; Alvarez, N.; Chin, K.L.; Bigi, F.; Tirado, Y.; García, M.A.; Anis, F.Z.; Norazmi, M.N.; Acosta, A. Tuberculosis vaccine candidates based on mycobacterial cell envelope components. *Tuberculosis* **2019**, *115*, 26–41. [CrossRef]
71. Wang, C.; Lu, J.; Du, W.; Wang, G.; Li, X.; Shen, X.; Su, C.; Yang, L.; Chen, B.; Wang, J.; et al. Ag85b/ESAT6-CFP10 adjuvanted with aluminum/poly-IC effectively protects guinea pigs from latent *Mycobacterium tuberculosis* infection. *Vaccine* **2019**, *37*, 4477–4484. [CrossRef]
72. Watson, A.; Li, H.; Ma, B.; Weiss, R.; Bendayan, D.; Abramovitz, L.; Ben-Shalom, N.; Mor, M.; Pinko, E.; Bar Oz, M.; et al. Human antibodies targeting a Mycobacterium transporter protein mediate protection against tuberculosis. *Nat. Commun.* **2021**, *12*, 602. [CrossRef]
73. Brodin, P.; Eiglmeier, K.; Marmiesse, M.; Billault, A.; Garnier, T.; Niemann, S.; Cole, S.T.; Brosch, R. Bacterial artificial chromosome-based comparative genomic analysis identifies Mycobacterium microti as a natural ESAT-6 deletion mutant. *Infect. Immun.* **2002**, *70*, 5568–5578. [CrossRef]
74. Herrmann, J.L.; O'Gaora, P.; Gallagher, A.; Thole, J.E.; Young, D.B. Bacterial glycoproteins: A link between glycosylation and proteolytic cleavage of a 19 kDa antigen from *Mycobacterium tuberculosis*. *EMBO J.* **1996**, *15*, 3547–3554. [CrossRef]
75. Kaufmann, S.H. How can immunology contribute to the control of tuberculosis? *Nat. Rev. Immunol.* **2001**, *1*, 20–30. [CrossRef]
76. Kaufmann, S.H. The contribution of immunology to the rational design of novel antibacterial vaccines. *Nat. Rev. Microbiol.* **2007**, *5*, 491–504. [CrossRef]
77. Skeiky, Y.A.; Sadoff, J.C. Advances in tuberculosis vaccine strategies. *Nat. Rev. Microbiol.* **2006**, *4*, 469–476. [CrossRef]
78. Cheng, C.H.; Lee, W.C. Protein solubility and differential proteomic profiling of recombinant Escherichia coli overexpressing double-tagged fusion proteins. *Microb. Cell Fact.* **2010**, *9*, 63. [CrossRef]

79. Rosano, G.L.; Ceccarelli, E.A. Recombinant protein expression in Escherichia coli: Advances and challenges. *Front. Microbiol.* **2014**, *5*, 172. [CrossRef]
80. Su, H.; Peng, B.; Zhang, Z.; Liu, Z.; Zhang, Z. The *Mycobacterium tuberculosis* glycoprotein Rv1016c protein inhibits dendritic cell maturation, and impairs Th1/Th17 responses during mycobacteria infection. *Mol. Immunol.* **2019**, *109*, 58–70. [CrossRef]
81. Yuan, C.; Qu, Z.L.; Tang, X.L.; Liu, Q.; Luo, W.; Huang, C.; Pan, Q.; Zhang, X.L. *Mycobacterium tuberculosis* Mannose-Capped Lipoarabinomannan Induces IL-10-Producing B Cells and Hinders CD4(+)Th1 Immunity. *iScience* **2019**, *11*, 13–30. [CrossRef]
82. Espitia, C.; Mancilla, R. Identification, isolation and partial characterization of *Mycobacterium tuberculosis* glycoprotein antigens. *Clin. Exp. Immunol.* **1989**, *77*, 378–383. [PubMed]
83. Willcocks, S.; Wren, B.W. Shared characteristics between *Mycobacterium tuberculosis* and fungi contribute to virulence. *Future Microbiol.* **2014**, *9*, 657–668. [CrossRef] [PubMed]
84. Pan, Q.; Wang, Q.; Sun, X.; Xia, X.; Wu, S.; Luo, F.; Zhang, X.L. Aptamer against mannose-capped lipoarabinomannan inhibits virulent *Mycobacterium tuberculosis* infection in mice and rhesus monkeys. *Mol. Ther.* **2014**, *22*, 940–951. [CrossRef] [PubMed]
85. Almagro-Moreno, S.; Boyd, E.F. Insights into the evolution of sialic acid catabolism among bacteria. *BMC Evol. Biol.* **2009**, *9*, 118. [CrossRef]
86. Giri, J.; Tang, J.M.; Wirth, C.; Peneff, C.M.; Eisenberg, B. Single-channel measurements of an N-acetylneuraminic acid-inducible outer membrane channel in *Escherichia coli. Eur. Biophys. J.* **2012**, *41*, 259–271. [CrossRef]
87. Li, Y.; Chen, X. Sialic acid metabolism and sialyltransferases: Natural functions and applications. *Appl. Microbiol. Biotechnol.* **2012**, *94*, 887–905. [CrossRef]
88. Vimr, E.R.; Kalivoda, K.A.; Deszo, E.L.; Steenbergen, S.M. Diversity of microbial sialic acid metabolism. *Microbiol. Mol. Biol. Rev.* **2004**, *68*, 132–153. [CrossRef]
89. Sweet, L.; Zhang, W.; Torres-Fewell, H.; Serianni, A.; Boggess, W.; Schorey, J. Mycobacterium avium glycopeptidolipids require specific acetylation and methylation patterns for signaling through toll-like receptor 2. *J. Biol. Chem.* **2008**, *283*, 33221–33231. [CrossRef]
90. Wang, B.; Henao-Tamayo, M.; Harton, M.; Ordway, D.; Shanley, C.; Basaraba, R.J.; Orme, I.M. A Toll-like receptor-2-directed fusion protein vaccine against tuberculosis. *Clin. Vaccine Immunol.* **2007**, *14*, 902–906. [CrossRef]

biomolecules

MDPI

Brief Report

Lack of Ecto-5′-Nucleotidase Protects Sensitized Mice against Allergen Challenge

Elisabetta Caiazzo [1], Ida Cerqua [1], Roberta Turiello [2,3], Maria Antonietta Riemma [1], Giacomo De Palma [1], Armando Ialenti [1], Fiorentina Roviezzo [1], Silvana Morello [2] and Carla Cicala [1,*]

1 Department of Pharmacy, School of Medicine and Surgery, University of Naples Federico II, 80131 Naples, Italy; elisabetta.caiazzo@unina.it (E.C.); ida.cerqua@unina.it (I.C.); mariaantonietta.riemma@unicampania.it (M.A.R.); giacomo.depalma@unina.it (G.D.P.); armando.ialenti@unina.it (A.I.); fiorentina.roviezzo@unina.it (F.R.)

2 Department of Pharmacy, University of Salerno, 84084 Fisciano, Italy; rturiello@unisa.it (R.T.); smorello@unisa.it (S.M.)

3 PhD Program in Drug Discovery and Development, University of Salerno, 84084 Fisciano, Italy

* Correspondence: cicala@unina.it

Abstract: Ecto-5′-nucleotidase (CD73), the ectoenzyme that together with CD39 is responsible for extracellular ATP hydrolysis and adenosine accumulation, regulates immune/inflammatory processes by controlling innate and acquired immunity cell functions. We previously demonstrated that CD73 is required for the assessment of a controlled allergic sensitization, in mice. Here, we evaluated the response to aerosolized allergen of female-sensitized mice lacking CD73 in comparison with their wild type counterpart. Results obtained show, in mice lacking CD73, the absence of airway hyperreactivity in response to an allergen challenge, paralleled by reduced airway CD23$^+$B cells and IL4$^+$T cells pulmonary accumulation together with reduced mast cells accumulation and degranulation. Our findings indicate CD73 as a potential therapeutic target for allergic asthma.

Keywords: airways; allergy; CD73; ecto-5′-nucleotidase; mast cells; hyperresponsiveness; B cells; CD23; sensitization

1. Introduction

Sensitization to an allergen is the condition for an allergic reaction to occur. Usually, humans recognize that they are sensitized to an allergen only when they come in contact with the allergen and experience the allergic reaction, which can even be fatal. Innate and adaptive immune systems talk to each other to assess allergic sensitization [1]. Sensitized subjects undergo different allergic manifestations when they encounter the allergen, depending on individual, genetic, and environmental factors [2].

Exploring host factors controlling the assessment of sensitization and those that control effector mechanisms of allergy is needed to identify potential therapeutic targets. Ecto-5′-nucleotidase (CD73) is the ectoenzyme responsible for tissue extracellular adenosine accumulation [3]. There is evidence that extracellular adenosine accumulates in asthmatic airways and plays a role in bronchial inflammation and hyperreactivity. An association between exposure to an allergen and airway responsiveness to adenosine and AMP has been demonstrated in experimental animals and humans [4–8]. CD73 regulates immune/inflammatory processes by controlling innate and acquired immunity cell functions [9–13]. By using a model of ovalbumin (OVA)-sensitized female mice, we previously demonstrated that CD73 is required for the assessment of a controlled sensitization. Indeed, sensitized mice lacking CD73 develop exacerbated airway inflammation, characterized by increased levels of Th2 cytokines and reduced frequency of regulatory T cells into the lung; nonetheless, they did not develop airway hyperreactivity [14]. The aim of the present work was to investigate the role of CD73 in the effector phase induced by an allergen challenge in the model of sensitized mice. We evaluated the response to aerosolized allergen of female-sensitized mice lacking CD73. Results obtained show, in mice lacking CD73, the absence of airway hyperreactivity in response to an allergen challenge, paralleled by reduced airway $CD23^{+}B$ cells and $IL4^{+}T$ cell pulmonary accumulation together with reduced mast cell accumulation and degranulation. Although further in-depth analysis is required, our findings suggest that targeting CD73 might be a therapeutic strategy to control allergic airway hyperresponsiveness.

2. Materials and Methods

2.1. Animals

Female C57BL/6J mice (Charles River, Calco, Italy) and $CD73^{-/-}$ were bred and housed in the animal facilities laboratory of Department of Pharmacy at the University of Naples Federico II. All animal procedures were approved by the Italian Ministry of Health and were in accordance with Italian (DL 26/2014) and European (n.63/2010/UE) guidelines and regulations.

2.2. Allergic Sensitization and Challenge

Mice (8 weeks old) were sensitized with 100 μg of ovalbumin (OVA, grade V; Sigma Aldrich, Milan, Italy) emulsified with aluminium hydroxide ($Al(OH)_3$, 13 mg/mL) and administered subcutaneously (s.c) on days 1 and 8. From day 21, mice were challenged with aerosolized OVA (5% *w*/*v* in saline) or with saline for 2 consecutive days for 20 min each day. All mice were sacrificed in 24 h following the last OVA or saline challenge.

2.3. Immunoblot Analysis for CD73

Pulmonary tissue was homogenized using a FastPrep-24 instrument (MP Biomedicals, Santa Ana, CA, USA) as previously described [12]. Fifty micrograms of proteins were loaded on 8% SDS-PAGE gel and transferred onto nitrocellulose membrane. The membrane was blocked with 5% (*w*/*v*) non-fat dry milk in phosphate buffer saline (PBS) with 0.1% (*v*/*v*) tween 20, before incubation overnight at 4 °C with anti-mouse CD73 polyclonal goat antibody 1:200 (Santa Cruz Biotechnology, Dallas, TX, USA). Then the membrane was incubated for 2 h at room temperature with the secondary antibody anti-goat IgG 1:2000 (Dako, CA, USA) conjugated with peroxidase. Anti-β-actin antibody (Santa Cruz Biotechnology) was used as a loading protein control. Protein bands were visualized using the ECL detection kit (Bio-Rad, Milan, Italy) and the ChemiDoc Imaging System (Bio-Rad, Italy). Densitometry was obtained using Image Lab software (Bio-Rad, Milan, Italy).

2.4. Measurement of Airway Responsiveness

From each animal group, the main bronchi were collected and rapidly cleaned from fat and connective tissue. Rings of 1–2 mm length were cut and mounted in 2.5 mL isolated organ baths containing Krebs solution, at 37 °C, oxygenated (95% O_2 and 5% CO_2), and connected to an isometric force transducer (type 7006, Ugo Basile, Comerio, Italy) associated to a Powerlab 800 (AD Instruments). Bronchial reactivity to cumulative concentrations of carbachol (10^{-9}–3×10^{-6} M) was evaluated. Results were expressed as dyne per mg of tissue [15].

2.5. Measurement of AMPase Activity

AMPase activity was assessed in lung homogenates by the measurement of inorganic phosphate (Pi) released after incubation with AMP (2 mM) through a Malachite Green assay as previously described [12]. To confirm that Pi measured was dependent on CD73 enzymatic activity, each sample was also tested following incubation with the CD73 inhibitor, adenosine 5′- (α, β-methylene) diphosphate (APCP, 100 μM); the value of nonspecific Pi released was subtracted from the value obtained following incubation with AMP. Results were expressed as Pi released (pmol/min/μg protein).

2.6. Enzyme-Linked Immunosorbent Assay

Total IgE levels were quantified in plasma samples by ELISA kit (BD Pharmingen, Franklin Lakes, NJ, USA or R&D Systems, Inc., Minneapolis, MN, USA) according to the manufacturer's instructions.

2.7. Lung Histology

Lung lobes were removed, fixed in 4% formalin, and embedded in paraffin. The tissue was then sectioned into 7 µm slices and stained with haematoxylin and eosin (H & E) for morphological analysis. In addition, to detect mast cells in the lung, tissue sections were stained with toluidine blue. Toluidine blue-positive mast cell numbers and degranulated mast cells were determined as previously described [16].

2.8. Flow Cytometry Analysis

Lungs were digested with 1 U/mL collagenase (Sigma-Aldrich, Milan, Italy). The analysis of inflammatory cells was determined by flow cytometry (BD FacsCalibur Milan, Italy) using the following antibodies: B220-Phycoerythrin (PE), CD19-Phycoerythrin-Cyanine5 (Pe-Cy5), CD23-Allophycocyanin (APC), CD3-Peridinin chlorophyll (PerCP), and CD4 Allophycocyanin-Cyanine7 (APC-Cy7). Intracellular staining for IL-4 (IL4-PE) was performed on fixed/permeabilized T cells after the extracellular staining for CD3 and CD4.

2.9. Statistical Analysis

All results are presented as mean ± standard error (SE). Statistical analysis was performed through Student's *t*-test for unpaired data or by one- or two-ways ANOVA for multiple comparisons followed by Bonferroni post-hoc test, as appropriate. A p value < 0.05 was considered statistically significant.

3. Results

3.1. Pulmonary Tissues from Ovalbumin-Challenged Mice Show High Levels of CD73 Expression and Activity

In order to investigate the role of CD73 in controlling the airway response to an allergen challenge, we used an ovalbumin-sensitized/challenged mouse model (Figure 1a). Mice were subcutaneously sensitized with ovalbumin (OVA), at day 1 and 8, followed by aerosolized OVA-challenge (OVA-challenge) for 2 consecutive days (day 21 and day 22) to induce allergic airway inflammation (Figure 1a). Mice were sacrificed at day 23. We first analyzed the expression and activity of CD73 in whole-lung tissue samples. CD73 expression and AMPase activity were up-regulated in lung tissue samples following sensitization, without a further increase following OVA-challenge (Figure 1b,c, respectively). The AMPase activity was completely reversed in the presence of the selective inhibitor of CD73, APCP (data not shown).

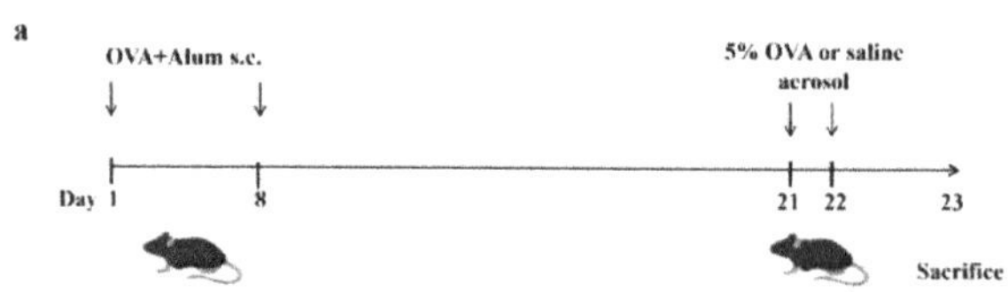

Figure 1. *Cont.*

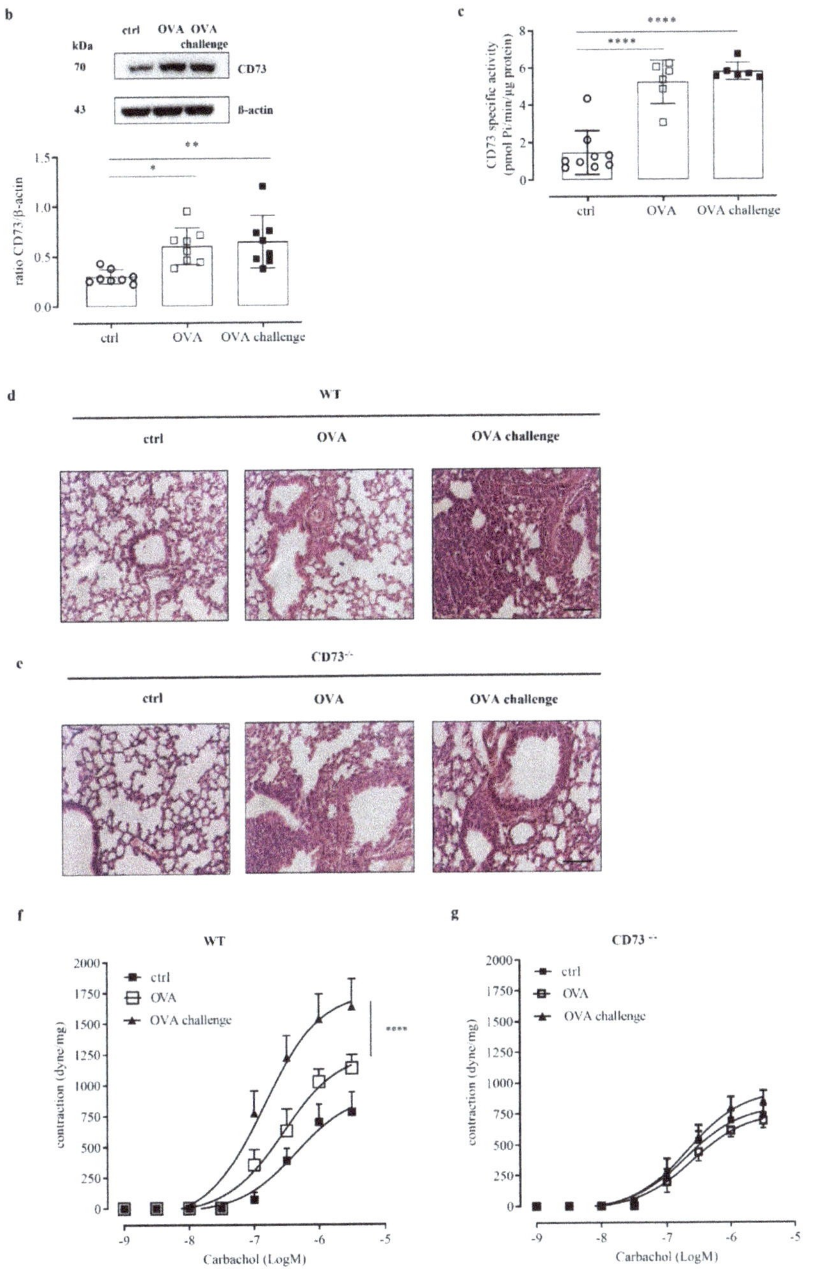

Figure 1. (**a**) OVA sensitization and aerosol challenge protocol in WT and CD73$^{-/-}$ mice. Mice were sensitized and challenged with OVA as described in Materials and Methods. (**b**) CD73 pulmonary expression in control and OVA-sensitized and challenged WT mice was evaluated by Western blots. Data are expressed as means ± SE of eight mice per group. * $p < 0.05$ and ** $p < 0.01$ (one-way ANOVA followed by Bonferroni's Multiple Comparison Test). (**c**) AMPase activity was determined by Malachite Green assay in lung tissue from control and OVA-sensitized and challenged WT mice. All data are expressed as mean ± SE of six to nine mice per group. **** $p < 0.0001$ (one-way ANOVA followed by

Bonferroni's Multiple Comparison Test). Histopathological changes in the lung tissues from control and sensitized and challenged WT (**d**) and CD73$^{-/-}$ mice (**e**) were assessed by hematoxylin and eosin (H&E) staining. Scale bar: 100 μm. Bronchial reactivity in response to carbachol was evaluated in control and OVA-sensitized and challenged WT (**f**) and CD73$^{-/-}$ (**g**) mice; * $p < 0.05$ and **** $p < 0.0001$; two-ways ANOVA plus Bonferroni.

3.2. Ovalbumin Challenge Does Not Increase Pulmonary Inflammatory Cell Infiltration in CD73$^{-/-}$ Mice

When we examined pulmonary morphology and inflammation by hematoxylin and eosin staining, we observed that the recruitment of inflammatory cells in the peribronchial region was markedly increased following OVA-challenge compared with lung sections from OVA-sensitized mice and control mice (Figure 1d) in WT mice. We noticed that the recruitment of inflammatory cells in the peribronchial region of lung from CD73-deficient mice (CD73$^{-/-}$) resulted as markedly increased after sensitization, as we have previously demonstrated [14], without a further increase following the OVA challenge (Figure 1e).

3.3. Ovalbumin Challenge Does Not Increase Bronchial Hyperreactivity to Carbachol in CD73$^{-/-}$ Mice

We evaluated the role of CD73 in the assessment of bronchial hyperreactivity triggered by an allergen challenge. Bronchial reactivity was evaluated as a response to carbachol of OVA-sensitized and challenged WT and CD73$^{-/-}$ mice. OVA sensitization increased bronchial reactivity to carbachol in WT mice, while it did not affect bronchial reactivity to carbachol in CD73$^{-/-}$ mice [14]. Here, we found that OVA challenge significantly increased bronchial reactivity to carbachol of sensitized WT mice (Figure 1f) but not in sensitized CD73$^{-/-}$ mice (Figure 1g).

3.4. Ovalbumin Challenge Enhances IgE Production in All Mice Groups

Analysis of total IgE levels was performed after ovalbumin challenge in plasma of WT and CD73$^{-/-}$ mice, as described in the Method section, and compared with plasma IgE levels determined in sensitized and control mice. The OVA challenge increased IgE levels in sensitized WT and CD73$^{-/-}$ mice, compared with their respective non-sensitized controls; notably, IgE levels following OVA challenge were significantly higher in sensitized CD73$^{-/-}$ mice than in their sensitized WT counterpart (122.2 ± 32.60 ng/mL, vs. 54.68 ± 20.86 ng/mL, n = 5; $p < 0.05$) (Figure 2a). Increased IgE levels in CD73$^{-/-}$ mice compared with WT mice were also observed before OVA challenge (Figure 2a).

a

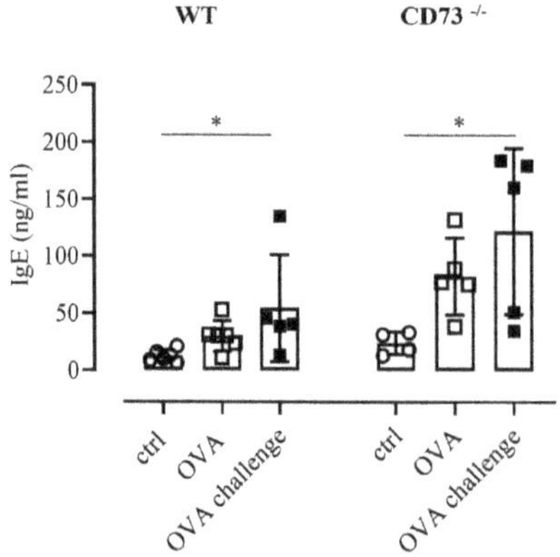

Figure 2. *Cont.*

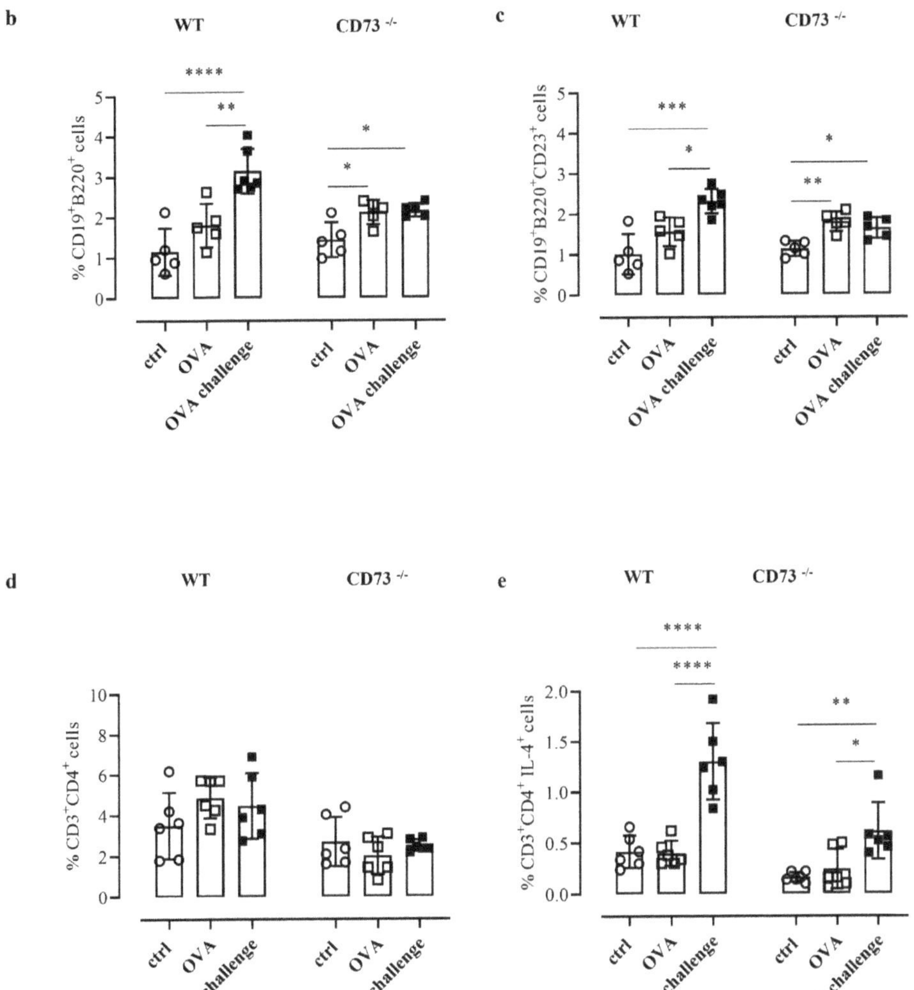

Figure 2. (**a**) Plasma levels of total IgE were measured in control and OVA-sensitized and challenged WT and CD73$^{-/-}$ mice by ELISA. All results are expressed as mean ± SE of n = 4–6 mice per group. * $p < 0.05$ (one-way ANOVA followed by Bonferroni's Multiple Comparison Test). The % of CD19$^+$B220$^+$ was evaluated in control and sensitized and challenged WT and CD73$^{-/-}$ mice (**b**) by flow cytometry as well as the % of CD19$^+$B220$^+$CD23$^+$ (**c**). Data are expressed as means ± SE of five to six mice per group. * $p < 0.05$, ** $p < 0.01$,*** $p < 0.001$, **** $p < 0.0001$ (one-way ANOVA followed by Bonferroni's Multiple Comparison Test). The % of CD3$^+$CD4$^+$ and CD3$^+$ CD4$^+$ IL4$^+$ in control and sensitized and challenged WT and CD73$^{-/-}$ mice was analyzed by flow cytometry (**d**,**e**). Data are expressed as means ± SE of five to six mice per group. * $p < 0.05$, ** $p < 0.01$, **** $p < 0.0001$ (one-way ANOVA followed by Bonferroni's Multiple Comparison Test).

3.5. OVA Challenge Increases Pulmonary Infiltration of CD23$^+$B Cells and IL-4$^+$ T Cells

We then analyzed the percentage of B cells (as CD19$^+$ B220$^+$ cells) in pulmonary tissues from all groups of WT mice and CD73$^{-/-}$ mice by flow cytometry. We found that OVA challenge increased the percentage of B cells in tissue samples from sensitized WT mice compared with control mice (Figure 2b). In tissue samples from sensitized CD73 $^{-/-}$ mice we found an increased frequency of B cells, compared to control mice, which were not further increased following OVA-challenge (Figure 2b). When we analyzed the expression of the low-affinity receptor for IgE (CD23) that is crucial in the IgE-induced inflammatory responses under allergen exposure [17], we observed that B cells were almost exclusively $CD23^+$ (Figure 2d). Indeed, the percentage of $CD23^+$B cells in WT mice following OVA challenge was significantly increased compared with sensitized WT mice (Figure 2c); conversely, in $CD73^{-/-}$ mice following OVA challenge the percentage of $CD23^+$B cells was increased compared with control mice but similar to that observed in sensitized $CD73^{-/-}$ mice (Figure 2c).

CD23 is crucial in the IgE-allergen-induced activation of T cells toward a Th2-like phenotype [17]. Th2-derived IL-4 represents an important effector mechanism of type 1 allergy [1]. Therefore, we next analyzed the percentage $CD4^+$ T cells expressing IL-4 in pulmonary tissues from all groups of WT mice and $CD73^{-/-}$ mice by flow cytometry analysis. Analyzing the total levels of $CD3^+CD4^+$T cells, we did not observe any significant changes among the experimental groups (Figure 2d). Instead, the frequency of $CD3^+CD4^+IL\text{-}4^+$ cells was significantly increased in both WT and $CD73^{-/-}$ OVA-challenged mice groups (Figure 2e). Of note, we also observed that the percentage of $IL4^+$T cells in $CD73^{-/-}$ mice, although they were significantly increased after OVA challenge, did not reach the frequency triggered by OVA challenge in WT mice ($0.61 \pm 0.11\%$, n = 6 vs. $1.30 \pm 0.15\%$, n = 6; $p < 0.0001$ Bonferroni test) (Figure 2e).

3.6. Ovalbumin Challenge Does Not Increase Mast Cell Degranulation in the Lung of $CD73^{-/-}$ Mice

Pulmonary mast cell number and degranulation were then evaluated by means of toluidine blue staining in WT and $CD73^{-/-}$ mice following sensitization and challenge. OVA sensitization increased the number of mast cells in the lung of WT (Figure 3a,c) and $CD73^{-/-}$ mice (Figure 3b,c). However, OVA challenge further increased the number of pulmonary mast cells in sensitized WT mice, whereas it did not affect the number of pulmonary mast cells in sensitized $CD73^{-/-}$ mice (Figure 3c). Concomitantly, OVA challenge increased the percentage (%) of degranulated mast cells in sensitized WT mice but not in $CD73^{-/-}$ mice (Figure 3d).

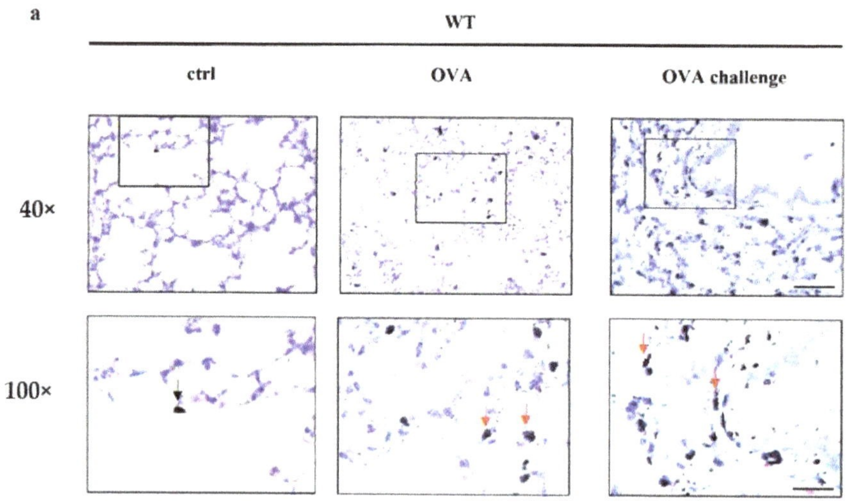

Figure 3. *Cont.*

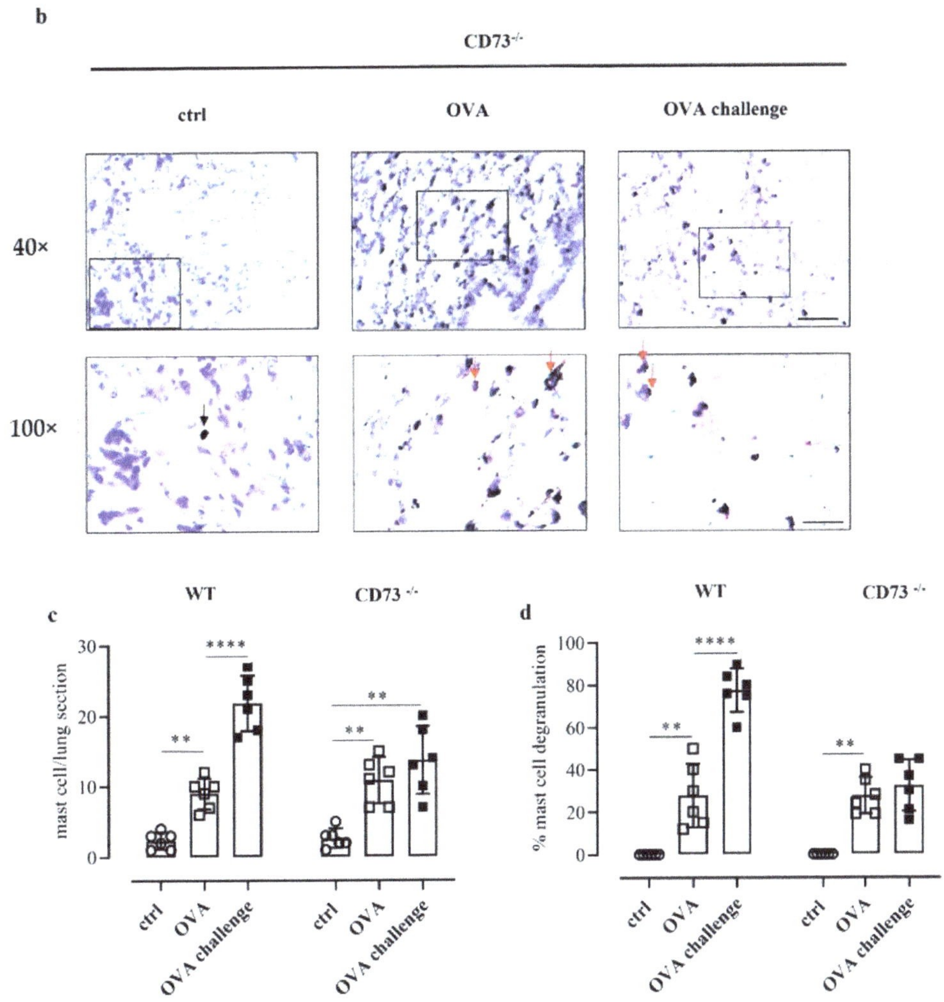

Figure 3. Representative photomicrographs of toluidine blue-stained lung sections from control and OVA-sensitized and challenged WT (**a**) and CD73$^{-/-}$ mice (**b**). The number of pulmonary mast cells as well as a differentiation between non-degranulated (deep blue, black arrows) and degranulated (light blue, red arrows) and the percentage of mast cell degranulation was performed (**c**,**d**). Scale bar: 50 μm and 20 μm. Data are expressed as means ± SE of six mice per group. ** $p < 0.01$ and **** $p < 0.0001$ (one-way ANOVA followed by Bonferroni's Multiple Comparison Test).

4. Discussion

We have previously shown that CD73 plays a role in the assessment of allergic sensitization, a requisite for an allergic reaction to occur. Indeed, in a model of OVA-sensitized CD73$^{-/-}$ mice we observed exacerbation of airway inflammation that was paralleled by increased pulmonary inflammatory cytokine levels and a reduced percentage of pulmonary Foxp3^{+}Tregs, suggesting that CD73 is functional to those surveillance mechanisms of the immune response [14].

Here, we investigated whether CD73 plays a role in the effector phase of allergic reaction. Thus, we developed the model of sensitized/allergen-challenged mice to compare functional and cellular response to the allergen challenge of sensitized WT and $CD73^{-/-}$ mice.

According to our previous results, we found that mice sensitization with OVA increased airway CD73 expression and activity [14]; however, there was no further increase following mice challenge with aerosolized OVA. The functional study showed hyperreactivity to carbachol of bronchi from sensitized mice and a further increase of reactivity of bronchi from allergen-challenged sensitized mice. Likely, CD73 up-regulation, following sensitization, primes bronchi to ova challenge, while hyperreactivity to carbachol, which was observed in sensitized WT but not $CD73^{-/-}$ mice, may be dependent upon adenosine availability into the airways [7].

Interestingly, in $CD73^{-/-}$ mice, neither sensitization alone nor sensitization plus an allergen challenge caused bronchial hyperreactivity, suggesting that the lack of CD73, despite causing allergic airway inflammation, as previously demonstrated, gives protection from hyperreactivity caused by an allergen challenge. Schreiber et al. [18] described the absence of airway hyperresponsiveness to an allergen in $CD73^{-/-}$ mice, of both sexes, due to inhibition of the tracheal Cl^- secretion stimulated by ATP and ADP following OVA challenge.

Here, we investigated cellular mechanisms that are differently activated by an allergen challenge in sensitized WT and $CD73^{-/-}$ mice. First of all, we found that in all groups of animals, circulating IgE levels were increased by an allergen challenge, indicating that the absence of CD73 did not affect the ability of B cells to produce IgE in response to an allergen challenge. This result suggests that the functional difference between sensitized WT and $CD73^{-/-}$ mice, following an allergen challenge, probably lies in effector mechanisms triggered by an allergen challenge in WT and $CD73^{-/-}$ mice.

It is worth noting that IgE levels from $CD73^{-/-}$ were higher than levels obtained from WT mice. The significance of such a difference needs to be further investigated; however, it was demonstrated that the couple CD39/CD73, through increasing adenosine levels, plays a role in Ig class switch in B cells [19]; on this basis, we could hypothesize that, under our experimental conditions, Ig class switch would be lost in B cells from $CD73^{-/-}$ mice, resulting in high IgE levels.

Besides the high affinity IgE receptor, FcεRI, on effector cells of type I allergy, another receptor for IgE, CD23 (FcεRII), represents an important modulator of immune response since it regulates IgE levels. CD23 on B cells is a low-affinity receptor for IgE, however, it can oligomerize, leading to enhanced IgE binding [17]. It is known that when allergen crosslinks the IgE-FcεRI receptor, effector cells are activated and an allergic response occurs. Evidence that there is a reciprocal inhibition between FcεRI and CD23 that blocks simultaneous binding by IgE suggested that these two receptors have different roles in the control of IgE [20,21]. We found that the OVA challenge increased the percentage of $CD23^+$ B cells into the pulmonary tissue of sensitized WT mice but not in sensitized $CD73^{-/-}$ mice.

It is known that the density of CD23 molecules on B cells correlates with IgE levels and facilitates allergen-specific T-cell activation [22]. Thus, there is a link between the Th2 response and CD23 density; in agreement, it has been shown that an anti-CD23 antibody inhibits Th2 response [23]. Increased levels of $CD23^+$B cells in patients with grass pollen allergy have been shown [22]. Thus, under our experimental conditions, we observed a reduced frequency of pulmonary $CD23^+$B cells in sensitized/OVA-challenged $CD73^{-/-}$ mice associated to a protection from allergen-induced airway hyperreactivity. It is worth noting that the difference in the frequency of pulmonary $CD23^+$B cells between WT and $CD73^{-/-}$ mice was observed following an OVA challenge but not before, in only sensitized mice, meaning that in $CD73^{-/-}$ mice, B cell response to an allergen challenge is defective, which is in agreement with an important role of CD73 in memory B cells development [24,25].

On this basis, we evaluated the frequency of pulmonary IL-4^+Tcells in our experimental groups, with or without an allergen challenge. We found that an allergen challenge increased the frequency of CD3$^+$CD4$^+$IL-4$^+$ cells in airways in both groups of OVA-sensitized WT and CD73$^{-/-}$ mice, with an increase in WT mice significantly higher than that achieved in CD73$^{-/-}$ mice. This result suggests that the lack of CD73 reduces Th2 response following an allergen challenge.

Finally, we also found that OVA challenge increased pulmonary mast cell accumulation and degranulation only in the WT-sensitized group but not in CD73$^{-/-}$-sensitized groups. This result requires further investigation to understand whether this is a key point. Mast cells represent effector cells of type I allergy, and their degranulation is mediated by allergens linking FcεRI-bound IgE and gives immediate allergic symptoms. The role of CD73/adenosine signalling in antigen-induced mast cell degranulation is still poorly understood; however, there is evidence that adenosine potentiates antigen-induced pulmonary mast cell activation [26].

In conclusion, our results demonstrate increased CD73 expression and activity in sensitized airways and the lack of airway response to an allergen challenge of CD73$^{-/-}$-sensitized mice characterized by reduced Th2 activation and reduced mast cell accumulation and degranulation. Although further investigation is needed, our results provide the basis to reflect on CD73 as a potential therapeutic target for type I allergy.

Author Contributions: Conceptualization, E.C., S.M. and C.C.; Methodology, E.C., S.M., F.R. and C.C.; Validation, E.C., S.M., F.R., R.T., I.C. and C.C.; Investigation, E.C., I.C., G.D.P., M.A.R., R.T., S.M. and C.C.; Writing—original draft preparation, E.C., S.M. and C.C.; Writing, review and editing, E.C., S.M., A.I., F.R. and C.C. All authors have read and agreed to the published version of the manuscript.

Funding: This research was funded by the Grant of University of Naples Federico II (Research Program 2017–2019; DR 409 del 7 Febbraio 2017).

Institutional Review Board Statement: All animal procedures were approved by the Italian Ministry of Health and were in accordance with Italian (DL 26/2014) and European (n.63/2010/UE) guidelines and regulations (authorization 5/2019 project code CF.43).

Data Availability Statement: The data that support the findings of this study are contained in the body of the manuscript. Any additional data used are available from the corresponding authors upon reasonable request.

Acknowledgments: We wish to thank the veterinarian Antonio Baiano and technicians Giovanni Esposito and Angelo Russo for animal care assistance and technical support given.

Conflicts of Interest: The authors declare no conflict of interest.

References

1. Holgate, S.T.; Wenzel, S.; Postma, D.S.; Weiss, S.T.; Renz, H.; Sly, P.D. Asthma. *Nat. Rev. Dis. Primers* **2015**, *1*, 15025. [CrossRef] [PubMed]
2. Ree, R.; Hummelshøj, L.; Plantinga, M.; Poulsen, L.K.; Swindle, E. Allergic sensitization: Host-immune factors. *Clin. Transl. Allergy* **2014**, *4*, 12. [CrossRef] [PubMed]
3. Zimmermann, H.; Zebisch, M.; Sträter, N. Cellular function and molecular structure of ecto-nucleotidases. *Purinergic Signal.* **2012**, *8*, 437–502. [CrossRef] [PubMed]
4. Currie, G.P.; Jackson, C.M.; Lee, D.K.C.; Lipworth, B.J. Allergen sensitization and bronchial hyper-responsiveness to adenosine monophosphate in asthmatic patients. *Clin. Exp. Allergy* **2003**, *33*, 1405–1408. [CrossRef]
5. Wilson, C.N. Adenosine receptors and asthma in humans. *J. Cereb. Blood Flow Metab.* **2008**, *155*, 475–486. [CrossRef]
6. Alfieri, A.; Parisi, A.; Maione, F.; Grassia, G.; Morello, S.; Ialenti, A.; Mascolo, N.; Cicala, C. Hyperresponsiveness to adenosine in sensitized Wistar rats over-expressing A1 receptor. *Eur. J. Pharmacol.* **2012**, *695*, 120–125. [CrossRef]
7. Cicala, C.; Ialenti, A. Adenosine signaling in airways: Toward a promising antiasthmatic approach. *Eur. J. Pharmacol.* **2013**, *714*, 522–525. [CrossRef]
8. Garcia-Garcia, L.; Olle, L.; Martin, M.; Roca-Ferrer, J.; Muñoz-Cano, R. Adenosine Signaling in Mast Cells and Allergic Diseases. *Int. J. Mol. Sci.* **2021**, *22*, 5203. [CrossRef]
9. Eckle, T.; Füllbier, L.; Wehrmann, M.; Khoury, J.; Mittelbronn, M.; Ibla, J.; Rosenberger, P.; Eltzschig, H.K. Identification of Ectonucleotidases CD39 and CD73 in Innate Protection during Acute Lung Injury. *J. Immunol.* **2007**, *178*, 8127–8137. [CrossRef]
10. Wang, L.-L.; Tang, H.-P.; Shi, G.-C.; Wan, H.-Y.; Tang, W.; Hou, X.-X.; Pan, L.-N.; Shi, B.-Y.; Tao, L.-Q. CD39/CD73 and the imbalance of Th17 cells and regulatory T cells in allergic asthma. *Mol. Med. Rep.* **2013**, *8*, 1432–1438. [CrossRef]

11. Neuberger, A.; Ring, S.; Silva-Vilches, C.; Schrader, J.; Enk, A.; Mahnke, K. Expression of CD73 slows down migration of skin dendritic cells, affecting the sensitization phase of contact hypersensitivity reactions in mice. *J. Dermatol. Sci.* **2017**, *87*, 292–299. [CrossRef] [PubMed]
12. Caiazzo, E.; Morello, S.; Carnuccio, R.; Ialenti, A.; Cicala, C. The Ecto-5′-Nucleotidase/CD73 Inhibitor, α,β-Methylene Adenosine 5′-Diphosphate, Exacerbates Carrageenan-Induced Pleurisy in Rat. *Front. Pharmacol.* **2019**, *10*, 775. [CrossRef] [PubMed]
13. Schneider, E.; Winzer, R.; Rissiek, A.; Ricklefs, I.; Meyer-Schwesinger, C.; Ricklefs, F.L.; Bauche, A.; Behrends, J.; Reimer, R.; Brenna, S.; et al. CD73-mediated adenosine production by CD8 T cell-derived extracellular vesicles constitutes an intrinsic mechanism of immune suppression. *Nat. Commun.* **2021**, *12*, 5911. [CrossRef] [PubMed]
14. Caiazzo, E.; Cerqua, I.; Riemma, M.A.; Turiello, R.; Ialenti, A.; Schrader, J.; Fiume, G.; Caiazza, C.; Roviezzo, F.; Morello, S.; et al. Exacerbation of Allergic Airway Inflammation in Mice Lacking ECTO-5′-Nucleotidase (CD73). *Front. Pharmacol.* **2020**, *11*, 589343. [CrossRef]
15. Morello, S.; Vellecco, V.; Roviezzo, F.; Maffia, P.; Cuzzocrea, S.; Cirino, G.; Cicala, C. A protective role for proteinase activated receptor 2 in airways of lipopolysaccharide-treated rats. *Biochem. Pharmacol.* **2005**, *71*, 223–230. [CrossRef]
16. Roviezzo, F.; Rossi, A.; Caiazzo, E.; Orlando, P.; Riemma, M.A.; Iacono, V.M.; Guarino, A.; Ialenti, A.; Cicala, C.; Peritore, A.; et al. Palmitoylethanolamide Supplementation during Sensitization Prevents Airway Allergic Symptoms in the Mouse. *Front. Pharmacol.* **2017**, *8*, 857. [CrossRef]
17. Engeroff, P.; Vogel, M. The role of CD23 in the regulation of allergic responses. *Allergy* **2021**, *76*, 1981–1989. [CrossRef] [PubMed]
18. Schena, F.; Volpi, S.; Faliti, C.E.; Penco, F.; Santi, S.; Proietti, M.; Schenk, U.; Damonte, G.; Salis, A.; Bellotti, M.; et al. Dependence of Immunoglobulin Class Switch Recombination in B Cells on Vesicular Release of ATP and CD73 Ectonucleotidase Activity. *Cell Rep.* **2013**, *3*, 1824–1831. [CrossRef]
19. Schreiber, R.; Castrop, H.; Kunzelmann, K. Allergen-induced airway hyperresponsiveness is absent in ecto-5′-nucleotidase (CD73)-deficient mice. *Pflügers Arch. -Eur. J. Physiol.* **2008**, *457*, 431–440. [CrossRef]
20. Conrad, D.H.; Ford, J.W.; Sturgill, J.; Gibb, D.R. CD23: An overlooked regulator of allergic disease. *Curr. Allergy Asthma Rep.* **2007**, *7*, 331–337. [CrossRef]
21. Cheng, L.E.; Wang, Z.-E.; Locksley, R.M. Murine B Cells Regulate Serum IgE Levels in a CD23-Dependent Manner. *J. Immunol.* **2010**, *185*, 5040–5047. [CrossRef] [PubMed]
22. Selb, R.; Eckl-Dorna, J.; Neunkirchner, A.; Schmetterer, K.; Marth, K.; Gamper, J.; Jahn-Schmid, B.; Pickl, W.F.; Valenta, R.; Niederberger, V. CD23 surface density on B cells is associated with IgE levels and determines IgE-facilitated allergen uptake, as well as activation of allergen-specific T cells. *J. Allergy Clin. Immunol.* **2017**, *139*, 290–299.e4. [CrossRef] [PubMed]
23. Poole, J.A.; Meng, J.; Reff, M.; Spellman, M.C.; Rosenwasser, L.J. Anti-CD23 monoclonal antibody, lumiliximab, inhibited allergen-induced responses in antigen-presenting cells and T cells from atopic subjects. *J. Allergy Clin. Immunol.* **2005**, *116*, 780–788. [CrossRef] [PubMed]
24. Anderson, S.M.; Tomayko, M.M.; Ahuja, A.; Haberman, A.M.; Shlomchik, M.J. New markers for murine memory B cells that define mutated and unmutated subsets. *J. Exp. Med.* **2007**, *204*, 2103–2114. [CrossRef]
25. Conter, L.J.; Song, E.; Shlomchik, M.J.; Tomayko, M.M. CD73 Expression Is Dynamically Regulated in the Germinal Center and Bone Marrow Plasma Cells Are Diminished in Its Absence. *PLoS ONE* **2014**, *9*, e92009. [CrossRef]
26. Gomez, G.; Zhao, W.; Schwartz, L.B. Disparity in FcεRI-Induced Degranulation of Primary Human Lung and Skin Mast Cells Exposed to Adenosine. *J. Clin. Immunol.* **2011**, *31*, 479–487. [CrossRef]

biomolecules

MDPI

Article

Deepening into Intracellular Signaling Landscape through Integrative Spatial Proteomics and Transcriptomics in a Lymphoma Model

Alicia Landeira-Viñuela [1], Paula Díez [1,2], Pablo Juanes-Velasco [1], Quentin Lécrevisse [1], Alberto Orfao [1], Javier De Las Rivas [3] and Manuel Fuentes [1,2,*]

1 Department of Medicine and General Cytometry Service-Nucleus, USAL/IBSAL, 37000 Salamanca, Spain; alavi29@usal.es (A.L.-V.); pauladg@usal.es (P.D.); pablojuanesvelasco@usal.es (P.J.-V.); quentin@usal.es (Q.L.); orfao@usal.es (A.O.)

2 Proteomics Unit, Cancer Research Centre (IBMCC/CSIC/USAL/IBSAL), 37007 Salamanca, Spain

3 Bioinformatics and Functional Genomics, Cancer Research Centre (IBMCC/CSIC/USAL/IBSAL), 37007 Salamanca, Spain; jrivas@usal.es

* Correspondence: mfuentes@usal.es; Tel.: +34-923-29-48-11

Abstract: Human Proteome Project (HPP) presents a systematic characterization of the protein landscape under different conditions using several complementary-omic techniques (LC-MS/MS proteomics, affinity proteomics, transcriptomics, etc.). In the present study, using a B-cell lymphoma cell line as a model, comprehensive integration of RNA-Seq transcriptomics, MS/MS, and antibody-based affinity proteomics (combined with size-exclusion chromatography) (SEC-MAP) were performed to uncover correlations that could provide insights into protein dynamics at the intracellular level. Here, 5672 unique proteins were systematically identified by MS/MS analysis and subcellular protein extraction strategies (neXtProt release 2020-21, MS/MS data are available via ProteomeXchange with identifier PXD003939). Moreover, RNA deep sequencing analysis of this lymphoma B-cell line identified 19,518 expressed genes and 5707 protein coding genes (mapped to neXtProt). Among these data sets, 162 relevant proteins (targeted by 206 antibodies) were systematically analyzed by the SEC-MAP approach, providing information about PTMs, isoforms, protein complexes, and subcellular localization. Finally, a bioinformatic pipeline has been designed and developed for orthogonal integration of these high-content proteomics and transcriptomics datasets, which might be useful for comprehensive and global characterization of intracellular protein profiles.

Keywords: affinity-based proteomics; human proteome project; LC-MS/MS; transcriptomics; size-exclusion-chromatography (SEC)

1. Introduction

Increased understanding of the events taking place during intracellular signaling has revealed a highly dynamic protein landscape for eukaryotic cells, especially in pathological settings (such as cancer, neurodegenerative, auto-immune diseases, etc.,) [1–3]. Most of the subcellular signaling pathways are led by protein complexes, which are constantly being formed and resolved. Furthermore, proteins are shuttling between different subcellular localizations to execute the expected and/or programmed biological processes according to the needs and requirements of the cell. These fluctuations could produce re-wiring of signaling networks for enabling phenotypic changes required for the adaptation to microenvironmental or external perturbations and/or stimuli [4–6]. Given these temporal and spatial variabilities, high-throughput biochemical methods are required to deepen the knowledge on the interacting biomolecules across subcellular localizations during the response to external stimulus, drug administration, and drug resistance mechanisms, etc., [1,2,4,6,7].

Bearing this in mind, the recent advances in next-generation sequencing (NGS) and high-resolution mass spectrometry (MS/MS) have opened up strategies methodologically which provide novel insights useful for understanding the complexity of intracellular processes [8,9]. Among other parameters in protein functionality, subcellular localization is one of the main determinants for a particular protein's intracellular dynamics and the final function; however, a global view of subcellular proteome organization remains relatively unknown, and it is currently studied by multipronged proteomics approaches. Despite some multiple integrations that have been recently performed (based on MS/MS and RNA-Seq, among others), there remains a huge interest in revealing the presence of proteoforms, multiprotein complexes, changes in transcript-protein, etc., [10–13]. Historically, subcellular localization of proteins has been determined by conventional biochemical approaches (mostly by targeting individual proteins); e.g., cell fractionation coupled with Western blot. Although additional development of large-scale GFP fusion protein-based [14], antibody-based assays [15,16], protein-metabolite interactions [17], Sequential Window Acquisition of All Theoretical Mass Spectra (SWATH-MS) characterization [18] has increased our knowledge of protein subcellular localization, these methods are labor-intensive, making them highly challenging and difficult to implement for high-content characterization in large cohort studies.

The development of mass spectrometry (MS/MS)-based proteomics coupled with protein extraction at subcellular level [19] as well as the high-throughput immunoassays coupled with size-exclusion chromatography (SEC) [20] have opened up new possibilities to query the spatial intracellular organization of the proteome on a larger scale [21]. More recently, the Aebersold group [22] has developed a strategy combining mass-spectrometry and SEC. Other similar attempts involve machine learning algorithms to assign subcellular localization based on protein quantification across multiple subcellular fractions by MS/MS. However, all these studies provide proteome-wide information but lack the multiple combinations between several proteomics characterizations and comprehensive integration with transcriptomics.

This study has two main purposes: i. Evaluation of protein extraction or complementary characterization with MS/MS-based proteomics and affinity proteomics (SEC-MAP), which may be useful for bioinformatic orthogonal integration with deep transcriptomics characterization; ii. generation of systematic pipeline for comprehensive orthogonal integration of protein subcellular localization, affinity proteomics (SEC-MAP), MS/MS data sets, and RNA-seq information in human lymphoma cell line, like a model for mapping cancer protein interactions. Moreover, the combination of these methodologies helps to deepen the knowledge about the architecture of the cells and the complexity of the spatial organization of the proteome, which can be a tool for interpreting a multiscale map of protein systems—which is relevant for deciphering cell signaling pathways linked to genetic perturbations/alterations, therapeutic interventions, or another external stimulus.

2. Materials and Methods

2.1. Cell Cultures

Human Ramos cell line (Burkitt's lymphoma, European Collection of Authenticated Cell Cultures—ECACC-Cat. no: 85030802) was cultured in RPMI 1640 medium supplemented with L-glutamine (Gibco, Whaltham, MA, USA), 10% (*v*/*v*) fetal bovine serum (Gibco, Whaltham, MA, USA) and 1% (*v*/*v*) penicillin/streptavidin (Gibco, Whaltham, MA, USA). The growth was monitored daily and cells were incubated at 37 °C/5% CO_2 with media renewal every 2–3 days.

2.2. Protein Extraction and Quantification

For protein extraction, 40×10^6 Ramos cells were pelleted by centrifugation and washed three times with PBS, and further centrifuged for 5 min (min.) at 1200 rpm. All the different protein extraction buffers were supplemented with phosphatase and protease inhibitors: tris [2-carboxyethyl]phosphine] (TCEP—Sigma-Aldrich, Burlington, MA, USA)

1 mM, phenylmethylsulfonyl fluoride (PMSF—Sigma-Aldrich, USA) 1 mM, sodium fluoride (NaF—Sigma-Aldrich, USA) 1 mM, sodium orthovanadate (Sigma-Aldrich, USA) 1 mM, β-glycerophosphate (Sigma-Aldrich, USA) 1 mM and sodium pyrophosphate tetrabasic decahydrate (Sigma-Aldrich, USA) 1 mM. All centrifugation and incubation steps were carried out at 4 °C. The extracted proteins from each strategy were stored at −20 °C until further use.

Protein quantification was performed by Coomassie Plus (Bradford) Assay Kit (ThermoFisher Scientific, Cat. no: 23236, USA) and Pierce™ BCA Protein Assay Kit (ThermoFisher Scientific, Cat. no: 23225, Waltham, MA, USA), following the manufacturer's instructions. Each sample (20 μg) was separated in 12% SDS-PAGE gel under reducing conditions. Subsequently, SDS-PAGE gels were stained with 0.1% (*w*/*v*) Coomassie brilliant blue and silver stains solutions according to standard laboratory procedure [23–25].

The different protein extraction procedures, which are evaluated herein, are described below and summarized in Table 1.

2.2.1. Protocol #1

Total of 375 μL of lysis buffer (140 mM NaCl, 50 mM EDTA, 10% (*v*/*v*) glycerol, 1% octylphenoxy poly(ethyleneoxy)ethanol—IGEPAL-, 20 mM Tris-HCl pH = 7 supplemented with proteases and phosphates inhibitors) per 1×10^7 cells was added, and cells were incubated on ice for 15 min. After that, the sample was centrifuged for 15 min at 15,000× *g* and the supernatant containing the protein content was stored at −20 °C until further analysis.

2.2.2. Protocol #2

As protocol #1, 1×10^7 cells were incubated with 375 μL of lysis buffer (20 mM HEPES pH = 8 and 9 M urea). Cell lysis was performed by sonication on ice (3 times 5 s bursts and 1 min break). Afterward, it was centrifuged for 15 min at 15,000× *g* and the supernatant containing the total protein content was stored at −20 °C until further analysis.

2.2.3. Protocol #3

Similar procedure as protocol #2 with a lysis buffer containing 7 M urea, 2 M thiourea, and 30 mM Tris-HCl pH = 8.5. The sample was then centrifuged for 15 min at 12,000× *g* and the supernatant was stored at −20 °C until further analysis.

2.2.4. Protocol #4

The same procedure as protocol #3 but with a different lysis buffer: 5 mM HEPES pH 8, 10 mM $MgCl_2$, 140 mM NaCl and 0.01% Tween 20.

2.2.5. Protocol #5

Total of 1×10^7 cells were incubated with 375 μL of hypotonic buffer (30 mM HEPES pH = 8, 15 mM KCl, 2 mM $MgCl_2$, 1 mM EDTA and 20% glycerol) supplemented with 10% (*v*/*v*) laurylmaltoside (n-Dodecyl-β-D-maltoside) for 30 min in rotation at 4 °C. Afterward centrifugation at 16,000× *g* for 5 min was carried out and the supernatant containing the proteins was collected.

2.2.6. Protocol #6

Similar to protocol #5 with a difference that the lysis buffer was supplemented with 1.5% (*v*/*v*) Triton X-100 instead of laurylmaltoside.

2.2.7. Protocol #7

In this case, 1×10^7 cells were incubated with 375 μL of buffer (5 mM HEPES pH 8, 10 mM $MgCl_2$, 140 mM NaCl and 0.01% Tween 20) and it was centrifuged for 15 min at 13,000× *g*. After centrifugation, the supernatant enrichment in cytoplasmic (Cyt) proteins was collected and the pellet was re-suspended in 375 μL of buffer supplemented with 38% (*v*/*v*) octyl-β-D-glucopyranoside. The sample was sonicated on ice (3 times 3 s bursts and

1 min break). After sonication, it was incubated for 30 min on ice and subsequently, it was centrifuged for 15 min at 13,000× *g*. Finally, the supernatant enriched in membrane proteins (Mem) was collected [26].

2.2.8. Protocol #8

Ramos cell pellets were re-suspended at a volume equal to 5 times that of the cell pellet in hypotonic buffer supplemented with protease and phosphatase inhibitors and 0.015% digitonin following the methodology published by Díez P. et al., 2015 [19].

Then, it was incubated on a rotatory shaker for 30 min and then it was centrifuged for 5 min at 500× *g*. After centrifugation, cytoplasmic (Cyt) proteins were found in the resulting supernatant 1. The pellet 1 was washed 3× with hypotonic buffer and it was centrifuged for 5 min at 500× *g* at 4 °C. The following fractions were processed similarly in a stepwise manner. For the supernatant 2 containing organelle (Org) proteins, a hypotonic buffer was supplemented with 0.5% Tween 20; for the supernatant 3 containing nuclear (Nuc) proteins, a hypotonic buffer supplemented with 14 mM NaCl was used; for the supernatant 4 containing membrane (Mem) proteins, a hypotonic buffer supplemented with 1% laurylmaltoside was used.

2.2.9. Protocol #9

Similar procedure as protocol #8 for cytoplasmic and organelle subcellular fractions. The membrane fraction was extracted with a hypotonic buffer supplemented with 0.5% octylphenoxy poly(ethyleneoxy)ethanol (IGEPAL) and centrifuged for 5 min at 3000× *g*. A hypotonic buffer supplemented with 1% laurylmaltoside was used for the extraction of the nuclear fraction.

2.3. Proteomics Analysis

2.3.1. Protein Digestion and LC-MS/MS Analysis

Each lane in SDS-PAGE gel (loaded with 15 µg protein extract) was cut into five equal fragments and digested using the method described by Olsen et al. with slight modifications (each piece was destained with 15 mM potassium ferrocyanide and for reduction and 50 mM sodium thiosulfate was used for the alkylation process) and incubated with 10 mM DTT at 56 °C for 45 min and then 55 mM iodoacetamide (IAA) was added and incubated at room temperature (RT) for 30 min, respectively. Trypsin (6.25 ng/mL) at 37 °C for 18 h was used for protein digestion, and the peptide solution was acidified with formic acid (FA) and desalted using C18-Stage-Tips columns) [19,27]. Samples were partially dried and stored at −20 °C until analyzed by LC-MS/MS.

A nanoUPLC system (nanoAcquity, Waters Corp., Milford, MA, USA) coupled to an LTQ-Velos-Orbitrap mass spectrometer (Thermo Fisher Scientific, San Jose, CA, USA) via a nanoelectrospray ion source (NanoSpray flex, Proxeon, Thermo) was used to study with LC-MS/MS. Peptides dissolution was carried out using 0.5% FA/3% acetonitrile (ACN). A trapping column (nanoACQUITY UPLC 2G-V/M Trap Symmetry 5 µm particle size, 180 µm × 20 mm C18 column, Waters Corp., Milford, MA, USA) was used to load. Separation was made with a linear gradient from 7% to 35% solvent B (ACN/0.1% FA) at a flow rate of 250 nL/min over 120 min in a nanoACQUITY UPLC BEH 1.7 µm, 130 Å, 75 µm × 250 mm C18 column (Waters Corp., Milford, MA, USA) [19,26,27]. Tandem mass spectra (MS/MS) acquisition and survey MS scan were applied to a data-dependent automatic switch using the positive ion mode of the nUPLC-LTQ-Orbitrap Velos. Acquisition scan was made with lock mass option enabled for the 445.120025 ion and mass range of m/z 400 to 1600. In the ion trap for fragmentation by collision-induced dissociation with 35% normalized energy, 10 ms activation time, q = 0.25, ± 2 m/z precursor isolation width and wideband activation, we selected the 20 peaks with the most intensity and with ≥2 charge state and above the 500 intensity threshold. Automatic gain control was 1×10^6 for MS and 5×10^3 for MS/MS scans and dynamic exclusion was enabled for 90 s [19,26,27].

Table 1. Summary of protein extraction procedures.

			Protocols									
		Buffer elements	#1	#2	#3	#4	#5	#6	#7	#8	#9	
Step 1	*Phosphatase Inhibitor (mM)*	***TCEP***	1	1	1	1	1	1	1	1	1	1
		PMSF	1	1	1	1	1	1	1	1	1	1
		NaF	1	1	1	1	1	1	1	1	1	1
		Sodium orthovanadate	1	1	1	1	1	1	1	1	1	1
		β-glycerophosphate	1	1	1	1	1	1	1	1	1	1
		Sodium pyrophosphate tetrabasic decahydrate	1	1	1	1	1	1	1	1	1	1
	Salt (mM)	***NaCl***	140	-	-	140	-	-	140	-	-	400
		KCl	-	-	-	-	15	15	-	15	15	-
	Nuclear envelope protector (mM)	***MgCl2***	-	-	-	10	2	2	10	2	2	2
	Metalloproteinase inhibitor (mM)	***EDTA***	50	-	-	-	1	1	-	1	1	1
	Chaotropic agent (M)	***Urea***	-	9	7	-	-	-	-	-	-	-
		Thiourea	-	-	2	-	-	-	-	-	-	-
	Buffer solution (mM)	***Tris/HCl***	20	-	30	-	-	-	-	-	-	-
		HEPES		20	-	5	30	30	5	30	30	30
	Thickening agent (%-v/v-)	***Glycerol***	10	-	-	-	20	20	-	20	20	-
	Non ionic detergent (%-v/v-)	***IGEPAL***	1	-	-	-	-	-	-	-	-	-
		Tween 20	-	-	-	0.1	-	-	0.1	-	-	-
		Laurylmaltoside	-	-	-	-	10	-	-	-	-	-
		Triton X-100	-	-	-	-	-	1.5	-	-	-	-
			Protocols									
		Buffer elements	#1	#2	#3	#4	#5	#6	#7	#8	#9	
Step 1	*Non ionic detergent (%-v/v-)*	***Digitonin***	-	-	-	-	-	-	-	0.015	0.015	-
		Octyl-β-D-glucopiranoside	-	-	-	-	-	-	38	-	-	-
Step 2	*Non ionic detergent (%-v/v-)*	***Tween 20***	-	-	-	-	-	-	-	5	5	-
Step 3	*Salt (mM)*	***NaCl***	-	-	-	-	-	-	-	14	-	-
	Non ionic detergent (%-v/v-)	***IGEPAL***	-	-	-	-	-	-	-	-	-	0.5
Step 4		***Laurylmaltoside***	-	-	-	-	-	-	-	1	-	1
	Centrifugation	***Time (min)***	15	15	15	15	5	5	15	5	5	5
		$10^3 \times g$	15	15	12	12	16	16	15	0.5	0.5	3/15

Legend

Phosphatase Inhibitor	Salt	Nuclear envelope protector	Metalloproteinase inhibitor
Chaotropic agent	Buffer solution	Thickening agent	Non ionic detergent
	Centrifugation		

2.3.2. Database Search

Raw data were converted into Mascot general file (.mgf) and target-decoy strategy was used to search in the neXtProt database (release 2016-02). Comet version (v.) 2015.01 rev.2 [28] was used to identify the peak list (.mgf file) obtained from MS/MS spectra and SearchGUI v. 1.30.1 [29] was used.

Concatenated target/decoy [30] version of the human complement of neXtProt release 2016-02 (41,992 sequences) was used for protein identification. SearchGUI v. 1.30.1 was used to design decoy sequences (the reverse target sequences) [29]. The process for the identification of the settings was as follows: trypsin with a maximum of two missed cleavages; 10.0 ppm as MS1 and 0.5 Da as MS2 tolerances; fixed modifications: carbamidomethylation of cysteine (+57.021464 Da). The variable modifications were acetylation of protein n-terminus (+42.010565 Da) and oxidation of methionine (+15.994915 Da) [19,26–30].

PeptideShaker v. 0.41.1 [31] was used to infer peptides and proteins from the identified spectra. A method, as the one published by our group in 2015, was used for the validation of decoy hit distribution (at a 1.0% False Discovery Rate—FDR—for Peptide Spectrum Matches –PSMs-, peptides and proteins) [19]. It is obtained at protein/peptide/PSM levels, the FDR (%), true positives, and false positives values for each replicated (1, 2, 3): Cyt 1 (0.99, 2899, 29/0.98, 10474, 104/1.0, 17627, 178); Cyt 2 (1.0, 3183, 32/1.0, 12897, 130/1.0, 20147, 203); Cyt 3 (0.99, 2802, 28/0,99, 10857, 109/1.0, 5666, 57); Mem 1 (0.93, 1166, 11/0.97, 2651, 26/0.99, 1303, 13); Mem 2 (0.97, 1122, 11/0.97, 1937, 19/0.99, 1607, 16); Mem 3 (1.0, 893, 9/0.98, 1523, 15/1.0, 1291, 13); Org 1 (0.97, 2543, 25/0.99, 7503, 75/0.99, 10305, 103); Org 2 (0.99, 2294, 23/0.99, 6919, 69/0.98, 4933, 49); Org 3 (0.96, 2465, 24/0.98, 7045, 70/1.0, 10507, 106); Nuc 1 (0.99, 2498, 25/0.98, 7145, 71/0.97, 915, 9); Nuc 2 (0.96, 2577, 25/1.0, 8938, 90/1.0, 4469, 45); Nuc 3 (1.0, 2872, 29/0.99, 10937, 109/0.99, 4585, 46). The protein-level FDR is an estimate and not all proteins that exceeded the threshold were "confidently identified".

The mass spectrometry data and the identification results have been deposited to the ProteomeXchange Consortium [32] via the PRIDE partner repository [33] with the dataset identifier PXD003939 and 10.6019/PXD003939. The data can be accessed with the following credentials upon login to the PRIDE website (http://www.ebi.ac.uk/pride/archive/login, accessed on 30 March 2021): Username: reviewer85106@ebi.ac.uk, Password: 0fVloZfQ.

2.3.3. Quantitative Analysis of MS/MS Datasets

Raw data were analyzed as reported by Paula et al. in 2021 [34]: i. For expression level measuring, the Label-Free Quantification method MaxLFQ [35] with the MaxQuant Suite v. 1.5.3.30 was used [36]. ii. PTXQC package v. 0.80.1 [37] in R v. 3.2.4 [38] was utilized for quality control analysis. iii. Perseus framework v. 1.5.3.2 was performed for ulterior analysis. Additionally, for each subcellular fraction (Mem, Cyt, Org and Nuc) total proteins and exclusive proteins were determined.

2.3.4. RNA-Sequencing Transcriptomics

RNA-Seq data from Ramos B-cell line was obtained with Illumina Genome Analyzer IIx with paired layout (experiment SRX105534: http://www.ncbi.nlm.nih.gov/sra/SRX105534, accessed on 30 March 2019 taken from the study SRP00931 (http://trace.ncbi.nlm.nih.gov/Traces/sra/?study=SRP009316, accessed on 30 March 2019) [39] from SRA (Sequence Read Archive) database. Gene expression analysis consisted of calculated values of FPKM fragment per kilobase of exon per million fragments mapped) on base for each gene on the following steps [19]: i. Use of SRA tools [40] to obtain the SRR387395 dataset from SRA database [40] and the subsequent conversion of the SRA file to paired-end fastq files; ii. trimming of the data with Trimmomatic [41]; iii. use the program STAR [42] to align the reads to ENSEMBL GRCh37 genome; iv. generation of a binary sequence alignment map (BAM) with SAMtools [43]; v. calculating the FPKM value for each gene with CuffLinks from the BAM files [44]; vi.-Use of the neXtProt ID mapping table to map ENSG_IDs within the neXtProt database v. 2016-02 (ftp://ftp.nextprot.org/pub/

current_release/mapping/, accessed on 30 March 2019). Finally, a total of 19,518 neXtProt IDs could be mapped within the RNA-Seq dataset, out of this 9523 neXtProt IDs had FPKM > 1.

2.4. Biotin Protein Labeling

Protein extracts obtained from protocols #1, #4, #5, #6, #7, #8, and #9 were biotin-labelled following the procedure described by Häggmark A. et al. 2013 [45]. All the cell lysates were incubated with two different concentrations (10 μg/μL—protocol A—and 1 μg/μL—protocol B-) of NHS-PEG_4-biotin (Thermo Scientific, no: 21363, USA) for 2 h at 4 °C. Biotin-labeling reactions were stopped with 4.5 μL for protocol A and 50% of the volume of the biotin-labeling solution for concentration for protocol B of 0.5 M Tris-HCl pH = 8.

After that, Amicon® Ultra-0.5 centrifugal filter 3K (Millipore-Merk, Damrstad, Germany) was used for removing the excess of biotin until a final volume of 110 μL (maximum volume of injection in column HPLC). In both approaches, 75 μg of protein was conjugated with biotin.

2.5. Size Exclusion Chromatography (SEC): Fast Protein Liquid Chromatography (FPLC)

Equipment HPLC 1100 series (Agilent) and column Superdex® 200 Increase 10/300 GL (GE Healthcare, Illinois, Sigma-Aldrich, USA) was used for protein fractionation based on molecular weight (MW). As MW standards, a mix of five purified proteins (Ferritin—440 kDa, aldolase—158 kDa-, conalbumin—75 kDa, ovalbumin—43 kDa and ribonuclease A—13.7 kDa) was used. Each sample was fractionated and collected in 24 aliquots at 0.5 mL/min. (flow rate) and PBS 1X Na^{2+}/K^{+}- tween 20 0.5% (*v*/*v*) as running buffer.

Then, all 24 collected fractions were merged in 8 fractions within defined MW ranges (fraction 1—166–473 kDa, fraction 2—121–142 kDa, fraction 3—74–103 kDa, fraction 4—54–63 kDa, fraction 5—33–46 kDa, fraction 6—24–28 kDa, fraction 7—17–21 kDa and fraction 8—11–15 kDa) by using Amicon® Ultra-0.5 centrifugal filter 3K (Millipore-Merk, Germany) until a final volume of 100 μL. All steps were performed at 4 °C. Each collected fraction was stored at −20 °C until incubation for protein microarray.

2.6. Protein Microarrays

Microarray preparation and performance evaluation were done following the procedures previously described by Sierra-Sánchez et al., 2017 [46]. The glass slide surface was activated by incubation with 2% (*v*/*v*) 3-(2-Aminoethylamino) propyldimethoxymethyl (MANAE) silane in acetone for 30 min and slight shaking at RT [47]. Then, activated glass slides were washed with acetone and Milli-Q water and dried with compressed filtered air. For antibody array printing, a non-contact inkjet printing technology (Arrayjet Inc., Edinburg, UK) was employed with a slide-out of 7 similar subarrays and 5 replicates per sample. The array content is described in Table S1 where 205 antibodies targeting 162 proteins, among positive and negative controls, were included (Table S2). All the samples were prepared at 1:1 (*v*/*v*) dilution with JetStar™ (Arrayjet Inc., Edingburgh, UK) printing buffer C.

2.7. Evaluation of Array Performance at Different MW Fractions

Protein microarrays were blocked with blocking solution PBS and 1% blocker BSA—10X, (Thermo Fisher Scientific, USA) for 1 h at RT with mild stirring. Then, blocked protein microarrays were thoroughly washed with distilled water. Regarding sample handling, 50% volume of each 24 collected fractions was processed with epitope retrieval treatment (30 min at 56 °C and 1 min at 20 °C) and at last, the remaining 50% samples were combined in one single solution. After that, 100 μL per subarray of each mix were incubated at 4 °C, with mild stirring. After overnight incubation, 100 μL of Cy3- Streptavidin (1:200 -*v*/*v*-) was added to each subarray, and it was incubated for 1 h in darkness in a humidified chamber at RT. Finally, the protein microarrays were washed and dried for further acquisition of array images. All steps were performed at RT unless otherwise specified.

2.8. Image Analysis and Data Acquisition

TIFF images obtained at different exposition times to achieve optimal images with SensoSpot® Fluorescence Microarray (Sensovation Gmbh, Radolfzell, Germany) were analyzed using GenePix Pro v. 6.0 software. Parameters were set to quantify light intensity values at Cy3 (λ = 532 nm) emission wavelength.

2.9. Protein Microarray Data Processing

Signal intensity values were processed by performing background subtraction, filtering, and housekeeping processes.

Normalization

To remove the background, the following was used:

$$\tau_{iMM_n} = (\widetilde{\delta} - k) - 2\sigma k$$

where:

τ_{iMM_n}: signal intensity value spot i containing MasterMix (MM_n).
$\widetilde{\delta}$: median intensity value per spot.
k: constant of subarray background intensity.
$2\sigma k$: variance of signal intensity background per subarray.

To remove the background effect of the mastermix (MM), as the same MM for each antibody was used, the following equation was used:

$$S_i = \tau_{iMM_n} - (\widetilde{\delta}_{MM_{nmax}} + \widetilde{\delta}_{MM_{nmax}} \cdot 0.05)$$

where:

S_i: signal intensity value of spot i without background.
$\widetilde{\delta}_{MM_{nmax}}$: maximum value of median intensity MM.

After background subtraction, the signal was normalized against a positive control (biotin). For further analysis, only the proteins detected with >50% of spotted antibodies displaying a normalized signal > 0 were included.

2.10. SEC-MAP Database

To combine the SEC-MAP data sets with transcriptomics and other proteomics characterization (LC-MS/MS), a database was designed and developed containing: i. Protein ratio at SEC-MAP for each MW fraction and protein extraction procedure; ii. Antibody info: type (monoclonal/polyclonal), supplier, and developed; iii. Protein ID, NextProt ID, Uniprot ID, MW (expected/observed/theoretical, etc.,), subcellular localization; iv. Detection by LC-MS/MS previously reported in the cell type of interest [19].

2.11. Integration of Transcriptomics, Proteomics, and SEC-MAP Datasets

A database called "complete protein mapping" was designed and developed from reported LC-MS/MS characterization of Ramos cell line in our previously reported studies [19]. Total of 5672 proteins were detected by unique tryptic peptides in three technical replicates and the subcellular localization (Cyt, Mem, Org, Nuc) was included (Table S3). The genes coding for each of the detected proteins were mapped to chromosomes with the R-package BiomaRt [48]. neXtProt IDs are used to merge the different datasets (RNA-Seq, LC-MS/MS, and SEC-MAP). In the case of proteins with more than one gene IDs, the gene ID with the highest FPKM value was selected, Table S4. In this study, a protein was considered to be fully observed if: i. Number of peptides $\geq$ 1; ii. FPKM $\geq$ 1; or iii. QAS value $\geq$ 1. Biological function analysis of observed proteins is based on DAVID and GeneTerm-Linker tools [49,50], which were used for functional enrichment analysis (FEA). Databases selected to find genes with annotated enriched terms were: (i) Gene Ontology (GO) using GO_BP, GO_CC and GO_MF; (ii) KEGG_PATHWAY; and (iii) the INTERPRO protein structural domain databases were used for generation of functional enrichment analysis (FEA). Moreover, a platform STING v. 11.0 [51] for the protein interactions obtained from SEC-MAP was used. For signaling pathways observed from the integration of proteomics and transcriptomics datasets, KEGG v. 93.0 [52] and Reactome v. 71 were used [53].

2.12. Visualization of Transcriptomics, Proteomics, and SEC-MAP Datasets

In this study, the software called Infinicyt™ 2.0 (Cytognos SL, Salamanca, Spain) has been employed for the visualization of multidimensional and multiparametric data sets. For that, a classification of each immunoassay was based on expected and theoretical subcellular localization and its correlation with the observed subcellular localization by the LC-MS/MS characterization. Hence, proteins were classified—according to subcellular localization—by positive selection in the following order: Nuc, Cyt, Mem, Org, Nuc-Org, Nuc-Cyt, Nuc-Mem, Org-Cyt, Org-Mem, Cyt-Mem, Nuc-Org-Cyt, Nuc-Org-Mem, Nuc-Cyt-Mem, Org-Cyt-Mem, and Nuc- Org-Cyt-Mem. Additionally, for each classified group, proteomics and transcriptomics integration were represented by APS, t-SNE plot, population burst, 2D (file number—y-axis- and each subcellular localization—x-axis-, and LFQ logarithmic—y-axis- and FPKM logarithmic—x-axis-). All these graphs (including 3D, tables, etc.,) are reported in Supplementary Materials Dataset I.

After that, a second classification was established according to proteomics values vs. transcriptomics values, represented by Log LFQ vs. Log FPKM, depicted in a 2D plot which allowed for classifying (according to proteomics vs. transcriptomics) five additional groups named: average population, average population, low Log LFQ vs. progressive Log FPKM, high Log LFQ vs. low Log FPKM, high Log LFQ vs. Log FPKM, and a group of outliers. Then, each group of proteins according to the subcellular localization was re-classified on these additional groups.

3. Results

In Figure 1, an overall representation of the experimental workflow performed for the systematic and multipronged multi-omics characterization and the multi-dimensional bioinformatics integration is shown.

3.1. Protein Extraction Strategies for Multi-Pronged Proteomics Characterization

Knowing and exploring the different cell lysis strategies is fundamental to know the compatibility of the different proteomics methodologies and the feasibility to perform an integration of multi-pronged proteomic strategies (LC-MS/MS and SEC-MAP, for example) (Figure 1). For that reason, this study compares nine different protein extraction methods (as described in materials and methods section -M&M-) on a cell line of interest (Ramos cell line, Burkitt's lymphoma, RA1—ATTC: CRL-1596-). Each method is different regarding the chemical composition (Table 1); therefore, to allow a better understanding and comparison between them, they are named from #1–#9. In addition, protocols #7–#9 allow protein separation at different subcellular localizations. Then, all the protein extraction strategies—and subsequent chemical biotin labeling—are evaluated for SEC-MAP performance, according to: i. Efficiency of protein extraction: direct correlation with protein abundance (Figure S1) and expected subcellular localization. ii. Compatibility with protein labeling: biotin is commonly used in protein microarrays; however, several chemical components of the lysate buffers can cause interferences in the biotin conjugation protocol and array performance must be evaluated under these conditions (Figure S2). iii. SEC: to correctly analyze multi-protein complexes, the array performance is evaluated at several MW fractions to confirm that protein abundance is not affected by protein size (MW) and subcellular localization.

Performance of SEC-MAP: Effect of Protein Extraction Procedures and Biotin Conjugation

The total amount of protein extracted with each of the strategies show wide variability. For the protein total extraction protocols, the amount of protein obtained ranged between 34 and 61 μg/protein total per 10^6 cells obtained, protocol #2 and #4, respectively. In the case of subcellular enrichment, the amount with membrane extraction protocol ranged from 169 to 270 μg/protein total per 10^6 cells with protocol #5 and #6; and subcellular fraction protocols (#9 & #7), the range is between 82 and 166 μg/protein total per 10^6 cells, respectively. These results confirm that the properties of chemical reagents—presented in the extraction buffers—are critical for the relative and/or absolute determination of protein abundance; but, it seems that protein distribution is not influenced as they remain comparable among all the protocols (Figure S1B).

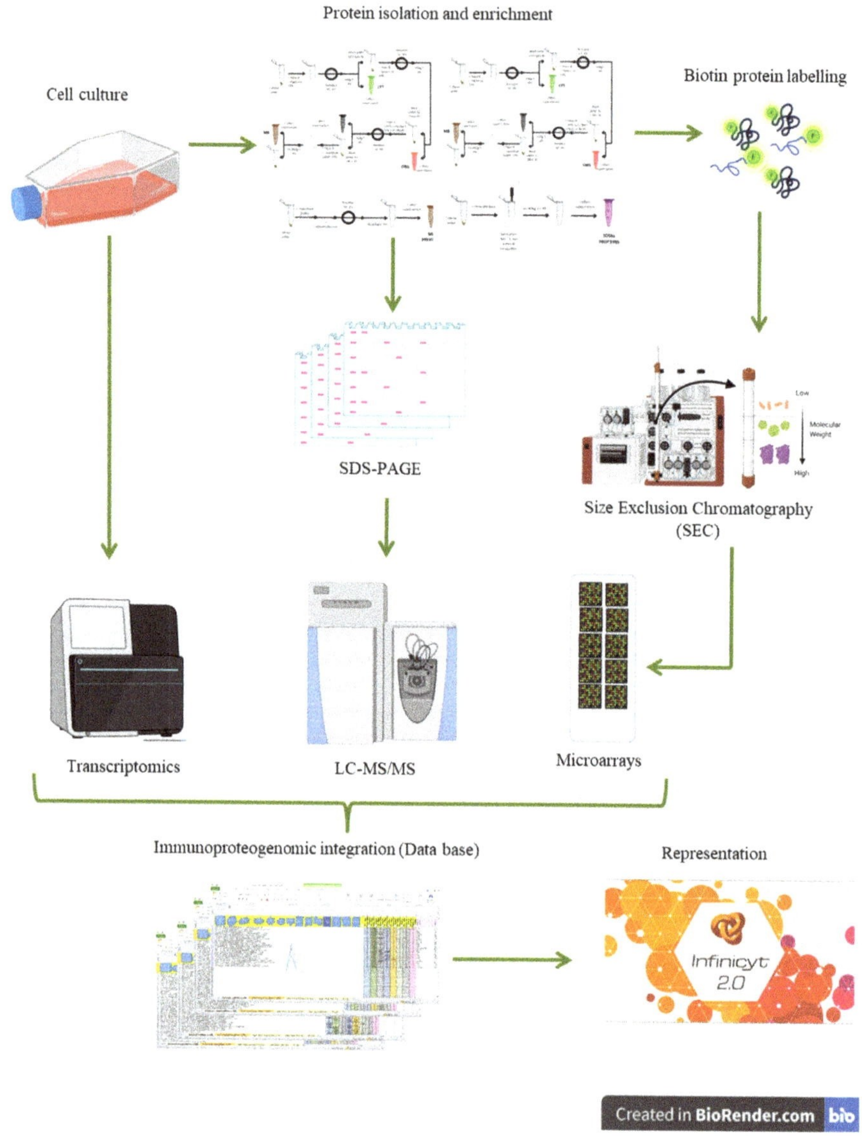

Figure 1. Schematic Representation of Experimental Design. Description of experimental steps involved for multi-omics integration.

About biotin conjugation, it does not show differences in all the analyzed MW by SEC (fraction 1—142–437 kDa –, fraction 2—74–121 kDa –, fraction 3—46–63 kDa –, fraction 4—26–39 kDa –, fraction 5—11–24 kDa). However, for antibody array performance, biotin-conjugated proteins present differences after SEC analysis (Figure S3). These are directly related to the effect on epitope recognition caused by chemical modification during biotin labeling. For SEC-MAP performance, it is also relevant that optimal conditions for SEC which were studied in all protein extractions in eight pre-defined MW ranges in the same antibody microarray. First, the number of identified proteins different is considered a parameter for comparison across the studied protein extraction strategies (Table S5). In this case, protocol #1 presents the highest number of identified total proteins, and protocol #6 the lowest. It is due to different detergent compositions in each protein extraction buffer (Figure S4A).

Moreover, a comparison of the observed subcellular localization of the proteins is made for each extraction procedure. A higher number of identified proteins are reached by protocol #1 and the lowest one by protocol #4. When the comparison is focused only on membrane proteins, protocol #5 displayed a higher number of identified proteins than protocol #6 (Table S6, Figure S4A). Likewise, protocols #8 and #9 for subcellular protein extractions (Cyt, Mem, Nuc, Org) gave similar numbers of identified proteins in all studied subcellular localizations (Table S6, Figure S4). As proof of concept, the SEC-MAP analysis focuses on BID, CASP 7, CASP 8, LYN, SYK, BTK, CDKN1B, DNAJB1, HSPD1, HSP90AA1, HIST1H4A, BAX, BAK1, PAK1 | 1, and CCNB1. Moreover, in the known subcellular localization are detected all of them (Table S7). Additionally, SEC-MAP has also detected proteins with more than one localization; for example, in 4 subcellular localizations (Cyt, Mem, Org, Nuc), BCL-2 is detected by protocol #8 and 3 subcellular localizations (Cyt, Org, Nuc) by protocol #9 (Table S7).

Furthermore, analysis of subcellular protein localization by SEC-MAP has evaluated the effect of different protein extraction procedures. One example is RELA, which is detected in Cyt by protocol #7 and in the 4 subcellular fractions by protocol #8 (corroborating the reported subcellular localization and observed in the LC-MS/MS datasets). Another illustrative example is MAPK1, which is detected in the Cyt by protocols #7 and 8; however, protocol #9 detects it in Nuc localization. The observed subcellular localizations also correlate with LC-MS/MS datasets. (Table S7).

These results show: i. Protein extractions strategies are critical in the identified protein number by SEC-MAP. ii. Extraction strategy may be critical in the orthogonal integration with other omics datasets.

3.2. *Deciphering Differential Protein Profiles by SEC-MAP*

3.2.1. Analysis of Intracellular Signaling Pathways by SEC-MAP

This study integrates multiplex antibody microarray detections (MAP) with subcellular protein localization by SEC analysis. At first glance, a normal distribution of identified proteins (in all the studied protein extracts) is observed in the analyzed MW range (437-11 kDa) (Table S5 and Figure S5), where proteins are also detected in MW fractions that correspond with large protein sizes (Figure 2).

SEC-MAP approach allows the analysis of intracellular signaling pathways by simultaneous detection of multiple proteins. For this reason, several well-characterized signaling pathways are explored in the Ramos cell line: i. apoptosis regulation, ii. apoptosis inhibition, iii. BCR signaling pathway, iv. cell cycle control, v. STING pathway, vi. Damage-associated molecular patterns (DAMPs), vii. MAPK signaling, viii. senescence signaling [54–57]. In fact, SEC-MAP approach simultaneously detects heat shock proteins (HSP), Histones (3, 4), BCL-2, BAX, BAK1, BID, CASP7-8-9, IFN-γ, IL10, IL6, TNF-β, NF-Kβ1, TP53, RELA, MAP2K1, MAPK1, FOS, among others. Figure 3A shows the summary of SEC-MAP analysis of these proteins at the expected molecular weight (MW and/or expected SEC—fraction), as well as the detection of several proteins in other MWs than the reported one (Table S7). Moreover, the SEC-MAP approach allows us to analyze the signaling pathways based on protein localization ($\geq$1 subcellular localization in one step by extraction protocols #7, #8, #9). Figure 3B shows a similar overall distribution for all SEC-MAP assays performed (Table S8).

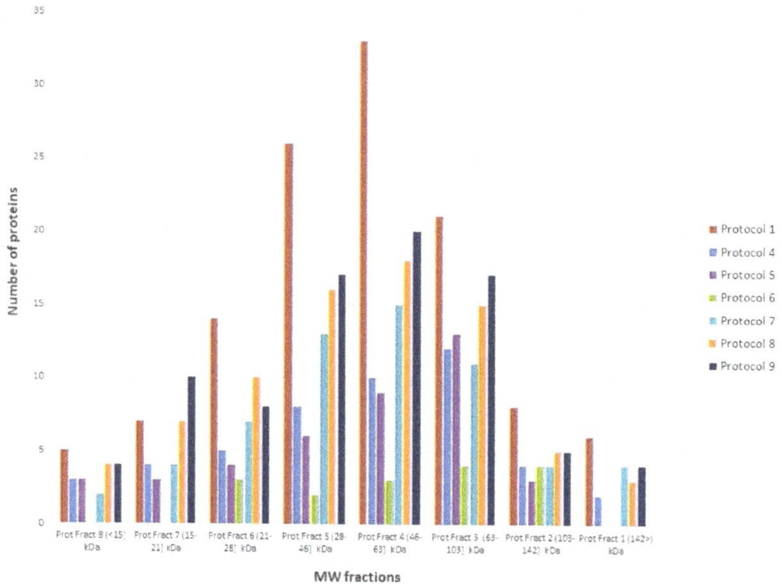

Figure 2. Effect of protein extraction protocols on protein identification by SEC-MAP approach.

3.2.2. Multi-Protein Complex Analysis by SEC-MAP

Nowadays, the inhibition of multi-protein complexes is one of the chemotherapeutical strategies in Burkitt's lymphoma, such as Venetoclax, Nutlin-3, Ibrutinib work by inhibiting the formation of multi-protein complexes (Figure 4) [57–59]. In this study, several multi-protein complexes are analyzed by SEC-MAP. These include BCL-2, BCL2L11, BID, BAD, BAK1, CASP3, and CASP8, which have been detected as multi-protein complexes (Figure 4A and Supplementary Materials Dataset II).

Moreover, the feasibility of SEC-MAP for the analysis of multi-protein complexes in only one intracellular signaling pathway. In Figure 4B, the BCR signaling pathway is analyzed by SEC-MAP, which reports protein complexes for PLCG2, SYK, CD19, ZAP70, LYN, BLNK, and BTK (Supplementary Materials Dataset II). Additionally, SEC-MAP analysis seems feasible to study protein complexes with discrimination at subcellular localization. A few illustrative examples are depicted in Figure 5A for apoptosis, Figure 5B for STING signaling and in Figure 5C for Ibrutinib targets (BTK, LYN, BLNK, ZAP70) at multiple subcellular localizations. Among these, SEC-MAP analysis also might be feasible to determine interacting partners, such as MDM2-TP53 (Figure 4C) and/or decipher novel protein complexes by the SEC-MAP analysis (Figure S6).

3.2.3. Analysis of Specific Protein Isoforms and/or Variants by SEC-MAP

In the content of protein microarray, there are spotted antibodies targeting specific motifs/amino acid residues with post-translational modifications (PTMs) of the same proteins (see M&M sections and Supplementary Materials Dataset II). Moreover, several antibodies target the same protein (i.e., monoclonal -mAb-, polyclonal -pAb-). Hence, whether SEC-MAP analysis could use to evaluate the specificity and/or selectivity of antibodies against a particular protein, as well as to evaluate protein states (such as monomer, complexed, degraded) or to provide info about isoforms/variants of a particular protein (Table S9). Here, FOS protein could be a representative example, which is detected by one antibody in two subcellular localizations (Protocol #8: Org, Cyt), and in three subcellular localizations (Protocol #9: Mem, Org, Nuc) by a different antibody (Table S9). As proof of concept, a few protein isoforms are highlighted (in all cases, all antibodies used are in the M&M section): i. 4E-BP1 (clone 53H11) Rabbit mAb (as described in the M&M) and phospho-4E-BP1 (Thr37/46) (clon 236B4) Rabbit mAb. The rabbit mAb (clone 4E-BP1) allowed the protein identification in protocols #1, #4, #5, #7 (Cyt), #8 (Mem, Org and Cyt) and #9 (Mem, Org, Cyt and Nuc), while phospho-4E-BP1 (Thr37/46) is detecting the isoform only by protocol #1. ii.-STAT3 phospho-protein isoforms are detected SEC-MAP analysis by both antibodies (pAb-clone C-20-to STAT3 and p-STAT3 mAb-clone B-7-) with all the protein extraction protocols (#1 to #9). iii.-p38 isoforms: Several antibodies against phospho-isoforms of p38 (Phospho-p38 MAPK—Thr180/Tyr182-, (Rabbit mAb clone D3F9) the native form (Rabbit mAb clone 9212) detect the phosphorylated isoforms by SEC-MAP, allowing the determination of the relative abundance.

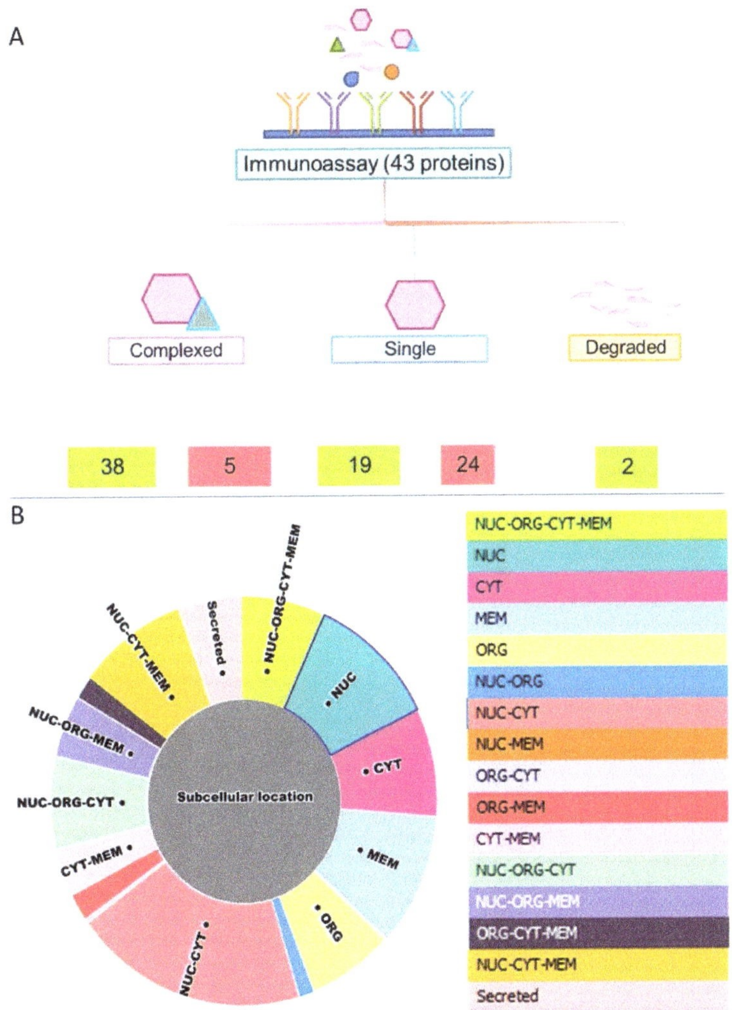

Figure 3. Summary of interest proteins selected for integration of SEC-MAP in the analysis of intracellular signaling pathways. (**A**): Schematic representation of proteins globally observed at different molecular isoforms (monomer, complexed, degraded). In green, the number of proteins detected, and in red, the number of non-detected proteins. (**B**): Summary pie chart of theoretical subcellular localization of proteins globally detected by all protein extraction protocols.

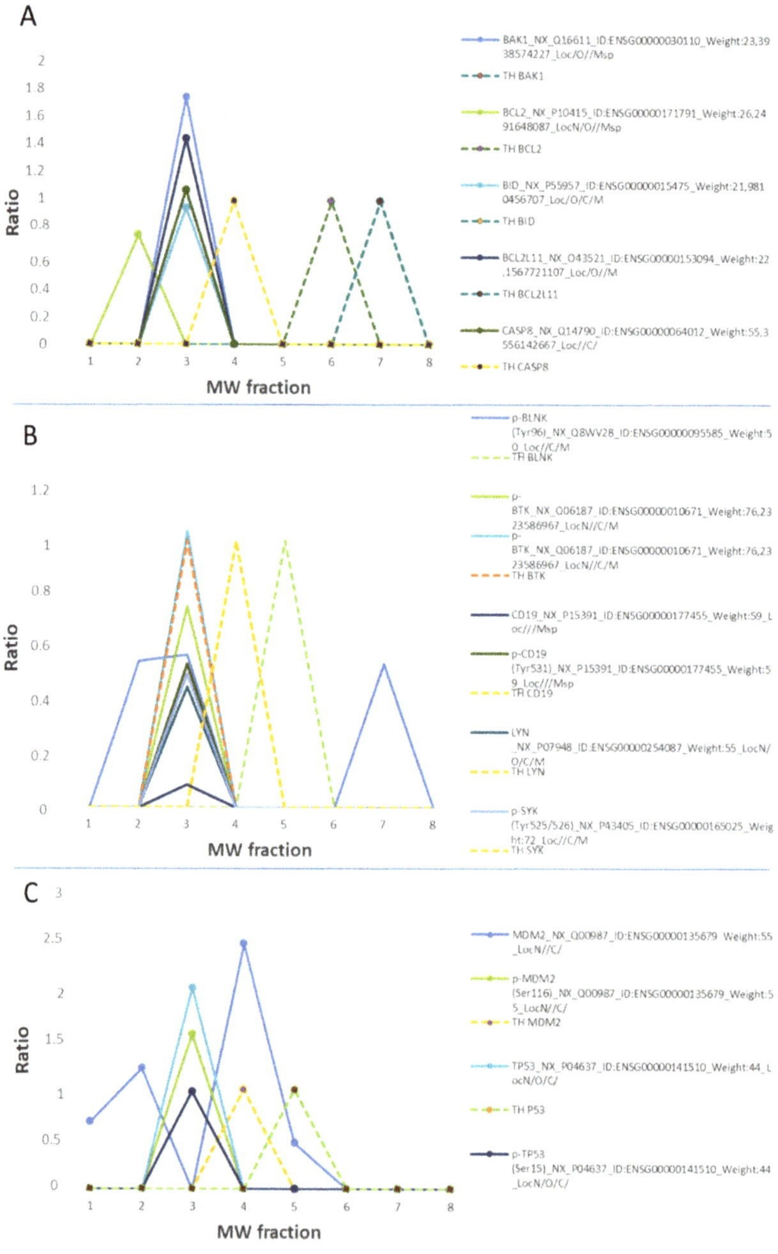

Figure 4. Multicomplex protein analysis by the SEC-MAP approach. Multiprotein complexes were detected at different protein extraction procedures. Peaks located at the same molecular weight (MW) fraction, and not found in their theoretical (TH) MW fraction, represent complexes formed by proteins when they are at a fraction above the expected one. When it is found in fractions below those expected, we speak of protein being hydrolyzed. If it is detected in theoretical fraction (dotted line), protein is found alone. (**A**). Proteins related to the BCL-2-BCL2L11 interaction signaling pathway. (**B**). Proteins related to the BTK pathway. (**C**). MDM2-TP53 interaction.

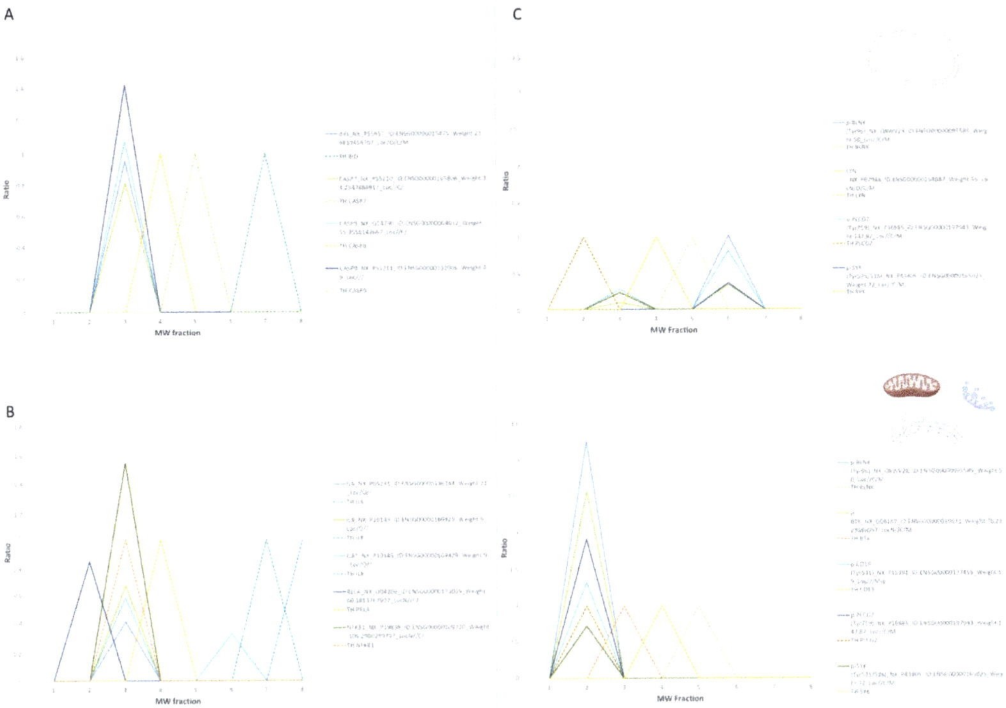

Figure 5. Multiprotein complex by SEC-MAP approach. Analysis of chosen multicomplexes involved in several cell signaling processes. (**A**): Proteins related to apoptosis inhibition. (**B**): Protein related to STING signaling. (**C**): Proteins related to B-cell receptor signaling in membrane and organelle subcellular localizations.

3.2.4. Orthogonal Integration of SEC-MAP with Multi-Omics Datasets (RNA-Seq & LC-MS/MS)

Regarding the results of SEC-MAP integration with RNA-seq and LC-MS/MS datasets for Ramos cell line, a pipeline for orthogonal integration of datasets has been designed (Figure 6). First, a systematic database containing the global information from SEC-MAP analysis (global proteome, multi-protein complexes, specific protein isoforms/state, etc.) (Table S5) was required. Subsequently, correlations between the protein array content, RNA-seq and LC-MS/MS data were explored. Accordingly, 162 proteins are studied by SEC-MAP, RNA-seq, and LC-MS/MS, showing a strong inter-relationship between the proteins detected by these three omics strategies. Taking into account the orthogonal integration, first as the starting point, two-by-two correlations could be reported: i. 55.90% matching proteins between SEC-MAP and LC-MS/MS characterization; ii. 50.93% matching proteins between SEC-MAP and RNA-Seq information. iii. 99.59% matching proteins characterization between RNA-Seq and LC-MS/MS (see Table S10). Furthermore, according to neXtProt release 2020.01-17, 518 identified proteins belong to PE1, 6 to PE2, and 6 to PE5 of MS/MS information only and, 5121 belongs to PE1, 3 to PE2, and 3 to PE5 corresponding to the integration of MS/MS and RNA-Seq data (Table S11).

In this study, aiming to provide an approach suitable for globally graphical visualization of the orthogonal integration of multi-omics datasets (SEC-MAP, RNA-seq, LC-MS/MS), a bioinformatics tool—named Infinicyt (https://www.cytognos.com/infinicyt/2.0, accessed on 30 March 2019; Cytognos SL, Salamanca, Spain)—has been successfully implemented. This software tool also allowed to study a quantitative correlation between these datasets (i.e., FPKM and LFQ, for RNA-seq and LC-MS/MS respective) (Figure 7).

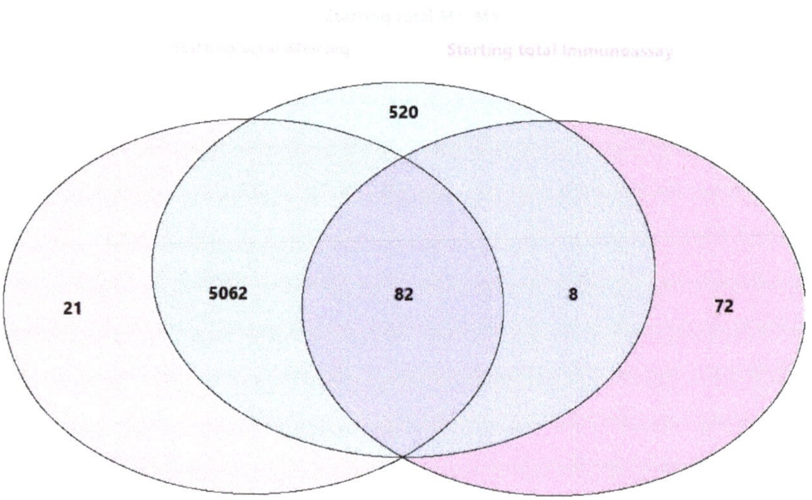

Figure 6. Venn diagram for orthogonal integration among all the datasets from SEC-MAP, LC-MS/MS, RNA-Seq (Number of proteins detected in each dataset).

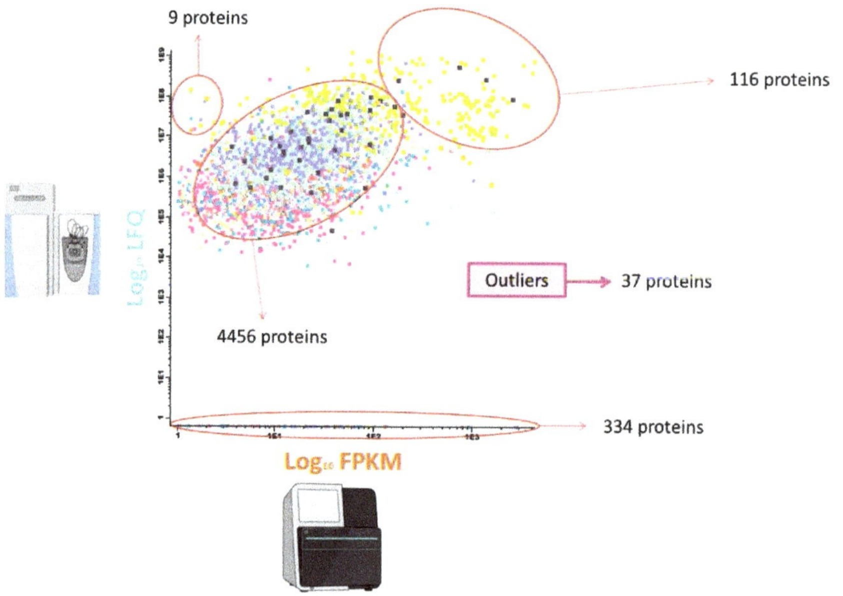

Figure 7. Dot-plot by Infinicyt 2.0 software displaying quantitative datasets for transcriptomics and LC-MS/MS characterization. Proteins/peptides are grouped according to different correlations between transcriptomics and proteomics datasets.

According to STRING and Reactome databases, it is observed that pathways related to the immune system and metabolism are significantly represented in 5 protein clusters. Metabolism of proteins and RNA pathways, cellular responses to external stimuli, transcription and DNA repair and cell cycle are represented in 4 proteins groups according to different high/low value log [LFQ] vs. high/low values of log [FPKM]. Finally, development biology and chromatin organization have been represented in three proteins groups according to log [LFQ] and log [FPKM] values (Table S12).

Bearing in mind similar patterns in the correlation between log[LFQ] and log[FPKM], the t-SNE plots provide insights from the functional point of view (within group of proteins): i. vesicle-mediated transport display high values on both parameters (Figure S7A), ii. organelle biogenesis and maintenance pathway displays high values in both parameters (Figure S7B); iii. hemostasis, DNA replication and programmed cell death for population averages appear as outliers on the correlation (Figure S7C,D). Deepening in this analysis by adding subcellular localization, several groups are observed on t-SNE plots according to the subcellular localization (Figure S7E–I, Table S12). For example, at Cyt localization, proteins related to cell–cell communication, programmed cell death, and hemostasis displayed low log[LFQ] and progressive log[FPKM] values. About proteins with ubiquitous localization in several subcellular localizations (Cyt + Mem + Nuc + Org) is observed as a homogeneous group at the t-SNE plot (with both high values, log[LFQ] & log [FPKM]) (Table S12).

4. Discussion

The synergistic integration of multi-omics datasets is highly dependent on multiple methodological aspects, which are also critical in the design and development of an algorithm for deepening the biological knowledge of intracellular signaling pathways. In this study, a simple approach has been designed and developed for the orthogonal integration of SEC-MAP, LC-MS/MS, and RNA-seq datasets corresponding to a Ramos cell line as a model. This orthogonal integration allowed—in a particular cellular situation—to decipher protein expression, isoforms, quantification at peptide and transcript levels, protein localization, protein interactions which also could be done at multiple cellular differentiation stages and/or physiological situations; thus, it could help uncover novel insights about the cellular dynamics and response to external stimulus.

Initially, sample preparation, as protein extraction procedures, has been revealed which is critical for orthogonal integration of multi-omics because of the expected influence in relative protein abundance, protein chemistry, protein stability, protein structure, tryptic digestion, protein solubility, and PTMs. It has been shown that protein extraction procedures have to be optimized for multi-omics integration because it could be an advantage/disadvantage in multi-pronged proteomics characterization (such as LC-MS/MS and SEC-MAP). Additionally, the specificity and selectivity of immunoassays are also affected by chemical labeling for detecting antigen-antibody interaction, which may hinder the epitopes and/or alter protein structure. Therefore, it is another critical factor for immunoassay validation of findings from RNA-seq and LC-MS/MS.

Overall, the importance of protein extraction procedures for further integration of proteomics with other omics datasets has been revealed. For example, protocols #2 and #3 have urea in the lysis buffer, which decreases the efficiency of biotin labeling and subsequent SEC-MAP analysis [60]. Another consideration is the chemical formulation of lysis buffers, such as the effect of non-ionic detergents (protocols #1, #5 and #6). Here, the role of critical micelle concentration (CMC) for each detergent is important in the protein extraction efficiency obtained for protocols #5 and #6. With this, it is shown that the detergent used in protocol #1 achieves a higher yield on protein extraction in comparison with other procedures. In this regard, among all protocols studied, protocol #1 is the one that gives us the best results; while protocol #6 produces the poorest results. Regarding subcellular localization, protocol #5 seems to be optimized for membrane protein extraction; meanwhile, for subcellular localizations, protocols #8 & #9 reported similar performance and effectiveness, with slight differences in membrane subcellular localization and organelle subcellular localization; which confirms the previously reported results by our group [19].

Regarding Human Proteome Project, it is a huge effort to detect "missing proteins"; which are classified into five groups (1–5) according to protein evidence (PE) [61]; then, protein enrichment is aligning a useful strategy to identify and detect "missing proteins" as it is shown in this study because a few proteins (reported as "missing proteins") have been detected (mainly on group PE2 and PE5).

Recently, several studies employing SEC (in *Escherichia coli*, HEK293, osteosarcoma cells, liver human tissue and others) focused mainly on analysis of protein interactions (protein-protein, protein-metabolite, protein-small molecules), the difference between transcriptomics-proteomics levels or in combination with spatial proteomics [17,18,21,26]. In this study, SEC is combined with multiplex protein arrays (MAP) such that, not only it can identify single proteins, but also detect isoforms and protein complexes. Such use of SEC-MAP has also been reported by Kirtwodd et al., in 2013 [62]. In this study, it is observed that proteins are identified in a wide MW range by SEC-MAP; where protocol #1 seems to yield the best performance in most of the MW range studied; however, few other protocols give a better identification only for a few of the MW fractions, which might be related with the subcellular localization, presence of protein complexes, and inherent chemical properties of the proteins.

Regarding the SEC-MAP integration, many proteins (i.e., PAK1 | P1, CCNB1, CASP, among others) are successfully detected at the expected MW and the subcellular localization as their confirmed existence by transcriptomics and MS/MS datasets. There is another group of proteins that are detected at the reported subcellular localization but at different MW, such as, BLC-2, RELA, MAPK1, MDM2, TP53, BLNK, SYK, CD19, LYN, ZAP70, PLCG2, suggesting the existence of protein–protein interactions and protein complexes. Thus, this approach is amenable for studying protein interaction networks and/or intracellular signaling dynamic as confirmed previously by our group in this study by correctly identifying the protein players in the previously reported pathways in this B lymphocyte cell line (such as senescence, evading growth suppressors, survival and death pathways, immune system evasion and immunoediting) [26,63–65]. This also opens the possibility to decipher interactions on newly reported pathways such as DAMPs (damage-associated molecular patterns), which might be useful for immunotherapies on lymphoma and leukemia because of their direct relation with the immunogenic cell death (ICD). Finally, similar to previously described by Díez P. et al. [19], it has been feasible to discriminate between protein groups in particular cell signaling pathways (such as immune system, metabolism, vesicle-mediated transport, organelle biogenesis and maintenance, homeostasis, etc.), monitoring subcellular localization and multi-omics correlation.

5. Conclusions

Multi-pronged proteomics characterization is highly dependent on protein extraction procedures, being a key step for functional outcomes from detection of proteins/multi-protein complexes in samples. Furthermore, it also reveals critical orthogonal integration of multi-omics data sets (such as SEC-MAP, LC-MS/MS, RNA-seq) and also provides complimentary info such as multi-protein complexes and subcellular localization. Moreover, the reported pipeline for multi-omics integration is useful for the HPP, as it systematically explores the compatibility of multi-omics data sets. Currently, the RNA-Seq technique provides us with information on the presence-absence of a particular protein of interest; meanwhile, MS/MS and SEC-MAP add data on protein localization, isoforms, and also, protein complexation status (complex-monomer-degraded). Consequently, SEC-MAP analysis seems to be a useful tool for orthogonal multi-omics integration, as it provides detailed complementary information about MW, subcellular localization, isoforms, PTMs, and protein complexes. Furthermore, SEC-MAP analysis seems feasible for determining predicted or unknown protein complexes and/or protein interactions in any sample (either from cell culture or clinical specimens).

With multi-omics orthogonal integration, it seems feasible to identify if a drug could target a protein complex of interest. It might help in the selection of drugs for a particular pathology.

Supplementary Materials: The following are available online at https://www.mdpi.com/2218-273X/11/12/1776/s1. Figure S1: Comparison of efficiencies according to the extraction protein processing used. Panel A: Amount protein obtained. Panel B: SDS-PAGE 12% Coomassie blue and silver staining. Figure S2: Comparison of efficiencies in differing concentrations of biotin. Panel A: Number of spots detected for the fraction. Panel B: Number of spots not detected for fraction. Figure S3: Graphic of molecular weight markers used after performing fast protein liquid chromatography (FPLC) technique. Figure S4: Descriptive analysis of behavior for each extraction protocol. Panel A: Global number of spots detected. Panel B: Number of spots detected for each subcellular fraction by four fraction protocol. Figure S5: Test of normality for protein extraction protocols. Panel A: Protocol 1. Panel B: Protocol 4. Panel C: Protocol 5. Panel D: Protocol 6. Panel E: Protocol 7. Panel F: Protocol 8. Panel G: Protocol 9. Figure S6: Multicomplex analysis for peaks by protocol 1. Panel A: Interaction detected in Fraction 2. Panel B: Interaction detected in Fraction 3. Panel C: Interaction detected in Fraction 2 and 4. Figure S7: Functional interaction (Reactome) and t-SNE plots (TS) of Infinicyt for groups obtained after analyses by values quantitative LFQ and FPKM. Panel A: Low LFQ and Progressive FPKM in Cyt for vesicle-mediated transport. Panel B: High LFQ and FPKM in Org for Organelle biogenesis and maintenance. Panel C: Outlier with High LFQ and Medium FPKM value (Hemostasis and Extracellular matrix organization). Panel D: Outlier with High LFQ and High FPKM value (Hemostasis). Panel E: Low LFQ and Progressive FPKM in Cyt. Panel F: Low LFQ and Progressive FPKM in Mem. Panel G: Low LFQ and Progressive FPKM in Org. Panel H: Low LFQ and Progressive FPKM in Nuc & Org. Panel I: Low LFQ and Progressive FPKM in Nuc & Cyt. Panel J: Low LFQ and Progressive FPKM in Org & Cyt. Table S1. Designer of subarray printing (Antibodies and controls—green: positive controls; and yellow: negative controls-). Table S2: List of protein target included in protein microarray. Table S3: Mapping of proteins with unique peptide. Table S4: Mapping of proteins with unique peptide and the highest FPKM value. Table S5: Descriptive statistic and summary information about protein extraction protocols by immunoassay and integration with RNA-Seq and MS/MS. Table S6: Z values results of comparisons between protein extraction protocols, both globally and by molecular weight fraction. Table S7: Deep analysis about state of protein, theoretical and assay subcellular localization (MS/MS and microarray) and RNA-Seq information of selection proteins according to relevance and interesting reported in literature. Table S8: Number of proteins detected in each subcellular localization and possible combinations. Table S9: Proteins detected with different antibodies and post-translational modification (PTM) detection. Table S10: Percentage two to two relationship for MS/MS, RNA-Seq and immunoassay techniques, Table S11: Orthogonal integration and missing proteins detected. Table S12: Functional interaction (Reactome) and biological enrichment processes (KEGG) present in groups obtained after analyses and visualization in Infinicyt by values quantitative LFQ and FPKM. Supplementary Materials Dataset I: Visualization of integration information from transcriptomic, proteomic and immunoassay. Supplementary Materials Dataset II: Dataset Immunoassay, proteomic and transcriptomic data integration.

Author Contributions: Conceptualization, P.D. and M.F.; methodology, A.L.-V., J.D.L.R. and P.D.; software, Q.L. and A.O.; validation, J.D.L.R., P.J.-V. and Q.L.; formal analysis, A.L.-V., J.D.L.R., P.D. and M.F.; investigation, A.L.-V., P.D. and M.F.; data curation, A.L.-V., P.J.-V., J.D.L.R. and P.D.; writing—original draft preparation, A.L.-V.; writing—review and editing, A.L.-V. and M.F.; visualization, A.L.-V. and Q.L.; supervision, M.F.; project administration, A.O., J.D.L.R. and M.F.; funding acquisition, M.F. All authors have read and agreed to the published version of the manuscript.

Funding: We are financial support from the Spanish Health Institute Carlos III (ISCIII) for the grants: FIS PI14/01538, FIS PI17/01930 and CB16/12/00400.

Institutional Review Board Statement: Not applicable.

Informed Consent Statement: Not applicable.

Data Availability Statement: MS/MS data are available via ProteomeXchange with identifier PXD003939 at EBI-EMBL.

Acknowledgments: We gratefully acknowledge Fondos FEDER (EU) and Junta Castilla-León (grant SA198A12-2 and COVID-19 grant COV20EDU/00187). Fundación Solórzano FS/38-2017. The Proteomics Unit belongs to ProteoRed, PRB3-ISCIII, supported by grant PT17/0019/0023, of the PE I + D + I 2017-2020, funded by ISCIII and FEDER. A. Landeira-Viñuela is supported by VIII

Centenario-USAL PhD Program, P. Juanes-Velasco is supported by the JCYL PhD Program "JCYL Nos Impulsa" and scholarship JCYL-EDU/601/2020. P.D. is supported by JCYL-EDU/346/2013 Ph.D. scholarship.

Conflicts of Interest: The authors declare no conflict of interest.

References

1. Lundberg, E.; Borner, G.H.H. Spatial proteomics: A powerful discovery tool for cell biology. *Nat. Rev. Mol. Cell Biol.* **2019**, *20*, 285–302. [CrossRef]
2. Wang, Z.; Deisboeck, T.S. Dynamic Targeting in Cancer Treatment. *Front. Physiol.* **2019**, *10*, 96. [CrossRef] [PubMed]
3. Koplev, S.; Longden, J.; Ferkinghoff-Borg, J.; Blicher Bjerregård, M.; Cox, T.R.; Erler, J.T.; Pedersen, J.T.; Voellmy, F.; Sommer, M.O.A.; Linding, R. Dynamic Rearrangement of Cell States Detected by Systematic Screening of Sequential Anticancer Treatments. *Cell Rep.* **2017**, *20*, 2784–2791. [CrossRef] [PubMed]
4. Comandante-Lou, N.; Fallahi-Sichani, M. Models of Cancer Drug Discovery and Response to Therapy. In *Reference Module in Biomedical Sciences*; Elsevier: Amsterdam, The Netherlands, 2019; pp. 269–276.
5. Nussinov, R.; Jang, H.; Tsai, C.-J.; Cheng, F. Precision medicine review: Rare driver mutations and their biophysical classification. *Biophys. Rev.* **2019**, *11*, 5–19. [CrossRef] [PubMed]
6. Aslam, B.; Basit, M.; Nisar, M.A.; Khurshid, M.; Rasool, M.H. Proteomics: Technologies and Their Applications. *J. Chromatogr. Sci.* **2017**, *55*, 182–196. [CrossRef] [PubMed]
7. Park, S.-M.; Hwang, C.Y.; Choi, J.; Joung, C.Y.; Cho, K.-H. Feedback analysis identifies a combination target for overcoming adaptive resistance to targeted cancer therapy. *Oncogene* **2020**, *39*, 3803–3820. [CrossRef] [PubMed]
8. Bhat, A.R.; Gupta, M.K.; Krithivasan, P.; Dhas, K.; Nair, J.; Reddy, R.B.; Sudheendra, H.V.; Chavan, S.; Vardhan, H.; Darsi, S.; et al. Sample preparation method considerations for integrated transcriptomic and proteomic analysis of tumors. *Proteom. Clin. Appl.* **2017**, *11*, 1600100. [CrossRef] [PubMed]
9. Buescher, J.M.; Driggers, E.M. Integration of omics: More than the sum of its parts. *Cancer Metab.* **2016**, *4*, 4. [CrossRef] [PubMed]
10. Dimitrakopoulos, L.; Prassas, I.; Diamandis, E.P.; Charames, G.S. Onco-proteogenomics: Multi-omics level data integration for accurate phenotype prediction. *Crit. Rev. Clin. Lab. Sci.* **2017**, *54*, 414–432. [CrossRef] [PubMed]
11. Eicher, T.; Patt, A.; Kautto, E.; Machiraju, R.; Mathé, E.; Zhang, Y. Challenges in proteogenomics: A comparison of analysis methods with the case study of the DREAM proteogenomics sub-challenge. *BMC Bioinform.* **2019**, *20*, 669. [CrossRef] [PubMed]
12. Ruggles, K.V.; Krug, K.; Wang, X.; Clauser, K.R.; Wang, J.; Payne, S.H.; Fenyö, D.; Zhang, B.; Mani, D.R. Methods, Tools and Current Perspectives in Proteogenomics. *Mol. Cell Proteom. MCP* **2017**, *16*, 959–981. [CrossRef]
13. Subbannayya, Y.; Pinto, S.M.; Gowda, H.; Prasad, T.S.K. Proteogenomics for understanding oncology: Recent advances and future prospects. *Expert Rev. Proteom.* **2016**, *13*, 297–308. [CrossRef]
14. Huh, W.-K.; Falvo, J.V.; Gerke, L.C.; Carroll, A.S.; Howson, R.W.; Weissman, J.S.; O'Shea, E.K. Global analysis of protein localization in budding yeast. *Nature* **2003**, *425*, 686–691. [CrossRef]
15. Sikorski, K.; Mehta, A.; Inngjerdingen, M.; Thakor, F.; Kling, S.; Kalina, T.; Nyman, T.A.; Stensland, M.E.; Zhou, W.; de Souza, G.A.; et al. A high-throughput pipeline for validation of antibodies. *Nat. Methods* **2018**, *15*, 909–912. [CrossRef]
16. Thul, P.J.; Åkesson, L.; Wiking, M.; Mahdessian, D.; Geladaki, A.; Ait Blal, H.; Alm, T.; Asplund, A.; Björk, L.; Breckels, L.M.; et al. A subcellular map of the human proteome. *Science* **2017**, *356*, eaal3321. [CrossRef] [PubMed]
17. Piazza, I.; Kochanowski, K.; Cappelletti, V.; Fuhrer, T.; Noor, E.; Sauer, U.; Picotti, P. A Map of Protein-Metabolite Interactions Reveals Principles of Chemical Communication. *Cell* **2018**, *172*, 358–372.e23. [CrossRef] [PubMed]
18. Heusel, M.; Bludau, I.; Rosenberger, G.; Hafen, R.; Frank, M.; Banaei-Esfahani, A.; van Drogen, A.; Collins, B.C.; Gstaiger, M.; Aebersold, R. Complex-centric proteome profiling by SEC-SWATH-MS. *Mol. Syst. Biol.* **2019**, *15*, e8438. [CrossRef] [PubMed]
19. Díez, P.; Droste, C.; Dégano, R.M.; González-Muñoz, M.; Ibarrola, N.; Pérez-Andrés, M.; Garin-Muga, A.; Segura, V.; Marko-Varga, G.; LaBaer, J.; et al. Integration of Proteomics and Transcriptomics Data Sets for the Analysis of a Lymphoma B-Cell Line in the Context of the Chromosome-Centric Human Proteome Project. *J. Proteome Res.* **2015**, *14*, 3530–3540. [CrossRef] [PubMed]
20. Wu, W.; Slåstad, H.; de la Rosa Carrillo, D.; Frey, T.; Tjønnfjord, G.; Boretti, E.; Aasheim, H.-C.; Horejsi, V.; Lund-Johansen, F. Antibody array analysis with label-based detection and resolution of protein size. *Mol. Cell. Proteom. MCP* **2009**, *8*, 245–257. [CrossRef]
21. Florinskaya, A.V.; Ershov, P.V.; Mezentsev, Y.V.; Kaluzhskiy, L.A.; Yablokov, E.O.; Buneeva, O.A.; Zgoda, V.G.; Medvedev, A.E.; Ivanov, A.S. The Analysis of Participation of Individual Proteins in the Protein Interactome Formation. *Biochem. Mosc. Suppl. Ser. B Biomed. Chem.* **2018**, *12*, 241–246. [CrossRef]
22. Rosenberger, G.; Heusel, M.; Bludau, I.; Collins, B.C.; Martelli, C.; Williams, E.G.; Xue, P.; Liu, Y.; Aebersold, R.; Califano, A. SECAT: Quantifying Protein Complex Dynamics across Cell States by Network-Centric Analysis of SEC-SWATH-MS Profiles. *Cell Syst.* **2020**, *11*, 589–607.e8. [CrossRef]
23. Laemmli, U.K. Cleavage of structural proteins during the assembly of the head of bacteriophage T4. *Nature* **1970**, *227*, 680–685. [CrossRef]
24. Bradford, M.M. A rapid and sensitive method for the quantitation of microgram quantities of protein utilizing the principle of protein-dye binding. *Anal. Biochem.* **1976**, *72*, 248–254. [CrossRef]

25. Smith, P.K.; Krohn, R.I.; Hermanson, G.T.; Mallia, A.K.; Gartner, F.H.; Provenzano, M.D.; Fujimoto, E.K.; Goeke, N.M.; Olson, B.J.; Klenk, D.C. Measurement of protein using bicinchoninic acid. *Anal. Biochem.* **1985**, *150*, 76–85. [CrossRef]
26. Díez, P.; Ibarrola, N.; Dégano, R.M.; Lécrevisse, Q.; Rodriguez-Caballero, A.; Criado, I.; Nieto, W.G.; Góngora, R.; González, M.; Almeida, J.; et al. A systematic approach for peptide characterization of B-cell receptor in chronic lymphocytic leukemia cells. *Oncotarget* **2017**, *8*, 42836–42846. [CrossRef]
27. Olsen, J.V.; de Godoy, L.M.F.; Li, G.; Macek, B.; Mortensen, P.; Pesch, R.; Makarov, A.; Lange, O.; Horning, S.; Mann, M. Parts per million mass accuracy on an Orbitrap mass spectrometer via lock mass injection into a C-trap. *Mol. Cell. Proteom. MCP* **2005**, *4*, 2010–2021. [CrossRef] [PubMed]
28. Eng, J.K.; Jahan, T.A.; Hoopmann, M.R. Comet: An open-source MS/MS sequence database search tool. *Proteomics* **2013**, *13*, 22–24. [CrossRef] [PubMed]
29. Vaudel, M.; Barsnes, H.; Berven, F.S.; Sickmann, A.; Martens, L. SearchGUI: An open-source graphical user interface for simultaneous OMSSA and X!Tandem searches. *Proteomics* **2011**, *11*, 996–999. [CrossRef]
30. Gaudet, P.; Michel, P.-A.; Zahn-Zabal, M.; Cusin, I.; Duek, P.D.; Evalet, O.; Gateau, A.; Gleizes, A.; Pereira, M.; Teixeira, D.; et al. The neXtProt knowledgebase on human proteins: Current status. *Nucleic Acids Res.* **2015**, *43*, D764–D770. [CrossRef]
31. Vaudel, M.; Burkhart, J.M.; Zahedi, R.P.; Oveland, E.; Berven, F.S.; Sickmann, A.; Martens, L.; Barsnes, H. PeptideShaker enables reanalysis of MS-derived proteomics data sets. *Nat. Biotechnol.* **2015**, *33*, 22–24. [CrossRef]
32. Vizcaíno, J.A.; Deutsch, E.W.; Wang, R.; Csordas, A.; Reisinger, F.; Ríos, D.; Dianes, J.A.; Sun, Z.; Farrah, T.; Bandeira, N.; et al. ProteomeXchange provides globally coordinated proteomics data submission and dissemination. *Nat. Biotechnol.* **2014**, *32*, 223–226. [CrossRef] [PubMed]
33. Martens, L.; Hermjakob, H.; Jones, P.; Adamski, M.; Taylor, C.; States, D.; Gevaert, K.; Vandekerckhove, J.; Apweiler, R. PRIDE: The proteomics identifications database. *Proteomics* **2005**, *5*, 3537–3545. [CrossRef]
34. Díez, P.; Pérez-Andrés, M.; Bøgsted, M.; Azkargorta, M.; García-Valiente, R.; Dégano, R.M.; Blanco, E.; Mateos-Gomez, S.; Bárcena, P.; Santa Cruz, S.; et al. Dynamic Intracellular Metabolic Cell Signaling Profiles During Ag-Dependent B-Cell Differentiation. *Front. Immunol.* **2021**, *12*, 637832. [CrossRef]
35. Cox, J.; Hein, M.Y.; Luber, C.A.; Paron, I.; Nagaraj, N.; Mann, M. Accurate proteome-wide label-free quantification by delayed normalization and maximal peptide ratio extraction, termed MaxLFQ. *Mol. Cell. Proteom. MCP* **2014**, *13*, 2513–2526. [CrossRef]
36. Cox, J.; Mann, M. MaxQuant enables high peptide identification rates, individualized p.p.b.-range mass accuracies and proteome-wide protein quantification. *Nat. Biotechnol.* **2008**, *26*, 1367–1372. [CrossRef] [PubMed]
37. Bielow, C.; Mastrobuoni, G.; Kempa, S. Proteomics Quality Control: Quality Control Software for MaxQuant Results. *J. Proteome Res.* **2016**, *15*, 777–787. [CrossRef]
38. RStudio Team. *R: A Language and Environment for Statistical Computing*; R. RStudio, PBC: Boston, MA, USA, 2011.
39. Schmitz, R.; Young, R.M.; Ceribelli, M.; Jhavar, S.; Xiao, W.; Zhang, M.; Wright, G.; Shaffer, A.L.; Hodson, D.J.; Buras, E.; et al. Burkitt lymphoma pathogenesis and therapeutic targets from structural and functional genomics. *Nature* **2012**, *490*, 116–120. [CrossRef]
40. Leinonen, R.; Sugawara, H.; Shumway, M. International Nucleotide Sequence Database Collaboration The sequence read archive. *Nucleic Acids Res.* **2011**, *39*, D19–D21. [CrossRef] [PubMed]
41. Bolger, A.M.; Lohse, M.; Usadel, B. Trimmomatic: A flexible trimmer for Illumina sequence data. *Bioinformatics* **2014**, *30*, 2114–2120. [CrossRef]
42. Dobin, A.; Davis, C.A.; Schlesinger, F.; Drenkow, J.; Zaleski, C.; Jha, S.; Batut, P.; Chaisson, M.; Gingeras, T.R. STAR: Ultrafast universal RNA-seq aligner. *Bioinformatics* **2013**, *29*, 15–21. [CrossRef]
43. Li, H.; Handsaker, B.; Wysoker, A.; Fennell, T.; Ruan, J.; Homer, N.; Marth, G.; Abecasis, G.; Durbin, R. 1000 Genome Project Data Processing Subgroup The Sequence Alignment/Map format and SAMtools. *Bioinformatics* **2009**, *25*, 2078–2079. [CrossRef]
44. Trapnell, C.; Williams, B.A.; Pertea, G.; Mortazavi, A.; Kwan, G.; van Baren, M.J.; Salzberg, S.L.; Wold, B.J.; Pachter, L. Transcript assembly and quantification by RNA-Seq reveals unannotated transcripts and isoform switching during cell differentiation. *Nat. Biotechnol.* **2010**, *28*, 511–515. [CrossRef] [PubMed]
45. Häggmark, A.; Byström, S.; Ayoglu, B.; Qundos, U.; Uhlén, M.; Khademi, M.; Olsson, T.; Schwenk, J.M.; Nilsson, P. Antibody-based profiling of cerebrospinal fluid within multiple sclerosis. *Proteomics* **2013**, *13*, 2256–2267. [CrossRef] [PubMed]
46. Sierra-Sánchez, Á.; Garrido-Martín, D.; Lourido, L.; González-González, M.; Díez, P.; Ruiz-Romero, C.; Sjöber, R.; Droste, C.; De Las Rivas, J.; Nilsson, P.; et al. Screening and Validation of Novel Biomarkers in Osteoarticular Pathologies by Comprehensive Combination of Protein Array Technologies. *J. Proteome Res.* **2017**, *16*, 1890–1899. [CrossRef] [PubMed]
47. González-González, M.; Bartolome, R.; Jara-Acevedo, R.; Casado-Vela, J.; Dasilva, N.; Matarraz, S.; García, J.; Alcazar, J.A.; Sayagues, J.M.; Orfao, A.; et al. Evaluation of homo- and hetero-functionally activated glass surfaces for optimized antibody arrays. *Anal. Biochem.* **2014**, *450*, 37–45. [CrossRef]
48. Durinck, S.; Spellman, P.T.; Birney, E.; Huber, W. Mapping identifiers for the integration of genomic datasets with the R/Bioconductor package biomaRt. *Nat. Protoc.* **2009**, *4*, 1184–1191. [CrossRef]
49. Dennis, G.; Sherman, B.T.; Hosack, D.A.; Yang, J.; Gao, W.; Lane, H.C.; Lempicki, R.A. DAVID: Database for Annotation, Visualization, and Integrated Discovery. *Genome Biol.* **2003**, *4*, P3. [CrossRef]
50. Fontanillo, C.; Nogales-Cadenas, R.; Pascual-Montano, A.; De las Rivas, J. Functional analysis beyond enrichment: Non-redundant reciprocal linkage of genes and biological terms. *PLoS ONE* **2011**, *6*, e24289. [CrossRef]

51. Szklarczyk, D.; Gable, A.L.; Lyon, D.; Junge, A.; Wyder, S.; Huerta-Cepas, J.; Simonovic, M.; Doncheva, N.T.; Morris, J.H.; Bork, P.; et al. STRING v11: Protein-protein association networks with increased coverage, supporting functional discovery in genome-wide experimental datasets. *Nucleic Acids Res.* **2019**, *47*, D607–D613. [CrossRef] [PubMed]
52. Kanehisa, M.; Goto, S. KEGG: Kyoto encyclopedia of genes and genomes. *Nucleic Acids Res.* **2000**, *28*, 27–30. [CrossRef]
53. Jassal, B.; Matthews, L.; Viteri, G.; Gong, C.; Lorente, P.; Fabregat, A.; Sidiropoulos, K.; Cook, J.; Gillespie, M.; Haw, R.; et al. The reactome pathway knowledgebase. *Nucleic Acids Res.* **2020**, *48*, D498–D503. [CrossRef]
54. Krysko, D.V.; Garg, A.D.; Kaczmarek, A.; Krysko, O.; Agostinis, P.; Vandenabeele, P. Immunogenic cell death and DAMPs in cancer therapy. *Nat. Rev. Cancer* **2012**, *12*, 860–875. [CrossRef] [PubMed]
55. Roh, J.S.; Sohn, D.H. Damage-Associated Molecular Patterns in Inflammatory Diseases. *Immune Netw.* **2018**, *18*. [CrossRef] [PubMed]
56. Berger, G.; Marloye, M.; Lawler, S.E. Pharmacological Modulation of the STING Pathway for Cancer Immunotherapy. *Trends Mol. Med.* **2019**, *25*, 412–427. [CrossRef] [PubMed]
57. Haseeb, M.; Anwar, M.A.; Choi, S. Molecular Interactions Between Innate and Adaptive Immune Cells in Chronic Lymphocytic Leukemia and Their Therapeutic Implications. *Front. Immunol.* **2018**, *9*, 2720. [CrossRef] [PubMed]
58. Kipps, T.J.; Stevenson, F.K.; Wu, C.J.; Croce, C.M.; Packham, G.; Wierda, W.G.; O'Brien, S.; Gribben, J.; Rai, K. Chronic lymphocytic leukaemia. *Nat. Rev. Dis. Primers* **2017**, *3*, 16096. [CrossRef] [PubMed]
59. Roberts, A.W.; Davids, M.S.; Pagel, J.M.; Kahl, B.S.; Puvvada, S.D.; Gerecitano, J.F.; Kipps, T.J.; Anderson, M.A.; Brown, J.R.; Gressick, L.; et al. Targeting BCL2 with Venetoclax in Relapsed Chronic Lymphocytic Leukemia. *N. Engl. J. Med.* **2016**, *374*, 311–322. [CrossRef] [PubMed]
60. Hermanson, G.T. Chapter 11—(Strept)avidin–Biotin Systems. In *Bioconjugate Techniques*, 3rd ed.; Hermanson, G.T., Ed.; Academic Press: Boston, MA, USA, 2013; pp. 465–505.
61. Adhikari, S.; Nice, E.C.; Deutsch, E.W.; Lane, L.; Omenn, G.S.; Pennington, S.R.; Paik, Y.-K.; Overall, C.M.; Corrales, F.J.; Cristea, I.M.; et al. A high-stringency blueprint of the human proteome. *Nat. Commun.* **2020**, *11*, 5301. [CrossRef]
62. Kirkwood, K.J.; Ahmad, Y.; Larance, M.; Lamond, A.I. Characterization of native protein complexes and protein isoform variation using size-fractionation-based quantitative proteomics. *Mol. Cell. Proteom. MCP* **2013**, *12*, 3851–3873. [CrossRef]
63. Hanahan, D.; Weinberg, R.A. Hallmarks of cancer: The next generation. *Cell* **2011**, *144*, 646–674. [CrossRef]
64. Kim, T.R.; Jeong, H.-H.; Sohn, K.-A. Topological integration of RPPA proteomic data with multi-omics data for survival prediction in breast cancer via pathway activity inference. *BMC Med. Genom.* **2019**, *12* (Suppl. 5), 94. [CrossRef] [PubMed]
65. Peng, W.; Zhu, R.; Zhou, S.; Mirzaei, P.; Mechref, Y. Integrated Transcriptomics, Proteomics, and Glycomics Reveals the Association between Up-regulation of Sialylated N-glycans/Integrin and Breast Cancer Brain Metastasis. *Sci. Rep.* **2019**, *9*, 17361. [CrossRef] [PubMed]

biomolecules

MDPI

Article

Attenuation of PITPNM1 Signaling Cascade Can Inhibit Breast Cancer Progression

Zihao Liu [1,2,†], Yu Shi [1,2,†], Qun Lin [1,2], Wenqian Yang [1,2], Qing Luo [1,2], Yinghuan Cen [1,2], Juanmei Li [1,2], Xiaolin Fang [1,2], Wen G. Jiang [3] and Chang Gong [1,2,*]

1 Breast Tumor Center, Sun Yat-Sen Memorial Hospital, Sun Yat-Sen University, Guangzhou 510120, China; liuzh75@mail2.sysu.edu.cn (Z.L.); shiyu9712@126.com (Y.S.); linq27@mail2.sysu.edu.cn (Q.L.); ywq20541@126.com (W.Y.); luoq63@mail2.sysu.edu.cn (Q.L.); cenyinghuan@163.com (Y.C.); lijm95@mail.sysu.edu.cn (J.L.); fangxiaolin2021@126.com (X.F.)

2 Guangdong Provincial Key Laboratory of Malignant Tumor Epigenetics and Gene Regulation, Sun Yat-Sen Memorial Hospital, Sun Yat-Sen University, Guangzhou 510120, China

3 Cardiff China Medical Research Collaborative, School of Medicine, Cardiff University, Heath Park, Cardiff CF14 4XN, UK; jiangw@cardiff.ac.uk

* Correspondence: gchang@mail.sysu.edu.cn or changgong282@163.com

† These authors contributed equally to this work.

Abstract: Phosphatidylinositol transfer protein membrane-associated 1 (PITPNM1) contains a highly conserved phosphatidylinositol transfer domain which is involved in phosphoinositide trafficking and signaling transduction under physiological conditions. However, the functional role of PITPNM1 in cancer progression remains unknown. Here, by integrating datasets of The Cancer Genome Atlas (TCGA) and Molecular Taxonomy of Breast Cancer (METABRIC), we found that the expression of PITPNM1 is much higher in breast cancer tissues than in normal breast tissues, and a high expression of PITPNM1 predicts a poor prognosis for breast cancer patients. Through gene set variation analysis (GSEA) and gene ontology (GO) analysis, we found PITPNM1 is mainly associated with carcinogenesis and cell-to-cell signaling ontology. Silencing of PITPNM1, in vitro, significantly abrogates proliferation and colony formation of breast cancer cells. Collectively, PITPNM1 is an important prognostic indicator and a potential therapeutic target for breast cancer.

Keywords: breast cancer; PITPNM1; proliferation; metastasis

1. Introduction

According to the 2020 cancer epidemiology statistics, breast cancer is the most common cancer worldwide among women with an estimated 2.3 million newly diagnosed cases [1,2]. Breast cancer becomes a major public health problem worldwide. Thus, uncovering new molecular mechanisms which are involved in the progression of breast cancer is of great clinical significance.

PITPNM1 (Phosphatidylinositol Transfer Protein Membrane Associated 1), also known as NIR2 (Pyk2 N-terminal domain-interacting receptor 2) and RDGB1 (Retinal Degeneration B Alpha 1), belongs to the subfamily of membrane-associated phosphatidylinositol transfer domain-containing proteins. PITPNM1 shares homology with Drosophila retina degeneration B (rdgB) protein [3–5]. PITPNM1, PITPNM2, and PITPNM3 were first found to be expressed in the retina and brain [6,7]. The genetic analysis of a small cohort of patients with autosomal dominant retinal dystrophy reveals that a point mutation c.1878G>C (p.Q626H) of PITPNM1′s homology, PITPNM3, will lead to retinal dystrophy disease [8]. These findings suggest that PITPNM1 might play a functional role in the phototransduction pathway and PITPNM1is considered as a candidate gene for human retina diseases. Recent studies suggest PITPNM1 can also be expressed in hair cells, adipose-derived stem cells, and hearing cells [9–11]. These studies, collectively, suggest that PITPNM1 plays a complex role in normal cells. Interestingly, cancer cells seem to hijack PITPNM1 to support their progression. It is reported that PITPNM1 is expressed in breast cancer cells and PITPNM1 promotes breast cancer migration, invasion, and epithelial-mesenchymal-transition (EMT) by activating the PI3K-AKT signaling pathway. These indicate that PITPNM1 potentially regulates the complex steps of cancer progression [12,13]. However, little attention has been drawn to PITPNM1 and its involvement in cancer progression. To reveal the interplay between PITPNM1 and the hallmark events of breast cancer, we performed data mining in The Cancer Genome Atlas (TCGA) and Molecular Taxonomy of Breast Cancer (METABRIC) to explore the association between the expression of PITPNM1 and clinicopathological data. Moreover, we carried out a series of in vitro tests by silencing PITPNM1 to determine whether PITPNM1 regulates breast cancer proliferation or not. Our results indicate that PITPNM1 plays an oncogenic role in breast cancer.

2. Materials and Methods

2.1. Cell Culture

Human breast cancer cell lines including SKBR3, MDA-MB-468, BT549, MDA-MB-231, T47D, ZR75-1, MCF-7, and BT474 were purchased from the American Type Culture Collection (ATCC, Manassas, VA, USA). All cells were cultured in Dulbecco's Modified Eagle Medium (DMEM) (Gibco, Life Technologies, Grand Island, NY, USA) supplemented with 10% fetal bovine serum under 5% CO_2 at 37 °C. All cells were confirmed by short tandem repeat DNA profiling and passaged for less than 6 months.

2.2. SiRNA Transfection

Breast cancer cells (5 × 10 per well) were seeded in a 6-well plate. Cells were cultured and transfected from 30–50% confluence. Two siRNAs targeting different sites of PITPNM1 (si-PITPNM1-1, sense: 5′-GCGGGCAAUACACACACAATT-3′, anti-sense: 5′-UUGUGUGUGUAUUGCCCGCTT-3′; si-PITPNM1-2, sense: 5′-GGAGAAAUUCUCCAUU GAATT-3′, anti-sense: 5′-UUCAAUGGAGAAUUUCUCCTT-3′, Genepharma, Shanghai, China) were, respectively, suspended in Opti-MEM and mixed with lipo3000. The mixture was added to plates and cultured for 48 h. After transfection, cells were harvested for further experiments.

2.3. CCK8 Cell Counting Assay

The CCK8 assay was used to assess the proliferation ability of cells. Cells (3 × 10^3 per well for MDA-MB-231, 5 × 10^3 per well for MCF-7 and SKBR3) were seeded in 96-well plates, incubated, and tested with CCK8. The CCK8 assay was tested daily for MDA-MB-231 cells and every other day for MCF-7 and SKBR3 cells. CCK8 reagent was added in a culture medium and incubated at 5% CO_2 at 37 °C for 1 h. Absorbance was measured by the TCAN plate reader (TECAN Spark 10M, Tecan Group Ltd., Zürich, Switzerland) at 450 nm. Data represent the mean of three independent experiments.

2.4. Colony Formation Assay

Cells were harvested after transfection, 1 × 10^3 to 2 × 10^3 cells were seeded in 6-well plates and cultured at 5% CO_2 at 37 °C for at least two weeks. After incubation, cells were fixed with 4% formaldehyde and stained with crystal violet. Data represent the mean of three independent experiments.

2.5. RNA Isolation and Real-Time Quantitative PCR

Total RNA was extracted using the RNA extraction kit (RC112-01, Vazyme, Nanjing, China) according to the manufacturer's protocol. Total RNA was used to synthesize complementary DNA (cDNA) through RNA reverse transcriptional reagents (Vazyme, Nanjing, China). Real-time quantitative PCR (qPCR) was performed by using SYBR green reagent (Vazyme, Nanjing, China) according to the manufacturer's protocol. The relative expression of genes was normalized to ACTB and calculated by the $2^{-\Delta\Delta Ct}$

method. Primers are listed as follows: ACTB sense: 5′-AGCGGGAAATCGTGCGTGAC-3′, ACTB anti-sense: 5′-CAGGAAGGAAGGCTGGAAGAGT-3′; PITPNM1 sense: 5′-GAGGAGTCTAGTGGTGAG
GGC-3′, anti-sense: 5′-TTCGGGTGTAGGGGTAGGC-3′.

2.6. Data Extraction and Bioinformatics Analysis

The transcriptome data of the gene count matrix was downloaded from the TCGA data portal (https://portal.gdc.cancer.gov/, accessed on 8 March 2020). The METABRIC datasets were downloaded from cBioPortal (http://www.cbioportal.org/, accessed on 13 February 2021).The Broad Institute Cancer Cell Line Encyclopedia (CCLE) datasets were downloaded from the official website (https://portals.broadinstitute.org/ccle, accessed on 8 May 2020). Estrogen receptor (ER) and progesterone receptor (PR) status were determined by immunohistochemistry. Human epidermal growth factor receptor 2 (HER2) status was determined by immunohistochemistry and/or fluorescence in situ hybridization (FISH). The survival data of the TCGA breast cancer cohort were downloaded from the TCGA Pan-Cancer Clinical Data Resource. Survival analysis was performed by website tools (Kaplan-Meier plotter, http://kmplot.com/analysis/, accessed on 10 November 2020; bc-GenExMiner, http://bcgenex.ico.unicancer.fr/BC-GEM/, accessed on 23 April 2021). Differentially expressed genes were analyzed by DESeq2. The gene ontology of genes differentially expressed in the high PITPNM1 group and the low PITPNM1 group was analyzed by DAVID and gene set variation analysis (GSEA) [14,15]. Immune infiltration scores were analyzed by the CIBERSORTx website portal (https://cibersortx.stanford.edu/, accessed on 29 February 2021) [16].

2.7. Statistical Analysis

Statistical analysis was performed using IBM SPSS Statistical 25 and R Studio v5. The log2FC > 1.5 and $p < 0.05$ or log2FC < −1.5 and $p < 0.05$ were characterized as differentially expressed genes. A two-tail student t-test and two-tail one-way ANOVA with Tukey's post hoc test were used to determine statistical significance between groups. K-M survival analysis was used to determine the prognosis of the high PITPNM1 and the low PITPNM1 groups with the higher quartile regarded as the cut-off value and their significance was calculated by the log-rank test.

3. Results

3.1. PITPNM1 Is Significantly Higher in Breast Cancer and Correlates with Poor Prognosis in Patients with Breast Cancer

We extracted two large-scale transcriptome datasets of breast cancer to determine the expression of PITPNM1. The mRNA of PITPNM1 of TCGA breast cancer samples was significantly higher in triple-negative cancer (TNBC, 1.64-fold, $p < 0.0001$, $n = 160$) as well as HER2 over-expression in breast cancer (1.99-fold, $p < 0.0001$, $n = 33$) and luminal A/B (1.55-fold, $p < 0.0001$, $n = 711$), compared to normal breast tissue ($n = 113$, Figure 1A). In breast cancer tissues of the TCGA datasets, the expression of PITPNM1 barely correlates with the breast cancer stage, ER status ($p > 0.05$), PR status ($p > 0.05$) or HER2 status ($p > 0.05$, Figure 1B,C). In METABRIC datasets, the expression of PITPNM1 in breast tissues is consistent with that seen in the TCGA datasets. The expression of PITPNM1 in TNBC (1.48-fold, $p < 0.0001$, $n = 299$), HER2 over-expression in breast cancer tissues (1.59-fold, $p < 0.0001$, $n = 127$), and luminal A/B (1.25-fold, $p < 0.0001$, $n = 1464$) is higher than that in normal breast tissues ($n = 147$, Figure 1D). The expression of PITPNM1 in stage 3 ($p < 0.05$) and stage 4 ($p < 0.05$) breast cancers is significantly higher than in stage 1 (Figure 1E). Compared with other subtypes, the level of PITPNM1 was higher in the ER-negative group, PR-negative group, and HER2-positive group (Figure 1F). Since PITPNM1 is much higher in breast cancer tissues than in normal breast tissues, we further explored the prognostic role of PITPNM1. By using the higher quartile, we separated breast cancer patients into two groups: the high expression group and the low expression group. We found high levels of PITPNM1 marginally associated with short disease-free survival (DFS) in the TCGA cohort (log-rank $p = 0.1075$), but not significantly correlated with overall survival (OS, Figure 1G). To further confirm whether high levels of PITPNM1 are associated with poor prognosis, three other independent cohorts of breast cancer patients were separated into two groups by the higher quartile, respectively. We found that higher expression of PITPNM1 was associated with poor OS in the METABRIC, SCAN-B, and Kaplan–Meier Plotter cohort (Figure 1H–J).

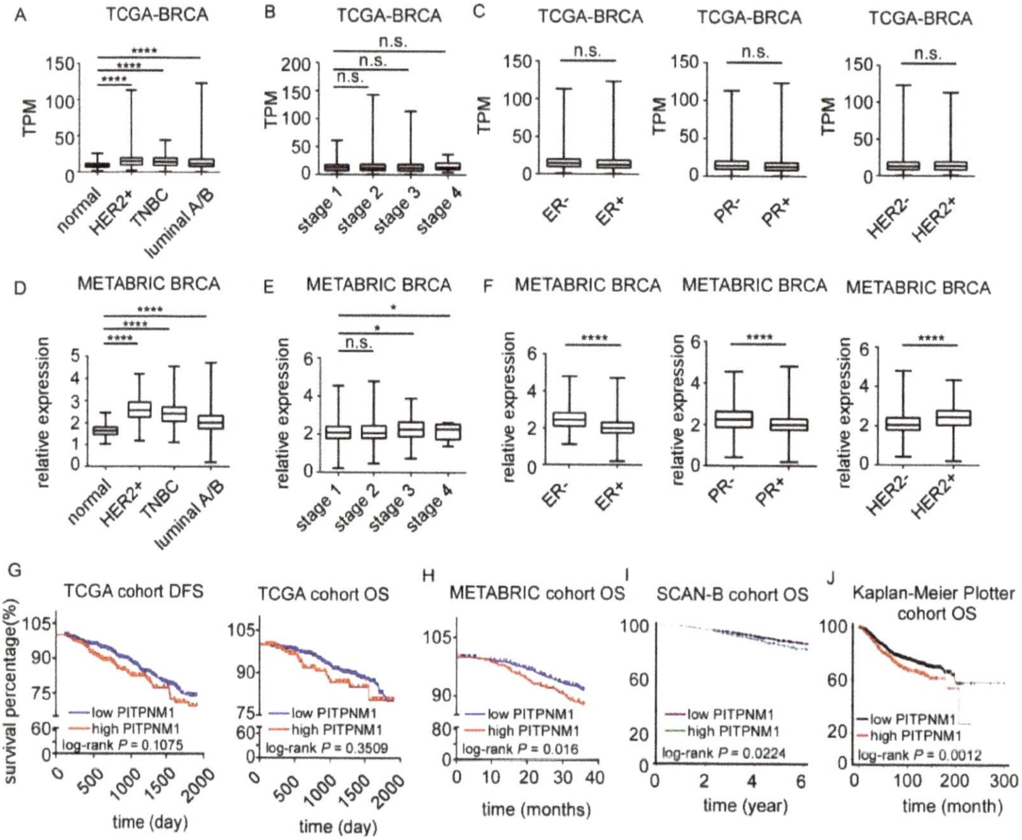

Figure 1. Prognostic value of PITPNM1 in breast cancer. (**A**) The expression of PITPNM1 in different molecular subtypes of breast cancer of TCGA cohort. Triple-negative cancer, TNBC, $n = 160$; human epidermal growth factor receptor 2 over-expression, HER2+, $n = 33$; luminal A/B, $n = 711$; normal breast tissue $n = 113$. **** $p < 0.0001$. (**B**) The expression of PITPNM1 in different stages of breast cancer of TCGA cohort. (**C**) The association of PITPNM1 with estrogen receptor (ER) status, progesterone receptor (PR) status, and HER2 status in TCGA cohort. n.s., not significant, $p > 0.05$. (**D**) The expression of PITPNM1 in different molecular subtypes of breast cancer of METABRIC cohort. TNBC, $n = 299$; HER2+, $n = 127$; luminal A/B, $n = 1464$; normal breast tissue $n = 147$. **** $p < 0.0001$. (**E**) The expression of PITPNM1 in different stages of breast cancer of METABRIC cohort. n.s., not significant, $p > 0.05$; * $p < 0.05$. (**F**) The association of PITPNM1 with estrogen receptor (ER) status, progesterone receptor (PR) status, and HER2 status in METABRIC cohort. **** $p < 0.0001$. (**G**) The prognosis of PITPNM1 in TCGA breast cancer cohort. Disease-free survival time, DFS; overall survival, OS. (**H**) The prognosis of PITPNM1 in METABRIC cohort. (**I**) The prognosis of PITPNM1 in SCAN-B cohort. (**J**) The prognosis of PITPNM1 in Kaplan–Meier cohort.

3.2. PITPNM1 Over-Expression Is Associated with Carcinogenesis Gene Ontology and Pathway

Since high expression of PITPNM1 predicts poor OS of patients with breast cancer as shown earlier, PITPNM1 could potentially promote breast cancer progression. In order to explore the potential molecular mechanisms underlying carcinogenesis of PITPNM1 in breast cancer, we employed GO term analysis, KEGG analysis, and GSEA analysis to determine the enrichment of cancer-related processes in the high PITPNM1 group and the low PITPNM1 group. Firstly, we analyzed the differentially expressed genes between the high PITPNM1 group and the low PITPNM1 group, defined by the higher quartile in the TCGA BRCA cohort. A total of 4543 differentially expressed genes were identified (Figure 2A, Supplementary Table S1).Through KEGG pathway analysis, we found that up-regulated genes in the high PITPNM1 group are mainly enriched in immune pathways such as: the T cell receptor signaling pathway, rheumatoid arthritis, primary immunodeficiency, and NK cell-mediated cytotoxicity, and cancer-related signaling pathways such as the NF-kappa B signaling pathway, the Jak-STAT signaling pathway, the chemokine signaling and cell adhesion molecules (Figure 2B, Supplementary Table S2). In contrast, down-regulated genes are mainly enriched in tight junction, taste transduction, steroid hormone biogenesis, salivary secretion, retinol metabolism, and so on (Figure 2C, Supplementary Table S3). Gene ontology analysis indicates that up-regulated genes are enriched in immune processes such as immune processes, inflammatory response, innate immune response, regulation of immune response, as well as T cell co-stimulation and cancer-related biological processes such as cell adhesion, positive regulation of cell proliferation, and cell differentiation (Figure 2D, Supplementary Table S2). As for the cellular counterpart, these genes are mainly enriched in the plasma membrane, the integral component of the membrane, and the extracellular region (Figure 2D, Supplementary Table S2). Molecular functions such as calcium ion binding, sequence-specific DNA binding, receptor binding, and so on, are mainly enriched by up-regulated genes. Down-regulated genes are not significantly associated with cancer progression since down-regulated genes are mainly enriched in multicellular organism development, the transport muscle contraction, and the chemical synaptic transmission for biological processes; the integral component of the membrane, the extracellular region, the extracellular space for the cellular counterpart, the G protein receptor activity, the olfactory receptor activity, actin-binding, and the structure molecule activity for molecular function (Figure 2E, Supplementary Table S3).

To strengthen the analysis of the DAVID database, we performed GSEA, which can reveal a great number of gene sets as well as their phenotypes enriched in the highPITPNM1group and the low PITPNM1 group. Interestingly, hallmarks of the T cell immune process such as the regulation of T cell activation and the activation of T cell proliferation, and hallmarks of cancer such as the regulation of cell-cell adhesion and the positive regulation of G1/S transition of the mitotic cell cycle, cell cycle check points are significantly enriched in the high PITPNM1 group, while negative regulation of the execution phase of apoptosis is enriched in the low PITPNM1 group (Supplementary Table S4).The top GSEA biological processes enriched in the high PITPNM1 group and the low PITPNM1 group are shown in Figure 2F.

Furthermore, we used the hallmarks of gene sets (h: hallmark gene sets, GSEA) which are gene signatures derived by many gene sets to represent well-defined biological states or processes to explore the potential biological processes enriched in the high PITPNM1 group as well as the low PITPNM1 group. Some cancer-related processes such as PI3K-AKT signaling, UV response signaling, the P53 pathway, and mitotic-spindle are enriched in the high PITPNM1 group while none of the gene sets are enriched in the low PITPNM1 group (Supplementary Table S5). The top one enrichment of GSEA (h: hallmark gene sets) analysis of the high PITPNM1 group is shown in Figure 2G.

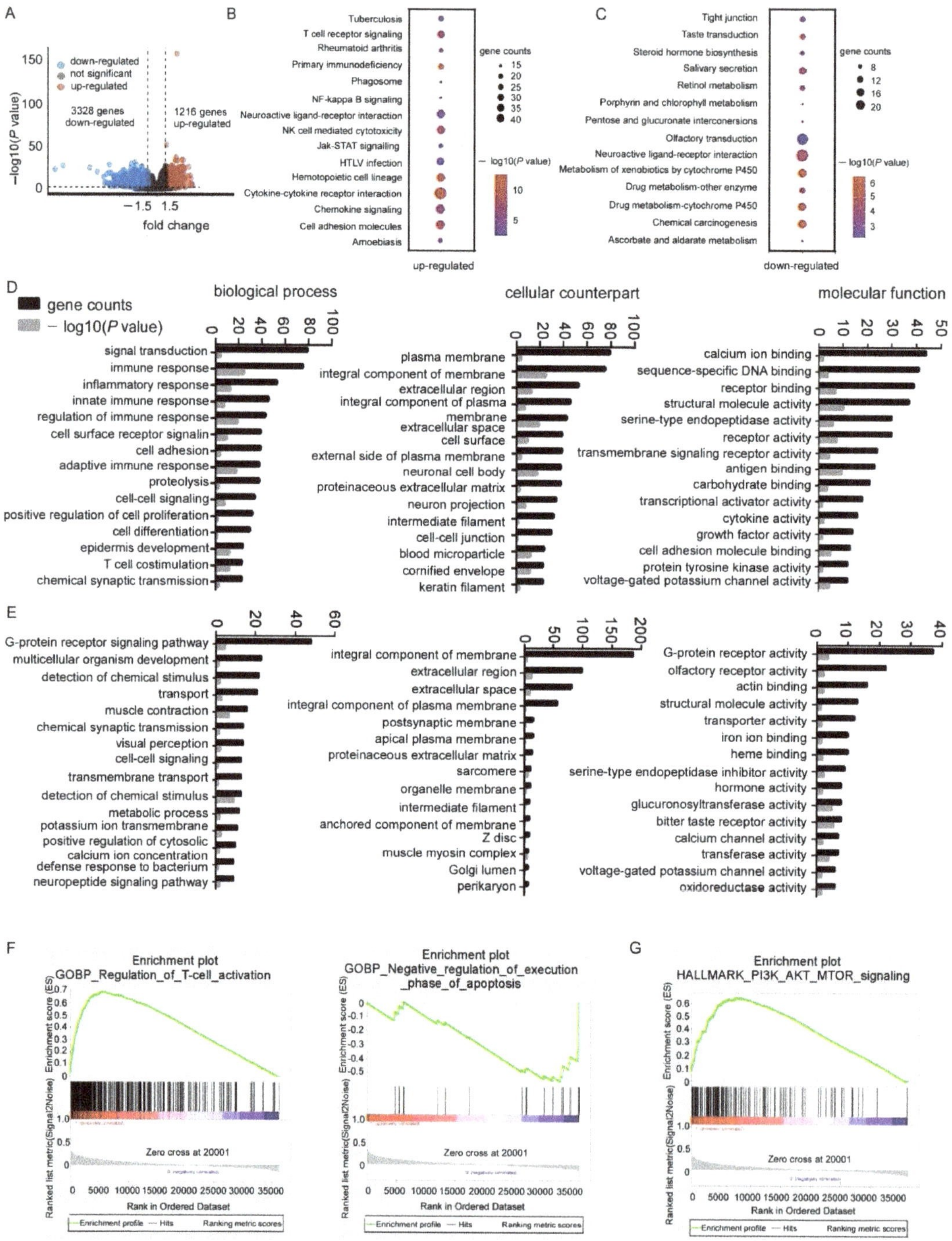

Figure 2. Gene ontology, KEGG pathway, and gene set enrichment analysis between PITPNM1 high expression group and PITPNM1 low group. (**A**) Volcano plot of differentially expressed genes between PITPNM1 high expression group and

PITPNM1 low expression group in TCGA cohort. (**B**) KEGG pathway of up-regulated genes between the high PITPNM1 group and the low PITPNM1 group. (**C**) KEGG pathway of down-regulated genes between the high PITPNM1 group and the low PITPNM1 group. (**D**) Biological process, cellular component, and molecular function ontology analysis of up-regulated genes. (**E**) Biological process, cellular component, and molecular function ontology analysis of down-regulated genes. (**F**) GSEA (C5 gene ontology sets) analysis of signatures of PITPNM1 high expression group and PITPNM1 low expression group. (**G**) GSEA (h: hallmark gene sets) analysis of signatures of PITPNM1 high expression group and PITPNM1 low expression group.

3.3. Silencing of PITPNM1 Inhibits Proliferation of Breast Cancer Cells

PITPNM1 predicts the poor prognosis of breast cancer and is associated with carcinogenesis gene ontology. The role of PITPNM1 in breast cancer progression was further determined. Firstly, the expression of PITPNM1 in different breast cancer cells was explored in CCLE datasets (Figure 3A). In different molecular subtypes, the mean value of PITPNM1 expression was much higher in HER2 over-expression and TNBC than that in the luminal A/B subtype. To validate the expression of PITPNM1 in different breast cancer cell lines and a non-tumorigenic immortalized breast epithelial cell, the mRNA levels of PITPNM1 were determined in nine cell lines including HER2 over-expression, TNBC, luminal subtypes, and non-tumorigenic MCF-10A (Figure 3B). PITPNM1 is highly expressed in most breast cancer cell lines. To confirm these results of transcript analyses, the protein level of PITPNM1 was also tested. Consistent with the qPCR results, the protein level of PITPNM1 was higher in breast cancer cell lines (Figure 3C). From the cell lines tested, SKBR3, MDA-MB-231, MCF-7, and MCF-10A cells were chosen. Two siRNAs targeting PITPNM1 were used to create PITPNM1-silencing cell models. Transfection with PITPNM1 siRNAs significantly reduced its expression from 60–70% in four different cell lines (Figure 3D). The CCK8 assay indicated that the silencing of PITPNM1 significantly inhibited MDA-MB-231 proliferation (Figure 3E). The colony formation ability of MDA-MB-231 was decreased after PITPNM1 siRNAs transfection (Figure 3E). In addition, the luminal subtype cancer cell line MCF7, as well as the HER2 over-expression cell line SKBR3, showed the same response to treatment of PITPNM1 siRNAs transfection (Figure 3F,G). On the other hand, silencing PITPNM1 in non-tumorigenic MCF-10A had little impact on cell proliferation or cell viability (Figure 3H). This indicates that PITPNM1 is essential for breast cancer proliferation but is not required by MCF-10A. PITPNM1 is a potential oncogene. Experiments in vitro suggest that PITPNM1 promotes breast cancer proliferation and maintains breast cancer cell viability.

3.4. The Relationship between PITPNM1 Expression and T Cell Populations

Through GSEA analysis, we found the high PITPNM1 group can enrich T cell immune process ontology such as T cell activation and T cell differentiation. In order to explore the potential involvement of PITPNM1 in the T cell immune process, firstly, we calculated the absolute abundance of T cell sub-populations in breast cancer by CIBERSORTx, which is a confident analytical tool to estimate the abundance of member cell types in cancer micro-environments based on bulk RNA sequencing. Activated CD4 positive memory T cells, CD8 positive follicular helper T cells, and regulatory T cells are significantly dysregulated in breast cancer tissues (Figure 4A). CD8 positive T cells, resting CD4 positive memory T cell, CD4 positive naive T cells, and gamma delta T cells showed no statistical difference between normal breast tissue and breast cancer tissue (Figure 4A). Since PITPNM1 is significantly up-regulated in breast cancer tissue (Figure 4B), we analyzed the correlation coefficient between PITPNM1 and the abundance of T cell sub-populations, which is significantly dysregulated in cancer tissues. PITPNM1 is correlated with regulatory T cells ($r = 0.28$, $p < 0.0001$, Figure 4C, Supplementary Table S3). These findings suggest PITPNM1 is potentially involved in breast cancer progression by regulating regulatory T cell infiltration and regulatory T cell function.

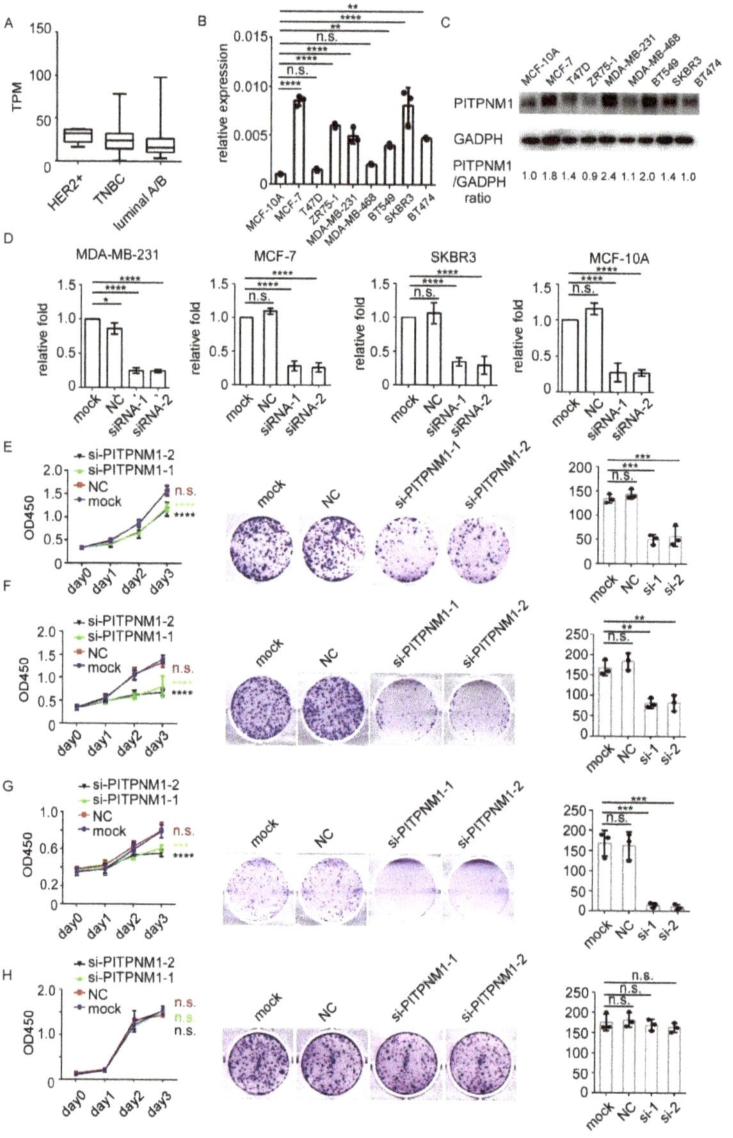

Figure 3. Silencing of PITPNM1 inhibits breast cancer proliferation. (**A**) The expression of PITPNM1 in different breast cancer subtypes. (**B**) The expression of PITPNM1 in breast cancer cell line validated by qPCR. n.s., not significant, $p > 0.05$; * $p < 0.05$; ** $p < 0.01$; *** $p < 0.001$; **** $p < 0.0001$. (**C**) The expression of PITPNM1 in breast cancer cell line validated by Western Blot. (**D**) siRNAs transfection efficiency validated by qPCR in four cell lines. n.s., not significant, $p > 0.05$, * $p < 0.05$; **** $p < 0.0001$. (**E**) The proliferation ability of MDA-MB-231 determined by the CCK8 assay and colony formation. n.s., not significant, $p > 0.05$; * $p < 0.05$; *** $p < 0.001$; **** $p < 0.0001$. (**F**) The proliferation ability of MCF-7 determined by the CCK8 assay and colony formation. n.s., not significant, $p > 0.05$; * $p < 0.05$; ** $p < 0.01$; **** $p < 0.0001$. (**G**) The proliferation ability of SKBR3 determined by the CCK8 assay and colony formation. n.s., not significant, $p > 0.05$; * $p < 0.05$; *** $p < 0.001$; **** $p < 0.0001$. (**H**) The proliferation ability of MCF-10A determined by the CCK8 assay and colony formation. n.s., not significant, $p > 0.05$.

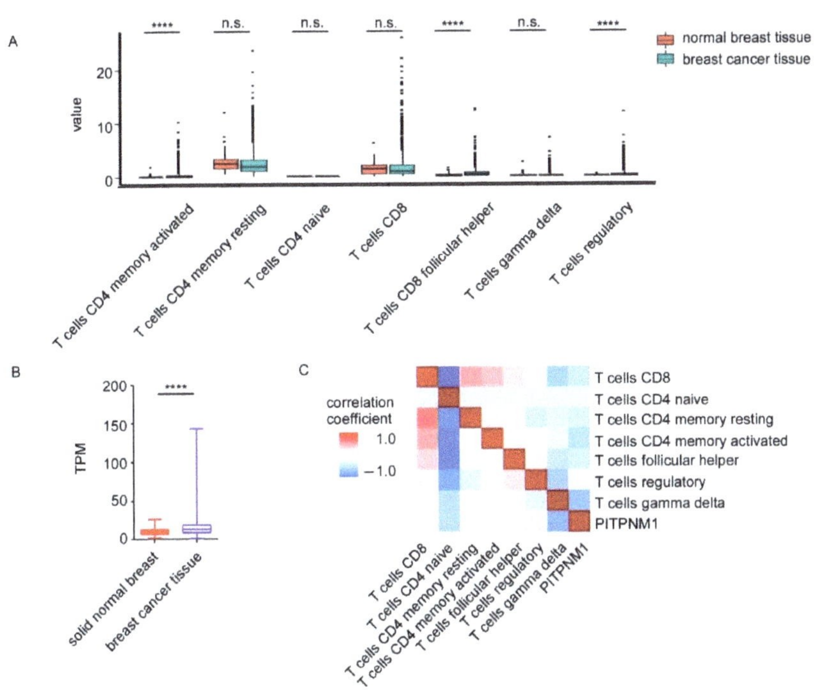

Figure 4. The relationship between PITPNM1 and T cell sub-population abundance. (**A**) The abundance of T cell subtypes in TCGA cohort estimated by CIBERSORTx. *t*-test; n.s., not significant, $p > 0.05$; **** $p < 0.0001$. (**B**) The expression of PITPNM1 in TCGA breast cancer cohort. Normal breast, $n = 113$; breast cancer tissue, $n = 1104$. **** $p < 0.0001$. (**C**) Heatmap of Spearman correlation coefficient between PITPNM1 and T cell subtypes.

4. Discussion

In the present study, we have revealed that PITPNM1 is associated with breast cancer prognosis. Through bioinformatics analysis, we found that PITPNM1 is associated with the carcinogenesis pathway and T cell immune process. The silencing of PITPNM1 inhibits the proliferation and colony formation ability of MDA-MB-231, MCF-7, and SKBR3. Moreover, PITPNM1 positively correlates with regulatory T cell infiltration. These results indicate that PITPNM1 could promote breast cancer progression by regulating breast cancer proliferation and the T cell immune process.

Previous studies have shown that PITPNM1 is an important regulator in maintaining the normal physiological functions of cells. PITPNM1 is expressed in retinal cells, adipose-derived stem cells, hearing cells, as well as hair cells, and is required for the normal function of these cells [7,9,10]. The dysregulation of PITPNM1 impairs signaling transduction and phosphatidylinositol transference, which in turn leads to abnormal functions of certain cells. It is reported that PITPNM1 is dysregulated in tissues from hearing-loss mouse models and Otx2 conditional knockout mice [11,17]. Moreover, PITPNM1 is regarded as a phosphoinositide trafficking and signaling transduction regulator [18,19]. It not only transfers phosphoinositide but also delivers phosphatidic acid between the endoplasmic reticulum and plasma membrane [18,20]. These findings highlight the important roles of PITPNM1 in homeostasis. However, few studies reveal the oncogenic roles of PITPNM1.

Of note, by integrating two large-scale breast cancer RNA sequence datasets, we found PITPNM1 was over-expressed in breast cancer tissue. We also found ahigh level of PITPNM1 can predict poor the prognosis of breast cancer patients. These results reveal the potential involvement of PITPNM1 in breast cancer progression. Interestingly, a recent study has provided some evidence for this link and showedPITPNM1 can promote breast metastasis by regulating breast cancer migration, invasion, and epithelial–mesenchymal-transition. The silencing of PITPNM1 will result in the impairment of migration as well as invasion [12]. In addition to regulating cell functions pertinent to metastasis, we found that PITPNM1 could regulate cell proliferation as well.

PITPNM1 was also shown to regulate the infiltration of regulatory T cells. Regulatory T cells play a tumor-promotional role in the micro-environment since regulatory T cells can migrate into cancer and suppress the antitumor effects of lymphocytes such as CD8 positive cytotoxic T cells [21–23]. Interestingly, through GSEA analysis, we found that the high PITPNM1 group could enrich T cell immune process ontology such as T cell activation and T cell differentiation. It raises the possibility that PITPNM1 potentially promotes breast cancer progression by regulating T cell immune processes. By using a highly confident immune abundance estimation tool, CIBERSORTx, we analyzed T cell subset abundance in bulk RNA sequence datasets. In addition, we found PITPNM1 was significantly correlated with regulatory T cells infiltration but is barely correlated with the infiltration of naive CD4 positive T cells or gamma delta T cells. Consistent with PITPNM1′s cancer-promotional roles, PITPNM1 promotes breast cancer progression by regulating regulatory T cell infiltration and inhibiting antitumor T cell immune processes.

5. Conclusions

PITPNM1, a phosphoinositide trafficking and signal transduction regulator, can promote breast cancer progression by regulating breast cancer proliferation and regulatory T cell infiltration. Therefore, PITPNM1 is proven to be an actionable target for breast cancer treatment.

Supplementary Materials: The following are available online at https://www.mdpi.com/2218-273X/11/9/1265/s1, Table S1: Differentially expressed genes in the high PITPNM1 group and the low PITPNM1 group in TCGA cohort, Table S2: GO term, KEGG pathway of up-regulated genes in the high PITPNM1group and the low PITPNM1 group, Table S3: GO term, KEGG pathway of down-regulated genes in the high PITPNM1group and the low PITPNM1 group, Table S4: GSEA BP analysis of the high PITPNM1group and the low PITPNM1 group in TCGA cohort, Table S5: GSEA h: hallmark gene sets analysis of the high PITPNM1group and the low PITPNM1 group in TCGA cohort, Table S6: Correlation analysis of PITPNM1 and immune cell infiltration.

Author Contributions: Investigation, data analysis and visualization, writing—original draft preparation, Z.L. and Y.S.; investigation, data analysis, W.Y. and Q.L. (Qun Lin); investigation, writing and editing, Q.L. (Qing Luo) and Y.C.; investigation, writing and editing, J.L. and X.F.; investigation, methodology, supervision, writing and editing, W.G.J.; conceptualization, methodology, investigation, supervision, writing and editing, C.G. All authors have read and agreed to the published version of the manuscript.

Funding: This work is supported by the National Key R&D Program of China (2017YFC1309103 and 2017YFC1309104); the Natural Science Foundation of China (81672594, 81772836, 81872139 and 82072907); Clinical Innovation Project of Bioland Laboratory (Guangzhou Regenerative Medicine and Health Guangdong Laboratory, 2018GZR0201004); Sun Yat-sen memorial hospital cultivation project for clinical research (SYS-C-201805); Key Projects of The National Natural Science Foundation of China(51861125203); Project of The Beijing Xisike Clinical Oncology Research Foundation(Y-Roche2019/2-0078); Technology Development Program of Guangdong province(2021A0505030082); Project of The Guangdong Provincial Key Laboratory of Malignant Tumor Epigenetics and Gene Regulation (2020B1212060018OF007).

Institutional Review Board Statement: Not applicable.

Informed Consent Statement: Not applicable.

Data Availability Statement: The data presented in this study are available upon request from the corresponding authors.

Conflicts of Interest: The authors declare no conflict of interest.

References

1. Sung, H.; Ferlay, J.; Siegel, R.L.; Laversanne, M.; Soerjomataram, I.; Jemal, A.; Bray, F. Global cancer statistics 2020: Globocan estimates of incidence and mortality worldwide for 36 cancers in 185 countries. *CA A Cancer J. Clin.* **2021**, *71*, 209–249. [CrossRef]
2. Siegel, R.L.; Miller, K.D.; Fuchs, H.E.; Jemal, A. Cancer statistics, 2021. *CA A Cancer J. Clin.* **2021**, *71*, 7–33. [CrossRef] [PubMed]
3. Trivedi, D.; Padinjat, R. Rdgb proteins: Functions in lipid homeostasis and signal transduction. *Biochim. Biophys. Acta (BBA)-Mol. Cell Biol. Lipids* **2007**, *1771*, 692–699. [CrossRef] [PubMed]
4. Woodard, C.; Alcorta, E.; Carlson, J. The rdgb gene of drosophila: A link between vision and olfaction. *J. Neurogenet.* **2007**, *21*, 291–305. [CrossRef] [PubMed]
5. Aikawa, Y.; Kuraoka, A.; Kondo, H.; Kawabuchi, M.; Watanabe, T. Involvement of pitpnm, a mammalian homologue of drosophila rdgb, in phosphoinositide synthesis on golgi membranes. *J. Biol. Chem.* **1999**, *274*, 20569–20577. [CrossRef] [PubMed]
6. Ocaka, L.; Spalluto, C.; Wilson, D.; Hunt, D.; Halford, S. Chromosomal localization, genomic organization and evolution of the genes encoding human phosphatidylinositol transfer protein membrane-associated (pitpnm) 1, 2 and 3. *Cytogenet. Genome Res.* **2005**, *108*, 293–302. [CrossRef]
7. Lev, S.; Hernandez, J.; Martinez, R.; Chen, A.; Plowman, G.; Schlessinger, J. Identification of a novel family of targets of pyk2 related to drosophila retinal degeneration b (rdgb) protein. *Mol. Cell. Biol.* **1999**, *19*, 2278–2288. [CrossRef] [PubMed]
8. Reinis, A.; Golovleva, I.; Köhn, L.; Sandgren, O. Ocular phenotype of cord5, an autosomal dominant retinal dystrophy associated with pitpnm3 p. Q626h mutation. *Acta Ophthalmol.* **2013**, *91*, 259–266. [CrossRef]
9. Carlisle, F.; Pearson, S.; Steel, K.; Lewis, M. Pitpnm1 is expressed in hair cells during development but is not required for hearing. *Neuroscience* **2013**, *248*, 620–625. [CrossRef]
10. Li, H.; Zhao, X.; Wen, X.; Zeng, A.; Mao, G.; Lin, R.; Hu, S.; Liao, W.; Zhang, Z. Inhibition of mir-490-5p promotes human adipose-derived stem cells chondrogenesis and protects chondrocytes via the pitpnm1/pi3k/akt axis. *Front. Cell Dev. Biol.* **2020**, *8*. [CrossRef]
11. Lewis, M.A.; Quint, E.; Glazier, A.M.; Fuchs, H.; de Angelis, M.H.; Langford, C.; van Dongen, S.; Abreu-Goodger, C.; Piipari, M.; Redshaw, N. An enu-induced mutation of mir-96 associated with progressive hearing loss in mice. *Nat. Genet.* **2009**, *41*, 614–618. [CrossRef]
12. Keinan, O.; Kedan, A.; Gavert, N.; Selitrennik, M.; Kim, S.; Karn, T.; Becker, S.; Lev, S. The lipid-transfer protein nir2 enhances epithelial-mesenchymal transition and facilitates breast cancer metastasis. *J. Cell Sci.* **2014**, *127*, 4740–4749. [CrossRef]
13. Hanahan, D.; Weinberg, R.A. Hallmarks of cancer: The next generation. *Cell* **2011**, *144*, 646–674. [CrossRef]
14. Subramanian, A.; Tamayo, P.; Mootha, V.K.; Mukherjee, S.; Ebert, B.L.; Gillette, M.A.; Paulovich, A.; Pomeroy, S.L.; Golub, T.R.; Lander, E.S. Gene set enrichment analysis: A knowledge-based approach for interpreting genome-wide expression profiles. *Proc. Natl. Acad. Sci. USA* **2005**, *102*, 15545–15550. [CrossRef] [PubMed]
15. Dennis, G.; Sherman, B.T.; Hosack, D.A.; Yang, J.; Gao, W.; Lane, H.C.; Lempicki, R.A. David: Database for annotation, visualization, and integrated discovery. *GenomeBiol.* **2003**, *4*, 1–11.
16. Newman, A.M.; Steen, C.B.; Liu, C.L.; Gentles, A.J.; Chaudhuri, A.A.; Scherer, F.; Khodadoust, M.S.; Esfahani, M.S.; Luca, B.A.; Steiner, D. Determining cell type abundance and expression from bulk tissues with digital cytometry. *Nat. Biotechnol.* **2019**, *37*, 773–782. [CrossRef]
17. Omori, Y.; Katoh, K.; Sato, S.; Muranishi, Y.; Chaya, T.; Onishi, A.; Minami, T.; Fujikado, T.; Furukawa, T. Analysis of transcriptional regulatory pathways of photoreceptor genes by expression profiling of the otx2-deficient retina. *PLoS ONE* **2011**, *6*, e19685. [CrossRef]
18. Kim, Y.J.; Guzman-Hernandez, M.-L.; Wisniewski, E.; Balla, T. Phosphatidylinositol-phosphatidic acid exchange by nir2 at er-pm contact sites maintains phosphoinositide signaling competence. *Dev. Cell* **2015**, *33*, 549–561. [CrossRef] [PubMed]
19. Lev, S. The role of the nir/rdgb protein family in membrane trafficking and cytoskeleton remodeling. *Exp. Cell Res.* **2004**, *297*, 1–10. [CrossRef] [PubMed]
20. Chang, C.-L.; Liou, J. Phosphatidylinositol 4, 5-bisphosphate homeostasis regulated by nir2 and nir3 proteins at endoplasmic reticulum-plasma membrane junctions. *J. Biol. Chem.* **2015**, *290*, 14289–14301. [CrossRef] [PubMed]
21. Togashi, Y.; Shitara, K.; Nishikawa, H. Regulatory t cells in cancer immunosuppression—Implications for anticancer therapy. *Nat. Rev. Clin. Oncol.* **2019**, *16*, 356–371. [CrossRef] [PubMed]
22. Liu, Y.-N.; Zhang, H.; Zhang, L.; Cai, T.-T.; Huang, D.-J.; He, J.; Ni, H.-H.; Zhou, F.-J.; Zhang, X.-S.; Li, J. Sphingosine 1 phosphate receptor-1 (s1p1) promotes tumor-associated regulatory t cell expansion: Leading to poor survival in bladder cancer. *Cell Death Dis.* **2019**, *10*, 1–11. [CrossRef] [PubMed]
23. Xu, R.; Wu, M.; Liu, S.; Shang, W.; Li, R.; Xu, J.; Huang, L.; Wang, F. Glucose metabolism characteristics and tlr8-mediated metabolic control of cd4+ treg cells in ovarian cancer cells microenvironment. *Cell Death Dis.* **2021**, *12*, 1–11.

biomolecules

MDPI

Article

A Thermodynamic Analysis of the Binding Specificity between Four Human PDZ Domains and Eight Host, Viral and Designed Ligands

Eva S. Cobos [1], Ignacio E. Sánchez [2], Lucía B. Chemes [3], Jose C. Martinez [1] and Javier Murciano-Calles [1,*]

1 Departamento Química Física, Unidad de Excelencia de Química Aplicada a Biomedicina y Medioambiente, Facultad de Ciencias, e Instituto de Biotecnología, Universidad de Granada, 18071 Granada, Spain; evasan@ugr.es (E.S.C.); jcmh@ugr.es (J.C.M.)

2 Laboratorio de Fisiología de Proteínas, Facultad de Ciencias Exactas y Naturales, Consejo Nacional de Investigaciones Científicas y Técnicas, Instituto de Química Biológica de la Facultad de Ciencias Exactas y Naturales (IQUIBICEN), Universidad de Buenos Aires, 1428 Buenos Aires, Argentina; isanchez@qb.fcen.uba.ar

3 Instituto de Investigaciones Biotecnológicas (IIBiO-CONICET), Universidad Nacional de San Martín, 1650 Buenos Aires, Argentina; lchemes@iib.unsam.edu.ar

* Correspondence: jmurciano@ugr.es

Abstract: PDZ domains are binding modules mostly involved in cell signaling and cell–cell junctions. These domains are able to recognize a wide variety of natural targets and, among the PDZ partners, viruses have been discovered to interact with their host via a PDZ domain. With such an array of relevant and diverse interactions, PDZ binding specificity has been thoroughly studied and a traditional classification has grouped PDZ domains in three major specificity classes. In this work, we have selected four human PDZ domains covering the three canonical specificity-class binding mode and a set of their corresponding binders, including host/natural, viral and designed PDZ motifs. Through calorimetric techniques, we have covered the entire cross interactions between the selected PDZ domains and partners. The results indicate a rather basic specificity in each PDZ domain, with two of the domains that bind their cognate and some non-cognate ligands and the two other domains that basically bind their cognate partners. On the other hand, the host partners mostly bind their corresponding PDZ domain and, interestingly, the viral ligands are able to bind most of the studied PDZ domains, even those not previously described. Some viruses may have evolved to use of the ability of the PDZ fold to bind multiple targets, with resulting affinities for the virus–host interactions that are, in some cases, higher than for host–host interactions.

Keywords: PDZ domain; binding specificity; virus; thermodynamics; isothermal titration calorimetry; differential scanning calorimetry

1. Introduction

Protein–protein interactions are crucial in biological functionality and are usually organized in networks that comprise two distinctive features: the occurrence of hubs and the repetitiveness of modules across the interactome. Precisely, hub proteins are commonly modular, which confers them the faculty to interconnect multiple processes by interacting with diverse partners. One of the most frequent modules present in hub proteins, and, possibly, the best studied to date, is the PDZ domain. More than 270 PDZ domains are part of human proteins and their principal roles are related to cell–cell communication and signaling and cell polarity [1]. These domains are especially relevant in the synaptic trafficking [2], as well as the tight junction formation and the maintenance of tissue integrity [3]. These functions make the PDZ domain very appealing to viruses, due to the importance for entrance and/or egression [4,5]. Indeed, plenty of viruses act through a PDZ-mediated interaction and PDZ domains are considered as a common route used by pathogenic viruses [6].

Another plausible good reason for virus to interact with the PDZ module is the relatively simple mode of interaction of PDZ binding. PDZ domains typically bind the C-terminal regions of their protein targets, which are often structurally disordered. In this canonical interaction, the residue comprising the free carboxyl terminus is accommodated in a conserved region of the PDZ domain. The residue at the C-terminus is frequently hydrophobic and is numbered as position 0 in the regular PDZ ligand notation (the rest of the residues are numbered backwards: -1, -2, etc.). The binding affinity and, therefore, specificity, in a given PDZ domain, is determined by the amino acid sequence of the ligand. Indeed, traditionally, PDZ domains have been classified by the sequence patterns that are able to recognize:

- Class I: X-(S/T)-X-ϕ_{-COOH};
- Class II: X-ϕ-X-ϕ_{-COOH};
- Class III: X-(D/E)-X-ϕ_{-COOH}.

where X is any amino acid, ϕ is a hydrophobic residue and the subscript $_{-COOH}$ means the C-terminus with the free carboxylate. This classification may be called "classic" or "traditional", since other attempts have been conducted in order to expand and make more precise the specificity classes. For instance, Tonikian et al. described up to 16 specificity classes based on peptide phage display screening, where specificity was determined down to the residue at position -5 [7]. Among them, Luck et al. reviewed the cases where a more extended surface and other structural regions different to the binding pocket modulate PDZ binding [8]. Moreover, Amacher et al. have summarized all the existing PDZ-binding data in a recent review [9]. In any case, the determinants of binding specificity are yet to be uncovered and the more basic and traditional classification has still validity.

Due to the magnitude and significance of viral infections worldwide and taking into account that plenty of diverse viruses bind PDZ domains as a part of their mechanisms of infection, proliferation and/or egression, we decided to perform a comparative study on PDZ binding to viral and host targets, along with designed ligands. Host PDZ motifs and viral mimics differ in many cases in sequence beyond the motif-determining residues. These may lead to differences in the specificity and thermodynamics of virus–host and host–host, protein–protein interactions [10]. Herein we have selected four PDZ domains, three of them known to be hijacked by viruses. Our selected domains comprise the three "classic" binding specificity classes. We have chosen natural ligands from the host and also from viruses that either are described to bind our selected PDZ domains or not, as well as a designed binder for one of the selected domains. Covering the whole set of cross interactions between the domains and the ligands using calorimetric techniques, we have obtained thermodynamic information for the cognate interactions and for non-cognate partners that also present measurable binding. Putting together all our information and gathering additional thermodynamic information from the literature, we shed some light on the intricate jungle of PDZ specificity.

2. Materials and Methods

2.1. PDZ Domains Purification and Peptide Samples

The DNA encoding the sequences of PSD95-PDZ2 (residues 155–249, according to PSD95 numbering), PSD95-PDZ3 (residues 302–415, according to PSD95 numbering), nNOS-PDZ (residues 1–127, in nNOS numbering) and Erbin-PDZ (given generously by Prof. Sachdev Sidhu, University of Toronto, Toronto, ON, Canada), were sub-cloned into pBAT4 vector (EMBL Core Purification Facility, Heidelberg, Germany) and expressed in Escherichia coli BL21-DE3 cells. Protein purification was carried out as previously described [11,12]. Briefly, cultures were over-expressed overnight and centrifuged. After cell lysis with a French press, the lysates were centrifuged and the pH of the supernatant was lowered down to a pH of 3 with diluted HCl. Upon centrifugation, ammonium sulfate was added to the samples, which were then injected into an AKTA FPLC for preparative size exclusion chromatography. Roughly 15 milligrams of protein per liter of LB culture were obtained and purity was determined by SDS-Page electrophoresis.

MALDI-TOF spectrometry, performed at the Scientific Instrumentation Services (CIC) of the University of Granada (Granada, Spain), confirmed the molecular mass for each PDZ domain. Extinction coefficients, determined as described [13], were used to calculate sample concentration. Experimental samples were always prepared by extensive dialysis against 50 mM potassium phosphate buffer at a pH of 7.5 and 4 °C.

Peptide samples were purchased both in JPT peptide technologies (Berlin, Germany) and in ChinaPeptides (Shanghai, China). In the cases where no Tryptophan or Tyrosine residue were present in the peptides, a quantitative analysis was performed and the extinction coefficients were experimentally obtained at 200 nm: 6458 $M^{-1}{\cdot}cm^{-1}$ (for FRETEV), 3272 $M^{-1}{\cdot}cm^{-1}$ (for IAESSV), 4160 $M^{-1}{\cdot}cm^{-1}$ (for KIATLV), 5251 $M^{-1}{\cdot}cm^{-1}$ (for GLDVPV), 9108 $M^{-1}{\cdot}cm^{-1}$ (for HRNTVV), 6184 $M^{-1}{\cdot}cm^{-1}$ (for LLDEIAV) and 9342 $M^{-1}{\cdot}cm^{-1}$ (for VVKVDSV).

2.2. Isothermal Titration Calorimetry (ITC)

Calorimetric titrations of the PDZ domains with the peptides were undertaken at 25 °C on an ITC-200 titration microcalorimeter (Microcal Inc., Northampton, MA, USA), as previously described [14]. Protein concentrations in the cell were routinely 70 µM and peptides solutions were prepared using the dialysis buffer to have a final concentration of 900 µM. Titrations were all made by injecting 2 µL volumes of ligand solution, with a final number of 19 injections per experiment. As a control and reference, independent experiments were carried out with only buffer in the calorimeter cell with the same ligand solutions to determine the corresponding heat dilution. The dilution isotherms were then subtracted from that obtained for protein titration. The binding stoichiometry, n, the dissociation constant, K_d and the corresponding enthalpy change, ΔH, were determined by fitting the resulting titration isotherm to a model considering a single ligand binding site using ORIGIN 7.0 (OriginLab Corporation, Northampton, MA, USA, 2002).

2.3. Differential Scanning Calorimetry (DSC)

Experiments were carried out in a VP-DSC instrument from Microcal INC. as described elsewhere [11]. The buffer conditions were the same as with ITC experiments, 50 mM buffer potassium phosphate at a pH of 7.5. For experiments including the peptides, the final concentration ratio was 1:3 (protein:peptide). Final data were obtained after treatment with ORIGIN 7.0 (OriginLab Corporation, Northampton, MA, USA, 2002).

3. Results

3.1. Rationale of the Choice of the PDZ Domains and Binding Partners

Among the plethora of the virus-interacting PDZ domains, we decided to select four domains that would allow us to address selectivity and specificity in terms of their respective partners and the canonical specificity classes. Firstly, we chose the highly studied PSD95 s and third PDZ domains. PSD95 is well known to act as a hub protein that interacts with multiple partners, allowing the synaptic trafficking [2]. PSD95 is just hijacked by two viruses: Human Papillomavirus (HPV) and Human T-cell leukemia/lymphotropic virus (HTLV). There is a homolog of PSD95, DLG1 or SAP97, which is bound by multiple viruses (more than 5 kinds, see [6]). We therefore opted for PSD95, to confirm a more built-in selectivity. We have selected its second and third PDZ domains, which are known to have similar specificities, both belonging to class I, although each one exerts diverse functions [15]. Our selection pursues to know whether the interaction energetics influences or reflects selectivity differences among the domains and between the viral ligands. Accordingly, we decided to pick the HTLV virus, since HPV has been demonstrated to be more promiscuous, as it targets several PDZ-containing proteins (at least 13 proteins have been described thus far [6]). There are several HTLV virus species and we opted to choose HTLV-1 and HTLV-3, since both are known to interact with PSD95 [16,17]. Additionally, we selected Ad E4 ORF adenovirus as a plausible negative control, since it should not bind PSD95, nor the other PDZ domains chosen in this study [6].

Secondly, we chose the Erbin PDZ domain, which, interestingly, has been shown to bind both class I [18] and class II [19] ligands. In this case, our choice was based more on the host partners than the viral, although just one virus is known to interact with Erbin, HTLV-1 [20], the one that we selected for PSD95 interaction. Regarding host interactions, we just mentioned that PSD95 regulates neuronal signaling. An important class of PSD95 partners are the tyrosine kinases that trigger several signaling cascades upon activation by neuregulin [21]. These tyrosine kinases, which act as receptors, are, among others, ERBB2 and ERBB4 [22]. Remarkably, ERBB2, whose C-terminus belongs to PDZ specificity class II, does not interact with PSD95, but does bind Erbin [19], whereas ERBB4 (specificity class I) interacts with PSD95 but not with Erbin [19,22]. We then considered to be interesting choosing the C-terminal regions of both receptors. In addition, we also selected a designed peptide for Erbin-PDZ, with the sequence TGWETWV. This peptide was obtained after phage display selections [23] and has served for other specificity studies [7,24,25]. As it can be inferred by the sequence, the designed peptide belongs to specificity class I.

Thirdly, we selected the nNOS PDZ domain, since it may be considered as one of the most versatile PDZ domains in terms of binding [26]. In the early studies of PDZ binding, the characteristic nNOS-PDZ interaction was its internal recognition by both the PDZ domain of syntrophin [24] and PSD95-PDZ2 [25,27]. Nevertheless, more interactions have been shown for nNOS-PDZ, comprising all the three PDZ classes ligands. As cognate nNOS partners we selected a described class II peptide, the C-terminus of CAPON, a cytoplasmic protein that regulates the PSD95/nNOS interaction by competing with PSD95 for nNOS binding [28]. In addition, we chose a class III peptide known to bind nNOS-PDZ, MelR, the C-terminal region of a melatonin receptor [29]. Interestingly, nNOS has not been described to be hijacked by any virus, so its ability to bind a significant diversity of targets, with its important neuronal role, together with PSD95 (with whom it indeed interacts) and Erbin, prompted us to choose this PDZ domain as well.

Thus, our selection comprises all three canonical and traditional binding classes of PDZ domains, including viral, host and designed ligands. All our selected PDZ domains and partners are summarized in Figure 1. Remarkably, basically all known interactions between PDZ domains and viruses are based on class I interactions (see Table 1 in [4]). All our selected PDZ domains also share relevant roles in neurons and synaptic trafficking.

Source		Sequence	Class	PSD95 PDZ2 1	PSD95 PDZ3 1	Erbin PDZ 1, 2	nNOS PDZ 1, 2, 3
Host	ERBB4	HRNTVV	1	✓ 13.9 μM	✓ 41.6 μM	✕	✕
	ERBB2	GLDVPV	2	✕	✕	✕	✓ 2 °C
	CAPON	LLDEIAV	2	✕	✕	✕	✓ 3 °C
	MelR	VVKVDSV	3	✕	✕	✕	✓ 3 °C
Virus	HTLV1 Tax	FRETEV	1	✓ 1.5 μM	✓ 3.8 μM	✓ 7.8 μM	✓ 3 °C
	HTLV3 Tax	IAESSV	1	✓ 20 μM	✓ 43 μM	✓ 23 μM	✕
	AdV E4 ORF1	KIATLV	1	✓ 11.5 μM	✕	✕	✓ 2.5 °C
Designed	Erbin PDZ binder	TGWETWV	1	✓ 5.7 μM	✕	✓ 0.24 μM	✓ 2.5 °C

Figure 1. The PDZ domains and the ligands used in this study. The ligands are grouped by their sources (host, virus or design) and are noted with both their respective names and amino acid sequences. The color fill means that the interaction has been previously described (green), or has also been demonstrated not to occur (red). If there is no fill, the interaction has not been studied according to bibliography, to the best of our knowledge (the references for the described interactions are in the text). The tick (✓) and the cross (✕) signs mean whether there is interaction or not, respectively, as demonstrated in this study. We have included the K_d in μM or the ΔT_m in °C, whenever an interaction has been shown by ITC or DSC, respectively (see Sections 3.2 and 3.3 for details).

Table 1. Thermodynamic parameters of the interactions between the PDZ domains and ligands that have been determined by ITC [1].

Source	PDZ	K_d (μM)	ΔH (kcal·mol^{-1})	−T·ΔS (kcal·mol^{-1})	ΔG (kcal·mol^{-1})
Host					
ERBB4	PSD95-PDZ2	13.9 ± 1.4	−9.7 ± 0.3	3.1	−6.6 ± 0.6
	PSD95-PDZ3	41.6 ± 10.2	−5.5 ± 0.1	−0.5	−6.0 ± 1.2
Virus					
HTLV1 Tax	PSD95-PDZ2	1.5 ± 0.1	−10.6 ± 0.1	2.6	−8.0 ± 0.5
	PSD95-PDZ3	3.8 ± 0.2	−9.5 ± 0.1	2.1	−7.4 ± 0.5
	Erbin-PDZ	7.8 ± 0.9	−9.3 ± 0.3	2.3	−7.0 ± 0.8
HTLV3 Tax	PSD95-PDZ2	20.0 ± 3.8	−5.5 ± 0.5	−0.9	−6.4 ± 1.3
	PSD95-PDZ3	43 ± 10	−3.5 ± 0.7	−2.4	−5.9 ± 1.9
	Erbin-PDZ	23 ± 9	−2.4 ± 0.3	−3.8	−6.2 ± 1.9
AdV E4 ORF1	PSD95-PDZ2	11.5 ± 0.5	−9.4 ± 0.2	2.6	−6.8 ± 0.3
Designed					
Erbin PDZ binder	PSD95-PDZ2	5.7 ± 0.2	-9.6 ± 0.4	2.5	−7.1 ± 0.3
	Erbin-PDZ	0.24 ± 0.08	-14.9 ± 0.1	5.8	−9.1 ± 2.0

[1] Stoichiometric (n) values ranged from 0.8 to 1.1.

3.2. ITC Experiments Between the PDZ Domains and Their Ligands

We performed ITC experiments at 25 °C between all the selected PDZ domains and peptides. For the sake of comparison, the experiments were all conducted in the same buffer conditions, 50 mM potassium phosphate and pH of 7.5, which presents the advantage of having an insignificant protonation heat, meaning that the experimentally determined enthalpy changes should just account for the net heat of reaction [14,30]. A table summarizing all interactions is shown in Figure 1, whilst we have just included the successful experiments graphs, both in Figure 2 and in the SI (Figure S1). As expected, many of the experiments showed no binding, since each PDZ domain has their corresponding partners previously described (see Figure 1).

PSD95-PDZ2 showed binding with all the peptide ligands that belong to class I specificity. K_d ranged from 1.5 to 20 μM and ΔH values were between -5.5 and -11.9 kcal·mol^{-1} (Table 1, see Section 3.3 for a dissection of the thermodynamics of interaction). Seemingly, according to these results, PSD95-PDZ2 is quite promiscuous, since only peptides belonging to classes II and III are unable to bind this domain. Maybe the most striking ligand partner of PSD95-PDZ2 from those analyzed is the designed binder for Erbin-PDZ. The interaction with it is moderately strong (5.7 μM), just one order of magnitude below than that for Erbin-PDZ (0.24 μM, Table 1).

Interestingly, results are different for PSD95-PDZ3. The K_d values are slightly higher than those obtained for PSD95-PDZ2 (range 3.8–43.8 μM), with smaller enthalpy changes as well, but, more importantly, binding has not been detected for two of the class I peptides (Figure 1), including the designed peptide for Erbin-PDZ. Again, no class II or III peptides were able to bind PSD95-PDZ3 either. A similar selectivity is found for Erbin-PDZ, with the notorious difference of its designed ligand (the tightest binder in this study, K_d = 0.24 μM and ΔH = −14.9 kcal·mol^{-1}). Instead, the striking result of Erbin-PDZ is the lack of binding with its canonical host partner, ERBB2, which actually gives the name to Erbin: ERBB2-interacting protein [31]. Our result may be due to the length of our peptide. We selected six residues as a standard for the recognition of the C-terminus of a PDZ domain partner, since, according to Tonikian et al., after position −5 (position 0 is the C-terminus and then numbering goes backward), PDZ domains are not specific enough [7]. In the case of ERBB2, it was already reported that the C-terminal tail comprising 9 residues (EYLGLDVPV) only binds with a K_d of around 50 μM and the same peptide with the tyrosine phosphorylated has a K_d of around 128 μM [32]. Thus, noting the relevance of the tyrosine residue at position −7, we postulate that shortening of the peptide might have caused the absence of detectable binding through ITC.

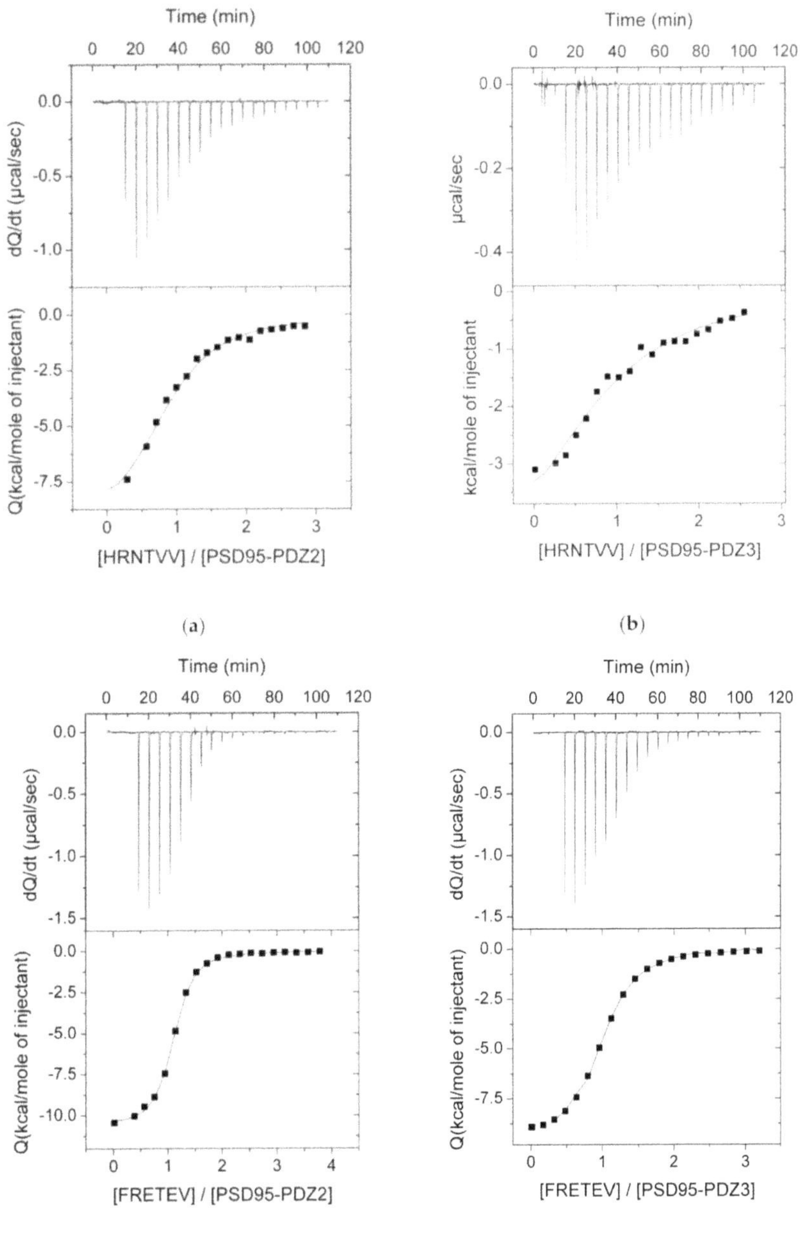

Figure 2. ITC experiments showing previously described interactions between the PDZ domains of PSD95 and both host (ERBB4) and viral (HTLV-1) ligands. Each panel corresponds to: (**a**) HRNTVV (ERBB4) with PSD95-PDZ2; (**b**) HRNTVV (ERBB4) with PSD95-PDZ3; (**c**) FRETEV (HTLV-1) with PSD95-PDZ2; (**d**) FRETEV (HTLV-1) with PSD95-PDZ3.

In the case of nNOS, we did not find binding using ITC with any of the peptides of our study. For all the non-binding ITC experiments, we then performed DSC in the search for weaker affinities binders (see following section).

3.3. DSC Experiments for the ITC That Did Not Show Binding

With those binding experiments that did not reveal an interaction by ITC, we decided to test them with DSC, both with and without the ligands, in the same buffer conditions. The final ratio of PDZ:peptide was always 1:3. DSC is also a calorimetric technique but that allows finding a more subtle binding, weaker than that from ITC [33]. Undoubtedly, ITC presents greater advantages, since it gives an exhaustive characterization of the binding interaction, granting the whole thermodynamics of the system; nevertheless, if the interactions are out of the ITC range, usually when they are weak (Kd values in the mM range), it is not sensitive enough to detect binding. By contrast, DSC can detect weak binding by a displacement of the melting temperature (T_m) due to the stabilization that occurs upon binding. Essentially, when a ligand is bound to the protein, energy is needed to decouple the interacting pair before protein denaturalization. In this way, the T_m is increased and the ΔT_m is proportional to the interaction energy.

We therefore tested our failed-ITC PDZ-ligand experiments using DSC. Remarkably, no binding was shown for any of the PDZ domains, except for nNOS-PDZ (Figure 3). Thus, PSD95-PDZ2, PSD95-PDZ3 and Erbin-PDZ seem to present stronger binding with their partners and no binding with others, reflecting a more stringent selectivity.

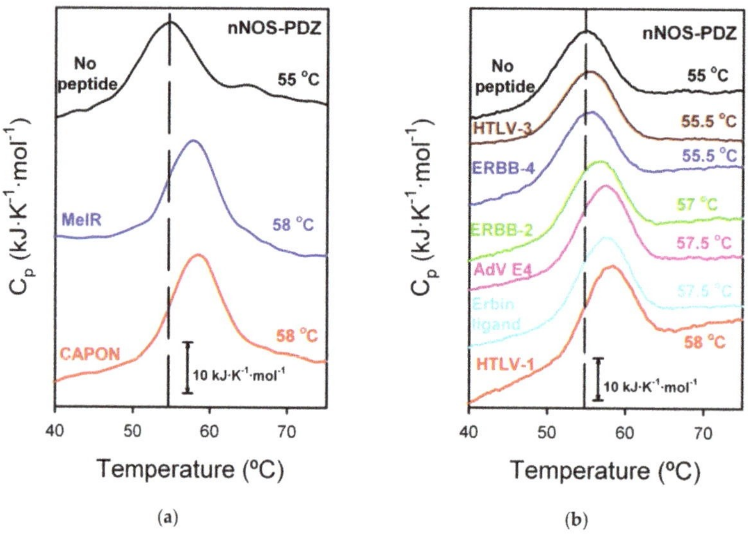

Figure 3. DSC experiments of nNOS-PDZ with the eight peptides of this study. In both panels, the nNOS-PDZ denaturalization without any peptide is also shown as reference, in black. We made two sets of experiments, (**a**) with cognate ligands and (**b**) with non-cognate ligands, including, in both sets, the nNOS-PDZ alone. The T_m is noted next to each calorimetric trace in the same color code ("Erbin ligand" refers to the phage-display designed ligand for Erbin). A bar indicating 10 kJ·K^{-1}·mol^{-1} is depicted as a reference.

DSC traces showed a T_m displacement in the case of nNOS-PDZ, with five of the tested ligands of around 2–3 °C. Unfortunately, the calorimetric traces for nNOS-PDZ are highly irreversible [12], which prevents an analysis that would potentially yield a K_d value. Instead, we can approximately estimate a K_d range based on previous cases in PDZ binding. We have taken as a reference two PDZ-peptide interactions. Firstly, the binding of PSD95-PDZ3 with the peptide KKETAV formerly described by us [11,34].

We had previously performed both ITC and DSC with PSD95-PDZ3 and KKETAV. The concentrations in the ITC experiments were the same than those used in this study and the DSC concentrations were similar, with the PDZ:peptide ratio also being 1:3. For the PSD95-PDZ3/KKETAV interaction, a ΔT_m of about 9 °C corresponds to a K_d of 1.5 μM. In the current study, we have also performed DSC for the interaction between PSD95-PDZ2 and TGWETWV and a ΔT_m of around 6 °C corresponds to a K_d of 5.7 μM (Table 1 and Figure S2). Considering that they are different PDZ domains, but have a somewhat similar ΔT_m-K_d correspondence, a ΔT_m of 2–3 °C would then imply a K_d of high micromolar or low millimolar, the range where ITC is not sensitive enough to detect binding. Lower ΔT_m values would denote even weaker affinities, out of the commonly accepted values for a physiological or biological interaction. nNOS-PDZ would therefore have affinity in the order of biological interaction with its natural counterparts (CAPON and MelR, specificity class II and III, respectively, ΔT_m = 3 °C for both peptides) and, surprisingly, with two C-terminal tails of virus proteins (HTLV1 Tax, ΔT_m = 3 °C and AdV E4 ORF1, ΔT_m = 2.5 °C, both belonging to class I), host partners (ERBB2, ΔT_m = 2 °C, class II) and even the designed peptide for the Erbin-PDZ (ΔT_m = 2.5 °C, class I) (see Figure 1). We considered no binding with HTLV-3 and ERBB-4 peptides, since the ΔT_m is too low (0.5 °C) and it may be within the experimental error, since these traces are highly sensitive to protein concentration, which allowed to reveal an intermediate in the conformational equilibrium [12]. These results imply that, although not with high affinities, nNOS-PDZ seems to be considerably promiscuous, comprising binding with ligands belonging to all the three traditional specificities in PDZ domains.

3.4. Analysis of the Energetic Balance of the Interactions

We have analyzed the relationship between enthalpy, entropy and free energy of interaction in our ITC results. Since we did not have enough data points for PSD95-PDZ2 or Erbin, we just performed our analysis with PSD95-PDZ3 and included previous data for mutant PSD95-PDZ3 domains measured under the same conditions [35]. Figure 4A shows a linear relationship between both -TΔS and ΔH (R^2 = 0.92, p = 0.03) and ΔG and ΔH (R^2 = 0.76, p = 0.04). Moreover, the slope of the relationship between $-T\Delta S$ and ΔH is far from unity (-0.67 ± 0.07). Altogether, these observations point at a bona fide enthalpy–entropy compensation phenomenon in the interactions of PSD95-PDZ3 with its targets [36–38]. The absence of outliers in the two linear relationships supports a robust mode of interaction that is not fundamentally altered by changes in the sequence of PSD95-PDZ3 or the peptides [39,40]. The slope of the relationship between ΔG and ΔH is 0.33 $\pm$ 0.07, indicating that for each energy unit of an increase in interaction enthalpy, one third translates into interaction free energy and two thirds are compensated by entropy [41]. We also analyzed ITC data by Spaller and coworkers for the interaction of PSD95-PDZ3 with six-residue peptides [42,43] and found similar results (Figure S3). In all, the interaction of peptides with PSD95-PDZ3 is robust and shows bona fide enthalpy–entropy compensation.

3.5. Comparison With Other PDZ–Ligand Interactions in Bibliography

It has been suggested that viral proteins displace other partners of their host targets by evolving for high affinity [10], so we compared the values of K_d depending on the nature and source of the ligand. In Figure 5, we have represented the K_d value in a logarithmic scale with respect to the ligand nature (host, virus or design). In our set of PDZ–ligand interactions, the strongest binding is with the designed peptide, then followed by viral ligand interactions and the host interactions as the weakest. This is, in principle, the more intuitive result and we wanted to check if it does happen in other PDZ examples. All the previously described cases, to the best of our knowledge, are represented in graphs, organized by PDZ domain, in the different panels of Figure S4. The trend observed by us happens in PDZ2 and PDZ3 from Scribble interacting with HTLV-1 Tax and host proteins [44], as well as PTPN4-PDZ and MAST2-PDZ with rabies and host proteins [45–48]. The opposite case appears in MUPP1-PDZ with HPV [49] and

PALS1-PDZ with SARS-CoV-2 [50]; the binding affinity between the viral protein and the PDZ domain is weaker than their respective PDZ–host interactions. Finally, for MAGI1-PDZ interaction with HPV E6, HTLV Tax, AdV E4 ORF1 and two host proteins [51], there is no well-defined trend in affinities differences. As a consequence, no clear rule in binding affinity governs the plausible competition between a viral and a host protein, other causes are surely influencing viral infection and proliferation. We speculate the following two scenarios to rationalize this result. Some viral proteins targeting host PDZ proteins show high affinity and function by displacing other host protein targets, thus "hijacking" the host PDZ protein for some specific purpose. Other viral proteins show affinities comparable to other host targets and function by integrating themselves in the existing protein interaction network, thus co-opting the incumbent partners of the host PDZ protein. Additionally, the relative in vivo effective concentrations of virus and host PDZ ligands can also strongly influence binding. Relative effective concentrations may vary during the virus life cycle due to changes in protein expression, localization and conformation. Thus, additional factors should be considered in the future for a definitive description of the competition between virus and host ligands PDZ ligands.

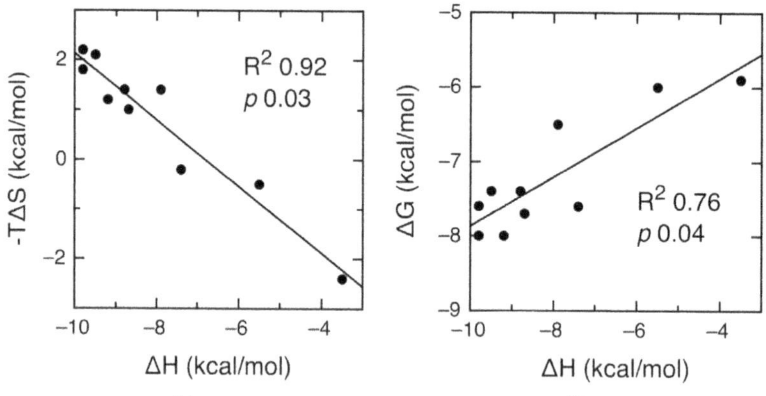

Figure 4. Enthalpy–entropy compensation in the PSD95-PDZ3 interactions with data from this study and from [35]. (**a**) Correlation between -TΔS and ΔH, showing the R^2 and p values. The experimental data are shown in circles and the linear regression as a line, with the equation y = −4.6 − 0.67x. (**b**) Correlation between ΔG and ΔH, showing the R^2 and p values. The experimental data are shown in circles and the linear regression as a line, with the equation y = −4.6 + 0.33x.

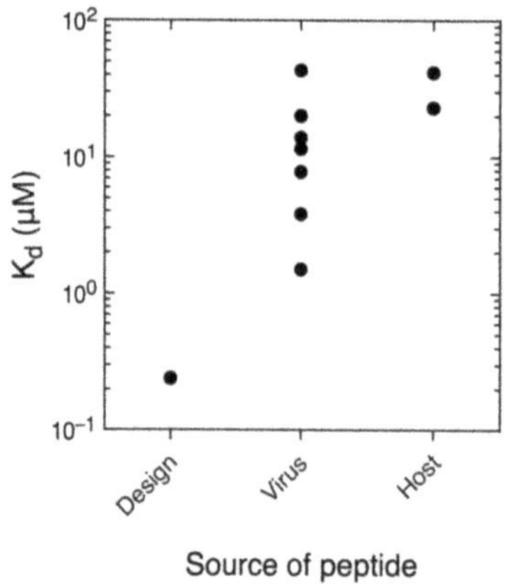

Figure 5. Logarithmic representation of the K_d vs. the peptide ligand source in our results.

4. Discussion

Our results indicate a basic and intrinsic specificity for a single PDZ domain, since all our four examples present binding with some ligands and not with others. However, it is also clear that the degree of such specificity differs in each individual domain. PSD95-PDZ2, PSD95-PDZ3 and Erbin-PDZ bind only class I ligands. Thus, elemental selectivity is present in all three of them. On the other hand, the specificity of the three domains for different class I ligands is different. PSD95-PDZ2 binds all five class I ligands, while PSD95-PDZ3 and Erbin-PDZ bind different sets of three peptide class I ligands. The nNOS-PDZ case is quite interesting, since it does not bind strongly to any of the tested ligands, but it weakly interacts with ligands belonging to the three traditional specificity classes. An alternative would be that the described binding by DSC might be weak enough to be considered as unspecific. Nevertheless, Merino-Gracia et al. dissected the binding specificity in nNOS-PDZ, revealing the molecular determinants of its ability to bind the three described specificity classes [26]. Hence, these authors demonstrated the versatility of this domain to bind multiple and diverse partners by showing at least three slightly different binding modes, including one able to recognize internal motifs and not the free carboxylate of the C-terminus. In several of the binding modes, residues outside the nNOS-PDZ binding pocket are involved and, considering that nNOS is a bigger protein, other additional allosteric pathways may also influence its PDZ binding. Thus, we can assume that our found interactions by DSC are specific and, perhaps, nNOS-PDZ interactions with its partners in vivo might be regulated not by the feeble intrinsic specificity of the domain, but, rather, by other elements, such as allosteric activators or the presence or absence of natural partners due to other signaling cascades.

PSD95-PDZ3 shows, however, a stronger specificity; it is for instance the only PDZ domain in this study that does not bind the phage-display designed binder for Erbin-PDZ. PSD95-PDZ3 has been exhaustively studied and two main factors influence binding: allostery [52,53] and post-translational modifications [35]. In both cases, the $\alpha 3$ helix of PSD95-PDZ3 [54] has a prominent role in the binding regulation, which is extended to the stability and folding of the domain [34,55–57]. Interestingly, PSD95 has three PDZ domains, in which PDZ1 and PDZ2 often act as a tandem with a very short linker between them [58–60], and PDZ3 alone has its specific functions, where the $\alpha 3$ helix is actually the linker with the next modules of the protein, an SH3 and guanylate-kinase domains. Maybe, the intrinsic binding regulation for PSD95-PDZ3 has then been evolved to be tighter than for PSD95-PDZ2, since the latter relies on its counterpart PSD95-PDZ1 to regulate their functionality. The presence of enthalpy–entropy compensation in the interaction between PSD95-PDZ3 and its targets may be a consequence of this evolutionary constraint.

Specificity regarding the host ligands is well defined. The ERBB4, CAPON and MelR peptides bind only to their respective PDZ domains. As discussed above, the host ERBB2 peptide binds to nNOS-PDZ, but not to its known partner Erbin-PDZ, most likely due to the length of the peptide. Thus, it seems that, at least in our set of PDZ interactions, the host and/or natural PDZ binders are best suited to their own particular domain. The designed and viral ligands have also shown a much larger promiscuity than the host ligands, with just a few cross interactions not taking place. The peptide designed to bind Erbin-PDZ also binds to PSD95-PDZ2 and nNOS-PDZ, which suggests that optimizing affinity can lead to decreased specificity, as discussed before for other PDZ-mediated interactions [61,62].

The viral ligands are also more promiscuous than the host ligands. The HTLV1 Tax peptide binds all four PDZ domains, while the HTLV3 Tax and AdV E4 ORF1 peptides bind three and two out of four PDZ domains, respectively. Our positive results include AdV E4 ORF1, which had not been reported to interact with any of the PDZ domains studied here. In addition, two of the three viral ligands interact with nNOS-PDZ, which has not been described to be the target of any virus up to date. Our results agree with previous high-throughput studies that report the interaction of individual viral proteins with up to dozens of host PDZ domains [63–68]. It may be the case that viruses have evolved to sequester a diverse range of host PDZ partners, exploiting the intrinsic interaction plasticity of the PDZ fold [69]. The interaction with multiple functionally related host proteins may increase robustness of the viral life cycle. This is the case for the budding-mediating Ebola virus VP40 protein, which entraps the host ESCRT pathway through several modular domains [70]. We also find it interesting that the homologous Tax proteins from HTLV1 and HTLV3 show differences in binding. This is in agreement with the differences in binding found for homologous papillomavirus E6 proteins and may be related to their clinical phenotypes [67].

Another important consequence of the specificity ligand results in this study is the property of the phage-display designed peptide to bind three of the four PDZ domains of our study and two of them tightly. Certainly, side effects are difficult to avoid, when targeting host proteins that are hijacked by virus. More problems along the same lines can be encountered if a PDZ domain is targeted, since the designed peptide/drug may bind other PDZ domains. Undeniably, designing strong and, more particularly, specific inhibitors for a given PDZ domain poses a significant challenge.

Supplementary Materials: The following are available online at https://www.mdpi.com/2218-273X/11/8/1071/s1, Figure S1: ITC experiments of seven PDZ-ligand interactions, Figure S2: DSC experiments of PSD95-PDZ2 alone and with TGWETWV, Figure S3: Enthalpy–entropy compensation in PSD95-PDZ3 interactions with ligands from the literature, Figure S4: Logarithmic representation of K_d values of PDZ interactions with virus, host or designed ligands found in bibliography.

Author Contributions: Conceptualization, I.E.S., L.B.C. and J.C.M.; methodology, E.S.C. and J.M.-C.; formal analysis, E.S.C., J.C.M. and J.M.-C.; writing—original draft preparation, J.M.-C.; writing—review and editing, all authors. All authors have read and agreed to the published version of the manuscript.

Funding: This research was funded by the Andalusian Regional Government (http://www.juntadeandalucia.es, accessed on 14 June 2021), grant number CVI-5915, and from ANPCyT, grant numbers PICT 2012-2550 and PICT 2015-1213. I.E.S. and L.B.C. are CONICET career investigators. J.M.-C. currently acknowledges funds from a reincorporation contract from the University of Granada.

Institutional Review Board Statement: Not applicable.

Informed Consent Statement: Not applicable.

Data Availability Statement: The data presented in this study are available in both this article and its supplementary information.

Acknowledgments: The authors acknowledge the CIC staff for their help in measuring the masses of the purified proteins.

Conflicts of Interest: The authors declare no conflict of interest.

References

1. Nourry, C.; Grant, S.G.N.; Borg, J.-P. PDZ Domain Proteins: Plug and Play! *Sci. STKE* **2003**, *2003*, 1–12. [CrossRef]
2. Kim, E.; Sheng, M. PDZ domain proteins of synapses. *Nat. Rev. Neurosci.* **2004**, *5*, 771–781. [CrossRef] [PubMed]
3. Heinemann, U.; Schuetz, A. Structural Features of Tight-Junction Proteins. *Int. J. Mol. Sci.* **2019**, *20*, 6020. [CrossRef] [PubMed]
4. James, C.D.; Roberts, S. Viral Interactions with PDZ Domain-Containing Proteins—An Oncogenic Trait? *Pathogens* **2016**, *5*, 8. [CrossRef]
5. Torres-Flores, J.M.; Arias, C.F. Tight Junctions Go Viral! *Viruses* **2015**, *7*, 5145–5154. [CrossRef] [PubMed]
6. Javier, R.T.; Rice, A.P. Emerging Theme: Cellular PDZ Proteins as Common Targets of Pathogenic Viruses. *J. Virol.* **2011**, *85*, 11544–11556. [CrossRef]
7. Tonikian, R.; Zhang, Y.; Sazinsky, S.L.; Currell, B.; Yeh, J.-H.; Reva, B.; A Held, H.; Appleton, B.; Evangelista, M.; Wu, Y.; et al. A Specificity Map for the PDZ Domain Family. *PLoS Biol.* **2008**, *6*, e239. [CrossRef] [PubMed]
8. Luck, K.; Charbonnier, S.; Travé, G. The emerging contribution of sequence context to the specificity of protein interactions mediated by PDZ domains. *FEBS Lett.* **2012**, *586*, 2648–2661. [CrossRef]
9. Amacher, J.F.; Brooks, L.; Hampton, T.H.; Madden, D.R. Specificity in PDZ-peptide interaction networks: Computational analysis and review. *JSBX* **2020**, *4*, 100022–100040. [CrossRef]
10. Chemes, L.B.; Gay, G.P.; Sánchez, I.E. Convergent evolution and mimicry of protein linear motifs in host–pathogen interactions. *Curr. Opin. Struct. Biol.* **2015**, *32*, 91–101. [CrossRef]

11. Murciano-Calles, J.; Cobos, E.S.; Mateo, P.L.; Camara-Artigas, A.; Martinez, J.C. An Oligomeric Equilibrium Intermediate as the Precursory Nucleus of Globular and Fibrillar Supramacromolecular Assemblies in a PDZ Domain. *Biophys. J.* **2010**, *99*, 263–272. [CrossRef]
12. Murciano-Calles, J.; Guell-Bosch, J.; Villegas, S.; Martinez, J.C. Common features in the unfolding and misfolding of PDZ domains and beyond: The modulatory effect of domain swapping and extra-elements. *Sci. Rep.* **2016**, *6*, 19242–19248. [CrossRef] [PubMed]
13. Gill, S.C.; von Hippel, P.H. Calculation of protein extinction coefficients from amino acid sequence data. *Analytical biochemistry* **1989**, *182*, 319–326. [CrossRef]
14. Murciano-Calles, J.; Coello, A.; Camara-Artigas, A.; Martinez, J.C. PDZ/PDZ interaction between PSD-95 and nNOS neuronal proteins. *J. Mol. Recognit.* **2019**, *33*, 1–11. [CrossRef]
15. Sheng, M.; Sala, C. PDZ Domains and the Organization of Supramolecular Complexes. *Annu. Rev. Neurosci.* **2001**, *24*, 1–29. [CrossRef]
16. A Chevalier, S.; Meertens, L.; Pise-Masison, C.; Calattini, S.; Park, H.; A Alhaj, A.; Zhou, M.; Gessain, A.; Kashanchi, F.; Brady, J.N.; et al. The tax protein from the primate T-cell lymphotropic virus type 3 is expressed in vivo and is functionally related to HTLV-1 Tax rather than HTLV-2 Tax. *Oncogene* **2006**, *25*, 4470–4482. [CrossRef]
17. Rousset, R.; Fabre, S.; Desbois, C.; Bantignies, F.; Jalinot, P. The C-terminus of the HTLV-1 Tax oncoprotein mediates interaction with the PDZ domain of cellular proteins. *Oncogene* **1998**, *16*, 643–654. [CrossRef]
18. Laura, R.P.; Witt, A.S.; Held, H.A.; Gerstner, R.; Deshayes, K.; Koehler, M.F.; Kosik, K.S.; Sidhu, S.S.; Lasky, L.A. The Erbin PDZ domain binds with high affinity and specificity to the carboxyl termini of delta-catenin and ARVCF. *J. Biol. Chem.* **2002**, *277*, 12906–12914. [CrossRef] [PubMed]
19. Huang, Y.Z.; Wang, Q.; Xiong, W.C.; Mei, L. Erbin Is a Protein Concentrated at Postsynaptic Membranes That Interacts with PSD-95. *J. Biol. Chem.* **2001**, *276*, 19318–19326. [CrossRef]
20. Song, C.; Wang, W.; Li, M.; Liu, Y.; Zheng, D. Tax1 enhances cancer cell proliferation via Ras-Raf-MEK-ERK signaling pathway. *IUBMB Life* **2009**, *61*, 685–692. [CrossRef]
21. Gassmann, M.; Lemke, G. Neuregulins and neuregulin receptors in neural development. *Curr. Opin. Neurobiol.* **1997**, *7*, 87–92. [CrossRef]
22. Garcia, R.A.G.; Vasudevan, K.; Buonanno, A. The neuregulin receptor ErbB-4 interacts with PDZ-containing proteins at neuronal synapses. *Proc. Natl. Acad. Sci. USA* **2000**, *97*, 3596–3601. [CrossRef] [PubMed]
23. Skelton, N.J.; Koehler, M.F.; Zobel, K.; Wong, W.L.; Yeh, S.; Pisabarro, M.T.; Yin, J.P.; Lasky, L.A.; Sidhu, S.S. Origins of PDZ domain ligand specificity. Structure determination and mutagenesis of the Erbin PDZ domain. *J. Biol. Chem.* **2003**, *278*, 7645–7654. [CrossRef] [PubMed]
24. Hillier, B.J.; Christopherson, K.S.; Prehoda, K.E.; Bredt, D.S.; Lim, W.A. Unexpected modes of PDZ domain scaffolding revealed by structure of nNOS-syntrophin complex. *Science* **1999**, *284*, 812–815. [CrossRef]
25. Brenman, J.E.; Chao, D.; Gee, S.H.; McGee, A.W.; E Craven, S.; Santillano, D.R.; Wu, Z.; Huang, F.; Xia, H.; Peters, M.F.; et al. Interaction of Nitric Oxide Synthase with the Postsynaptic Density Protein PSD-95 and α1-Syntrophin Mediated by PDZ Domains. *Cell* **1996**, *84*, 757–767. [CrossRef]
26. Gracia, J.M.; Costas-Insua, C.; Canales, A.; Rodríguez-Crespo, I. Insights into the C-terminal Peptide Binding Specificity of the PDZ Domain of Neuronal Nitric-oxide Synthase. *J. Biol. Chem.* **2016**, *291*, 11581–11595. [CrossRef]
27. Tochio, H.; Mok, Y.K.; Zhang, Q.; Kan, H.M.; Bredt, D.S.; Zhang, M. Formation of nNOS/PSD-95 PDZ dimer requires a preformed beta-finger structure from the nNOS PDZ domain. *J. Mol. Biol.* **2000**, *303*, 359–370. [CrossRef]
28. Jaffrey, S.; Snowman, A.M.; Eliasson, M.J.; Cohen, N.; Snyder, S.H. CAPON: A Protein Associated with Neuronal Nitric Oxide Synthase that Regulates Its Interactions with PSD95. *Neuron* **1998**, *20*, 115–124. [CrossRef]
29. Tochio, H.; Zhang, Q.; Mandal, P.; Li, M.; Zhang, M. Solution structure of the extended neuronal nitric oxide synthase PDZ domain complexed with an associated peptide. *Nat. Genet.* **1999**, *6*, 417–421. [CrossRef]
30. Palencia, A.; Cobos, E.S.; Mateo, P.L.; Martínez, J.C.; Luque, I. Thermodynamic Dissection of the Binding Energetics of Proline-rich Peptides to the Abl-SH3 Domain: Implications for Rational Ligand Design. *J. Mol. Biol.* **2004**, *336*, 527–537. [CrossRef]
31. Borg, J.-P.; Marchetto, S.; Le Bivic, A.; Ollendorff, V.; Jaulin-Bastard, F.; Saito, H.; Fournier, E.; Adélaïde, J.; Margolis, B.; Birnbaum, D. ERBIN: A basolateral PDZ protein that interacts with the mammalian ERBB2/HER2 receptor. *Nat. Cell Biol.* **2000**, *2*, 407–414. [CrossRef]
32. Birrane, G.; Chung, J.; Ladias, J.A.A. Novel Mode of Ligand Recognition by the Erbin PDZ Domain. *J. Biol. Chem.* **2003**, *278*, 1399–1402. [CrossRef]
33. Jelesarov, I.; Bosshard, H.R. Isothermal titration calorimetry and differential scanning calorimetry as complementary tools to investigate the energetics of biomolecular recognition. *J. Mol. Recognit.* **1999**, *12*, 3–18. [CrossRef]
34. Murciano-Calles, J.; Marin-Argany, M.; Cobos, E.S.; Villegas, S.; Martinez, J.C. The Impact of Extra-Domain Structures and Post-Translational Modifications in the Folding/Misfolding Behaviour of the Third PDZ Domain of MAGUK Neuronal Protein PSD-95. *PLoS ONE* **2014**, *9*, e98124. [CrossRef] [PubMed]
35. Murciano-Calles, J.; Corbi-Verge, C.; Candel, A.M.; Luque, I.; Martinez, J.C. Post-Translational Modifications Modulate Ligand Recognition by the Third PDZ Domain of the MAGUK Protein PSD-95. *PLoS ONE* **2014**, *9*, e90030. [CrossRef]
36. Jen-Jacobson, L.; Engler, L.E.; Jacobson, L.A. Structural and Thermodynamic Strategies for Site-Specific DNA Binding Proteins. *Structure* **2000**, *8*, 1015–1023. [CrossRef]

37. Krug, R.R.; Hunter, W.G.; Grieger, R.A. Enthalpy-entropy compensation. 1. Some fundamental statistical problems associated with the analysis of van't Hoff and Arrhenius data. *J. Phys. Chem.* **1976**, *80*, 2335–2341. [CrossRef]
38. Krug, R.R.; Hunter, W.G.; Grieger, R.A. Enthalpy-entropy compensation. 2. Separation of the chemical from the statistical effect. *J. Phys. Chem.* **1976**, *80*, 2341–2351. [CrossRef]
39. Takeda, Y.; Ross, P.D.; Mudd, C.P. Thermodynamics of Cro protein-DNA interactions. *Proc. Natl. Acad. Sci. USA* **1992**, *89*, 8180–8184. [CrossRef]
40. Yang, C.; Mori, T.; Inoue, Y. Supramolecular Enantiodifferentiating Photocyclodimerization of 2-Anthracenecarboxylate Mediated by Capped γ-Cyclodextrins: Critical Control of Enantioselectivity by Cap Rigidity. *J. Org. Chem.* **2008**, *73*, 5786–5794. [CrossRef] [PubMed]
41. Leffler, J.E. The Enthalpy-Entropy Relationship and its Implications for Organic Chemestry. *J. Org. Chem.* **1955**, *20*, 1202–1231. [CrossRef]
42. Saro, D.; Klosi, E.; Paredes, A.A.; Spaller, M.R. Thermodynamic Analysis of a Hydrophobic Binding Site: Probing the PDZ Domain with Nonproteinogenic Peptide Ligands. *Org. Lett.* **2004**, *6*, 3429–3432. [CrossRef] [PubMed]
43. Saro, D.; Li, T.; Rupasinghe, C.; Paredes, A.; Caspers, N.; Spaller, M.R. A Thermodynamic Ligand Binding Study of the Third PDZ Domain (PDZ3) from the Mammalian Neuronal Protein PSD-95. *Biochemistry* **2007**, *46*, 6340–6352. [CrossRef] [PubMed]
44. Ivarsson, Y.; Arnold, R.; McLaughlin, M.; Nim, S.; Joshi, R.; Ray, D.; Liu, B.; Teyra, J.; Pawson, T.; Moffat, J.; et al. Large-scale interaction profiling of PDZ domains through proteomic peptide-phage display using human and viral phage peptidomes. *Proc. Natl. Acad. Sci. USA* **2014**, *111*, 2542–2547. [CrossRef] [PubMed]
45. Babault, N.; Cordier, F.; Lafage, M.; Cockburn, J.; Haouz, A.; Prehaud, C.; Rey, F.; Delepierre, M.; Buc, H.; Lafon, M.; et al. Peptides Targeting the PDZ Domain of PTPN4 Are Efficient Inducers of Glioblastoma Cell Death. *Structure* **2011**, *19*, 1518–1524. [CrossRef] [PubMed]
46. Delhommel, F.; Chaffotte, A.; Terrien, E.; Raynal, B.; Buc, H.; Delepierre, M.; Cordier, F.; Wolff, N. Deciphering the unconventional peptide binding to the PDZ domain of MAST2. *Biochem. J.* **2015**, *469*, 159–168. [CrossRef] [PubMed]
47. Khan, Z.; Terrien, E.; Delhommel, F.; Lefebvre-Omar, C.; Bohl, D.; Vitry, S.; Bernard, C.; Ramirez, J.; Chaffotte, A.; Ricquier, K.; et al. Structure-based optimization of a PDZ-binding motif within a viral peptide stimulates neurite outgrowth. *J. Biol. Chem.* **2019**, *294*, 13755–13768. [CrossRef]
48. Terrien, E.; Chaffotte, A.; Lafage, M.; Khan, Z.; Préhaud, C.; Cordier, F.; Simenel, C.; Delepierre, M.; Buc, H.; Lafon, M.; et al. Interference with the PTEN-MAST2 Interaction by a Viral Protein Leads to Cellular Relocalization of PTEN. *Sci. Signal.* **2012**, *5*, 1–12. [CrossRef]
49. Sharma, S.C.; Rupasinghe, C.N.; Parisien, R.B.; Spaller, M.R. Design, Synthesis, and Evaluation of Linear and Cyclic Peptide Ligands for PDZ10 of the Multi-PDZ Domain Protein MUPP1. *Biochemistry* **2007**, *46*, 12709–12720. [CrossRef]
50. Cascio, E.L.; Toto, A.; Babini, G.; De Maio, F.; Sanguinetti, M.; Mordente, A.; Della Longa, S.; Arcovito, A. Structural determinants driving the binding process between PDZ domain of wild type human PALS1 protein and SLiM sequences of SARS-CoV E proteins. *Comput. Struct. Biotechnol. J.* **2021**, *19*, 1838–1847. [CrossRef]
51. Fournane, S.; Charbonnier, S.; Chapelle, A.; Kieffer, B.; Orfanoudakis, G.; Travé, G.; Masson, M.; Nominé, Y. Surface plasmon resonance analysis of the binding of high-risk mucosal HPV E6 oncoproteins to the PDZ1 domain of the tight junction protein MAGI-1. *J. Mol. Recognit.* **2010**, *24*, 511–523. [CrossRef] [PubMed]
52. Petit, C.M.; Zhang, J.; Sapienza, P.J.; Fuentes, E.; Lee, A.L. Hidden dynamic allostery in a PDZ domain. *Proc. Natl. Acad. Sci. USA* **2009**, *106*, 18249–18254. [CrossRef]
53. Reynolds, K.; McLaughlin, R.; Ranganathan, R. Hot Spots for Allosteric Regulation on Protein Surfaces. *Cell* **2011**, *147*, 1564–1575. [CrossRef]
54. Doyle, D.A.; Lee, A.; Lewis, J.; Kim, E.; Sheng, M.; MacKinnon, R. Crystal Structures of a Complexed and Peptide-Free Membrane Protein–Binding Domain: Molecular Basis of Peptide Recognition by PDZ. *Cell* **1996**, *85*, 1067–1076. [CrossRef]
55. Calosci, N.; Chi, C.; Richter, B.; Camilloni, C.; Engstrom, A.; Eklund, L.; Travaglini-Allocatelli, C.; Gianni, S.; Vendruscolo, M.; Jemth, P. Comparison of successive transition states for folding reveals alternative early folding pathways of two homologous proteins. *Proc. Natl. Acad. Sci. USA* **2008**, *105*, 19241–19246. [CrossRef]
56. Gautier, C.; Visconti, L.; Jemth, P.; Gianni, S. Addressing the role of the α-helical extension in the folding of the third PDZ domain from PSD-95. *Sci. Rep.* **2017**, *7*, 12593–12600. [CrossRef]
57. Murciano-Calles, J.; Martinez, J.C.; Marin-Argany, M.; Villegas, S.; Cobos, E.S. A thermodynamic study of the third PDZ domain of MAGUK neuronal protein PSD-95 reveals a complex three-state folding behavior. *Biophys. Chem.* **2014**, *185*, 1–7. [CrossRef]
58. Kalia, L.V.; Salter, M.W. Interactions between Src family protein tyrosine kinases and PSD-95. *Neuropharmacology* **2003**, *45*, 720–728. [CrossRef]
59. Long, J.-F.; Tochio, H.; Wang, P.; Fan, J.-S.; Sala, C.; Niethammer, M.; Sheng, M.; Zhang, M. Supramodular structure and synergistic target binding of the N-terminal tandem PDZ domains of PSD-95. *J. Mol. Biol.* **2003**, *327*, 203–214. [CrossRef]
60. Sainlos, M.; Tigaret, C.; Poujol, C.; Olivier, N.B.; Bard, L.; Breillat, C.; Thiolon, K.; Choquet, D.; Imperiali, B. Biomimetic divalent ligands for the acute disruption of synaptic AMPAR stabilization. *Nat. Chem. Biol.* **2010**, *7*, 81–91. [CrossRef]
61. Ernst, A.; Gfeller, D.; Kan, Z.; Seshagiri, S.; Kim, P.M.; Bader, G.; Sidhu, S.S. Coevolution of PDZ domain–ligand interactions analyzed by high-throughput phage display and deep sequencing. *Mol. BioSyst.* **2010**, *6*, 1782–1790. [CrossRef]

62. Karlsson, O.A.; Sundell, G.N.; Andersson, E.; Ivarsson, Y.; Jemth, P. Improved affinity at the cost of decreased specificity: A recurring theme in PDZ-peptide interactions. *Sci. Rep.* **2016**, *6*, 1–11. [CrossRef]
63. Bavagnoli, L.; Dundon, W.G.; Garbelli, A.; Zecchin, B.; Milani, A.; Parakkal, G.; Baldanti, F.; Paolucci, S.; Volmer, R.; Tu, Y.; et al. The PDZ-Ligand and Src-Homology Type 3 Domains of Epidemic Avian Influenza Virus NS1 Protein Modulate Human Src Kinase Activity during Viral Infection. *PLoS ONE* **2011**, *6*, e27789. [CrossRef]
64. Belotti, E.; Polanowska, J.; Daulat, A.M.; Audebert, S.; Thomé, V.; Lissitzky, J.-C.; Lembo, F.; Blibek, K.; Omi, S.; Lenfant, N.; et al. The Human PDZome: A Gateway to PSD95-Disc Large-Zonula Occludens (PDZ)-mediated Functions. *Mol. Cell. Proteom.* **2013**, *12*, 2587–2603. [CrossRef]
65. Genera, M.; Quioc-Salomon, B.; Nourisson, A.; Colcombet-Cazenave, B.; Haouz, A.; Mechaly, A.; Matondo, M.; Duchateau, M.; König, A.; Windisch, M.P.; et al. Molecular basis of the interaction of the human tyrosine phosphatase PTPN3 with the hepatitis B virus core protein. *Sci. Rep.* **2021**, *11*, 1–17. [CrossRef] [PubMed]
66. Melik, W.; Ellencrona, K.; Wigerius, M.; Hedström, C.; Elväng, A.; Johansson, M. Two PDZ binding motifs within NS5 have roles in Tick-borne encephalitis virus replication. *Virus Res.* **2012**, *169*, 54–62. [CrossRef] [PubMed]
67. Thomas, M.; Myers, M.P.; Massimi, P.; Guarnaccia, C.; Banks, L. Analysis of Multiple HPV E6 PDZ Interactions Defines Type-Specific PDZ Fingerprints That Predict Oncogenic Potential. *PLoS Pathog.* **2016**, *12*, e1005766. [CrossRef] [PubMed]
68. Vincentelli, R.; Luck, K.; Poirson, J.; Polanowska, J.; Abdat, J.; Blémont, M.; Turchetto, J.; Iv, F.; Ricquier, K.; Straub, M.-L.; et al. Quantifying domain-ligand affinities and specificities by high-throughput holdup assay. *Nat. Methods* **2015**, *12*, 787–793. [CrossRef] [PubMed]
69. Murciano-Calles, J. The Conformational Plasticity Vista of PDZ Domains. *Life* **2020**, *10*, 123. [CrossRef] [PubMed]
70. Gordon, T.B.; Hayward, J.A.; Marsh, G.A.; Baker, M.L.; Tachedjian, G. Host and Viral Proteins Modulating Ebola and Marburg Virus Egress. *Viruses* **2019**, *11*, 25. [CrossRef] [PubMed]

MDPI AG
Grosspeteranlage 5
4052 Basel
Switzerland
Tel.: +41 61 683 77 34

Biomolecules Editorial Office
E-mail: biomolecules@mdpi.com
www.mdpi.com/journal/biomolecules

www.ingramcontent.com/pod-product-compliance
Lightning Source LLC
Chambersburg PA
CBHW061818210926
51850CB00006B/736
9783725868827